Acoustics of Nanodispersed Magnetic Fluids

Acoustics of Nanodispersed Magnetic Fluids

V. Polunin

CISP

CRC Press
Taylor & Francis Group
Boca Raton London New York

CRC Press is an imprint of the
Taylor & Francis Group, an **informa** business

CRC Press
Taylor & Francis Group
6000 Broken Sound Parkway NW, Suite 300
Boca Raton, FL 33487-2742

First issued in paperback 2019

ISBN-13: 978-1-4987-3393-9 (hbk)
ISBN-13: 978-0-367-37765-6 (pbk)

Visit the Taylor & Francis Web site at
http://www.taylorandfrancis.com

and the CRC Press Web site at
http://www.crcpress.com

Contents

Preface

Recently, the number of publications concerned with the study of the nanodispersed media by acoustic methods has been increasing. It may be assumed that a new scientific direction is being formed – the acoustics of nanodispersed media which is to a certain extent the extension of the classic methods of molecular acoustics to the study of the nanoscale structure of matter.

The need for detailed explanation of the materials obtained in investigating the acoustic properties of products of nanotechnologies – magnetic colloids, referred to as magnetic fluids in the scientific and technical literature – results from the fact that this science should be available to a wide range of scientists and investigators. At the same time, the generalisation of the results of investigations in this area makes it possible to form more fundamental representation of the physical nature of unique acoustic effects in magnetic fluids and widens the possibilities of the diagnostics of the acoustic properties of actual magnetic fluids, prepared using various ferromagnetic materials and the dispersed media.

Studying together acoustic and electromagnetic fields, we can obtain information on the properties of both colloid solutions and individual nanoparticles, nanoaggregates and nanoclusters, included in their composition. The unique special feature of the acoustomagnetic method of investigation of the mechanism of magnetisation of the dispersed medium, based on the application of the acoustomagnetic effect, in comparison with other available methods, including the wave methods, is the possibility of probing baric and thermal dependences of these mechanisms.

In addition to solving 'purely measurement' tasks, the acoustometry of nanodispersed magnetic fluids may be useful in solving a large number of physical problems concerned with the examination of quantum constraints (minimum dimensions of the domains) and investigation of the special physical properties of the nanosized crystals (magnetostriction of the single-domain nanosized particle, baric and thermal dependence of its magnetic moment).

The monograph 'Acoustics of nanodispersed magnetic fluids' presented here to the reader includes in the concentrated form a large amount of very important information on the acoustic properties of magnetic fluids, obtained as a result of investigations, carried out by the author with colleagues at the Department of Physics of the Southwest State University (Russian Federation, Kursk), and also a large number of publications in the Russian and foreign literature for the period since 1969 until now.

This monograph is an English translation of the book by V.M. Polunin 'Acoustic properties of nanodispersed magnetic fluids', published in 2012 by Fizmatlit (Moscow). The two added sections (9.8 and 10.8) provide further information on the subject. The text of the translated manuscripts includes corrections of misprints found in the Russian edition of the book.

Previously, the same publishing house published in 2008 the book by V.M. Polunin 'Acoustic effects in magnetic fluids' which was also included in the current book. Except for these monographs by V.M. Polunin there are no books in the scientific literature of the world dealing with the acoustics of magnetic fluids.

The author is very pleased to present to the English readers experiments carried out in these fields in Russia and throughout the world. When writing the monograph, I took into account the studies by English physicists in acoustics, especially the work by J.W. Strutt – Lord Rayleigh, and the results obtained by these authors are used directly for explaining the observed acoustic effects in magnetic fluids. In order to present an easy to understand book, I have used, where possible, the relatively simple mathematical expressions.

The monograph includes a wide range of subjects in acoustics: special features of the propagation of sound oscillations in the magnetised fluids and the model theories describing them; acoustomagnetic and magnetoacoustic effect in magnetic fluids; the oscillatory system with the magnetic fluid inertial element; the results of investigation of the temperature dependence of the speed of sound in the magnetic fluid and the interpretation of the results using the additive model of elasticity; the results of investigation of the dispersion of the speed of sound in the *fluid–cylindrical shell* system on the basis of the unique experimental procedure using magnetic fluids which makes it possible to carry out analysis of the physical adequacy of the theoretical models; examination of the vibration–rheological effect in the oscillatory system in which the magnetic fluid place the role of the inertial–viscous element. Analysis of the experimental data obtained on the basis of the model of the rotational viscosity makes it possible to evaluate the geometrical parameters of the magnetic nanoaggregates and non-magnetic microparticles, dispersed in the sample of the actual magnetic colloid; investigation by the optical methods of the kinetic – strength properties of the magnetic fluid membrane in the form of the model of a magnetic fluid seal used in different devices; acoustic granulometry of the magnetic nanoparticles, dispersed in the carrier fluid, and construction of the size distribution curve of the particles. Acoustic granulometry offers new approaches to the examination of matter in the nanocrystalline state and widens the possibilities of the currently available experimental methods.

The monograph includes mostly material relating to magnetic colloids, but the actual magnetic fluids may also contain microsize particles in addition to the nanoparticles. Therefore, it is important to discuss the distinguishing characteristics of the ferrosuspensions. Therefore, the monograph also discusses the following questions: comparative analysis of the equilibrium magnetisation of the magnetic fluid and the ferrosuspensions; the results of investigation of the special features of passage of ultrasound through the ferrosuspensions; recommendations for the optimisation of the acoustic parameters of the magnetic fluids and ferrosuspensions.

The book consists of the Introduction and 10 chapters: chapter 1. Equilibrium magnetisation of magnetic fluids; chapter 2. Perturbation of magnetic induction by sound; chapter 3. The speed of propagation of sound; chapter 4. Absorption and scattering of sound; chapter 5. The ponderomotive mechanism of electromagnetic excitation of sound; chapter 6. Magnetoacoustic effect in the megahertz frequency range; chapter 7. Magnetic fluid sealant as the oscillatory system; chapter 8. Acoustomagnetic spectroscopy; chapter 9. Acoustic granulometry; chapter 10. Acoustometry of the form of magnetic nanoaggregates and non-magnetic microaggregates. A large number of literature references and also a list of symbols and abbreviations are also included.

The description of the experimental procedure, included in all sections of the book, makes it possible (if necessary) to repeat experiments. The additional material, included at the end of the book in the form of Appendix, can be used for evaluating the acoustic parameters of the magnetic fluids based on various dispersed media, in different physical conditions. The large number of examples of calculating the physical quantities can be used as a training material for obtaining further knowledge and self-control.

The material in the book should be studied in the proposed sequence because of the importance of the problem of perturbation of magnetisation relaxation of this process in discussing the acoustic parameters of magnetic fluids.

The book is intended for scientists, engineers and investigators in the area of physical acoustics, hydroacoustics, magnetic hydrodynamics, RF physics, rheological physics, developers of equipment using magnetic colloids. It is also hoped that the monograph will be useful to the students of universities and post-graduates of physical sciences.

Acknowledgements

The inspiring atmosphere in the team at the Southwest State University (formerly the Kursk Polytechnical Institute) fruitfully influenced the development of physical research of one of the first products of nanotechnology – magnetic fluids.

It is pleasant for me to note that my wife Galina Dmitrievna Polunina is also involved in the research of the acoustic properties of magnetic fluids. In 1973 she found an articles in the Izvestiya newspaper and said: ‘Look here! How interesting this fluid is!’

I am grateful to my friends and colleagues with whom I worked before and continue to work. Among them, Gennady T. Sychev, who gave me considerable and diverse help in my many years of activity as the Head of the Department of physics, through which it was possible to focus more on research and work on the book.

I also express my gratitude to graduate students with whom I worked as a scientific supervisor and took active part in studies of the acoustic properties of magnetic fluids. Many of my students have co-authored conference papers and articles in various journals and their names are present in the bibliographic list of books.

I am especially grateful to my colleagues – professors: Alexander Pavlovich Kuzmenko, Alexander Fedorovich Pshenichnikov, Nikolay Sergeevich Kobelev, Eugene Borisovich Postnikov. Communication with them gave me the opportunity to better understand the range of issues raised in the book.

My thanks to Anastasia Mikhailovna Storozhenko, who kindly helped me in preparing the manuscript.

I am sincerely grateful to Victor Riecansky, the publisher of Cambridge International Science Publishing, who also translated the book and provided me with the opportunity to offer our readers this English translation.

A number of studies, the results of which are presented in the book, were carried out with financial support of the Federal Target Program 'Scientific and scientific-pedagogical personnel of innovative Russia' (grants FAO NK-410P, NK-387P, No. 14.V37.21.0906). Publication of the books in 2008 and 2012 was carried out with the support of the Russian Foundation for Basic Research (projects 07-02-07004 and 11-02-07030). This work is supported by the Ministry of Education and Science of the Russian Federation; project code 3.1941.2014/K.

V.M. Polonin

Symbols and abbreviations

B – magnetic induction vector
c – the speed of propagation of sound waves
c_{SS} – adiabatically–adiabatic speed
c_{ST} – adiabatic–isothermal speed
C_p – specific heat at constant pressure
C_v – specific heat at constant volume
e – electromotive force
G – the gradient of the magnetic field strength
H – magnetic field vector
J – the intensity of the sound wave
K_a – anisotropy constant
k – the wave vector
k_o – Boltzmann constant
$L(\xi) = \text{cth}\,\xi - \xi^{-1}$ – Langevin function
$\xi = \mu_0\, m_* H / k_0 T$
M – magnetization vector of material
M_S – saturation magnetization of the magnetic fluid
M_{S0} – saturation magnetisation of ferroparticles
M_0 – magnetization of the medium in the unperturbed state
M_n – concentration factor of magnetization
M_T – temperature coefficient of magnetization
m_* – the magnetic moment of the particle
N – demagnetization factor
N_d – dynamic demagnetizing factor
N_k – number of turns in the coil
n – concentration of ferroparticles
p – pressure
Q – Q-factor
q – coefficient of thermal expansion
R, r – radius, coordinate of the cylindrical system
S – entropy, the surface area

T, T_c – absolute temperature, the temperature in Celsius
t – time
V_f – volume of the magnetic core of particles
z – wave resistance, coordinate
α – the coefficient of absorption of sound waves (mostly)
α_c – temperature coefficient of the speed of sound
β – the relative value, damping coefficient
β_S – adiabatic compressibility
γ – the ratio of specific heats
δ – the thickness of the stabilizing shell
ε – the dielectric constant of the medium
ε_0 – electric constant
η – coefficient of viscosity, total viscosity
η_s – the shear viscosity coefficient
η_v – the bulk viscosity coefficient
λ – wave length
λ_0 – the length of the standing wave
μ – magnetic permeability of material
μ_0 – magnetic constant
ν – frequency of oscillation
ρ – density
ρ_1 – density of fluid carrier
ρ_2 – density of solid particles
σ – surface tension coefficient
τ_H – Néel relaxation time
τ_B – Brownian rotational motion time of particles
φ – the volume concentration of the dispersed phase
χ – magnetic susceptibility
ω – angular frequency of oscillation
Γ_n – nonlinearity parameter
AME – acoustomagnetic effect
MAE – magnetoacoustic effect
MF – magnetic fluid
MFST – magnetic–sound transmitter
MFM – magnetic fluid membrane
MFS – magnetic fluid seal
MMF – magnetized magnetic fluid
NMMF – non-magnetized magnetic fluid
SAS – surfactant
FS – ferrosuspension
FP – ferroparticles

Introduction

Investigations of the physical properties of liquid magnetising media already started in the first half of the 20th century with the examination of the magnetic and the rheological properties of fluids with ferromagnetic particles suspended in them, with the size of these particles in the range from several micrometres to tens of micrometres. These dispersed systems are referred to as ferrosuspensions (FS). Application of these systems in practice is based on the very strong dependence of viscosity on the strength of the magnetic field. The ferrosuspensions and paste-like compositions are used for visualisation of domain boundaries, in braaking devices, in magnetic flaw inspection, in the manufacture of recording tapes, in the technology of separation of iron ores and in some other areas.

However, when solving a number of other problems in practice, the anomalous magnetic dependence of the viscosity is an interfering factor. Another shortcoming of this type of dispersed system is that they are unstable, together with the irreversible separation of the magnetic and non-magnetic phases under the effect of gravitational force or an inhomogeneous magnetic field.

A quantitative jump in the development of stable fluid magnetically controlled media with the high stability of the structure and the almost complete independence of viscosity on the magnetic field was made in the 60s as a result of the development of magnetic fluids (MF) – one of the first products of new nanotechnologies.

The magnetic fluids have the form of a colloidal solution of single-domain ferri- and ferromagnetic particles in the fluid carrier. To make sure that the disperse system has the required aggregate stability, the magnetic particles are coated with a mono-molecular layer of a stabilising agent. The shape of the particles is almost spherical; the mean radius and thickness of the stabilising shell have the order of several nanometres. The high intensity thermal motion of such small particles of the disperse phase determines the high macroscopic homogeneity of the entire system. In the final analysis, the solution of the problem of development of materials with the required physical properties not available in nature has been made in the adjacent area of knowledge – physics of ferromagnetism, colloidal chemistry and magnetic hydrodynamics.

As a result of the combinations of these 'mutually excluding' properties, typical only of the magnetic fluids, such as fluidity, the compressibility of the liquid medium and the capacity to magnetise to saturation in relatively weak magnetic fields – of the order of $\sim 5 \cdot 10^5$ A/m and at the same time high magnetisation of $\sim 10^5$ A/m, these materials are used in various areas of science and technology: magnetic fluid seals (hermetizers), magnetically controlled lubrication in friction sections and supports, separators of non-magnetic materials, cleaning of the water surface to remove oil products, the sensors of the angle of inclination and acceleration, and fillers of the gaps of magnetic heads of loudspeakers.

The application of the magnetic fluids opens new possibilities of utilising the unique acoustic effects: perturbation of the electromagnetic fields by the sound; stabilisation of moving acoustic contact; electromagnetic excitation of elastic oscillations in a liquid medium; magnetic stabilisation of the oscillatory system with a liquid inertial – viscous element, 'spring-loaded' by the gas cavity and the elasticity of the ponderomotive type; self-restoration of the broken magnetic fluid membrane, accompanied by the generation of low-frequency oscillations; control of the cross-section and direction of the sound beam; pneumatic–acoustic modulation of the magnetic flux.

The magnetic fluid, being a colloidal system, with the particles of the disperse phase having the size larger than the dimensions of the molecules of the disperse medium by only one or two orders of magnitude, intensively interact with each other and with the external magnetic field. Consequently, the magnetic fluids are of considerable interest for the physical acoustics of heterogeneous media and maybe use as a highly suitable object for verification of different theoretical models of propagation of sound in the disperse media with the nanosized particles of the disperse phase.

On the other hand, the requirements in practice stimulate interest in the detailed examination of the acoustic properties of liquid magnetised media which may be used as a basis for the development of new devices and systems. Insufficient information on the processes of kinetics of ferroparticles, interphase exchange during the propagation of sound way is in the magnetic fluids, the effect of aggregation of the ferroparticles and the processes of the magnetic hydrodynamic nature on the elastic and dissipative properties of the ferrocolloids greatly delay the development of applied investigations.

The material in the book is divided into 10 chapters.

The first chapter presents general information on the magnetic fluids: formation of the magnetic fluids, the conditions for stability of the fluids, the main physical properties. The method for measuring the magnetic parameters of the magnetic fluids and ferrosuspensions is briefly described. Attention is given to the processes of equilibrium magnetisation using the examples of magnetisation of magnetic fluids and ferrosuspensions. The effect of deformation of a cylindrical sample of the magnetic fluid on his magnetisation in both the static mode and in the dynamic deformation mode is studied. The concept of the static deformation demagnetising factor is introduced. The equation of relaxation of magnetisation for a compressed magnetic fluid is derived.

The second chapter describes the special features of perturbation of magnetisation of the magnetic fluid by sound. A new parameter is introduced for the magnetism of the non-deformed magnetics for calculating the sound-induced demagnetising field: 'the dynamic demagnetising factor', and the algorithm for calculating the parameter is proposed. Information is given on the acoustomagnetic effect (AME), based on the 'concentration' model theory. The method for the experimental examination of the AME is described.

The third and fourth chapters discuss in detail the characteristics of the acoustic parameters of the magnetic fluids – the speed of propagation and the coefficient of absorption of sound waves in the magnetic fluid; the existing model theories describing the special features of the propagation of sound waves in the non-magnetised and magnetised magnetic fluids are described, the conclusions of the theoretical considerations are compared with the experimental results; a method is described for measuring the acoustic parameters of the magnetic fluids and ferrosuspensions in the process of magnetisation; some special features of the passage of ultrasound through the ferrosuspensions and the possibilities of optimising the acoustic parameters of the magnetic fluids and ferrosuspensions are discussed.

The fifth chapter generalises the ponderomotive mechanism of electromagnetic excitation of sound in the magnetic fluid (magnetoacoustic effect) and examines a number of specific cases of the generation of elastic oscillations by magnetic fluid transducers based on this mechanism.

The six chapter is concerned with describing the results of investigation of the magnetoacoustic effect in the megahertz frequency range. Analysis of the experimental data indicates the existence of a mechanism of electromagnetic excitation of elastic oscillations in the magnetic fluid of the non-ponderomotive nature which is associated with the special features of the nanodispersed magnetic components of the magnetic fluid, i.e., it is determined by the structural factor.

At present, the magnetic fluids are used mostly in magnetic fluid hermetizers (MFH) also magnetic fluid seals (MFS). Of considerable importance are the strength and kinetic characteristics of these devices. However, investigation of these characteristics is a complicated experimental task. Therefore, the seventh chapter describes the results of experimental and theoretical studies of the elastic and strength properties of the magnetic fluid membrane which can be used, on the one side, as a model of magnetic fluid sealing and, on the other side, it can be used as an independent device.

The eighth chapter includes the material for acoustomagnetic spectroscopic. The chapter describes the unique possibilities of spectral analysis of oscillation mode is excited in the fluid–cylindrical shell system, based on the application of magnetic fluids.

The ninth chapter presents the theoretical and experimental justification of the new methods of examining the physical parameters of the magnetic nanoparticles, dispersed in the fluid carrier – acoustic granulometry.

The tenth chapter discusses the problems of the rheology of the magnetised magnetic fluid in the vibration flow conditions. The model of rotational viscosity is used for the analysis of experimental data obtained in the investigation of the oscillatory system with a magnetic fluid inertial – viscous element which can be used to determine the shape of the magnetic nanoaggregates and non-magnetic microparticles, dispersed in a sample of the actual magnetic colloid.

It is important to note that the development of the methods of acoustic granulometry and acoustometry of the shape of the nano- and microaggregates is fully consistent with the existing trends in the development of fluid magnetised media. If the task in the initial stage of production of the magnetic fluids in the 60s and 70s of the previous century was two obtain the aggregate stability and, in the ideal state, the development of a system with a monoparticle disperse phase, the next task is the organisation of the control process of aggregation in which the integrity of the solution is not disrupted and the system itself acquires new properties and application possibilities.

The main stage in the investigation of the acoustic properties of the magnetic fluids was the period of 10 years between 1975 and 1985. Up to this period, only a study by B.B. Cary and F.H. Fenlon was published in 1969 [1]. This study was concerned with the evaluation of the conversion effect of a new material. Another study was published by I.E. Tarapov in 1973 [2], describing the phenomenon logic theory of the propagation of sound in the magnetised system.

Experimental studies of the speed of propagation of sound in a non-magnetised magnetic fluid (NMMF) were started in 1976 by B.I. Pirozhkov, YU.M. Pushkarev and I.V. Yurkin [7]. The measurements show that the speed of sound in the magnetic fluid is lower than the speed of sound in a pure dispersed medium. However, in some cases, the experimental data contradicted each other, for example, in the problem of the displacement of the temperature maximum of the speed of sound in the magnetic fluid based on water with the concentration of the solid phase.

The most contradicted where the experimental data for the field dependence of the speed of sound. For example, according to the data obtained by B.I. Pirozhkov, Yu.M. Pushkarev and I.V. Yurkin, the speed of sound in the measurement error range of 0.3% in a magnetic fluid based on kerosene does not depend on the strength of the magnetic field, whereas D.Y. Chung and W.E. Isler [10] reported the anomalous dependence of the speed of sound in the magnetic fluid based on water. In the magnetic field of up to ~80 kA/m the speed of sound increase non-monotonically by 30–50%. From the purely physical viewpoint, the 'defect' of this type is unlikely to occur because it is difficult to explain the reason for more than twofold change of the compressibility of the fluid under the effect of a very moderate, homogeneous magnetic field.

Theoretical studies of the field dependence of the speed of sound in fluid magnetised media within the framework of the magnetohydrodynamics of solids were carried out by I.E. Tarapov in 1973 [2] and M.I. Shliomis and B.I. Pirozhkov in 1977 [12]. In 1975 J.D. Parsons introduced [13] the equation of movement of a unit vector to the system of equations for calculating the speed of sound in the magnetic fluid. The direction of the vector coincides with the direction of the magnetic circuit.

Experimental investigations of the dissipation of acoustic energy in the magnetic fluids were started in 1978 by W.E. Isler and D.Y. Chung [10, 14], with the determination of the dependence of the coefficient of absorption of ultrasound on the strength and direction of the magnetic field. A strong dependence of the absorption coefficient of this parameters was found. However, analysis of the results did not take into account the conclusions of the theory of propagation of sound in inhomogeneous media available at that time.

V.V. Sokolov et al published in 1984 [17] the data on the frequency dependence of the speed of propagation in the coefficient of absorption of ultrasound in the magnetic fluid in the frequency range ~3–50 MHz.

In 1985, V.V. Gogosov et al [18] showed theoretically that the change of the dimensions and form of the aggregates in the magnetic field may strongly influence the acoustic parameters of the medium.

Theoretical studies of the magnetoacoustic effect (electromagnetic excitation of sound waves) in an unlimited magnetic fluid were carried out in 1974 by V.G. Bashtovoi anf M.S. Krakov [19, 20].

Experimental studies of a magnetic fluid converter – *the source of sound* – were carried out in 1978 by A.R. Baev and P.P. Prokhorenko [21]. The experiments carried out at frequencies from 16 to 26.7 kHz and showed the relatively high efficiency of the converter.

Subsequently, in 1985–2000, considerable attention was paid to the low-frequency range of sound oscillations. For example, in 2001 B.R. Mace, R.W. Jones and N.R. Harland published studies [27] describing the oscillatory systems with controlled magnetic fluid inserts. In 2002 Yu.K. Bratukhin and A.V. Lebedev performed theoretical and experimental studies of force oscillations of the shape of the droplet of a magnetic fluid in a magnetic field [31].

Starting in 1978, studies were published in the domestic and foreign literature dealing with the acoustic properties of the magnetic fluids carried out by the author of the present book together with colleagues at the Department of physics of the Southwest State University.

In 1978, the theory of a cylindrical magnetic fluid converter was proposed [22, 23], and in 1981–1982 experimental and theoretical studies were carried out into the special features of the converter functioning in the megahertz frequency range [24–26]. The conversion of the elastic oscillations in a magnetised fluid to electromagnetic oscillations – the acoustomagnetic effect (AME) –

was investigated for the first time theoretically in 1982 [3], and was confirmed by experiments in [4, 5].

In 1978, the additive model of the formation of compressibility of the magnetic colloid was proposed in [8]. Later, in 1983, the special features of propagation of sound waves in the magnetic fluid on the basis of the considerations of the micro-in homogeneity of its structure, was published [9].

The article in [11] discusses the anisotropy of the speed of ultrasound in a magnetic fluid magnetised to saturation,, and for the collinear and orthogonal positions of the wave vector and the vector of the strength of the magnetic field the values of the speed differential by only 2 m/s. The experimental studies in [15, 16] were used to make conclusions regarding the presence of the bulk viscosity in the magnetic fluid and special features of the absorption of sound in an inhomogeneous magnetic field.

Studies [28–30] were published in 2001 and 2002 and discuss the effect of self-restoration of the magnetic fluid membrane and its elastic properties. The strength and kinetic properties of the magnetic fluid membrane were studied in 2005 [32, 33]. In 2010, the strength and kinetic parameters of the magnetic fluid membrane were investigated by optical methods [255].

The low-frequency oscillations of the oscillatory system with a magnetic fluid inertial-viscous element were studied in 2009 [256, 257, 264]. The model of rotational viscosity was used to explain the results and obtain information on the geometry of the dispersed nanoparticles.

The results of evaluation of the magnetic moment and the size of the magnetic nanoparticles of the magnetic fluids, obtained by acoustic granulometry, were discussed in the studies published in 2010– 2012 [263, 289, 303, 307, 316, 319]. The results are in satisfactory agreement with the data obtained in atomic force microscopy, magnetic relaxometry and magnetic granulometry.

The generalisation of the studies concerned with the acoustics of nanodispersed magnetic fluids will make it possible to propose new directions of solving problems in the area of the physics of nanocrystals and dispersed media based on them.

1

Equilibrium magnetisation of magnetic fluids

1.1. General information on magnetic fluids

The liquid ferromagnetics, synthesised in the middle of the 60s of the 20th century – magnetic fluids (MF) – are colloidal solutions of various ferro- or ferrimagnetic substances in conventional fluids [34–40, 314]. The production of the magnetic fluids makes it possible to solve one of the most important tasks of colloidal chemistry – the production of nanoparticles of the solid material and dispersion of the material in the fluid–carrier [39–41]. As a result of these small dimensions of the particles, they become single-domain particles [34, 41]. In the absence of the magnetic field and in fields in which the para-process is not significant [41, 42], the single-domain particles can be regarded as magnetised to saturation. The magnetic moment of these particles is $m_* = VM_{S0}$, where V is the volume of the particles. Saturation magnetisation M_{S0} depends on the particle size and decreases with the reduction of the particle size; at the particle size typical of the magnetic colloids, M_{S0} is equal to ~50% of the appropriate value of the multi-domain material. The reduction of M_{S0} is attributed to the deficit of the neighbours in the volume interaction in the surface layer or to chemical changes in the surface layer of the particles [34, 35, 39].

Van der Waals forces operate between the particles in the colloidal solution. These forces act over a short range and are considerable only in the distance between the particles is small. They belong in the group of the surface forces [43, 44]. In addition to the surface forces, there are forces between the particles of the magnetic colloids determined by the presence of a constant magnetic moment at the particles. The energy of the dipole interaction of the pair of identical ferroparticles (FP) can be described by the following equation [43]:

$$U = -2\mu_0^2 m_*^4/(3r^6 k_0^T)$$

where μ_0 is the magnetic constant; m_* is the magnetic moment of the particle; r is the distance between the particle; k_0 is the Boltzmann constant; T is absolute temperature.

The energy of magnetic interaction decreases with distance at a rate which is considerably lower than that of the energy of van der Waals interaction, i.e., magnetic forces act over long distances (long-range forces). When the particles come together, the magnetic forces cause bonding of the particles, aggregation of the dispersed phase and, consequently, the stability of the colloidal solution is impaired. The condition of existence of the MF as a stable colloidal solution is reduced to the fact that the energy of the magnetostatic interaction of the magnetic dipoles U is a small fraction of the thermal energy of the particles k_0T.

The aggregate stability of the colloids is ensured by the formation, on the surface of the particles, of protective shells preventing the bonding of the particles into aggregates. Initially, the production technology was developed in the direction of refining the coarse-grained ferromagnetic particles [37, 38]. This method makes it possible to produce magnetite, ferrite fluids, and also the fluids based on classic ferromagnetics of Fe, Ni and Co. An alternative method with respect to both reducing the expenditure and greatly increasing the productivity is the technology of production of the magnetic fluids by chemical condensation [40]. Chemical condensation is based on the precipitation of magnetite particles from an aqueous solution of bivalent and trivalent iron by an excess of the concentrated alkali solution:

$$2FeCl_3 + FeCl_2 + 8NaOH \xrightarrow{\text{Surplus of NaOH}} Fe_3O_4 \downarrow + 8NaCl + 4H_2O.$$

The resultant deposit of the colloidal particles is transferred into the fluid–carrier by peptisation based on the formation of a layer of molecules of a surface-active substance (SAS) on the particle surface. This results in the separation of the particles and subsequent dispersion in the fluid–carrier.

Peptisation is carried out by adding a solution of the fluid–carrier and the SAS to the magnetite deposit during heating to

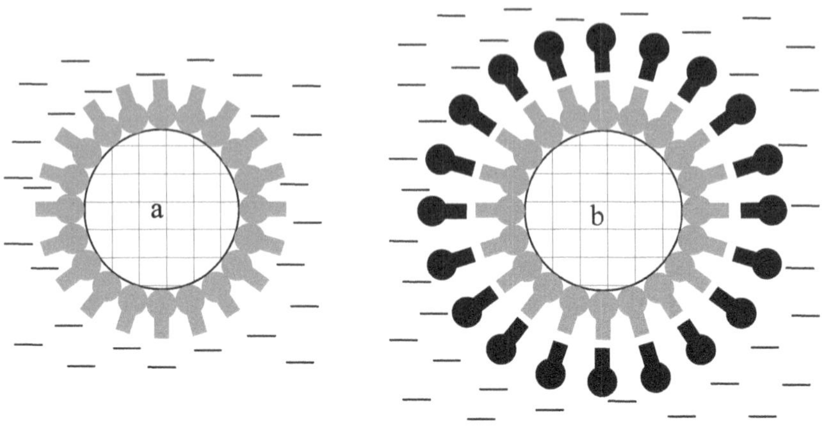

Fig. 1.1. A nanoparticle in a non-polar solvent (a) and in a polar solvent (b).

70–100°C and stirring. Magnetite and oleic acids, used most frequently for the production of the magnetic fluids, are the optimum components [40, 44] because of the maximum magnetic saturation and simple technology. The oleic acid $C_8H_{17}CH = CH(CH_2)_7CO{-}O^-H^+$ is attracted by its polar end O^-H^+ to the surface of the solid phase forming on it a dense monomolecular layer with a thickness of $\delta \approx 2$ nm (Fig. 1.1a). In the non-polar dispersion media (oil, kerosene, dodecane, octane, etc), the flexible non-polar ends of the SAS, related to the fluid–carrier, are directed from the particle to the liquid.

The stability of the dispersed particles in the polar fluid, for example, in water, is achieved by the characteristic position of two SAS layers (Fig. 1.1b); the first layer consists of oleic acid molecules, the second layer consists of sodium oleate molecules and the polar ends of the second SAS layer, related to the polar fluid–carrier, are directed from the particle to the fluid. In this case, the thickness of the protective shell is twice the thickness of the protective magnetite shell, stabilised in hydrocarbon media. There are magnetic fluids based on vacuum, transformer, vaseline and some other oils. Electrically conducting MFs are also produced using fluids such as mercury or an indium–gallium–tin (ingas) eutectic alloy in which the particles of Fe, Ni, and Co, stabilised with tin, bismuth and lithium, are dispersed.

The value of saturation magnetisation and stability of the MFs are strongly affected by the special features of the process of synthesis of the MF: the rate of supply of the solution of iron salts to the alkali and the mixing rate of the reaction mixture, the selection of the precipitation agent, the temperature conditions. In addition to this, the properties of the final product are also influenced by the 'internal' factors of the technological process: the type and concentration of the stabiliser, the composition of the fluid–carrier, the presence of various additions. The number of factors and also special features of the synthesization technology, influencing the properties of the actual magnetic fluids, are so large that they cannot be controlled in order to produce the fluid with the required properties.

The interaction of the particles in the magnetic colloids under specific conditions results in the formation of the structure from FP – floccules, granules, chains, clusters, spatial networks, droplet-shaped aggregates [34, 43, 45–47], which has a strong effect on the magnetic properties of the magnetic fluids. Two mechanisms result in the coalescence of the magnetic colloids – magnetic attraction between the suspended particles and the dipole–dipole interaction specific for the FP. In [34] the authors introduced the 'pairing' constant of the particles with the diameter *d*:

$$\Pi = \mu_0 m_*^2 / (d^3 k_0 T).$$

For the single-domain particles, the value Π is proportional to the particle volume. At $\Pi < 1$ the van der Waals forces play a certain role. With the increase of the particle size the contribution of magnetic interaction to the overall balance of the interparticle forces become stronger. At $\Pi \geq 1$ the magnetic attraction of the particles results in the formation of spatial structures – *chains and clusters* – as a result of the appearance of the minimum of the total energy of interaction of the particles at large distances between them.

The magnetic susceptibility of the MF increases with the increase of the size of the magnetic particles and the volume concentration of the particles in the colloid. The particle size of $d \sim 10$ nm is optimum because this is the largest size at which the particles do not yet aggregate as a result of the dipole–dipole interaction at room temperature (bonding is prevented by the thermal motion of the particles). The maximum concentration of the magnetic substance φ_m in the colloid depends on the ratio δ/d and on the size distribution of the particles. If the particles were identical spheres with the diameter *d*, then in the case of the dense hexagonal or face-centred cubic packing, the value $\varphi_m = (\pi / 3\sqrt{2})[d_0 / (d_0 + 2\delta)]^3$ would be ≈0.27 at $d_0 = 10$ nm and $\delta = 2$ nm. Usually, the particle size of the magnetic fluids differs and they can be packed with greater density. The concentration of the magnetic phase in the MF may reach 0.3 but in most cases in the magnetic colloids $\varphi_m \approx 0.1$–0.2, and the excess of viscosity above the viscosity of the fluid-carrier is in the range from several percent to several times.

The repulsion force, referred to as steric force, forms between the particles coated with the layer of long chain-like molecules when they come into contact.

Steric repulsion forms [43, 44] as a result of the distortion of the long molecules and the increase of the local concentration of these molecules in the zone of interaction of solvate layers. In long-term contract of the particles, the excess of the flexible chains of the SAS molecules can be redistributed in a large volume or in the entire adsorption layer at a sufficiently high surface mobility of the adsorbed molecules. At the same time, the repulsion forces depend on the contact time of the particles indicating that these forces are not purely potential forces. As regards the nature of resistance to the particles coming together, the adsorption layers should be regarded as elastoviscous shells with the elasticity modulus which depends on the repulsion potential, and with the stress relaxation time, determined by the rate of establishment of the equilibrium distribution of the molecules in the adsorption layer [43, 48].

In the field, the minimum interaction energy of the particles corresponds to the angle $m_*\mathbf{r} = 0$ and, consequently, the particles should be distributed in chains along the field. In a strong magnetic field U may increase by several orders of magnitude in comparison with the interaction energy outside the field. Therefore, in the colloids in which there is no spontaneous agglomeration of the particles, the external magnetic field may cause reversed agglomeration. This is indicated by, for example, the effect of the field on the optical transparency of the magnetite colloids or on the anisotropy of light scattering [34, 49, 50] which change when a field is applied and restore their initial value when the field is switched off.

In the presence of the aggregate stability of the system the particles of the dispersed phase are maintained, because of their small size, by the thermal Brownian motion in the volume of the fluid-carrier.

The nanoparticles with the mass m at room temperature move at thermal velocities $v = \sqrt{2k_0T/m} \approx 1.7\,\text{m/s}$, , and the characteristic time during which the particle changes its direction of movement is $t \sim m/3\pi\eta d_0 \sim 10^{-10}$ s. During this time, the particle travels the distance

~0.1 nm. Carrying out fast random motion with a 'step' of ~0.1 nm, the particle slowly diffuses, travelling on average the distance $(2Dt)^{1/2}$ during the time t, where $D = k_0T/3\pi\eta d_0$ is the diffusion coefficient. The establishment of the equilibrium distribution of the particle concentration takes place in the finite time τ_*. The order of this time is determined by the characteristic diffusion time [53]:

$$\tau_* = 6\pi\eta R k_0 T / f^2,$$

where f is the force acting on the particle (in this case $f = mg$); η is the viscosity of the fluid–carrier; R is the particle radius.

If it is assumed that $\eta = 1.38 \cdot 10^{-3}$ Pa · s (viscosity of kerosene), $R = 5$ nm, $T = 300$ K, $\rho = 5240$ kg/m^3 (density of magnetite), then $\tau_* \approx 7 \cdot 10^7$ s ≈ 23 years.

In a gravitational force field, such a system remains homogeneous for as long as necessary.

The magnetic fluids show also high stability in the magnetic fields with high heterogeneity. In this case, the FP is subjected to the effect of the ponderomotive force

$$f = \mu_0 m_* G,$$

where G is the gradient of the strength of the magnetic field, m_* is the magnetic moment of the particle.

Assuming that $m_* \approx 10^{-19}$ A·m^2, $G = 10^6$ A/m^2, we obtain $\tau_* \approx 6 \cdot 10^6$ s ≈ 60 days.

The magnetic fluids are almost completely opaque liquids. Experiments with clarification can be carried out either if the thickness of the layer is small (~10 μm), or in the case of a low concentration ($\leq 10^{-2}$) at a layer thickness of ~1 mm.

In electrical or magnetic fields, the magnetic fluids become similar to uniaxial crystals. They show the anisotropy of thermal and electrical conductivity, viscosity, and also of the optical properties: beam birefringence, dichroism, scattering anisotropy. These effects are associated with the orientation along the external magnetic field **H** or electrical field **E** of the non-spherical colloidal particles, and also with straightening of these particles into dense chains, directed along the field. The characteristic values of the strength of the electrical and magnetic fields, at which the orientation effects become significant, can be estimated by equating the electrostatic magnetostatic energy for the particle with the mean volume $V_f = 5 \cdot 10^{-23}$ m^{-3} to the energy of its thermal motion: $\mu_0 m_* H \approx k_0 T$ or

$\varepsilon_0 V_f E^2 \approx k_0 T$. From here, we obtain $H \approx k_0 T/\mu_0 m \approx 1.46 \cdot 10^4$ A/m and $E_0 \approx (k_0 T/Vf)^{1/2} \approx 3 \cdot 10^6$ V/m.

The values of the electro- and magnetooptic effects in the magnetic fluids are six orders of magnitude higher than the identical values in the conventional fluids, because the volume of the colloidal particles is ~10^6 times greater than the volume of the molecules. In the crossed electrical and magnetic fields, the magnetic fluids are similar to biaxial crystals in which the optical anisotropy can be varied both in magnitude and direction. At a specific ratio between **H** and **E**, directed normal to each other, we observe the effect of compensation of optical anisotropy. This takes place at $H/E \approx 5 \cdot 10^{-3}\ \Omega^{-1}$.

In a study by V.V. Chekanov *et al* [49], the investigated specimens of the MF were divided into three groups. The first group included the MFs with no anisotropy of light scattering in the fields of up to ~800 kA/m. The second group includes the MFs in which the appearance of the anisotropy of scattering with the application of the external magnetic field is of the threshold character (~80 kA/m). The third group contains the MFs in which the anisotropy of scattering is recorded in weak fields of <10^2 A/m and increases with increase of the strength of the field. The magnetic fluids with no scattering anisotropy are stable in the aggregate state to ~800 kA/m. The threshold value of the scattering anisotropy indicates the threshold nature of aggregation. The MFs in which the scattering anisotropy is detected in weak fields are unstable in the aggregate state. The width of the aggregates stretched along the field in one of the specimens at H = 800 kA/m, calculated by analogy with [45], was equal to 10 μm. These large aggregates are detected directly using optical microscopes [50].

The magnetic fluids, produced using different bases and different technologies, can be subdivided into three types.

The first type includes stable fluids with the linear dimensions of the particles of the dispersed phase of ~10 nm in which there are no aggregates in homogeneous magnetic fields with the strength of 500–800 kA/m.

The second type includes the stable MFs in which the aggregation of the magnetic particles continues, starting from a more or less defined threshold value of the strength of the magnetic field.

The third type basically combines low-stability fluids forming a significant deposit as a result of short-term holding (from several hours to several days); in these fluids, the aggregates are present also in the absence of the magnetic field.

1.2. The equation of the magnetic state

The equation of the magnetic state is the analytical dependence of the magnetisation of a substance on the strength of the magnetic field and temperature, i.e. $M(H, T)$. The magnetic nanoparticles, responsible for the magnetisation process in the MFs, are in the superparamagnetic condition, i.e., their magnetic momentum performs Brownian motion. The application of the external magnetic field results in rapid saturation of the magnetisation of the MF in weak and medium strength magnetic fields $M \sim 100$ kA/m, since the magnetic moment of the single-domain particle is many times greater than the magnetic moments of the individual atoms which additionally stresses the suitability of using the term 'superparamagnetic' for these media.

The energy of interaction of the magnetic moment with the external field and the body of the particle in the case of the anisotropy of the 'easy axis' is determined by the relationship [53]:

$$U = -\mu_0 \mathbf{m}_* \mathbf{H} - K_a V_f (\mathbf{m}_* \cdot \mathbf{n})^2 / \mathbf{m}_*^2$$

where K_a is the anisotropic constant; n is the unit vector, specifying the direction of the anisotropy axis.

The mechanism of rotation of the magnetic moment of the particle depends on the ratios of the terms in this equation.

If $\mu_0 m_* H \ll K_a V_f$, the magnetic moment is rigidly connected with the axis of easy magnetisation, i.e., it is 'frozen' into the body of the particle. In this case, the mechanism determining the rotation of the magnetic moment is the rotation of the particle. The orientation of the magnetic moment along the axis of easy magnetisation is established during the period of attenuation of the Larmor precession of the magnetic moment

$$\tau_\gamma = (\beta \omega_\gamma)^{-1},$$

where $\omega_\gamma = \mu_0 \gamma_e H_a$ is the frequency of ferromagnetic resonance and is equal to 10^9 Hz; $\gamma_e = 1.76 \cdot 10^{11}$ C/kg is the hydrodynamic ratio for the electron; β is the dimensionless parameter of damping, with the order of magnitude of 10^{-2}; $H_a = 2K_a / \mu_0 M_s$ is the strength of the anisotropic field.

The thermal rotational fluctuations of the moment result in weakening of the bond of the moment with the body of the particle. This mechanism is characterised by the dimensionless parameter

$$\sigma_* = K_a V_f / k_0 T,$$

equal to the ratio of the anisotropy energy to the energy of thermal fluctuations. The anisotropic constant for the ferromagnetics changes in a wide range $K_a \approx (10^6–10^2)$ J/m^3. At room temperature $k_0 T \approx 10^{-21}$ J and the particle radius of $R \approx 4$ nm we obtain $\sigma_* \approx 10^2 \div 10^{-2}$.

Thus, σ_* may have the values both considerably higher or lower than unity.

If $\mu_0 m_* H \gg K_a V_f$, the orientation of the magnetic moment of the particle is close to the orientation of the magnetic field. The relaxation time depends on the frequency of the ferromagnetic resonance determined in this case by the strength of the external field.

In non-rigid dipoles, when the ratio $\mu_0 m_* H / k V_f$ may have arbitrary values, and the orientation of the vectors **H** and **n** is arbitrary but fixed, the magnetic moment is established along the effective field [53]

$$\mathbf{H}_e = \mathbf{H} + \mathbf{H}_a,$$

where $\mathbf{H}_a = \dfrac{2K_a}{\mu_0 M_s m_*}(\mathbf{m}_* \cdot \mathbf{m})\mathbf{n}$.

In order to take into account the thermal fluctuations of the moment, it is necessary to determine more accurately the conditions of existence of the rigid dipoles. The fluctuation energy can be considerably lower than the binding energy of the moment with the body of the particle, i.e. $\sigma_* \gg 1$. However, even if this condition is fulfilled in the time period characteristic of the given FP τ_N (Néel relaxation time), the magnetic reversal of the particle takes place. Therefore, the particle can be regarded as a rigid dipole over the period of time satisfying the condition $t_* \ll \tau_N$.

It is assumed that the vector the magnetisation of the magnetic fluid **M** and the vector of the strength of the magnetic field of the medium **H** are parallel in the quasi-static approximation. The relationship between the moduli M and H is given by the equation of equilibrium magnetisation. At low and moderate concentrations of the ferrocolloid the dispersion of the FP in the latter can be regarded as

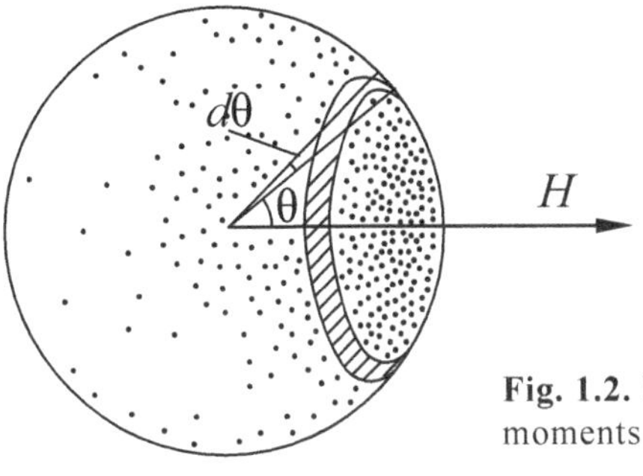

Fig. 1.2. Distribution of directions of the magnetic moments in the magnetic field.

non-interacting Brownian particles, taking part in the random thermal motion with the energy k_0T. As a result, the set of these particles assuming the rigid bonding of the magnetic moment with the body of the particle can be regarded as a gas and used to describe the process of its magnetisation by the theory of magnetisation of the paramagnetic gas [34, 51].

As a result of the frequent use of the Langevin model of magnetisation in the analysis of different processes taking place in magnetic fluids, we will derive the equation of the magnetic state of the paramagnetic gas, presented in a book by S. Tikadzuma [244] with the notation is used in the present book and with small abbreviations.

As shown in Fig. 1.2, to describe the angular distribution of the magnetic moments of the particles we use unit vectors drawn from the origin of the coordinates to the surface of the unit sphere. At H = 0, the magnetic moments are oriented in all directions and, therefore, the distribution of the ends of the unit vectors on the surface of the unit sphere is uniform. When the field H is applied, the distribution is displaced to the directions close to the direction H. When the magnetic moment of the particle forms the angle θ with the field V_f, the probability of orientation of the moment in this direction is proportional to the Boltzmann vector

$$\exp\left(-\frac{U}{k_0T}\right)=\exp\left(\mu_0\frac{m_*H}{k_0T}\cos\theta\right).$$

The proportion of the magnetic moments, forming with the field the angles in the range from θ to $\theta + d\theta$, is proportional to the area of the crosshatched ring in Fig. 1.2 which is equal to $2\pi\cdot\sin\theta\cdot d\theta$. Consequently, we obtain the following equation for the probability of orientation of the magnetic moments in the angle range between

θ and $\theta + d\theta$:

$$p(\theta)d\theta = \frac{\exp\left(\mu_0 \dfrac{m_* H}{k_0 T}\cos\theta\right)\sin\theta d\theta}{\int_0^{\pi} exp\left(\mu_0 \dfrac{m_* H}{k_0 T}\cos\theta\right)\sin\theta d\theta}.$$

If the magnetic moments forms the angle with the field H, it's component along the field is equal to $m_*\cdot\cos\theta$ and, therefore, the magnetisation, determined by all magnetic moments, present in the unit volume, is

$$M = nm_* \cos\theta = nm_* \int_0^{\pi} \cos\theta\, p(\theta)d\theta = nm_* \frac{\int_0^{\pi} \exp\left(\mu_0 \dfrac{m_* H}{k_0 T}\cos\theta\right)\cos\theta \sin\theta d\theta}{\int_0^{\pi} \exp\left(\mu_0 \dfrac{m_* H}{k_0 T}\cos\theta\right)\sin\theta d\theta}$$

Carrying out integration in the numerator and denominator of the equation, and assuming $\mu_0 m_* H/k_0 T \equiv \xi$, we obtain

$$M = nm_* \left(\frac{e^{\xi} + e^{-\xi}}{e^{\xi} - e^{\xi}} - \frac{1}{\xi}\right) = nm_* \left(\operatorname{cth}\xi - \frac{1}{\xi}\right).$$

The expression in the brackets is the Langevin function and is denoted by $L(\xi)$.

Thus, the theory leads to the magnetisation law, described by the Langevin function:

$$M = nm_* L(\xi),\ L(\xi) = \operatorname{cth}\xi - \frac{1}{\xi},\ \xi = \frac{\mu_0 m_* H}{k_0 T}. \qquad (1.1)$$

The thermal fluctuations determine the stochastic rotations of the magnetic moment in relation to the direction of the field. The effect of this mechanism on the orientation of the magnetic moment is determined by the Langevin argument ξ. The conventional boundary between the 'weak' and 'strong' magnetic fields in the case of the magnetic fluids may be represented by the value $H_T = k_0 T/\mu_0 m_*$, derived from the conditions $\xi = 1$.

We estimate H_T. For the magnetite particles with the characteristic volume $V_f = 5\cdot 10^{-23}\ m^3$, the magnetic moment is $m_* = M_{S0} V_f = 2.25\cdot 10^{-17}$ A·m². At the energy $k_0 T = 4.5\cdot 10^{-21}$ J we have $H_T = 1.46\cdot 10^4$ A/m.

With the increase of the strength of the magnetic field the curve $L(\xi)$ approaches asymptotically unity and this corresponds to the saturation magnetisation of the medium $M_s = nm^*$, i.e., total orientation of the magnetic moments of all particles along the field. In strong magnetic fields where $H \gg k_0T/\mu_0m$, equation (1.1) has the form:

$$M = M_S - \frac{3M_S k_0 T}{4\pi\mu_0 M_{S0} HR^3}, \tag{1.2}$$

where M_{S0} is the saturation magnetisation of the dispersed ferromagnetic; R is the ferroparticle radius.

In weak fields in the expansion of the Langevin equation into a Taylor series, we obtain

$$\lim_{\xi \to 0} L(\xi) = \xi / 3,$$

and, consequently, the initial magnetic susceptibility $\chi_0 = \dfrac{M}{H}$ does not depend on the strength of the field

$$\chi_0 = \frac{4\pi\mu_0 M_S M_{S0} R^3}{9K_0 T}. \tag{1.3}$$

The equations (1.2) and (1.3) show that the superparamagnetism of the magnetic fluid is interesting not only as a specific magnetic phenomenon but also as a non-destructive method of determination of the dimensions and magnetic moment of the magnetic nanoparticles dispersed in the colloid.

The numerical volume of the initial magnetic susceptibility of a concentrated magnetic fluid (the volume concentration of magnetite ~0.2) at room temperature reaches ~10, which is ~10^4 times greater than the susceptibility of the conventional fluids.

The value χ_0 decreases with increasing temperature. When the temperature approaches the Curie point T_c of the magnetic from which the colloid is produced, its spontaneous magnetisation also shows a strong temperature dependence. Heating the magnetic fluid above T_c may greatly reduce its magnetic susceptibility which is the basis of the phenomenon of thermomagnetic convection. The layers of the magnetic fluid with $T < T_c$ have higher magnetic susceptibility and are pulled into the region with the higher strength of the magnetic field, displacing the layers with $T > T_c$. The intensity of thermomagnetic convection may be many times higher than that of gravitational convection.

Susceptibility increases in accordance with the Curie–Weiss law with the reduction of temperature T, but this increase does not take place without limits and at some temperature T_g the dependence $\chi_0(T)$ reaches a maximum and subsequently χ_0 decreases. The numerical value of T_g is not associated with the solidification temperature of the fluid–carrier and depends on the concentration φ of the magnetic substance of the MF and the frequency of the measuring field. The system of interacting magnetic dipoles – single-domain colloidal particles – forms with a reduction of temperature a random structure of the entangled and branched dipole chains. For example, at $T > T_g$, the MFs are liquid superparamagnetics, and at $T < T_g$ they change to the disordered gel-like state.

Equation (1.1) shows that in short periods of changes of temperature, the strength of the magnetic field, and the concentration the equilibrium value of the magnetisation of the compressed magnetic fluid M_e can be represented by a linear dependence:

$$M_e = M_0 + M_n \cdot \delta n + M_T \cdot \delta T + M_H \cdot \delta H, \tag{1.4}$$

where M_0, $M_n \equiv \left(\dfrac{\partial M}{\partial n}\right)_0$, $M_T \equiv \left(\dfrac{\partial M}{\partial T}\right)_0$, $M_H \equiv \left(\dfrac{\partial M}{\partial H}\right)_0$ relate to the unperturbed medium.

In addition to M_H, we also use the concept of the 'total' or 'integral' magnetic susceptibility $\chi = M/H$.

The temperature dependence of the magnetisation of the MF is determined by two factors: in the explicit form and by the dependence of the magnetic moment of the particles $m_*(T)$. Therefore

$$M_T = \left(\partial M/\partial T\right)_{H,n,m} + \left(\partial M/\partial T\right)_{H,n,T} \cdot \left(\partial M_*/\partial T\right)_n. \tag{1.5}$$

Away from the Curie temperature (for magnetite – at room temperature), the dependence of the magnetic moment of the particles on temperature is very weak $\left(\partial m_*/\partial T\right)_n \approx 0$. However, in the vicinity of the Curie point, the value of this term may greatly increase [42, 52].

M_T does not include a factor of thermal expansion of the fluid. In this case, it is found in the second term of equation (1.4), which is objectively reflected by the dependence of the concentration of the dispersed medium on temperature $n(T)$.

1.3. The method of measurement of the magnetic parameters of magnetic fluids and ferrosuspensions

Only the methods of measurement of the magnetic parameters which have been reliably verified will be discussed.

The magnetic field in vacuum is characterised by the vector **H** – the strength of the magnetic field, $[H]$ = A/m. The magnetic field in a substance is determined by the magnetic induction vector **B**, $[B] = T$ (Tesla). In the case of weak magnetic materials, and also the ferromagnetics with a low coercive force, these two quantities are linked together by the linear dependence

$$\mathbf{B} = \mu_0 \mu \mathbf{H}$$

where $\mu_0 = 4 \cdot 10^{-7}$ A/m; μ is the magnetic permittivity of the substance. For vacuum (and, approximately, also for air) $\mu = 1$.

The magnetic moment of the unit volume of the substance is represented by the magnetisation M, $[M]$ = A/m. Taking this parameter into account, we can write:

$$\mathbf{B} = \mu_0 (1 + \chi)\mathbf{H} = \mu_0 (\mathbf{H} + \mathbf{M}),$$

where χ is the magnetic susceptibility of the substance; $M = \chi H$ is the equation of the magnetic state.

The parameter χ changes its numerical value in the process of magnetisation of the substance. In weak fields (at the start of the magnetisation curve), χ represents the so-called initial susceptibility χ_0. The initial magnetic susceptibility χ_0 of the specimens of the magnetic fluid can be determined from the initial slope of the magnetisation curve:

$$\chi_0 = M / H.$$

With increase of **H**, the ferromagnetic is magnetised to saturation, i.e., M becomes maximum M_S (saturation magnetisation) for the given material.

To characterise the substance in magnetic fields of different strength, we introduce the concept of the differential magnetic susceptibility: $\chi_\partial = \Delta M / \Delta H$. In the fields close to saturation $\chi_\partial \to 0$.

The degree of magnetisation of strongly magnetic substances depends not only on the value of the magnetic susceptibility but also on the geometrical form of the substance. In magnetisation of a strongly magnetic body with finite dimensions and placed in the external field, magnetic poles form at both end surfaces of the body ('magnetic charges' with the opposite sign) and this determines the formation of a field in the substance with the opposite direction (demagnetising field H').

The strength of the field H' is proportional to the magnetisation M and, consequently, we can write

$$H' = N \cdot M,$$

where N is the coefficient of proportionality referred to as the demagnetising factor. The resultant field in the substance H_i is:

$$H_i = H_e - NM.$$

In a general case, the coefficient N is a tensor but its value for the anisotropic magnetic depends only on the shape of the magnetic. For example, in magnetisation of a very long thin bar along its axis the coefficient N is equal to almost zero and, conversely, in the case of short and thick specimens $N \approx 1$.

In magnetisation of bodies of 'irregular shape' the distribution of the demagnetising field in them is non-uniform, i.e. the strength and direction of the field change from point to point. In such cases it is difficult to calculate the demagnetising factor.

Exact and accurate calculations can be carried out only for magnetics in the form of ellipsoids. The calculated numerical values of the demagnetising factor for prolate and oblate ellipsoids of rotation and also the experimentally determined values of N for a circular bar for different values of k, presented in [244], are listed in Table 1.1.

In the case of an ellipsoid of general type, the equation for N has a complicated form, but there is a simple relationship between the demagnetising factors along the three main axes of the ellipsoid x, y, z:

$$N_x + N_y + N_z = 1.$$

Taking this relationship into account, it is possible to find quite easily the value of N for several partial cases of highly symmetric ellipsoids. For example, in a sphere all three axes are equivalent ($N_x = N_y = N_z$) and, consequently, we obtain

Table 1.1

Ratio of dimensions	Circular bar	Prolate ellipsoid of revolution	Oblate ellipsoid of revolution
1	27	0,333 3	0,333 3
2	14	0,173 5	0,236 4
5	0.040	0,055 8	0,124 8
10	0,017 2	0,020 3	0,069 6
20	0,006 17	0,006 75	0,036 9
50	0,001 29	0,001 44	0,014 72
100	0,000 36	0,000 430	0,007 76
200	0,000 090	0,000 125	0,003 90
500	0,000 014	0,000 023 6	0,001 567
1000	0,000 003 6	0,000 006 6	0,000 784
2000	0,000 000 9	0,000 001 9	0,000 392

$$N = \frac{1}{3}.$$

In transverse magnetisation of a long circular bar the demagnetising factor along its axis N_z will be equal to 0, and from the evident condition $N_x = N_y$ we obtain

$$N = \frac{1}{2}.$$

In magnetisation of the plane in the direction of its normal, with the axis z selected along the normal, we have N_x=N_y=0, from which

$$N = 1.$$

Assuming $M(H) = \chi H_i$, we obtain $H_i = H_e - NM = H_e - N\chi H_i$, hence

$$H_i = \frac{H_e}{1+N\chi} \text{ and } M = \frac{\chi H_e}{1+N\chi}.$$

In this case, it is assumed that the parameter χ was obtained from the magnetisation curve in the conditions in which the demagnetising field can be ignored, for example, if the length of the vessel with the magnetic fluid in the longitudinal field is considerably greater than its diameter.

For a cylindrical column of the MF in the magnetic field transverse to the column N = 0.5 and therefore:

$$H_i = \frac{H_e}{1+0.5\chi} \text{ and } M = \frac{\chi H_e}{1+0.5\chi}.$$

In the initial section of the magnetisation curve of the magnetic fluid $\chi = \chi_0 = \text{const}$. In the non-linear region of the magnetisation curve it should be taken into account that $\chi = \frac{M}{H_i}$ – the ratio of total (final) values of M and H_i. At $H_e >> NM$ we have $H_i = H_e$ and, consequently, demagnetising fields in relatively strong magnetic fields can be ignored.

The strength of the magnetic field, the gradient of the strength, and the magnetisation of the magnetic fluid and FP and also the remanent magnetisation of the FP are measured by different variants of the induction method. This method is based on the law of electromagnetic induction according to which the EMF of the induction, formed in the conducting circuit, is numerically equal to the rate of variation of the magnetic flux penetrating this circuit.

To measure magnetisation, the fluid is poured into a cylindrical vessel. The longitudinal magnetisation of the samples is carried out either inside a solenoid, calibrated in advance with respect to current, or between the pole terminals of a laboratory electromagnet. Magnetisation or remanent magnetisation is determined by the ballistic method by recording the change of the magnetic flux penetrating the turns of the measuring coil when the magnetised specimen is pulled out from the coil.

In the simplest case, measurements are taken of the magnetic flux penetrating the turns of the coil with activation or deactivation of the magnetic field with and without the specimen, Φ_M and Φ_{MO}. In this case, magnetisation is calculated from the equation

$$M = \left(\Phi_M - \Phi_{MO}\right) / \mu_0 S N_K \left(1 - N\right),$$

where S is the area of the circuit, N_K is the number of turns in the coil, N is the demagnetising factor.

The variation of the magnetic flux in the inductance measuring coil (with and without the specimen) is achieved also by rotating the coil through 180°C around the axis, normal to the lines of the strength of the magnetic field.

The highest sensitivity is typical of the measurement method based on the application of the system consisting of two inductance coils connected anti-parallel. A vessel with the sample is placed in one of the coils.

The diagram of experimental equipment for measuring magnetisation of the magnetic fluids by this method is shown in Fig. 1.3. The vessel 1 with the sample of the magnetic fluid is placed inside one of the two identical inductance coils 2 and 3 connected in the opposite direction and placed on the rotating bar 4. Subsequently, the measuring cell is placed between the pole terminals of the laboratory electromagnet 5 and connected to the micro-webermeter 6.

When rotating the bar of the cell by 180°, the magnetic flux changes:

$$\Delta\Phi = \mu_0(M-(-M))S = 2\mu_0 MS$$

where $S = \pi d^2/4$, d is the inner diameter of the vessel.

Magnetisation is calculated from the equation:

$$M = 2\Delta\Phi / \pi\mu_0 d^2 N_{K1}(1-N),$$

where N_{K1} is the number of turns in a coil.

If the condition $d/l \ll 1$ is fulfilled, the demagnetising factor $N \approx 0$, and the relative error of the measurements ε_M:

$$\varepsilon_M = \sqrt{(\delta\Delta\Phi / \Delta\Phi)^2 + (\Delta d / d)^2 + \Delta N_{K1} / N_{K1})^2}$$

does not exceed 5%.

In the calculation of the constant of the small measuring coils it may be necessary to take into account the correction associated

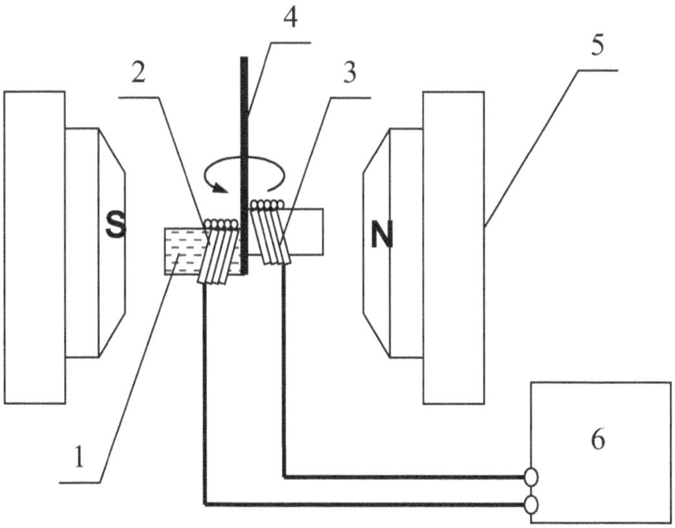

Fig. 1.3. The diagram of equipment for measuring magnetisation.

with the cylindrical shape of the winding conductor and with its finite thickness

$$\Delta S / S = 2r^2 / 9R^2,$$

where r, R are the radii of the sections of the conductor and the coil.

The saturation magnetisation of the magnetic fluid M_s can be determined from the limiting value of magnetisation which is obtained using relatively strong magnetic fields. The reliability and accuracy of measurement of M_S will be higher if in accordance with equation (1.2) we use the resultant data for constructing the graph of the dependence M ($1/H$) and carry out linear approximation in the range $1/H \approx 0$.

The measurement of the magnetisation of the magnetic fluid and ferrosuspensions by the weighing method is described below in section 1.4.

1.4. Comparison of equilibrium magnetisation of magnetic fluid and ferrosuspension

We present a number of experimental data for the equilibrium magnetisation of magnetic fluids and compare them with the appropriate data for ferromagnetic suspensions (FS) produced by the same method [245].

The nature of the force effect of the inhomogeneous magnetic field on the ferrosuspension is strongly affected by remanent magnetisation. When the direction of the magnetic field is changed to opposite in some strength range, the specimen of the ferrosuspension is pushed out from the magnetising solenoid.

In [245] attention was given to the hysteresis loop in cycling variation of H from 0 to 10 kA/m, the coercive force H_c and remanent magnetisation M_r in the range of the strength of the external magnetic fields 10 to 75 kA/m, the dependence of remanent magnetisation on the delay time t and also some possibilities of demagnetising the specimen of the ferrosuspension. The investigated ferrosuspension was produced by thorough mixing of F-600 ferrite powder with castor oil. The volume concentration of the ferromagnetic was increased to 30%. The viscosity of the suspension, determined from the rate of discharge from the tube, was (10 ± 0.5) Pa·s, i.e. an order of magnitude higher than that of the castor oil. Taking into account the Newtonian nature of the flow of the FS, the above result is regarded only as an estimate of its static shear viscosity.

The disperse phase in the investigated sample of the magnetic fluid was represented by magnetite particles of the single-domain size, and the dispersion medium – by PES-2 polyethylsiloxane liquid. The density and viscosity of the magnetic fluid were equal to 1.23 g/cm^{3}s and 1 Pa·s, respectively.

Magnetisation and coercive force were measured in the weighing equipment.

There are two variants of the classic 'weighing' method used for measuring the magnetic parameters of weakly magnetic substances [55], – weighing a sample in the form of a sphere with a small radius and a sample in the form of a long cylinder, with one end placed in the area of the maximum field and the other one in the region in which the field is equal to almost zero (GUI method), – which cannot be used for studying the concentrated magnetic fluid and ferrosuspensions. The first method – because weighing is accompanied by the continuous evaporation of the liquid droplet and the appropriate change of its volume, mass and concentration, and the second – due to the fact that the magnetisation of such a system is a non-linear function of the strength of the magnetic field and its value is very high so that the demagnetising factor cannot be ignored.

As regards the sufficiently high stability of the suspension and the magnetic fluid, the above shortcomings are not experienced in the variant of the weighing method in which a relatively short cylindrical specimen is placed in the region of the magnetising field with a small field strength gradient. Consequently, at a minimum difference of the average and local characteristics, it is quite easy to take into account the demagnetising factor and calculate the values of the strength of the field 'in the substance', H_i. This method will now be described.

The magnetic weighing method is based on measuring the force acting on the investigated specimen in a heterogeneous magnetic field. Magnetic weighing equipment, used in [245], is shown schematically in Fig. 1.4. The analytical damping balance 1 is designed for measuring the force acting on the cylindrical vessel with the sample placed on the axis of the magnetising solenoid 3 in the section with a small field strength gradient. The fixed position of the specimen is achieved using the electromagnetic device with zero compensation 2. The power source includes the accumulator, a set of rheostats, ammeter 4 and the current direction switch. The equation, expressing the magnetisation of the investigated sample M through the quantities, obtained by direct measurements, has the following form

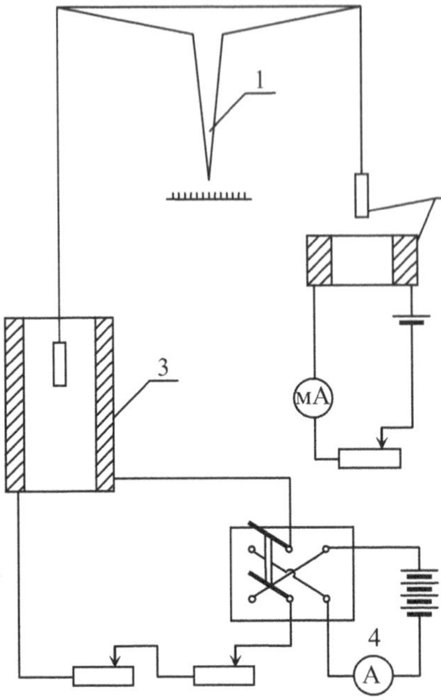

Fig. 1.4. Magnetic weighing equipment.

$$M = gND^2lm / \Phi hld^2,$$

where N and D is the number of the turns and the diameter of the winding of the coil used for calibrating the gradient of the strength of the field, respectively; Φ is the variation of the magnetic flux penetrating the turns of the calibrating coil with the length l at the current intensity of the magnetising solenoid of 1 A; h and d are the height and inner diameter of the vessel; m is the weight difference of the sample obtained at the current of the magnetising solenoid I; g is freefall acceleration.

The error of measurement by this method is $\Delta M/M = 5\%$.

The dependence $M(H)$ and remanent magnetisation of the sample of the ferrosuspension were determined by the ballistic method. In this case, measurements are taken of the magnetic flux, penetrating the turns of the inductance coil ensuring withdrawal of the magnetised specimen from the coil. All measurements are taken at a temperature of 20–25°C.

Figure 1.5 shows the hysteresis of the specimens of the FS, constructed taking the demagnetising factor into account. The magnetisation values, indicated by the solid circles on the graph, were obtained in the process of initial magnetisation (branch 0–a),

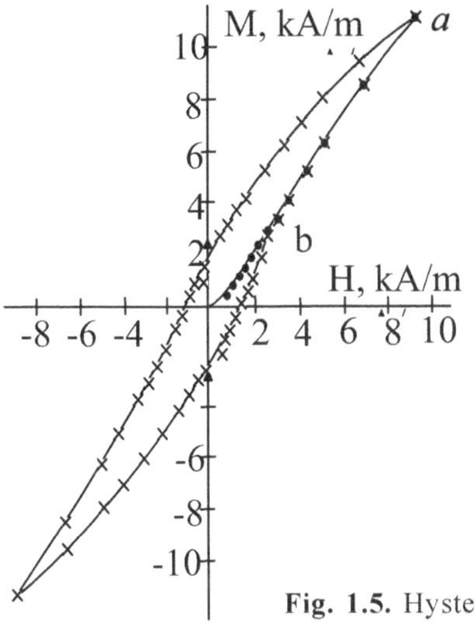

Fig. 1.5. Hysteresis loop of the FS.

and the values indicated by the crosses were obtained in the processes of cyclic changes of the field strength from 10 to –10 kA/m and in the reverse direction. The dark triangles indicate the values of the remanent magnetisation, measured by the ballistic method. In the vicinity of $H = 0$, the measurement error rapidly increases as a result of the increase of the error of fixing $\Delta\Phi$.

In contrast to the classic hysteresis loop, characteristic of solid ferromagnetics, the curve is not closed at the point *a* belonging to the tip of the loop and it is closed at the point *b* situated in the intermediate section of the initial magnetisation branch. Evidently, to explain this special feature, it is necessary to take into account the fact that magnetic reversal of PS takes place to a large extent as a result of the rotation of magnetised particles. The effect of this factor (together with the processes of intradomain nature) in transition through the demagnetising state may result in a steep curvature of the curve of the dependence $M(H)$.

Another special feature of magnetic reversal of the FS is the considerable (in this case ≈0.5 min) delay of establishment of M in relation to H in the vicinity of the point $H = \pm H_c$. This delay is also associated with the rotation of the ferrite particles. The duration of reorientation of the particles should depend strongly on the local viscosity which, as a result of the dipole–dipole interaction between them, can greatly exceed the viscosity of the dispersion medium. This is one of the factors causing the delay. The second factor is associated with the sequence of development of the process: the reorientation of the magnetic dipoles takes place mostly in the vicinity of the base of the measuring cylinder–vessel situated in the area with the highest strength, and gradually propagates throughout the entire specimen.

From the graph, we obtain directly the values of the coercive force, magnetic susceptibility (determined as the ratio of the maximum values of M and H) χ in the investigated field strength range:

$$H_c = 1.3 \text{ kA/m and } \chi = 1.2.$$

Figure 1.6 shows the dependence M (H) for a magnetic fluid with the cyclic variation of H in the range 0 to 10 kA/m. The dark circles indicate the direct course curve, corresponding to the increase of the strength in both the direct and reversed direction. The crosses indicate the reversed course curve, during which the strength of the field decreases to 0.

The data presented in Fig. 1.6 show that within the limits of the measurement error, the curves of the direct and reversed course are identical. The results obtained on the basis of the ballistic method and the weighing method indicate the complete absence in the investigated MF sample of the remanent magnetisation in the range of the strength of the outer magnetising field from 0 to 75 kA/m. In this respect, the magnetically treat is regarded as an ideal magnetically soft material. Within the framework of these considerations, the absence of any manifestation of the magnetic hysteresis in the magnetic fluid in the static measurement conditions is explained by the short relaxation time of rotational diffusion of the magnetic moment of the magnetic particles of the single-domain dimensions, suspended in the liquid and carrying out thermal Brownian motion.

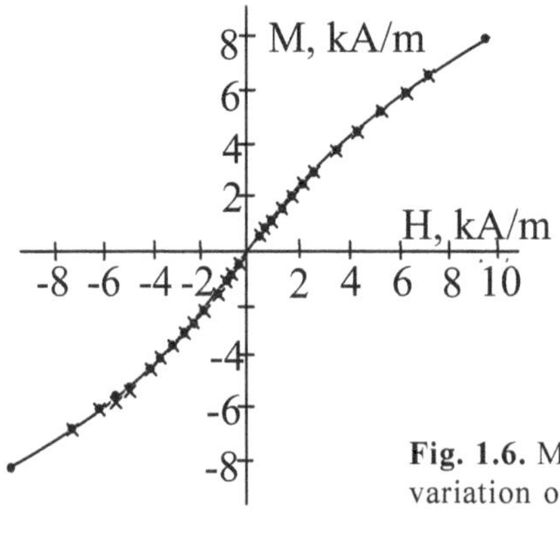

Fig. 1.6. Magnetisation of the magnetic fluid at cyclic variation of the strength of the field H.

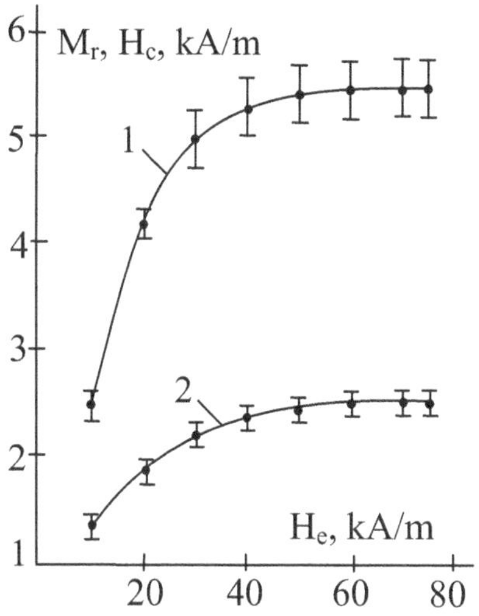

Fig. 1.7. The dependences M_r (H_e) and H_c (H_e). 1) M_r; 2) H_c.

According to the data in Fig. 1.6, the magnetic susceptibility of the investigated magnetic fluid is 0.86.

Figure 1.7 shows the dependence of remanent magnetisation M_r and coercive force H_c on the strength of the outer magnetising field H_p for the ferromagnetic suspension. The values of M_r were obtained by the ballistic method, the values of H_e by the weighing method. The magnetisation time was 10 s. Prior to every magnetisation cycle, the specimen was returned to the initial condition by slow withdrawal from the alternating magnetic field followed by weighing. An increase of H_e is accompanied by a gradual reduction of the rate of increase of both parameters, and the ratio H_c/M_r in the error range remains constant and equal to 0.5 which, possibly, indicates the existence of a physical relationship between them.

An important applied aspect is the question of the magnetic ageing of the material based on the reduction of remanent magnetisation and the change of its main magnetic parameters with time. The variation of the remanent magnetisation of the FS sample during holding is shown in Fig. 1.8.

In the experiment period (≈ 6000 h) M_r decreased by 16% from the initial value, recorded 15 seconds after switching off the magnetising field. It is noteworthy that in the first 70–80 h M_r decreased by ≈ 8%, and by the same amount in the following 1000 h. This was followed by the stabilisation of magnetisation.

The time required for the passage of a spherical ferrite particle with a radius of 1.5 μm in descent in castor oil of 1/3 height of

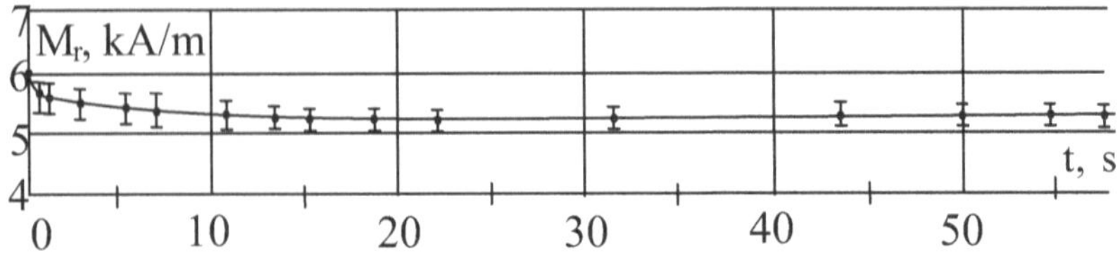

Fig. 1.8. Variation of M_r of a ferrosuspension sample during holding.

the column of the suspension (i.e. the duration of complete phase separation of the system) is ≈800 h. At the same time, after holding for 6000 h there was only a thin (≈ 0.5 mm) film of the liquid phase on the surface of the sample. This fact and also the relative stability of remanent magnetisation indicate that the FS contains a more or less continuous spatial structure formed by the magnetised particles interacting over a distance.

The possibilities of demagnetising the material are also of considerable practical interest. A demagnetisation method is available in which the sample is gradually withdrawn from the alternating magnetic field whose strength amplitude is slightly higher than the strength of the magnetising field; in this method the magnetisation of the ferromagnetic suspension is reduced by at least a factor of 100. The demagnetisation of this magnitude can also be achieved by careful stirring of the FS. Naturally, the latter of these two possibilities of demagnetisation is suitable exclusively for liquid and paste-like systems.

As a result of these data, the given FS sample can be regarded as a 'liquid magnet' because, being fluid (under the condition of constant form), is capable of retaining for a long time the remanent magnetisation on a level close to the initial level. A completely different behaviour is observed in the process of static magnetisation for the specimen of the magnetic fluid. At switching off the field the specimen is completely demagnetised without any external effect thus confirming the validity of the model of the superparamagnetic used for magnetic colloids.

It is also important to note a large difference between the magnetic fluid and the ferromagnetic suspension as regards their stability in the non-uniform magnetic field. The dimensions of the particles of the dispersed phase in the FS are approximately $\sim 10^3$ times greater than the dimensions of the particles in the magnetic fluid, and their volume and magnetic moment differ by a factor of $\sim 10^9$. For such large particles of the dispersed phase we can ignore the factor of thermal Brownian motion and calculate the value τ_* from the equation

$$\tau_* = \ell / \upsilon = 6\pi\eta R\ell / (\mu_0 m_* G) = 9\eta\ell / (2\mu_0 R^2 M_{S0} G),$$

where ℓ is the linear dimensions of the system (height of the vessel); M_{S0} is the saturation magnetisation of magnetite.

At R = 5 μm, η = 1.38·10^{-3} Pa, M_{S0} = 4.77·10^5 A/m, G = 10^6 A/m^2, ℓ = 0.1 m, we obtain $\tau_* \approx$ 40 s. It should be mentioned that in the identical situation the value obtained for the magnetic fluid was $\tau_* \approx$ 60 days.

Taking into account the presence in the real magnetic fluids of aggregates in the form of magnetic fluid chains, it should be mentioned that the orientation of the magnetic changes in the ultrasound and magnetic fields is determined by three factors: magnetic field, thermal motion and the speed of the carrier liquid.

The body around which a uniform flow of an ideal, incompressible liquid flows, is subjected to the effect of the force moment equal to [18]:

$$M_r = -\frac{1}{2}(\lambda_{||} - \lambda_{\perp})U^2 \sin 2\theta,$$

where $\lambda_{||}$ and $\lambda_{\perp}$ are the components of the tensor of the attached masses of the ellipsoid; U is the speed of the flow around the body; θ is the angle between the direction of the speed U and the major axis of the ellipsoid.

The efficiency of the rotational effect of the fluid flow on the aggregates in comparison with the effect of thermal Brownian motion and the magnetic field is relatively low. This is indicated by the evaluation of the strength of the magnetic field resulting in the rotational effect characteristic of the ultrasound wave of medium power, and by the comparison of the energy of the rotational effect of the flow with the thermal energy of the particles.

Using the expression $\lambda_{||}$ in $\lambda_{\perp}$ for the ellipsoid of rotation with the major and minor half-axes ℓ and d, we obtain for the order of magnitude [18]: $M_r \sim (4.3)\pi\rho_1 \ell d^2 U^2$. The magnetic moment of the ellipsoid is determined from the equation

$$m = m_* N_{ag} = M_s' 4\pi\ell d^2 / 3,$$

where N_{ag} is the number of ferromagnetic particles in the aggregate,

m^* is the magnetic moment of a single ferromagnetic particle, M'_s is the magnetisation of the ferromagnetic particles.

We estimate the strength of the magnetic field at which the $\frac{M_r}{\mu_0 mH} = 1$ equality is fulfilled. We derive the relationship

$$\frac{M_r}{\mu_0 mH} = \frac{(4/3)\pi \ell d^2 U^2 \rho_1}{(4/3)\mu_0 \pi \ell d^2 M'_s H} = \frac{U^2 \rho_1}{\mu_0 M'_s H}.$$

For ultrasound with the power of 1 W/cm^2 at the frequency ν = 1 MHz, the displacement amplitude is approximately $2 \cdot 10^{-8}$ m. This corresponds to the amplitude of vibrational speed U = 0.13 m/s. Assuming that $M'_S \sim 4.7 \cdot 10^5$ A/m, we obtain that $M_r = \mu_0 mH$ at $H \approx 20$ A/m.

We estimate the volume of an ellipsoidal particle having the energy in the flow comparable with the energy of thermal Brownian motion ($U \sim 0.02$ m/s, $T \sim 300$ K)

$$V_{ag} = \frac{4}{3}\pi \ell d^2 = \frac{k_0 T}{\rho_1 U^2} \approx 1.3 \times 10^{-20} \text{m}^3.$$

Assuming that $V_{ag} = N_{ag} \cdot 4\pi R_m^3/3$ (R_m is the radius of a single ferromagnetic particle), we obtain the estimate of the number of particles in the aggregate N_{ag}: $N_{ag} \approx 2 \cdot 10^4$.

Consequently, the Brownian motion can be ignored only in the case of relatively large aggregates.

1.5. Magnetisation of the specimen in quasi static deformation

Attention will begin to a number of electromagnetic effects, caused by elastic quasi-static deformation of the specimen of the magnetised liquid filling a cylindrical vessel [54]. The vessel is made of a non-magnetic and non-conducting material with the magnetic permittivity μ = 1. Since any deformation is completed in the finite time Δt, the quasi-static condition will be represented by fulfilment of the inequality $\Delta t \geq L/c$, where L is the length of the cylinder, c is the speed of propagation of sound in the fluid.

We are interested either in the 'purely longitudinal' deformation of the liquid cylinder when its length changes and the diameter d remains constant, or 'purely radial' deformation when only the diameter (radius R_c) of the cylinder changes. In both cases, the liquid is magnetised along the axis of the cylinder.

The increment of the magnetic field δH, present in the equation (1.4), can be the result of the change of the strength of the external magnetic field and may be associated with the change of the demagnetising field. The demagnetising field of the magnetised specimens can be expressed using the demagnetising factor N [55]. In the theoretical study [56], the authors derived the expression for the demagnetising factor averaged over the central circular cross-section, for the cylinder uniformly magnetised along the axis. The demagnetising factor, obtained by this procedure, is referred to as 'ballistic'. The ballistic factor is the function of the form parameter P ($P \equiv L/d$):

$$N_\delta = 1 - \frac{2P}{\pi k_1}[K_e(\mathrm{k}_1) - E_e(k_1)], \tag{1.6}$$

where $k_1 = (1 + P^2/4)^{-0.5}$; K_e and E_e are the total elliptical integrals of the first and second kind.

The equation (1.6) holds for any values of P. The approximation relationships for $P \ll 1$ and $P \gg 1$ have following form

$$N_\delta \approx 1 - \frac{2P}{\pi}\left[\ln\frac{8}{P} - 1\right]; \tag{1.7}$$

$$N_\delta \approx \frac{1}{2}P^{-2}\left[1 - \frac{3}{2}P^{-2} + \frac{25}{8}P^{-4}\right]. \tag{1.8}$$

The effect of the perturbation of the demagnetising field, caused by the deformation of the cylinder, is determined by the change of its linear dimensions and magnetisation:

$$\delta H = M \cdot \delta N + N \cdot \delta M.$$

In the linear approximation, the effect of the the demagnetising field in the central circular cross-section of the uniformly magnetised cylinder is proportional to the perturbation of magnetisation:

$$\delta H_p = N_c \cdot \delta M. \tag{1.9}$$

The values of the proportionality coefficient N_c for the longitudinal and radial deformation of the cylinder differ:

$$N_{c\ell} = N_\delta - P\frac{\partial N_\delta}{\partial P}; \tag{1.10}$$

$$N_{cr} = N_\delta + 0.5P\frac{\partial N_\delta}{\partial P}. \tag{1.11}$$

Coefficient N_c is the static deformation demagnetising factor. Equations (1.10) and (1.11) shows that $N_{c\ell} > N_\delta$, $N_{cr} < N_\delta$.

Table 1.2 gives the values of the ballistic demagnetising factor N for a number of values of P in the range from 0 to 10, taken from [56]. The table also gives the values of the parameters $\Delta N \equiv P(\partial N_\delta/\partial P)$ $N_{c\ell}$ and N_{cr}. ΔN was calculated from the equations

$$\Delta N = -\frac{4P}{\pi} - \frac{2P}{\pi}\cdot\ln\frac{P}{8} (\text{at } 0 \le P \le 0.4),$$

$$\Delta N = P^{-2}\left(1 - 3P^{-2} + \frac{75}{8}P^{-4}\right)(\text{at } 2 \le P \le 10),$$

derived using the approximation expressions (1.7) and (1.8). In the range $0.4 \le P \le 2$ the values $\partial N_\delta/\partial P$ are determined on the basis of the angle of inclination of the tangent to the curve N_δ (P).

As indicated by the equations (1.4) and (1.4), perturbation of the magnetisation in isothermal deformation within the framework of

Table 1.2

$P=L/d$	N_δ	ΔN	$N_{c\ell}$	N_{cr}
0	1.000	0.000	1.000	1.000
0.01	0.9638	0.0298	0.9938	0.9489
0.1	0.7845	0.1516	0.9366	0.7087
0.2	0.6565	0.2150	0.7202	0.549
0.4	0.4842	0.25	0.729	0.361
0.8	0.2905	0.24	0.532	0.170
1.0	0.2322	0.22	0.427	0.135
2	0.09351	0.21	0.240	0.025
3	0.04800	0.087	0.1262	0.0089
4	0.02865	0.053	0.0811	0.0026
6	0.01334	0.0257	0.0386	0.00049
8	0.007633	0.0149	0.0229	0.000183
10	0.004923	0.0097	0.0147	0.000073

the considered concentration model is determined by the variation of the concentration and perturbation of the demagnetising field. For the longitudinal and radial deformation the perturbation of magnetisation in the plane of the central circular section of the cylinder is expressed by the relationships

$$\delta M = nM_n(1+N_{c\ell}M_H)^{-1}\frac{\delta L}{L} \tag{1.12}$$

and

$$\delta M = -2nM_n(1+N_{cr}M_H)^{-1}\frac{\delta R_c}{R_c}. \tag{1.13}$$

The magnetisation increment can be expressed by the increment of static pressure ΔP by carrying out in the equations (1.12) and (1.13) substitution $\frac{\delta L}{L} = -\beta_T \cdot \delta P$ and $2\frac{\delta R_c}{R_c} = -\beta_T \cdot \delta P$, where $\beta_T = -V^{-1}\left(\frac{\partial V}{\partial P}\right)_T$ is isothermal compressibility.

For the adiabatic longitudinal and radial static deformation, the perturbation of magnetisation can be expressed as follows

$$\delta M = -(nM_n + \gamma_* M_T)(1+\mathrm{N}_{c\ell}\,M_H)^{-1}\frac{\delta L}{L}; \tag{1.14}$$

$$\delta M = -2(nM_n + \gamma_* M_T)(1+\mathrm{N}_{cr}\,M_H)^{-1}\frac{\delta R_c}{R_c}, \tag{1.15}$$

where $\gamma_* = qTc^2C_p^{-1}$.

The appearance of the term $\gamma_* Mt$ will be justified in section 2.1.

1.6. Dynamic deformation of the magnetised specimen

Dynamic deformation of a fluid column forms as a result of the propagation of the landing sound waves or the establishment of the system of standing sound waves.

We write the system of electrodynamic equations in which the Maxwell equations are presented in the integral form:

$$\oint_{r_1} \boldsymbol{E}^* dl = -\int_{S_0} (\partial \boldsymbol{B}/\partial t) d\boldsymbol{S}; \tag{1.16}$$

$$\oint_{r_2} \boldsymbol{H} dl = \int_{S_0} i d\boldsymbol{S} + \int_{S_0} (\partial \boldsymbol{D}/\partial t) d\boldsymbol{S}; \tag{1.17}$$

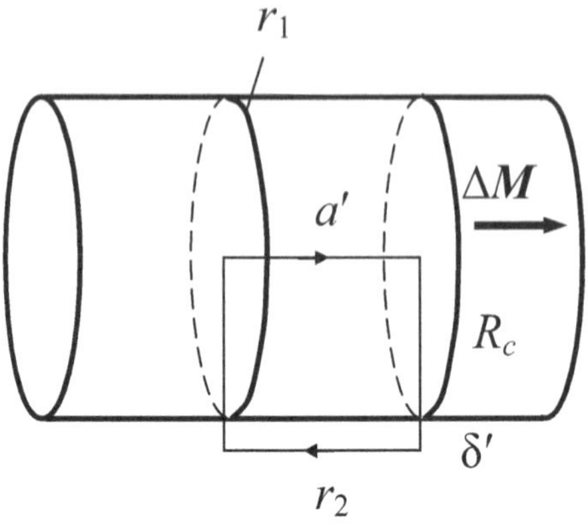

Fig. 1.9. Calculation model.

$$\mathbf{D} = \varepsilon\varepsilon_0 \mathbf{E}^*; \tag{1.18}$$

$$\mathbf{i} = \sigma \mathbf{E}^*; \tag{1.19}$$

$$\mathbf{B} = \mu_0(\mathbf{H} + \mathbf{M}) \tag{1.20}$$

Figure 1.9 shows the calculation model. According to the equation (1.16), the circular contour of the central section r_l is characterised by the formation of an electrical field with the strength

$$E^* = 0.5\left(R_c + \delta'\right)\frac{\partial B}{\partial t}.$$

Or, assuming that $\delta'/R_c \ll 1$,

$$E^* = 0.5 R_c \frac{\partial B}{\partial t}.$$

According to (1.17) and (1.18)

$$\oint_{r_2} \mathbf{H}_l \, dl = \varepsilon\varepsilon_0 \int_{S_0} (\partial \mathbf{E}^* / \partial t) d\mathbf{S},$$

here r_2 is the closed contour in the form of a rectangle in which the two sides are directed along the axis and two others in the radial direction; $S_0 = a'(R_c + \delta')$ is the area of this contour; H_l is the magnetic field induced by the alternating electrical field.

Taking into account equation (1.16) we can write

$$H_l a' = \varepsilon\varepsilon_0 \int_{S_0} \frac{\partial}{\partial t}\left(\frac{R_c + \delta'}{2} \cdot \frac{\partial B}{\partial t}\right) dS.$$

From this equation, after simple transformations, we obtain:

$$H_i = -\frac{\varepsilon\varepsilon_0 d^2}{16}\frac{\partial^2 B}{\partial t^2}.$$

In the evaluation calculations, it is permissible to assume that the only reason for the formation of the alternating magnetic field H_i is the variation of the magnetisation of the fluid due to its deformation and therefore $\Delta B = \mu_0 \cdot \Delta M$ and, consequently

$$H_i = -\frac{\varepsilon d^2}{16c_e^2}\frac{\partial^2 \Delta M}{\partial t^2},$$

where c_e is the speed of propagation of the electromagnetic wave.

In the case of harmonic oscillations we obtain

$$\frac{\partial^2(\Delta M)}{\partial^2} = -\omega^2 \cdot \Delta M$$

and

$$\frac{H_i}{\Delta M} = \frac{\pi^2 \varepsilon d^2}{4\lambda_e^2}.$$

Finally, we have

$$\frac{H_i}{\Delta M} \cong 2.5\left(\frac{d}{\lambda_e}\right)^2, \tag{1.21}$$

where λ_e is the length of the electromagnetic wave at the frequency of elastic oscillations.

This equation shows that $H_i \ll \Delta M$ in the entire frequency range up to hypersonic.

Thus, the distribution of the magnetic fields in the specimen at every moment of time can be described by the equations of the static magnetic field:

$$\operatorname{div}\boldsymbol{B} = 0, \ \operatorname{rot}\boldsymbol{H} = 0,$$

the relationships $B_{n1} = B_{n2}$ and $H_{\tau 1} = H_{\tau 2}$ are fulfilled on the surface of the specimen, and the effects, associated with the finite value of the speed of propagation of electromagnetic perturbations in the ultrasonic frequency range can be ignored.

At the same time, the variations of induction in the specimen, according to equation (1.16), lead to the appearance of an EMF which shows that the investigated field is quasi-stationary [57]. The condition of the quasi-stationary nature of deformation magnetisation is based on fulfilling the inequality

$$d / \lambda_e < 0.1. \tag{1.22}$$

The process of magnetisation of the magnetic fluid is determined mainly by two mechanisms of orientation of the magnetic moments of the FP along the magnetic field. One mechanism is associated with the Brownian rotational movement of the particles in the liquid matrix, the other mechanism is determined by the thermal fluctuations of the moment inside the very particle [34]. Each relaxation processes is characterised by the specific time. The duration of rotational Brownian diffusion of colloidal particles is expressed by the equation

$$\tau_B = \frac{3V\eta_{SO}}{k_0 T}, \tag{1.23}$$

where η_{SO} is the static shear viscosity of the fluid–carrier.

At $\eta_{SO} = 0.13 \cdot 10^{-2}$ Pa·s, $T = 300$ K, $V = 10^{-24}$ m^3, $\tau_e \approx 3 \cdot 10^{-7}$ s.

The mechanism, determined by the thermal fluctuation of the magnetic moment inside the particle, is typical of the small single-domain particles. With the reduction of the size of the single-domain particles to several nanometres, the coercive force, characterising these particles, rapidly decreases to 0. The behaviour of the ensemble of the small solid particles is similar to that of the paramagnetic atoms with a large magnetic moment. At the thermal fluctuations, vector $\mathbf{m}_*$ is oriented in different spatial directions – the axes of easy magnetisation, separated by the potential barriers.

In a uniaxial magnetic particle in the absence of the magnetic field, the magnetic moment is rotated to a specific side along the axis of easy magnetisation. To ensure that the direction of the magnetic moment changes to the opposite direction, the magnetic moment must overcome a potential barrier whose height is determined by the energy of the crystallographic magnetic anisotropy $K_a V_f$. This orientation mechanism $\mathbf{m}_*$ is referred to as the Néel's mechanism.

The processes of establishment of the equilibrium magnetic moment averaged over the ensemble of the identical particles is characterised by two times: the relaxation time of the magnetic moment at the specific direction of the two-sided axis of easy magnetisation τ_0 and the Néel relaxation time through the barrier of magnetic anisotropy, separating two equivalent directions of the magnetic moment τ_N [53]:

$$\tau_0 = \tau_\gamma \sigma_* = m_*/(2\beta\gamma_e k_0 T),\quad \tau_N = \tau_\gamma \sigma_*^{-3/2} \exp\sigma_*. \tag{1.24}$$

The range of variation of time τ_0 is determined only by the range of variation of the magnetic moment which is very narrow for the particles, used in the preparation of the magnetic fluids. The characteristic value is $\tau_0 = 10^{-7}$ s and, consequently, the average magnetic moment for many hydrodynamic processes can be regarded as established along the specific direction of the axes of magnetic anisotropy of the particle. The Néel time τ_N exponentially increases with the increase of σ and can vary in a wide range. Therefore, the dynamic properties of the magnetic fluids may depend strongly on the nature of Néel relaxation in the particles used in the preparation of the fluids.

The Néel time depends very strongly on the particle size. At the values of the diameter of the magnetite particle, dispersed in kerosene, 8, 10 and 12.5 nm at $T_c = 25°C$ τ_N has the values of respectively 10^{-18}, 10^{-9} and 1 s [58] and, at the same time, the time τ_e is equal to $3.8\cdot10^{-7}$, $7.6\cdot10^{-7}$ and $1.5\cdot10^{-6}$ s, i.e., changes only slightly.

The actual magnetic fluids are mixtures of particles of different sizes and, consequently, the size distribution of the particles is not always available. Of the two mechanisms of relaxation of magnetisation the important mechanism is the one characterised by the shorter duration of rotational diffusion. When $\tau_N >> \tau_e$ the equilibrium orientation of the magnetic moments is established mainly by the Brownian rotation of the particles.

M.M. Maiorov [60] carried out the measurements of the complex magnetic permittivity of seven MF specimens in the frequency range from 30 Hz to 100 kHz and determined the relaxation time of the magnetic moment. The results show that in the specimens of the magnetic fluids, prepared by the conventional procedure on the basis of magnetite and kerosene, $\tau = 3\cdot10^{-5}$ s, and the increase of the viscosity of the fluid-carrier by a factor of 3.5 results in a fivefold increase of the relaxation time. This result is explained by the dominant contribution of the Brownian motion of the ferroparticles to magnetisation relaxation. With the increase of the dimensions of the superparamagnetic particles, the contribution of the Néel mechanism to the relaxation of magnetisation decreases. The polydispersed nature of the ferrophase in the actual magnetic fluids, and also the presence of the aggregates determine the existence of the spectra of the magnetic relaxation time [34, 39, 60–62].

The time dependence of the process of establishment of the equilibrium magnetic state is described by the magnetisation relaxation equation [34]. This for a compressible magnetic fluid was proposed in [12]:

$$\frac{\partial \mathbf{M}}{\partial t} = -\tau_1^{-1}(\mathbf{M} - \mathbf{M}_e) - \mathbf{M}\,\mathrm{div}\,\frac{\partial \boldsymbol{u}}{\partial t}, \tag{1.25}$$

where τ_1 is the relaxation time of the magnetisation component, parallel to the vector of the strength of the magnetic field, u is the displacement of the particle of the medium from the equilibrium position.

The internal rotation of the magnetic particles, typical of the magnetic colloids, results in slightly different field dependences of the relaxation time for the components longitudinal and transverse to the external field. In particular, the equation obtained for the longitudinal component by M.I. Shliomis [34] has the following form

$$\tau_1 = \frac{d(\ln L(\xi))}{d(\ln \xi)}.$$

The first term in the right-hand part of the equation (1.25) characterises the delay of the increase of magnetisation, the second term is the instantaneous component of this increment. In the case of perturbation of the magnetisation of the fluid by a flat sinusoidal acoustic wave, the magnetisation relaxation equation has the form

$$(\mathbf{M} = \mathbf{M}_0 + \delta \mathbf{M})$$

$$i\omega\tau_1 \partial \mathbf{M} = \mathbf{M}_e - \mathbf{M} - i\omega\tau_1 \mathbf{M}\frac{\partial u}{\partial x}. \tag{1.26}$$

2

Perturbation of magnetic induction by sound

2.1. Perturbation of magnetisation of the magnetic fluid by sound

The expression for the perturbation of the magnetisation of the medium by a flat sound wave will be derived [3]. The perturbation of the magnetisation of the medium by a flat sound wave as the result of the dependence of magnetisation of the medium on its density was taken into account for the first time in a theoretical study by I.E. Tarapov [2] and subsequently by B.I. Pirozhkov and M.I. Shliomis [12]. They showed that the perturbation of magnetisation is accompanied by the perturbation of the strength of the magnetic field. In contrast to [12], in which the solutions of the magnetisation of the magnetic fluid were evaluated assuming that the sound wave is isothermal, here it will be assumed that the wave is adiabatic and, in contrast to [2], the phenomenon of relaxation of magnetisation will be taken into account.

Let it be that a flat monochromatic isoentropic sound wave with the circular frequency ω propagates in a magnetic fluid along the OX axis, and the vector **H** is directed along the axis OY, i.e., **H** and wave vector **k** are mutually orthogonal. At $\mathbf{H} \perp \mathbf{k}$ $\delta H = 0$, there is no perturbation of the demagnetizing field. This claim will be confirmed. Since the wave propagates in the direction normal to **B**, in the compression phase, in the vicinity $x = 0$ (Fig. 2.1), **B** increases (the density of the induction lines increases), and in the rarefaction phase, in the vicinity $x = \lambda/2$, **B** decreases. For the rectangular contour $ACDF$, whose sides AF and CD are parallel to **B**, we will use the theorem of the circulation of the vector **H** in the quasi stationary

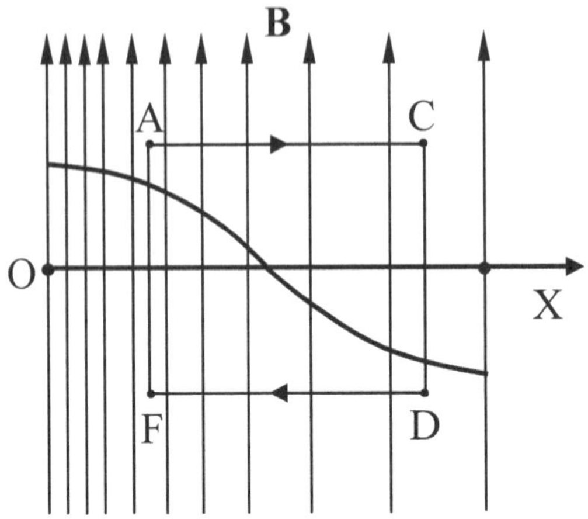

Fig. 2.1. Modulation of the induction of the magnetic field by the flat sound wave.

approximation: $\oint_L (\mathbf{H} \cdot d\mathbf{l}) = \sum_i I_i$. The right-hand part of the equation converts to 0 because there are no macroscopic currents I_i inside the circuit. Therefore, the strength of the magnetic field at all points of the sections AF and CD is the same. Because of the arbitrary length of AC and FD at any value x $H = H_0 =$ const.

If there is no perturbation of the demagnetising field, according to the equation (1.4) the equilibrium value of the magnetisation of the fluid can be expressed using the equation

$$M_e = M_0 + M_n \cdot \delta n + M_T \cdot \delta T. \quad (2.1)$$

The equations of conservation of energy and continuity lead to [63]:

$$\delta T = qTc^2 C_p^{-1} \rho^{-1} \cdot \delta\rho; \quad (2.2)$$

$$\frac{\delta n}{n} = \frac{\partial u}{\partial x}, \quad (2.3)$$

where $q \equiv -\rho^{-1} \frac{\partial \rho}{\partial T}$ is the temperature coefficient of expansion; ρ is the density of the liquid; c is the velocity of propagation of sound in the magnetic fluid in the absence of the magnetic field; C_p is the specific heat capacity at constant pressure and constant strength of the magnetic field; u is the displacement of the particles from the equilibrium position.

The condition of constancy of the strength of the magnetic field in the determination of its capacity becomes important because of the magnetocalorific effect.

Solving the system of equations (1.26), (2.1)–(2.3), and denoting $\gamma_* \equiv qTc^2C_p^{-1}$, we obtain the following equation:

$$\frac{\delta M}{M_0} = -\left[\begin{array}{l}\left(n\frac{M_n}{M_0} + \gamma_*\frac{M_T}{M_0} + \omega^2\tau^2\right)(1+\omega^2\tau^2)^{-1} \\ +i\omega\tau\left(1 - n\frac{M_n}{M_0} - \gamma_*\frac{M_T}{M_0}\right)\times(1+\omega^2\tau^2)^{-1}\end{array}\right]\frac{\partial u}{\partial x}. \tag{2.4}$$

Under the condition of the linear form of the dependence $M(n)$, i.e. at $M = C \cdot n$, where C = const, the real part δM can be written in the following form

$$\delta M = -[M_0 + \gamma_* M_T\left(1+\omega^2\tau^2\right)^{-1}]\frac{\partial u}{\partial x}. \tag{2.5}$$

The second term in the square brackets of the above expression is of the relaxation nature because of the variations of temperature in the adiabatic sound wave.

In the high-frequency range ($\omega\tau \gg 1$) from (2.5) we obtain

$$\left(\frac{\Delta M}{M_0}\right)_\infty = -\frac{\partial u}{\partial x},$$

At $\omega\tau \ll 1$

$$\left(\frac{\Delta M}{M_0}\right)_0 = -\left(n\frac{M_n}{M_0} + \gamma_*\frac{M_T}{M_0}\right)\frac{\partial u}{\partial x}. \tag{2.6}$$

Using the Langevin equation (1.1), the equation (2.6) can be transformed to the following form

$$\left(\frac{\Delta M}{M_0}\right)_0 = -\left[1 - qc^2\left(\xi^{-1} - \frac{\xi}{\text{sh}^2\xi}\right)C_p^{-1}\left(\text{cth}\,\xi - \xi^{-1}\right)^{-1}\right]\frac{\partial u}{\partial x}. \tag{2.7}$$

In the initial section of the magnetisation curve ($\xi \ll 1$) we have

$$\left(\frac{\Delta M}{M_0}\right)_0 = -\left[1 - qc^2C_p^{-1}\right]\frac{\partial u}{\partial x}.$$

and on approach to saturation ($\xi \gg 1$, $M \to M_s$)

$$\frac{\Delta M}{M_0} = -\left[\frac{1 - qc^2nk_0T}{\mu_0 C_p H M_s}\right]\frac{\partial u}{\partial x}.$$

Finally, if $\xi = 1$, then

$$\left(\frac{\Delta M}{M_0}\right)_0 = -\left(1-0.83qc^2C_p^{-1}\right)\frac{\partial u}{\partial x}.$$

Let it be that $C_p = 2\cdot10^3$ J/kg·K [35], $c = 1120$ m/s [64], $q = 0.53\cdot10^{-3}$ K^{-1} [159], then $qc^2C_p^{-1} = 0.33$ The second term in the square brackets of the expression (2.5), determined by the absence of heat exchange in the wave, is significant in the range of weak and moderate magnetic fields [66]. On reaching the magnetic saturation $\frac{\gamma_* M_T}{M_0} = 0$, and the oscillations of magnetisation become 'instantaneous'.

The oscillations of the magnetisation will be evaluated on the bases of the previously mentioned experimental data and assuming that the amplitude of deformation in the sound wave is 10^{-4} and, correspondingly, the intensity is $J = 10^4$ W/m^2. Consequently, for the amplitude of the oscillations $\left(\frac{\Delta M}{M_0}\right)_0$ and $\left(\frac{\Delta M}{M_0}\right)_\infty$ we obtain $0.5\cdot10^{-4}$ and 10^{-4}.

From equation (2.7) we obtain easily the expression for the amplitude of oscillations of magnetisation

$$\Delta M_m = M_S\left[\operatorname{cth}\xi - \xi^{-1} - qc^2C_p^{-1}\left(\xi^{-1} - \xi\operatorname{sh}^{-2}\xi\right)\right]ku_m, \qquad (2.8)$$

where u_m is the amplitude of the displacement of the particles of the fluid from the equilibrium position, k is the reciprocal number.

The dependence $\Delta M_m(H_0)$ is determined by the multiplier present in the square brackets. This multiplier will be denoted by $F(\xi)$. Accepting the values of the parameters $q = 0.64\cdot10^{-3}$ K^{-1}, $c = 1200$ m/s, $C_p = 2.1\cdot10^3$ J/(kg · K), we obtain

$$qc^2C_p^{-1} = 0.45$$

and

$$F(\xi) = L(\xi) - 0.45D(\xi),$$

where $D(\xi) = \xi^{-1} - \xi\operatorname{sh}^{-2}\xi$.

The graphs of the functions $F(\xi)$ and $L(\xi)$ are shown in Fig. 2.2. The solid line is $L(\xi)$, and the dashed line $F(\xi)$. The curve of the dependence $F(\xi)$ passes below the $L(\xi)$ curve and reaches saturation in the range of saturation of magnetisation.

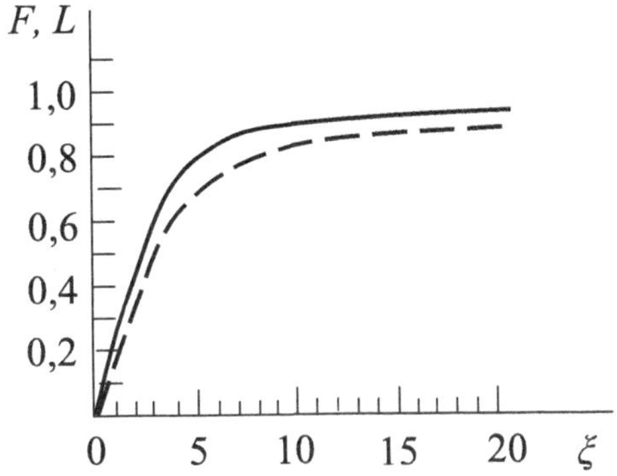

Fig. 2.2. Dependences $F(\xi)$ and $L(\xi)$: solid line – $L(\xi)$, dashed line – $F(\xi)$.

Let us now assume that the flat sound wave propagates in the magnetic fluid with along the axis OX in the direction parallel to the vector of the strength of the magnetic field, i.e. $\mathbf{H}||\mathbf{k}$. In the absence of sound $\mathbf{H}$ = const and $\mathbf{M} = \mathbf{M}_0$ = const. and, therefore, $\mathbf{B} = \mu_0 (\mathbf{H}_0 + \mathbf{M}_0)$. Due to the Maxwell equation div $\mathbf{B} = 0$, the increment of magnetisation involves the increment of the strength of the magnetic field [12]:

$$\delta H = -\delta M.$$

The equilibrium value of the magnetisation in this case has the form of (1.4) and the equations of energy and continuity remain without changes. Taking into account (2.2), (2.3), and the previously introduced notations, equation (1.4) will be transformed to the form

$$M_e = M_0 - (nM_n + \gamma_* M_T)\frac{\partial u}{\partial x} - M_H \cdot \delta M. \tag{2.9}$$

The nature of approach of M to M_e is determined by the same relaxation equation (1.25), and for the sinusoidal wave by (1.26). After substituting M_e into (1.26), from equation (2.9) we obtain

$$\delta M = -\frac{nM_n + \gamma_* M_T + i\omega\tau_1 M_0}{1 + M_H + i\omega\tau_1} \cdot \frac{\partial u}{\partial x}.$$

Denoting $\tau \equiv \tau_1 (1 + M_H)^{-1}$, we obtain

$$\delta M = -\frac{(nM_n + \gamma_* M_T)/(1 + M_H) + i\omega\tau M_0}{1 + i\omega\tau} \cdot \frac{\partial u}{\partial x}. \tag{2.10}$$

The equations of the oscillations of magnetisation (2.5) and (2.8) can be used when there is no slipping of the particles in relation to the liquid matrix. For the relatively large particles, for example, in a ferrosuspension (FS), this condition is fulfilled only approximately. The theory of propagation of sound in the dispersed systems contains the equation for the calculation of the relative velocity of the particles in the medium [67, 68] β□$_v$ (β□$_v$ is the ratio of the oscillatory velocity of the suspended particles to the velocity of the surrounding medium):

$$\tilde{\beta}_V = \frac{1+\sqrt{\Psi_v}+i\sqrt{\Psi_v}\left(1+2\sqrt{\Psi_v}/3\right)}{1+\sqrt{\Psi_v}+i\sqrt{\Psi_v}\left(1+b_2\sqrt{\Psi_v}\right)}, \tag{2.11}$$

where $\Psi_v = \omega\rho_1 R_p^2 / 2\eta_{s1}$; $b_2 \equiv 2/9(1+2\gamma_0)$; ρ_1 and η_{s1} is the density and shear viscosity of the fluid-carrier; R_p is the radius of the particles of the dispersed phase; ω is the circular frequency of the harmonic oscillation; ρ_2 is the density of the particles of the dispersed phase.

Separating the real part of the equation (2.11) we obtain

$$\beta_v = \frac{(1+\sqrt{\psi_v})^2+\psi_v\left(1+2\sqrt{\psi_v}/3\right)\left(1+b_2\sqrt{\psi_v}\right)}{(1+\sqrt{\psi_v})^2+\psi_v\left(1+b_2\sqrt{\psi_v}\right)^2}.$$

Figure 2.3 shows the graph of the dependence β_v (R_p) on the semi-logarithmic scale. It was accepted that ρ_1 = $0.8\cdot10^3$ kg/m^3, ρ_2 = $5.2\cdot10^3$ kg/m^3, η_{s1} = $1.3\cdot10^{-3}$ kg/m·s, v = 25 MHz. The crosshatched area indicates the range of the values of R_p typical of the magnetic fluid. The graph shows that the sleeping of the particles of the dispersed phase is detected starting at R_p equal to 450 nm and when R_p increases to 1–10 μm it becomes very large. The range of the size of the particles, corresponding to the stable magnetic fluid, is situated at the beginning of the horizontal section of the curve.

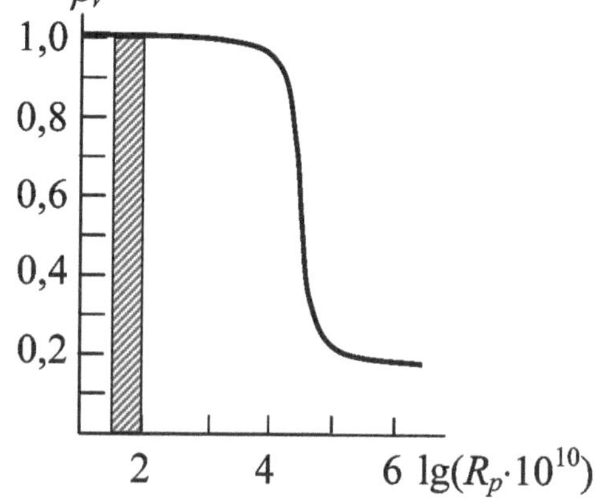

Fig. 2.3. The dependence β_V (R_p), [R_p] = m.

The magnetisation oscillations, determined by the oscillations of the concentration of the ferroparticle (FP) in the sound wave, are described by the equations (2.5) and (2.8).

Taking into account the possible slipping of the particles, the continuity equation assumes the following form $\frac{\delta n}{n} = -\frac{\partial u_k}{\partial x}$, where u_k is the displacement of the particles from the equilibrium position. Since in the harmonic process $\vartheta = i\omega u$, $u_k = \frac{\tilde{\beta}_V \vartheta}{i\omega}$, then $\frac{\partial u_k}{\partial x} = \frac{\partial(\tilde{\beta}_V u)}{\partial x}$. Therefore, the increment of the magnetisation δM can be written in the following form:

$$\delta M = -(nM_n\tilde{\beta}_V + \gamma_* M_T) \cdot \frac{\partial u}{\partial x}.$$

The real part of the last equation for the linear dependence $M(n)$ has the form

$$\Delta M = -(\beta_V M + \gamma_* M_T) \cdot \frac{\partial u}{\partial x}.$$

Taking into account the finite relaxation time, we obtain

$$\Delta M = -\left[\beta_V M + \frac{\gamma_* M_T}{1+\omega^2\tau^2}\right] \cdot \frac{\partial u}{\partial x}.$$

The amplitude of the oscillations of magnetisation is determined from the following expression:

$$\Delta M = -\left[\beta_V M + \frac{\gamma_* M_T}{1+\omega^2\tau^2}\right] \cdot u_m k.$$

Assuming that the static magnetisation is described by the Langevin equation, we obtain for the amplitude of oscillations of magnetisation

$$\Delta M_m = M_s\left[\beta_V L(\xi) - \frac{qc^2 D(\xi)}{C_p\left(1+\omega^2\tau^2\right)}\right] u_m k. \tag{2.12}$$

In the low frequency range ($\omega\tau \ll 1$)

$$\Delta M_m = M_s\left[\beta_V L(\xi) - \frac{qc^2 D(\xi)}{C_p}\right] u_m k. \tag{2.13}$$

The equations (2.12) and (2.13) differ from (2.8) by the multiplier β_v at the Langevin function.

The process of slipping of the particles in the magnetic fluid can be intensified as a result of the aggregation of the magnetic particles in the presence of the magnetic field [18, 45, 47, 126–130]. In cases in which the agglomeration results in the formation of magnetic chains, elongated preferentially along the field, the direction of flow of the liquid around these chains becomes very important.

2.2. Elastic oscillations of the magnetic fluid cylinder at the basic frequency

In propagation of the sound wave along the cylindrical column of the magnetised fluid, the perturbation of magnetisation is non-uniform along the length of the cylinder even when magnetic saturation is reached [54].

We examine the calculation model (Fig. 2.4). Both bases of the liquid cylinder are acoustically free; the magnetising field, the axis of the cylinder and the OX axis are parallel to each other; the point $x = 0$ is situated in the middle of the liquid; the displacement of the liquid particles is governed by the equation $u = u_0 \cdot \sin kx \cdot \sin \omega t$; the length of one standard wave is located along the length of the liquid cylinder. The cylinder is regarded as a set of pairs of discs with thickness dx and coordinates x and $-x$. The demagnetising field at the points of the central circular section, formed by the symmetric discs, is proportional to the increment of magnetisation in the disc thickness $dH_p = N \cdot dM$. The proportionality coefficient $N\left(\frac{2x}{d}\right)$ is assumed to be equal to $N_\delta(P)$ which is valid for the homogeneous cylinder magnetised to the value dM characterised by the shape parameter $P = \frac{2x}{d}$ [56].

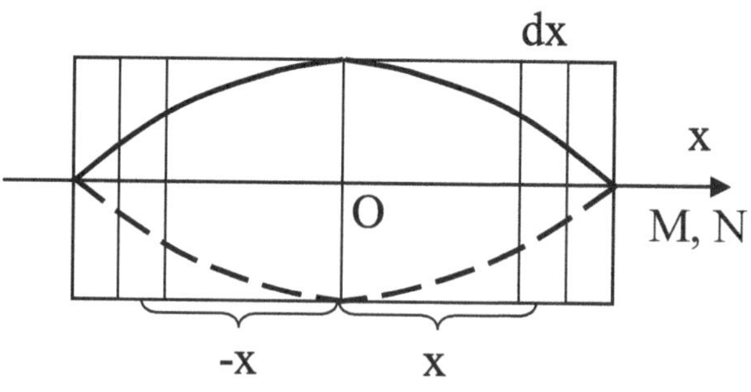

Fig. 2.4. Calculation model.

The perturbation of the demagnetising field in the central circular section of the cylinder can be described by the following equation:

$$\delta H_p = \left(N_{\delta d} + \Delta N_d\right)\delta M. \tag{2.14}$$

The second term in the brackets characterises the part of the perturbation of the demagnetising field determined by the oscillations of the length of the cylinder. It will be referred to as the parameter $N_d \equiv N_{\delta d} + \Delta N_d$, 'the dynamic demagnetising factor'. It may be shown that

$$\Delta N_d = -\frac{2P}{\pi}\frac{\partial N_\delta}{\partial P}.$$

The parameter $N_{\delta d}$ characterises the demagnetising field connected with the perturbation of magnetisation of the cylinder

$$N_{\delta d} = b\int_0^{\frac{\lambda}{2d}} N_\delta(P)\cdot \sin(bP)\cdot dP, \tag{2.15}$$

where $b = kd / 2$.

Substitution of the equation (1.6) into equation (2.15) causes that the integral can be computed through the elementary functions. Therefore, we use the interpolating function in the form

$$N_\delta(P) = N_\delta(P_n)\cdot \exp(-a_n P), \tag{2.16}$$

where $n = 1, 2, 3,\ldots$ is the number of the interpolation node; $P_1 = 0$ and $P_n \le P \le P_{n+1}$.

The unknown parameter a_n is determined from the expression

$$a_n = P_{n+1}^{-1}\ln\left(\frac{N_\delta(P_n)}{N_\delta(P_{n+1})}\right).$$

The following interpolation nodes are introduced: 0, 0.4, 1, 2, 3, 6.

The results of calculation of the parameters ΔN_d, $N_{\delta d}$ and $N_d \equiv N_{\delta d} + \Delta N_{\delta d}$ are presented graphically by the curves 1, 2 and 3 in Fig. 2.5. The function $\Delta N_{\delta d}$ (P) has the maximum at $P \approx 0.5$. With increase of P the parameter $N_{\delta d}$ decreases from 1 to 0, but the slope of the curve $N_{\delta d}$ (P) is less steep than that of the N_δ (P) curve, represented by the dashed line. The value N_d is also in the range from 1 to 0.

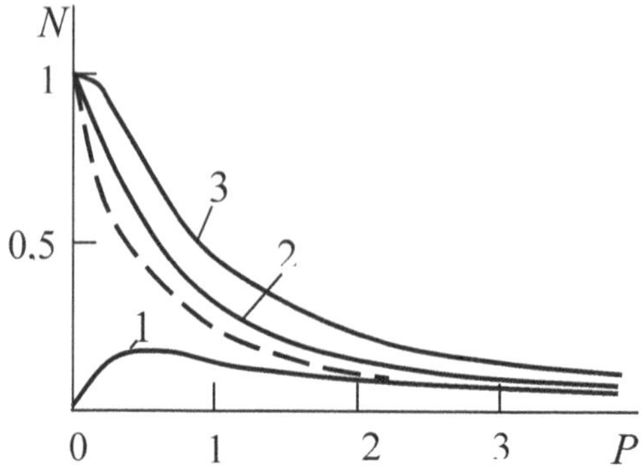

Fig. 2.5. Dependence of the demagnetising factor on the form parameter P: 1 – $N_{\delta d}$, 2 – $N_{\delta d}$, 3 – N_d, the dashed line N_δ.

Using a similar calculation procedure, we could obtain the expression for the parameter N_d for the mutually orthogonal position of the vector of the strength of the magnetising field and the axis of the cylinder. It is natural to expect in this case that with the variation of P from 0 to ∞, N_d changes from 0 to 0.5.

In the presence of a system of standing (or running) waves in the column of the magnetised fluid, the distribution of the contribution of the parameters $N_{\delta d}$ and ΔN_d in the value of the dynamic deformation demagnetising factor greatly changes.

Let it be that along the length of the cylinder L there are $(2m+1)$ standing waves ($m = 0, 1, 2,\ldots$). The presence of symmetry in the perturbation of magnetisation in relation to the central circular section makes is also possible in the present case to use the introduced calculation model

$$N_{\delta d} = \frac{kd}{2} \int_0^{(2m+1)\lambda/2d} N_\delta(P) \cdot \sin\left(\frac{kd}{2} P\right) dP. \qquad (2.17)$$

The same equation for the parameter $N_{\delta p}$ is obtained in the case of the propagation of the running wave in the liquid cylinder if the condition of symmetry of the perturbation in relation to the central cross-section is fulfilled: $N_{\delta d}$ is determined at the moment of time m/v in the bases of the column, $x = \pm L/2$; the phase of the oscillations is equal to $\pm(2m+1)\pi/2$. To obtain exclusively qualitative conclusions, the interpolation equation (2.16 in) will be simplified by restricting the equation to a single term: $N'(P) = \exp(-aP)$. Replacing N_d by N' in equation (2.17) we obtain

$$N_{\delta d} = \frac{1 - \frac{a\lambda}{\pi d} N'(P) \cdot \sin\left[(2m+1)\frac{\pi}{2}\right]}{1 + (a\lambda / \pi d)^2}$$

If $P \gg 1$, then $N' \approx 0$ and $\Delta N_d \approx 0$, and, therefore,

$$N_d \cong \left[1 - \left(\frac{a\lambda}{\pi d}\right)^2\right]^{-1}. \tag{2.18}$$

Under these conditions, the factor N_d does not depend on the form parameter P and its value is now determined by the ratio λ/d. With the reduction of wavelength $\lambda/d \to 0$, and $N_d \to 1$.

The ellipsoidal particles in the conventional fluids during propagation of the sound wave in them are influenced by the orientation mechanism, introduced in the monograph by Ya.I. Frenkel' [69]. In the areas in which the fluid is subjected to tensile loading, the chains of the ferroparticles should be oriented preferentially in the direction of the vector $\pm\mathbf{k}$, and in the compressed regions – in the transverse direction. The magnetic chain should be subjected to rotating oscillations whose amplitude, with the other conditions being equal, decreases with the increase of the strength of the external magnetic field. In the compression phase, the orientation effect results in a small reduction of the degree of magnetisation of the liquid, and in the tensile phase in a corresponding increase of the magnetisation. The amplitude of perturbation of magnetisation at the points of the central circular section of the cylindrical column of the demagnetising liquid in the mode of longitudinal oscillations of the column is calculated from the equation

$$\Delta M = \frac{M\left[M_\beta + \left(\frac{\pi c \tau}{L}\right)^2\right]\dot{u}_0}{c\left[1 + \left(\frac{\pi c \tau}{L}\right)^2\right]}, \tag{2.19}$$

where $\dot{u}_0$ is the amplitude of the oscillatory velocity;

$$\tau \equiv \tau_1\,(1+N_d + M_n)^{-1};$$

$$M_\beta \equiv (nM_n + \gamma_* M_T)[M(1+N_d M_H)]^{-1}$$

The sliping of the ferromagnetic particles of aggregates in the fluid–carrier with the relative velocity β_V is taken into account by replacing M_β in equation (2.19) by $M'_\beta = (\beta_V\, nM_n + \gamma_* M_t)[M(1+N_d M_H)]^{-1}$.

Equation (2.19) showed that in the region of magnetic saturation, where $M_H = 0$, the perturbation of magnetisation becomes maximum and independent of the perturbation of the demagnetising field. In the range of weak and moderate magnetic fields, the perturbation of the demagnetising field can be ignored in determination of δM only in the case of a magnetic fluid with a low concentration and for weakly magnetic fluids for which $M_H << 1$.

2.3. Acousto-magnetic effect

A very important electromagnetic effect, caused by the propagation of the flat soundwave in the magnetic fluid, is the induction of the alternating electrical fields and EMF in the conducting circuit which is referred to as the acousto-magnetic effect (AME).

We estimate the value of the induced EMF, assuming that the conducting circuit is rectangular with the one side equal to h and the other side equal to $\lambda/2$ [3]. Consequently, on the basis of the law of electromagnetic induction we obtain

$$e_m = 2\mu_0 h N_k M_s \omega u_m,$$

where N_k is the number of turns in the circuit.

Let it be that the intensity of sound is equal to 10^5 W/m^2 and, therefore, the amplitude of the speed of displacement in the sound wave $\omega u_m \approx 0.35$ m/s. At $h = 10^{-2}$ m, $N_k = 10$, $M_s = 30$ kA/m, we obtain $e_m = 2.5\cdot 10^{-3}$ V.

The problem of the application of the magnetic fluid as a transducer–receiver of sound oscillations was studied by B.B. Cary and F.H. Fenlon [1]. The study was concerned mainly with the problem of generation of sound oscillations using a magnetic fluid. The transducer, proposed in the study, has the form of a tablet; magnetisation is carried out by a magnetic field whose lines of induction are colinear in relation to the tablet axis. In the model which they examined, the magnetic field was homogeneous on both sides of the medium–magnetic fluid interface, and because of the constant value of the normal component of magnetic induction it was concluded that the AME can not operate in this case.

In isothermal longitudinal deformation, according to the Maxwell equation (1.16), the EMF forms in the circuit encircling the cross-section of the liquid cylinder and containing N_k turns

$$e = -0.25\mu_0(1 - N_{c\ell})\pi n M_n d^2 N_k \beta_T (1 + N_{c\ell} M_n)^{-1} \frac{\partial(\delta p)}{\partial t}. \qquad (2.20)$$

If the variable part of the pressure varies in accordance with the harmonic law with a frequency ω, the induced EMF also carries out harmonic oscillations with the amplitude:

$$e_0 = 0.25\mu_0(1 - N_{c\ell})\pi n M_n d^2 N_k \omega \beta_T (1 + N_{C\ell} M_n)^{-1} \delta p_0. \qquad (2.21)$$

In the adiabatic process of longitudinal deformation

$$e_0 = 0.25\mu_0(1 - N_{c\ell}) N_k \left[\pi d^2 (n M_n + \gamma_* M_T)\right] \omega \beta_s (1 + N_{c\ell} M_n)^{-1} \delta p_0 . \qquad (2.22)$$

For a sufficiently long cylinder $P>>1$, $N_{c\ell} \approx 0$.

In the case of radial deformation it is assumed that there is a gap between the circuit and the surface of the cylinder within which the oscillatory movement of the side surface of the cylinder takes place. In the adiabatic processes of radial deformation

$$\begin{aligned} e_0 = 2\mu_0 \pi N_k R_c (1 + N_{cr} M_H)^{-1} [n M_n - M + \gamma_* M_T (1 - N_{cr}) \\ - n M_n N_{cr} + M(N_\delta - M_H N_{cr}(1 - N_{cr}))] \delta R_0 \omega \end{aligned} \qquad (2.23)$$

For a disc ($P \ll 1$) the equation (2.23) gives $e_0 = 0$. If the condition $P >> 1$ is satisfied, then $N_{cr} = 0$, $N_\delta = 0$ and equation (2.23) assumes the following form:

$$e_0 = 2\mu_0 \pi N_k R_c (n M_n - M + \gamma_* M_T) \delta R_0 \omega.$$

If the magnetisation of the liquid is directly proportional to the concentration of the particles n, then

$$e_0 = 2\mu_0 \pi N_k R_c \gamma_* M_T \delta R_0 \omega \qquad (2.24)$$

Under these conditions, the EMF is induced by the oscillations of temperature, and in the absence of temperature variations (isothermal process) it is no longer possible.

Taking into account equation (2.19), we can write the expression for the amplitude of the EMF induced in the circuit under the effect of longitudinal oscillations of the cylinder of the magnetised liquid at the basic frequency ω:

$$e_0 = \mu_0 \cdot 0.25(1 - N_d) N_k \pi d^2 M \ddot{u}_0 \left[M_\beta + (\pi c \tau / L)^2\right] c^{-1} \left[1 + (\pi c \tau / L)^2\right]^{-1} \qquad (2.25)$$

where $\ddot{u}_0$ is the amplitude of the oscillatory acceleration.

If the relaxation time of magnetisation is large, i.e., $\tau >> \frac{L}{\pi c}$, then

$$e_0 = \mu_0 \cdot 0.25(1 - N_d)\pi d^2 N_k M \ddot{u}_0 c^{-1} \tag{2.26}$$

Substitution into equation (1.14), (2.19), (2.25), (2.26) of the values of $N_{c\ell}$, N_{cr} and N_d, shown in Table 1.2 and in Fig. 2.5, should be regarded as only the first approximation, because these parameters are calculated assuming the magnetic saturation of the magnetic fluid.

If the magnetisation of the magnetic fluid is instantaneous and is governed by the Langevin law (1.1), the relative amplitude of the EMF $\beta_e = e_0/e_{0max}$ is a function of ξ:

$$\beta_e = \left[1 + \frac{\mu_0 n m_*^2 N_d D(\xi)}{k_0 T \xi}\right]^{-1} F(\xi), \tag{2.27}$$

where $D(\xi) \equiv \xi^{-1} - \xi \cdot sh^{-2}\xi$;

$$F(\xi) \equiv L(\xi) - 0.45D(\xi).$$

Correspondingly, for the equation (2.24) for the radial deformation we obtain

$$\beta_e' = \frac{D(\xi)}{D_{\max}(\xi)} = 2.87D(\xi) \tag{2.28}$$

The curves 1, 2 and 3 in Fig. 2.6 represent the dependences $L(\xi)$, $\beta_e(\xi)$ and $\beta_e'(\xi)$. Calculations were carried out assuming that $N_d = 0.5$ – does not depend on ξ, $n = 10^{23}$ m^{-3}, $C_p = 2000$ kJ/kg·K,

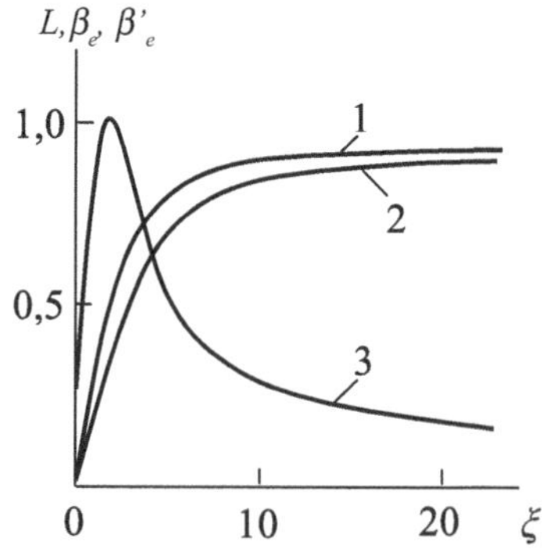

Fig. 2.6. Dependences: 1 – $L(\xi)$; 2 – $\beta_e(\xi)$; 3 – $\beta'_e(\xi)$.

ρ = 1200 kg/m^3, 20°C. Under these assumptions, the dependence β_e (ξ) has the form of a rising curve with a tendency to saturation, similar to the curve L (ξ), and the dependence β'_e (ξ) has a maximum at $\xi \cong 2$, $H \approx 25$ kA/m.

Thus, the exciting modes of the elastic oscillations can be characterised by the qualitatively different dependence $\beta_e(H)$. Equations (2.18) and (2.25) show that with the reduction of the wavelength, i.e., at $\lambda/d \rightarrow 0$, $N_d \rightarrow 1$ and $e_0 \rightarrow 0$, $\delta B \rightarrow 0$, in the limiting case of propagation of a flat infinite sound wave along the magnetising field the perturbation of the induction becomes equal to 0 [3].

2.4. Method for the experimental investigation of the acousto-magnetic effect

The most suitable procedure is the one based on the application of a cylindrical pipe produced from a non-magnetic and non-conducting material. The pipe, filled with a magnetic fluid, is placed partially or completely in the transverse or longitudinal magnetic field and, subsequently, a sound wave is introduced into the fluid. The alternating magnetic field, inducing the fluid, is received by the measuring inductance coil, and the variable EMF from the coil travels to the measuring device.

Various modes of normal waves can exist in the cylindrical pipe. If the frequency of oscillations is smaller than the critical frequency [70–72], only flat waves can exist in the pipe and propagate with the phase velocity c_T. The criterion of propagation of the flat wave in a cylindrical pipe is based on fulfilling the inequality

$$R_c \angle 0.61\lambda. \tag{2.29}$$

The introduction of the sound oscillations into the fluid takes place both through the free upper surface of the fluid column and through the membrane covering the pipe at the top. The very important circumstance, reflecting the specific properties of the investigated objects, is the fulfilment of the condition of reduction of the strength of the magnetic field in the vicinity of the free surface of the fluid to minimum. The component of the homogeneous magnetic field normal to the horizontal surface of the fluid must be smaller than the critical value H_{cr}. If this condition is not fulfilled, the flat form of the surface of the magnetic fluid is unstable in relation to small perturbations [34, 73, 74]. As a result of the large distortion of the surface of the liquid, the condition of formation of the standing waves is violated and the area of the active surface of the emitter changes in an uncontrollable manner [75]. For kerosene magnetic fluid $H_{cr} = 10 \div 15$ kA/m [39]. To fulfil this condition, it is necessary to use relatively long pipes and, consequently, the open surface of the fluid is far away from the working zone of the magnetic field in the area in which the strength of the field is almost insignificant.

In [75] the source of the constant magnetic field was represented by an electromagnet: an induction sensor with the number of turns $3000 \div 5000$ was used; the intensity of the sound emited to the fluid did not exceed 10^3 W/m^2.

Figure 2.7 shows the flow diagram of experimental equipment. The following devices are incorporated in the system: 1 – voltmeter, 2 – frequency meter, 3 – piezoelement (2 MHz); 4 – waveguide; 5 – generator; 6 – glass pipe; 7 – magnetic fluid; 8 – current source; 9 – electromagnet; 10 – inductance coil; 11 – oscilloscope.

This gives the relationship describing the dependence of the amplitude of the AME on the angle φ, formed between the direction of the magnetic field and the normal to the frame (Fig. 2.8), assuming that the circular frame is tightly pressed to the surface of the pipe and the axis of the pipe is normal to the vector of strength.

In static deformation (the pipe is absolutely rigid) of the fluid column

$$\delta B = \mu_0(\delta M - N.\delta M) = \mu_0(1 - N)\delta M.$$

In this case, $N = 0.5$. Because of the constant normal component of induction at the boundary of the magnetics we have:

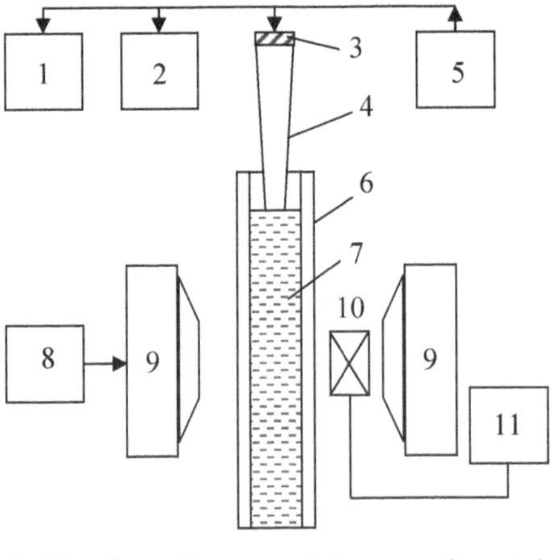

Fig. 2.7. The flow diagram of the experimental equipment.

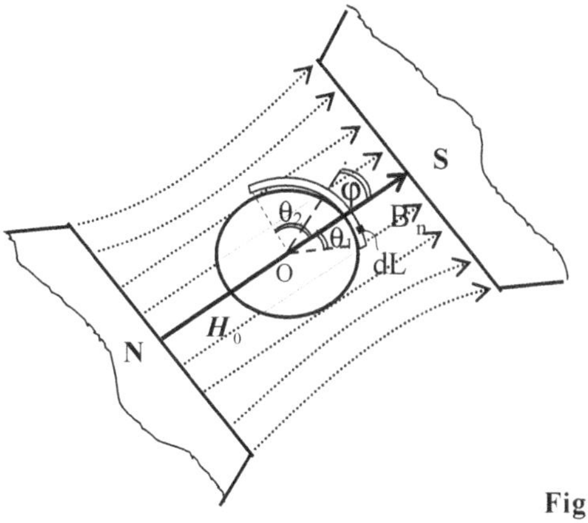

Fig. 2.8. The diagram of the problem.

$$\delta B_n^{(i)} = \delta B_n^{(e)},$$

where $\delta B_n^{(i)}$ and $\delta B_n^{(e)}$ are the normal components of the increment of the magnetic induction inside the pipe and on the pipe surface

$$\delta B_n^{(e)} = \delta B \cdot \cos\theta = \mu_0 (1 - N) \cdot \delta M \cdot \cos\theta,$$

where θ is the angle between the direction of the magnetic field and the beam restricting the frame.

The increment of the magnetic flux through the band with the width dL is:

$$\delta\Phi = N_B h \cdot \delta L \cdot \delta B_n^{(e)} = N_B h \cdot \delta L \cdot \mu_0 (1 - N) \cdot \delta M \cdot \cos\theta,$$

where L is the length of the frame; h is its height ($h \ll \lambda$); N_B is the number of turns.

On the other side

$$d\theta = 2dL / d,$$

where d is the pipe diameter.

Consequently

$$\delta\Phi = \mu_0 (1 - N) N_B \frac{d}{2} h \cdot \delta M . \cos\theta \cdot d\theta.$$

The magnetic flux, penetrating the contour of the frame, is:

$$\Delta\Phi = \Delta\Phi_1 + \Delta\Phi_2 = \mu_0 (1 - N) N_B \frac{d}{2} h \cdot \delta M \left[\int_0^{\theta_1} \cos\theta \cdot d\theta + \int_0^{\theta_2} \cos\theta \cdot d\theta \right].$$

In this case $\varphi = \frac{\theta_2 - \theta_1}{2}$, and, consequently,

$$\Delta\Phi = \mu_0(1-N)N_B \frac{d}{2} h \cdot \delta M \left[\int_0^{L/d-\varphi} \cos\theta . d\theta + \int_0^{L/d+\varphi} \cos\theta \cdot d\theta \right],$$

$$\Delta\Phi = \mu_0(1-N)N_B \cdot d \cdot h \cdot \delta M \cdot \sin\frac{L}{d} \cdot \cos\varphi.$$

The amplitude of the EMF induced in the circuit:

$$e = -\frac{d(\Delta\Phi)}{dt} = -\mu_0(1-N)N_B \cdot d \cdot h \cdot \frac{d(\delta M)}{dt} \cdot \sin\frac{L}{d} \cdot \cos\varphi.$$

At $L \ll d$ we obtain

$$e = -\mu_0(1-N)N_B \cdot L \cdot h \cdot \frac{d(\delta M)}{dt} \cdot \cos\varphi.$$

If $L < d$, should be replaced by N on N_d – the dynamic demagnetising factor (section 2.2).

In the equipment (Fig. 2.9) we can determine the field dependence of the EMF of induction for the mutually orthogonal and colinear position **H** and **k** on the same magnetic fluid and in the same mode of excitation of the ultrasonic oscillations. The identical conditions are produced by the application in the experimental equipment of the glass pipe 1 at a sound guide, with the pipe bent under the right angle. The pipe is filled with the magnetic fluid 2. The ultrasound emitter 3 is in contact with the open surface of the fluid column. 4 and 5 are the measuring inductance coils. The first of them is intended for measurement with the vectors **H** and **k** in the orthogonal position, the second one – for the collinear position of the vectors. At the bottom, the pipe is covered with a thin glass sheet. The pipe is situated between the poles of the electromagnet and can move in the vertical direction. The establishment of the system of the standing waves in the liquid column is confirmed by the presence of the maximum of EMF situated along the axis at the same distance from each other and observed in displacement of the measuring coils along the pipe [77].

To compare the dependences of the relative amplitude of the induced EMF on the strength of the field, obtained at $\mathbf{H} \perp \mathbf{k}$ and $\mathbf{H} \| \mathbf{k}$, the results of measurements at $\mathbf{H} \perp \mathbf{k}$ must be corrected taking into account the demagnetising field. In practice, this can be carried out conveniently by the following procedure: using the data of

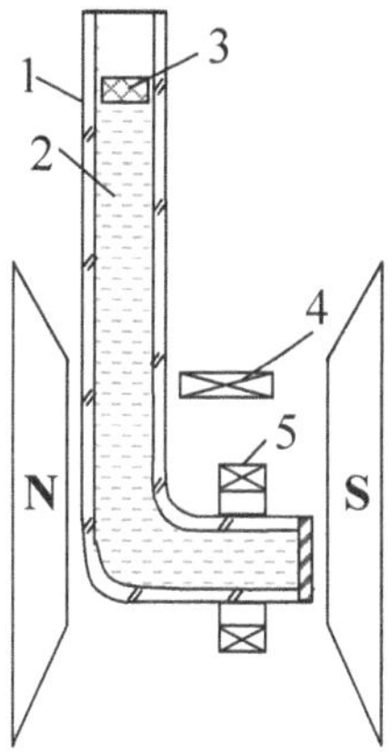

Fig. 2.9. The flow diagram of the experimental equipment.

the field dependence of magnetisation $M(H)$ in the longitudinal magnetisation of a relatively long cylindrical specimen of the investigated magnetic fluid and the results of calculations using the equation $H_e = H_i + 0.5M(H_i)$ it is possible to construct the table of the appropriate values of H_e, $= H_i$, $M(H_i)$; subsequently, the smooth curve of the dependence $H_i(H_e)$ gives the relevant values of H_i at **H** ⊥ **k**, and the 'corrected' curve is plotted.

The flow diagram of the experimental equipment, designed for comparative examination of the AME in the central lower (in the vicinity of the bottom plate) part of the pipe [78] and also for examining the spectrum of the modes of the oscillations in the magnetic fluid–cylindrical shell system [79, 92], is presented in chapter 8 in Fig. 8.1. The functioning of the individual units is also described there. The special feature of this equipment is that the elastic waves are introduced through the free surface of the magnetic fluid, filling the pipe, using the waveguide, and the magnetic head, probing the sound wave, activates the constant circular magnet, magnetised along the axis, and the inductance coil placed inside and rigidly connected with it.

Figure 2.10 shows the flow diagram of the experimental equipment designed for examining AME and measuring the speed of sound in the magnetic fluid, filling the pipe in the transverse heterogeneous magnetic field of a permanent magnet. The sound oscillations with a frequency of 20–70 kHz are introduced into the fluid through the plane-parallel bottom plate of the pipe. The signal from the generator of sound oscillations 1 travels in the parallel direction to the frequency meter 2, the voltmeter 3 and the piezosheet 4.

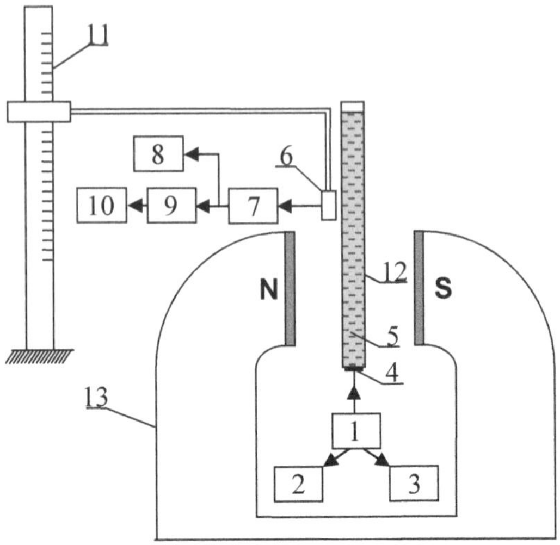

Fig. 2.10. Flow diagram of the experimental equipment based on AME.

Passing through the column of the magnetic fluid 5 and reflecting from its free surface, the sound signal produces a standing wave. The signal is recorded with the inductance coil 6 of cylindrical shape and adjacent to the site surface of the pipe. Subsequently, the signal travels to the selective amplifier 7 and, after amplification, to the oscilloscope 8 and the analog–digital transducer 9 connected with the computer 10 which records the amplitude of the received signal for subsequent processing. The inductance coil is rigidly secured on the kinematic section of the cathetometer 11. The magnetic fluid fills the glass pipe 12 situated between the poles of the permanent magnet 13. The inductance sensor is in the form of a winding of a copper conductor in the form of a circular frame (for encircling half the pipe).

The sensor is moved along the pipe using the cathetometer with the accuracy of 0.01 mm. The displacement section includes, in addition to the central part of the gap between the poles, also the adjacent areas of the heterogeneous field.

The NI LabView software is used for filtration of the resultant signal, expansion into a spectrum for controlling the level of interference, determination of frequency and the amplitude of the AME, and also for retaining the required data in the Excel format.

The composition of equipment includes the permanent magnet 13 whose magnetic field in the gap between the poles and in its vicinity has been studied sufficiently. The results of experimental and theoretical investigations of the heterogeneous magnetic field are presented in section 8.4.

2.5. Experimental results and analysis

In [80–82] investigations were carried out into the dependence of the relative amplitude of the EMF of the induction, determined by the AME, on the strength of the external magnetic field H, varying in the range 0–240 kA/m, and also on the relative amplitude of the alternating voltage, supplied to the piezotransducer, and on the displacement of the magnetic head along the column of the magnetic fluid. In the experiments, the vectors **H** and **k** were in the orthogonal position in relation to each other. Investigations were carried out on a stable sample of a concentrated magnetic fluid, prepared on the basis of magnetite and kerosene with the addition of oleic acid as a stabiliser. The density of the magnetic colloid, measured using a pycnometer, was $1.31 \cdot 10^3$ kg/m^3.

Figure 2.11 shows the dependence of the relative amplitude of the EMF β_v on the strength of the external field H_e, obtained at a frequency of 57.2 kHz. The black circles show results of measurement of β_e in the direct direction – with increasing H_e, and the open circles – in the reverse direction. β_e increases monotonically with increasing H_e, and with increasing H_e the rate of this increase decreases, and the dependence $\beta_e(H_e)$ tends to saturation.

The dependence of the amplitude of the induced EMF on the angle φ in the relative units, obtained in the experiments with the rotating magnetic field at $\mathbf{H} \perp \mathbf{k}$, is shown in Fig. 2.12. The points represent the experimental values, obtained in the process of rotation of the magnetic fields with a step of 5°. The thin line shows the graph of

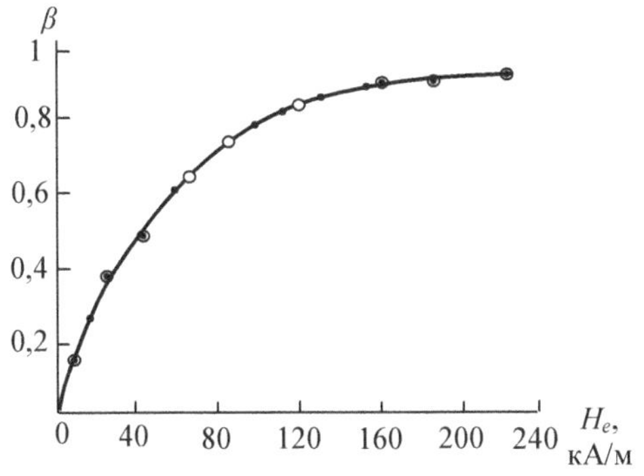

Fig. 2.11. Dependence of the relative amplitude β_e on H_e.

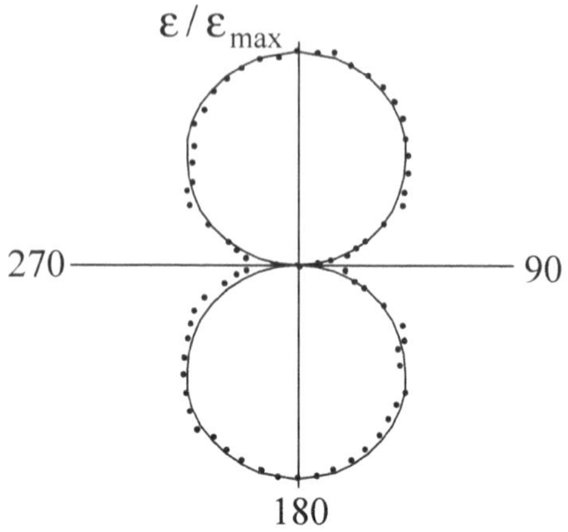

Fig. 2.12. Dependence of the relative amplitude of AME on the angle φ.

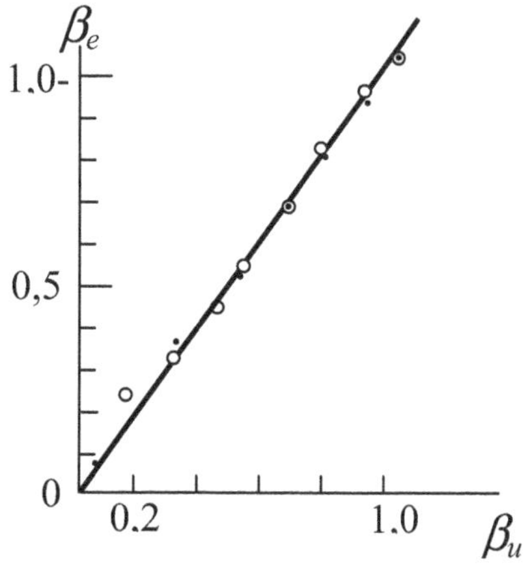

Fig. 2.13. Dependence of the relative amplitude $\beta_e(\beta_u)$.

cos φ. Thus, in the single rotation of the magnet, the amplitude, following the variation |cos φ|, assumes twice the maximum value and twice is equal to 0 [81].

Figure 2.13 shows the dependence of the amplitude of the induction EMF e_0 on the amplitude of the voltage of the alternating EMF *U*, supplied to the piezoelement, in the relative expression β_e (β_u). The experimental points are approximated quite efficiently by the straight line indicating the absence of cavitation processes in the investigated range of the amplitude of the ultrasonic field.

Investigations were carried out on magnetic fluids with a kerosene base with the density of $\rho = 1294$ kg/m^3 at a temperature of 20°C. The induced EMF was measured in the mode of external

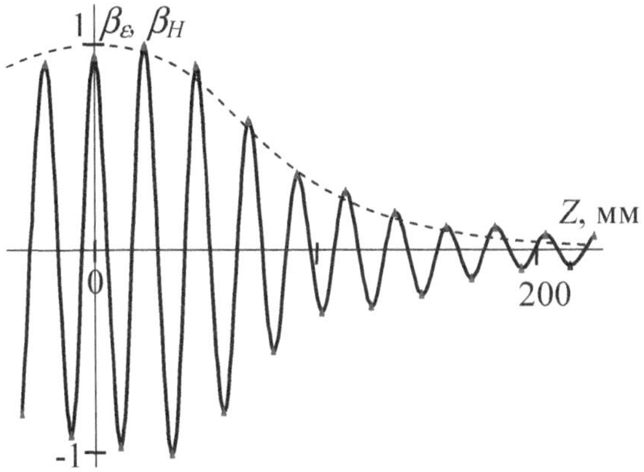

Fig. 2.14. Comparison of the dependences $\beta_e(Z)$ and $\beta_H(Z)$.

synchronisation of the oscilloscope from the generator. In transition through the boundary of the adjacent half-waves, the oscillation phase changed 'in a jump' by π.

Figure 2.14 shows the dependence of the relative amplitude of the oscillations of the EMF, induced in the circuit β_e on the coordinate Z (point with the coordinate $Z = 0$ coincides with the centre of the gap between the terminals of the permanent magnet). The same graph shows the dependence of the relative strength of the magnetic field $\beta_H(Z)$.

It may be seen that the dependences are in a qualitative agreement. This fact also corresponds to the introduced theoretical model. There are also some differences: firstly, at a large distance from the point $Z = 0$, the numerical value $\beta_e(Z)$ is greatly higher than $\beta_H(Z)$, secondly, with increasing distance, the monotonic variation of the adjacent maximum of the dependence $\beta_e(Z)$ is superposed with the small alternating 'steps'. The first special feature is explained by the non-linear form of the dependence $\beta_e(H)$, and the second one by the presence of a running wave, travelling away through the holder of the pipe, and of the long wave oscillation modes.

The coordinates of the maxima and the data on the strength of the field at these points were used to plot the graphs of the dependence of the AME amplitude on the strength of the magnetic field. One of these relationships is shown in Fig. 2.15. The points show the experimental data, the solid line is the approximation of the dependence in Excel software. The non-linear form of the dependence $\beta_\varepsilon(H)$ is clearly visible here.

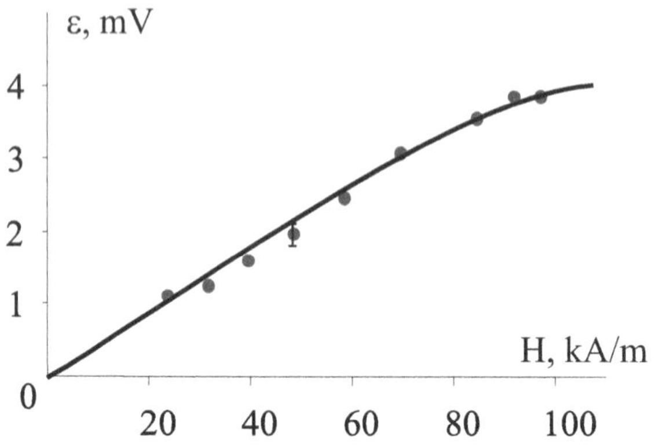

Fig. 2.15. Dependence of the AME amplitude on the strength of the magnetic field.

The investigation of the acousto-magnetic effects and the measurement of the speed of sound in the magnetic fluid using a permanent magnet under specific conditions may be a preferred procedure, because:

– using the heterogeneous magnetic field, we can obtain the field dependence of the amplitude of the induced EMF without any need to use the procedure of gradual increase of the strength of the field which is necessary when the electromagnet is used;

– the investigations of the AME in the heterogeneous magnetic field can be combined with the measurement of the speed of sound in the magnetic fluid:

– in the heterogeneous magnetic field, the concentration of the magnetic matter is redistributed and this may be recorded on the basis of the variation of the amplitude of perturbation of magnetisation which, in turn, can be used to characterise the stability of the magnetic fluid.

The determination of the reliable experimental results for the field dependence of the induction EMF is of considerable importance for justifying the physical model of AME in the magnetic fluid. Therefore, experiments were carried out to determine the dependence of the induction EMF on the strength of the magnetic field for the mutually orthogonal and collinear positions of **H** and **k** on the same fluid and in the same mode of excitation of ultrasonic oscillations (Fig. 2.9). The frequency of the ultrasonic oscillations, introduced into the fluid, is 61 kHz. Measurements were taken at a temperature of 22°C. Verification shows that the longitudinal oscillations are excited in the fluid column, as indicated by the presence of the

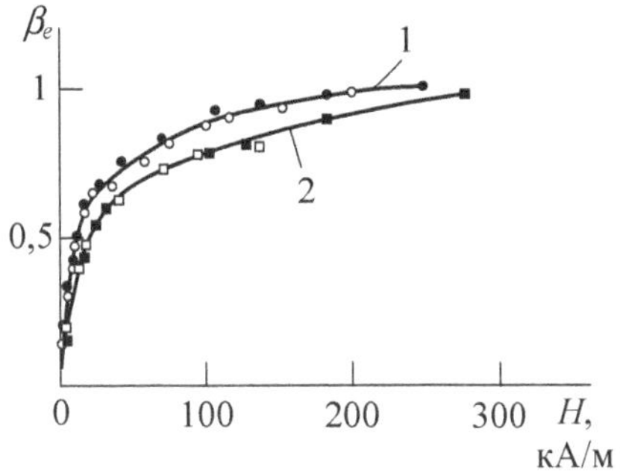

Fig. 2.16. Dependence $\beta_e(H)$ at **H** ⊥ **k** (1) and **H**||**k** (2).

maximum of the dependence $e_m(Z)$, observed in the displacement of the measuring coils along the pipe [77].

The specimen of the magnetic fluid was prepared on the basis of magnetite and kerosene and is density was $1.7 \cdot 10^3$ kg/m^3.

The graph of the dependence of the magnetisation of the magnetic fluid on the strength of the magnetising field $M(H)$, with its strength changing in the range from 0 to 820 kA/m, is similar to the 'Langevin' magnetisation curve. The curve of the dependence $M(H)$ was used to determine the values of initial permittivity χ and saturation magnetisation M_S: 6.7 and 82.5 kA/m.

Figure 2.16 shows the graph of the field dependence of the relative amplitude of EMF induced in the measuring coil, with the orthogonal position of **H** and **k** (curve 1) and for the collinear position (curve 2). The direct course of the dependence, obtained with increasing strength of the field, on the curve 1 is indicated by the crosshatched circles, and on the curve 2 by the crosshatched squares. The reverse course is indicated by the circles and squares without crosshatching. Both curves are corrected taking the demagnetising field into account.

Increasing the strength of the magnetising field increases the amplitude of the induced EMF, and in the fields of ~10^5 A/m there is a tendency for saturation of the dependence $\beta_e(H)$. The qualitative agreement of the curves 1 and 2 indicates the operation of a single mechanism of induction phenomena for two different orientations **H** and **k**, which is described by the given model.

2.6. Dependence of the amplitude of the acousto-magnetic effect on frequency along the length of the magnetic fluid cylinder

The investigation of the acousto-magnetic effect in the field

Table 2.1.

ν, kHz	20	30	42	50	59	63
A_C/A_B	2	1.6	1.25	1.0	0.8	0.5

of powerful ultrasound showed the increase of the amplitude of the induced EMF in the vicinity of the bottom of a cylindrical container [83].

Later, a similar effect was also observed in the field of low intensity ultrasonic waves [78]. Experiments were carried out in the equipment described in section 8.2 (Fig. 8.1) and in the frequency range 20–70 kHz because high frequencies resulted in complicated wave phenomena which in [79] were interpreted as a result of propagation of higher oscillation modes in the pipe with the liquid. Table 2.1 shows the results of measurement of the ratio of the maximum AME amplitude in the centre of the magnetic fluid column (A_C) and in the vicinity of its base (A_B), with the frequency of the excited oscillations ν being the parameter.

We examine the proposed mechanism of the increase of the amplitude of the oscillations in the immediate vicinity of the edge using the following model. It is assumed that the magnetic fluid is magnetised to saturation, both bases of the liquid cylinder are acoustically free, and the magnetising field, the axis of the cylinder and the axis OX are parallel to each other (Fig. 2.17). $(2m+1)$ lengths of the standing waves ($m = 0, 1,\ldots$) are placed along the length of the cylinder. The point $x = 0$ is situated in one of the bases.

The displacement of the liquid particles is governed by the equation

$$u = u_0 \cdot \sin(bP + \varphi_0) \cdot \sin \omega t,$$

where $P = 2x/d$, $b = kd/2$ are the dimensionless coordinate and the wave number; φ_0 is the initial phase; k is the wave number.

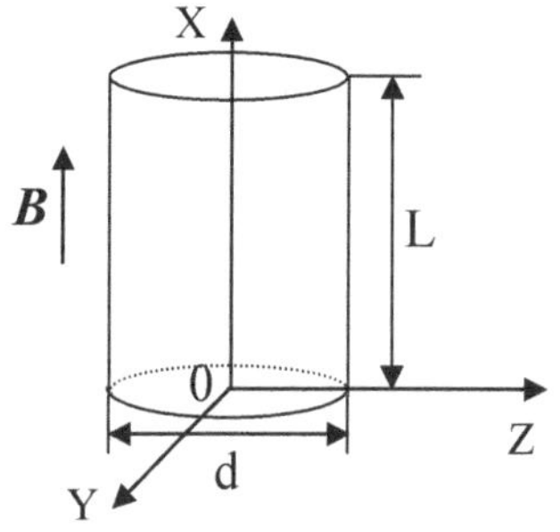

Fig. 2.17. Orientation of the cylinder in the field in the selected coordinate system.

The demagnetising factor, averaged over the cross-section of the cylinder with length L and diameter d, uniformly magnetised along the axis, at the distance $x = Pd/2$ from its end has the form [56]:

$$N(x)=1-\frac{2}{\pi}\left\{\frac{x}{\chi_1 d}[K(\chi_1)-E(\chi_1)]+\frac{L-x}{\chi_2 d}[K(\chi_2)-E(\chi_2)]\right\}, \quad (2.30)$$

where $\chi_1=1/\sqrt{1+(x/d)^2}, \chi_2=1/\sqrt{1+((L-x)/d)^2}$.

Equation (2.30) can be transformed to the following form (in dimensionless coordinates):

$$N(x)=\frac{1}{2}N(P)+\frac{1}{2}N\left(\frac{2L}{d}-P\right), \quad (2.31)$$

where $N(\xi)=1-\frac{2\xi}{\pi\chi}[K(\chi)-E(\chi)],\ \chi=1/\sqrt{1+(\xi/2)^2}$.

Equation (2.31) is the half sum of two equations coinciding with the ballistic demagnetising factors for a cylinder with the form parameter ξ.

Using the method proposed in [54] for calculating the ballistic factor in the dynamic deformation mode, and using the representation (2.31) we obtain

$$N_d=\frac{b}{2}\int_0^P N(P)\cdot\sin(bP+\varphi_0)\cdot dP+\frac{b}{2}\int_0^{(2L/d)-P} N(P)\cdot\sin(bP+\varphi_0)\cdot dP.$$

The value of the dynamic demagnetising factor in the arbitrary section of the magnetic fluid cylinder includes the contribution from the demagnetising factors, determined by the columns of the magnetic fluid on both sides of the given cross-section. The sound wave carries out the modulation of the concentration of the magnetic particles and, correspondingly, of the magnetisation of the liquid. In the case of the edge section, the perturbation of the magnetic field, associated with the effect of the sound waves on the magnetisation of the magnetic fluid, remains uncompensated. Consequently, demagnetising is weaker than in the central section. It should also be taken into account that the ballistic demagnetising factor is a rapidly decreasing function [56] and, therefore, the contribution to the behaviour of the magnetic field, recorded by the inductance sensor, comes only from several waves closest to the investigated section.

The dependence of the demagnetising factor on the wave length (and, consequently, frequency) will be evaluated using the approximation $N(P) = e^{-aP}$, proposed in [54] (a is a coefficient the highest accuracy of the approximation). For the central cross-section ($P = L/d$) and the section closest to the edge ($bP = \pi/2$) we have correspondingly

$$N_{dc} = \frac{1}{1+\left(\dfrac{a\lambda}{\pi d}\right)^2}; \tag{2.32}$$

$$N_{ab} = \frac{1-\alpha\dfrac{a\lambda}{\pi d}}{1+\left(\dfrac{a\lambda}{\pi d}\right)^2}, \tag{2.33}$$

where $\alpha = 0.5e^{-\frac{a\lambda n}{2d}} \sin(\pi n/2), n = 1, 2, \ldots.$

Taking into account the approximate nature of the resultant equations, to verify the qualitative agreement of these equations with the experiment, we consider the limiting case of small wave length (high-frequency), determined by the inequality $a\lambda/\pi d \ll 1$. In this case, the demagnetising factors (2.32) and (2.33) have the following form:

$$N_{dc} = 1-\left(\frac{a\lambda}{\pi d}\right)^2; \tag{2.34}$$

$$N_{db} = 1-\frac{a\lambda}{\pi d}\alpha. \tag{2.35}$$

The EMF, induced in the coil, is proportional to the multiplier $(1 - N_d)$ which at the small wave length decreases quadratically in the centre (2.34) and linearly in the vicinity of the pipe edge (2.35). In this case, it is important to note that the quadratic multiplier tends to 0 at a higher rate than the linear one.

The values of the relative amplitude β for the central and edge AME, relating to the appropriate values of $1/v$, are presented in Figs. 2.18 and 2.19. The points indicate the experimental data. The solid curve shows the results of computer approximation by polynomials. The form of the solid curve indicates the qualitative agreement of the proposed model with the observed 'edge' effect.

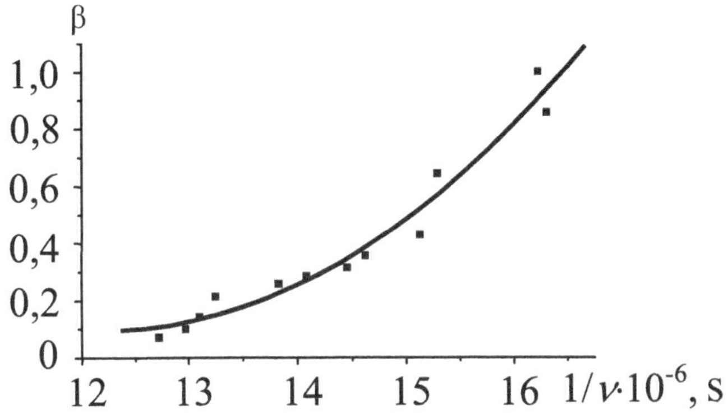

Fig. 2.18. Dependence β(1/ν) relating to the central part of the pipe.

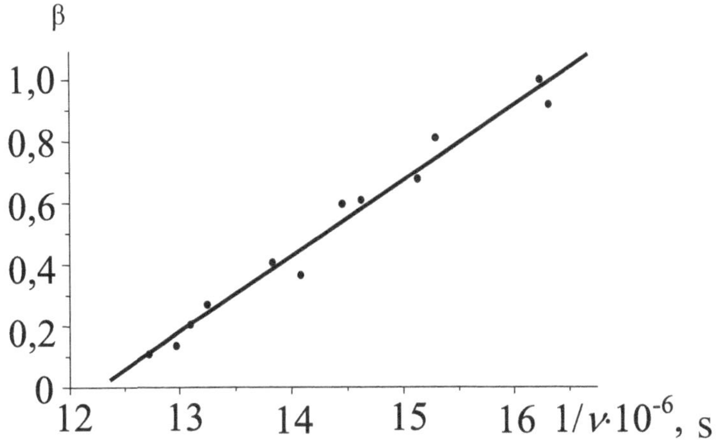

Fig. 2.19. Dependence β(1/ν) relating to the edge of the pipe.

2.7. Identification of oscillation modes

The speed of propagation of the elastic oscillations in the liquid, filling the container, the waveguide and the pipe, can greatly differ from the speed of sound in a free medium. The majority of investigations into the subject are theoretical. The number of experimental studies is very small and they are not systematic and there have been considerable shortcomings as regards the experimental procedure.

The qualitative improvement of the experimental procedure may be achieved using the acousto-magnetic identification of the modes of elastic oscillations. In the method, the magnetic fluid (not a conventional liquid) is poured into the pipe. The propagation in the

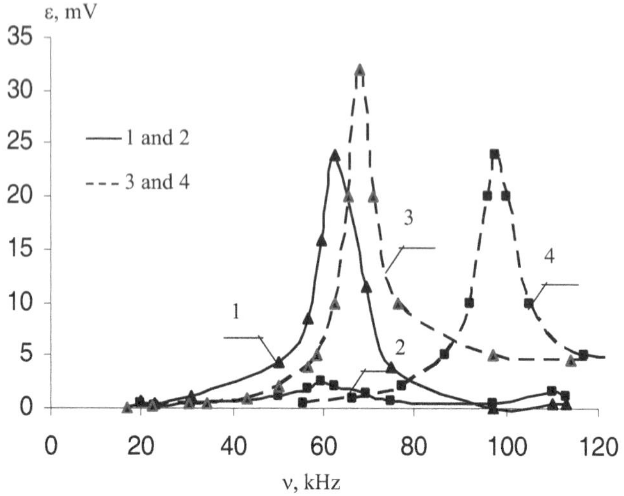

Fig. 2.20. The amplitude-frequency characteristic of the AME.

magnetised MF of the acoustic wave in the circuit, situated outside the pipe, results in the formation of a variable EMF [89, 91].

The 'scale factor', i.e., the ratio of the wave length of sound to the geometrical parameters of the induction circuit, is of considerable importance in the identification of the oscillation modes. If the circuit encircles the pipe with the MF, the thickness of the winding of the circuit should be smaller than the length of the standing wave of the investigated oscillation mode, but if the circuit is placed parallel to the side surface of the pipe, the height of the frame of the circuit should be equal to or smaller than the length of the standing wave. In both cases, the experiment conditions are optimum if the 'conventional' frequency matching for electronics and radio technology (setting the resonance frequency of the oscillatory circuit of the receiving device with respect to the frequency of ultrasound in the magnetic fluid) is fulfilled.

The amplitude–frequency characteristic of the receiving device, in which the right-angled circuit is parallel to the side surface of the pipe, is shown in Fig. 2.20. The solid curves 1 and 2 are the amplitude–frequency characteristics of the AME, and the dotted lines 3 and 4 are the amplitude–frequency characteristics of the receiving device which receives only the 'aiming' single, generated by the alternating voltage generator. The squares and the triangles show the data, obtained after activation of one or two inductance coils, respectively [232]. When two coils are used (with 1300 turns each), the resonance frequency decreases from 110 kHz to 62.4 kHz, and the maximum EMF increases ~10 times. Intensification of the AME at a frequency of 62.4 kHz is associated with frequency matching and the fulfilment of the condition of the scale factor.

The right section of the curve 2 is slightly raised because it relates to the relevant frequency of the circuit with a single inductance coil, i.e., 110 kHz. ~1.7 lengths of the standing wave are placed on the height of the conducting frame and, consequently, partial compensation of the magnetic fluxes, penetrating the circuit of the frame, takes place.

We determine the dependence of the amplitude of the variable EMF on the angles φ_n and $\hat{\varphi}'$, formed between the unit normal to the conducting flat frame **n** and the direction $\mathbf{H}_0$ from one side and the wave vector **k** from the other side. The magnetic flux through the element of the circuit, containing N_k terms, with the area $ds = h \cdot dr$, can be written in the form $d\Phi_M = N_k(\delta\mathbf{B}\cdot\mathbf{n})\cdot ds$. Since $\delta\mathbf{B} = \mu_0(\delta\mathbf{M} + \delta\mathbf{H})$ and $\delta H_x = -\delta M_x$, then $\delta\mathbf{B} = \mu_0(\mathbf{i}\delta M_x + \mathbf{j}\delta M_y - \mathbf{i}\delta M_x)$, i.e. $\delta\mathbf{B} = \mathbf{j}\mu_0\delta M_y$. Therefore, $\delta B = \mu_0\delta M \sin\hat{\varphi}$ and $d\Phi_M = \mu_0 N_k \delta M \hat{\varphi}' ds(\mathbf{j}\cdot\mathbf{n})$. Here $\delta M = \delta M_x \cdot \cos\hat{\varphi}' + \delta M_y \cdot \sin\hat{\varphi}'$, δM_y and δM_x are the perturbations of the magnetisation, determined by $\mathbf{j}H_y$ and $\mathbf{i}H_x$ are the components of the field. The dependence $\delta M(x)$ is included in the multiplier $\partial u/\partial x$ so that we can write $\partial M = C \cdot \partial u/\partial x$, where C = const. Therefore

$$\delta\Phi_M = \mu_0 C N_k \sin\hat{\varphi}' \cdot h \int_{-(L_p/2)\cdot\sin\varphi_n}^{(L_p/2)\cdot\sin\varphi_n} \partial u/\partial x \cdot dx$$

where L_p is the width of the frame.

For a harmonic flat wave $u = u_m \cos(\omega t - kx)$, $\partial u/\partial x = u_m k \sin(\omega t - kx)$. Integration in the previous equation in this case is carried out by the elementary procedure. On the basis of the law of electromagnetic induction, the following equation is obtained for the amplitude of the EMF

$$e_m = 2\mu_0 \cdot N_k \cdot C \cdot h \cdot c \cdot \sin\left((\pi L_p/\lambda)\sin\hat{\varphi}_n\right) \cdot \sin\varphi. \qquad (2.36)$$

The dependence of the amplitude of the EMF on the angle, formed by the vectors $\mathbf{H}_0$ and **k**, determines the directional characteristic of the receiver, shown in Fig. 2.21. The characteristic is distinguished by the distinctive directionality which could be used for investigating the ultrasound waves in the frequency range ~1÷10 MHz. If the

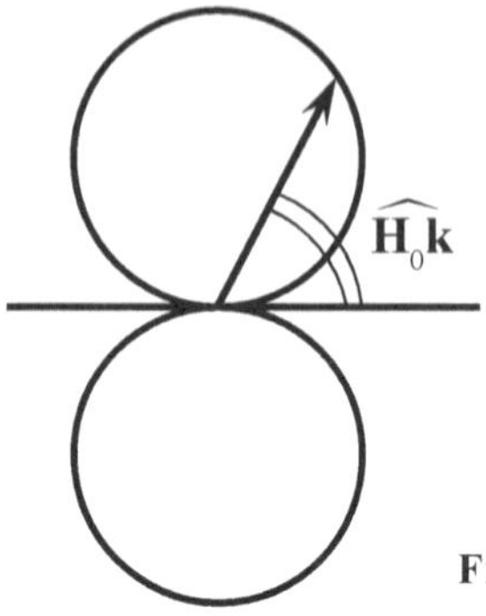

Fig. 2.21. The directional characteristic of the receiver.

angle $\hat{\varphi}'$ is fixed by making it equal to, for example, $\pi/2$, then at $\omega\tau \ll 1$ We obtain

$$e_m = 2\mu_0 \left(\beta_\upsilon M_0 + \gamma_* M_T\right) N_k u_m \cdot h \cdot \omega \cdot \sin\left(\left(\pi L_p / \lambda\right) \cdot \sin\varphi_n\right). \quad (2.37)$$

Rotating the frame from $\varphi_n = \pi/2$ to $\varphi_n = 0$, we observe a number of maxima e_m; for example, if $2L_p/\lambda = (2m + 1)$, where $m = 0, 1, 2,$ – the number of maximum equals $m + 1$. When rotating the frame by the angle 2π, the number of maxima is 2 $(2m + 1)$. The minimum angle of rotation of the frame φ_{n1}, at which the maximum e_m is observed is

$$\varphi_{n1} = \arcsin(\lambda / 2L_p). \quad (2.38)$$

At $\lambda/2L_p \ll 1$ $\varphi_{n1} \approx \lambda/2L_p$.

Thus, it is possible to determine the wave length of sound by rotating the frame.

It should be mentioned that in the experiments with AME the EMF is induced in the circuit as a result of the modulation of the magnetic flux by the sound waves in the magnetised magnetic fluid [4–6]. Therefore, the magnitude of the EMF depends on the magnetisation of the magnetic fluid which is reflected in particular in equation (2.20). In the experiments, based on the phenomenon of self-induction in the circuit, immersed in the magnetic fluid, the EMF is determined by the passage of alternating current through the circuit, and its magnitude depends on the inductance of the circuit and, consequently, on the magnetic permittivity of the magnetic fluid. Therefore, the dependences of the EMF on the strength of the field magnetising the magnetic fluid in these two methods will differ qualitatively.

3

Speed of sound

3.1. The additive model of elasticity

To obtain the functional dependence of the speed of propagation of sound waves in the magnetic fluid and the concentration of the solid phase, it is necessary to use the 'additive' model of formation of elasticity, available for the dispersed media [8, 93–96]. The calculation equation is derived using the following procedure: initially, assuming that there is no heat exchange between the components of a hypothetical dispersed system, we obtain the expression for c_{SS} and, subsequently, taking into account the interphase ('internal') heat exchange and the magnetic fluid, we obtain the expression for calculating c_{ST}. It is assumed that the density of the stabiliser and the carrier fluid are approximately equal to each other and, therefore, we use the relationship: $\rho = \rho_1(1-\varphi)+\rho_2\varphi$.

The smallness of the linear dimensions of the particles of the disperse phase in comparison with the length of ultrasound waves enables us to use for the magnetic fluids some conclusions of the solid state mechanics, in particular, use the equation for the speed of sound [97]:

$$c = (\rho\beta_s)^{-0.5}, \qquad (3.1)$$

where ρ is the density of the fluid, β_S is the adiabatic compressibility of the fluid.

The volume of the magnetic fluid consists of the volumes of the carrier fluid V_1, the solid phase V_2 and the stabiliser V_α. To obtain the highest stability of the system, the concentration of the stabiliser should be optimum [39]. Let the value $\alpha \equiv V_\alpha / V_2$ satisfy this requirement and remain constant for some class of the magnetic fluids.

With the quasi-static increase of external pressure by Δp, the volume of the system changes by ΔV, and $\Delta V = -\beta_T V \cdot \Delta p$, where β_T is the isothermal compressibility of the system.

The increase of the volume of the system should be equal to the sum of the increments of the volumes of each component:

$$\Delta V = \Delta V_1 + \Delta V_2 + \Delta V_\alpha,$$

where ΔV_1, ΔV_2, ΔV_α are the increments of the volume of the dispersion medium, the carrier fluid and the stabiliser.

Consequently, we obtain

$$\beta_T = (1 - \varphi - \alpha\varphi)\beta_{T1} + \alpha\varphi\beta_{T\alpha} + \varphi\beta_{T2}, \tag{3.2}$$

where β_{T1}, $\beta_{T\alpha}$, β_{T2} are the isothermal compressibilities of the dispersed medium, the stabiliser and the solid phase, $\varphi \equiv V_2 / V$ is the volume concentration of the solid phase.

Taking into account the relatively small compressibility of the solids, it is assumed that $\beta_{T2} = 0$. As regards the parameters $\beta_{T\alpha}$ and β_{T1}, two different assumptions were made in [64]: $\beta_{T\alpha} \ll \beta_{T1}$ and $\beta_{T\alpha} \approx \beta_{T1}$. The first case is realised at a sufficiently strong bond of the molecules of the stabiliser with the surface of the particles, and the second case – in the absence of such a bond. It would be more accurate to assume that $\beta_{T\alpha} = \gamma'\beta_{T1}$ and, consequently, equation (3.2) has the following form:

$$\beta_T = [1 - \varphi - (1 - \gamma')\alpha\varphi]\beta_{T1} \tag{3.3}$$

In the case of high-rate processes, the isothermal compressibilities β_{T1}, β_{T2} and $\beta_{T\alpha}$ change to adiabatic compressibilities β_{S1}, β_{S2} and $\beta_{S\alpha}$, which have the actual values in the absence of the dissipative processes.

If the mutual effect of the components is ignored, the adiabatic compressibility of the system is the sum of the specific adiabatic compressibilities:

$$\beta_S = [1 - \varphi - (1 - \gamma')\alpha\varphi]\beta_{S1} \tag{3.4}$$

When the agreement with the experiment is satisfactory, the resultant value of the parameter γ' provides information on the relative compressibility of the components of the actual magnetic fluid.

Substitution of (3.4) to (3.1) enables us to express the speed of propagation of sound in the magnetic fluid in the form

$$c_{SS} = c_1\rho_1^{0.5}\{\rho[1-\varphi-(1-\gamma')\alpha\varphi]\}^{-0.5}, \tag{3.5}$$

where $c_1 \equiv (\rho_1\beta_{S1})^{-0.5}$ is the speed of sound in a pure disperse medium.

The concentration of the solid phase in the magnetic fluid is calculated from the equation

$$\varphi = (\rho-\rho_1)/(\rho_2-\rho_1). \tag{3.6}$$

Therefore,

$$c_{SS} = c_1\left\{(1-\varphi+\varphi\rho_2/\rho_1)[1-\varphi-(1-\gamma')\alpha\varphi]\right\}^{-0.5}. \tag{3.7}$$

Replacing in (3.5) φ from equation (3.6) we obtain

$$c_{SS} = c_1\left\{\frac{(\rho_2-\rho_1)\dfrac{\rho_1}{\rho}}{\rho_2-\rho-\alpha(1-\gamma')(\rho-\rho_1)}\right\}^{0,5} \tag{3.8}$$

The medium with the small inhomogeneities and the distances between them in comparison with the wavelength is referred to as microheterogeneous [93, 94]. The magnetic fluid is a unique example of the microheterogeneous medium [9]. The magnetic particles, dispersed in the carrier fluid, are so small that the given condition is fulfilled in the entire ultrasound range.

The passage of the sound wave is accompanied by periodic compression and tensioning of the components of the magnetic fluid, and on the 'macroscopic' scale this process, as the majority of homogeneous liquids, is adiabatic. However, the variation of the temperature of the components of the system, determined by the varying sound pressure, differs. Because of the smallness and the relatively high heat conductivity of the ferromagnetic particles, the temperature of the particles manages to become equal to the temperature of the carrier fluid and, consequently, the process is 'microscopically' isothermal. The critical frequency below which the frequency range, corresponding to the given process, is situated, is determined from the equation [94]:

$$v_{cr} = \frac{\chi_2}{\pi\rho C_{p_2}R^2},$$

where χ_2 and C_{p2} is the heat conductivity and specific heat capacity of the solid particles at constant pressure; R is the radius of the particle; ρ is the density of the magnetic fluid.

For a low-concentration magnetic fluid of the first kind $\nu_{cr} \approx 10^{11}$ Hz. In conventional emulsions the process of propagation of sound can be 'microscopically' adiabatic and also isothermal [94]. In further considerations, the speed of sound in the case of adiabatic 'macroscopically' and adiabatic 'microscopic' process will be denoted by c_{SS}, and in the case of the adiabatic–isothermal process by c_{ST}.

The mechanism of equalisation of temperature between the opponents of the magnetic fluid influences the value of the adiabatic compressibility and the speed of propagation of sound. For the disperse systems with the condition $\nu \ll \nu_{cr}$ fulfilled, we can write the expression derived by M.A. Isakovich [94]:

$$c_{ST} = c_{SS}\left[1 - 0.5\varphi T c_{SS}^2 \rho\rho_2 C_{p2}\left(\frac{q_2}{\rho_2 C_{p2}} - \frac{q_1}{\rho_1 C_{p1}}\right)^2\right]. \qquad (3.9)$$

where ρ is the density of the magnetic fluid; C_{p1} and C_{p2} are the specific heat capacities; q_1 and q_2 are the coefficients of thermal expansion of the disperse medium and the disperse phase.

Substituting the expression for c_{SS} (3.7) to (3.9), we obtain

$$c_{ST} = c_1\left[\left(1-\varphi+\varphi\frac{\rho_2}{\rho_1}\right)(1-\varphi-\varepsilon\varphi\right]^{-0.5}$$
$$\times\left\{1 - \frac{0.5\varphi c_1^2 \rho\rho_2 C_{p_2} T}{(1-\varphi+\varphi\rho_2/\rho_1)(1-\varphi-\varepsilon\varphi)}\left(\frac{q_2}{\rho_2 C_{p_2}} - \frac{q_1}{\rho_1 C_{p_1}}\right)^2\right\}. \qquad (3.10)$$

where $\varepsilon \equiv (1 - \gamma')\alpha$.

In the case of low concentrations $\varphi \ll 1$ the second term in the square brackets of the equation (3.9) forms, as shown later, a small correction to 1 and, therefore, when a substituting in this equation c_{SS} by c_1 the magnitude of the correction does not change greatly. The expression for c_{SS} placed in front of the square bracket is presented taking into account the equation (3.5) and the given notation of ε in the form

$$c_{SS} = c_1\left(\frac{\rho_1}{\rho}\right)^{0.5}(1-\varphi-\varepsilon\varphi)^{-0.5} \approx c_1\left(\frac{\rho_1}{\rho}\right)^{0.5}\left[1+0.5(1+\varepsilon)\varphi\right].$$

After substituting c_{SS} into (3.9), retaining the terms linear with respect to φ, we obtain

$$c_{ST} = c_1\left(\frac{\rho_1}{\rho}\right)^{0.5}\left(1 + B^*\varphi\right), \tag{3.11}$$

where $B^* \equiv 0.5(1+\varepsilon) - A^*T; A^* \equiv 0.5c_1^2\rho\rho_2 C_{p_2}\left(\frac{q_2}{\rho_2 C_{p_2}} - \frac{q_1}{\rho_1 C_{p_1}}\right)^2$.

The value A^* greatly differs from zero in the condition $\frac{q_1}{\rho_1 c_{p1}} \rangle\rangle \frac{q_2}{\rho_2 c_{p2}}$. Consequently, the internal heat exchange in the investigated disperse system influences the elastic properties of the system mainly through the mechanism of thermal expansion of the carrier fluid. It should be mentioned that the correction in equation (3.9), determined by internal heat exchange, is small in comparison with 1 and the temperature dependence present in it is weak. Actually, accepting the values for ρ = 1230 kg/m³, c_1 = 1200 m/s [64], C_{p1} = 2 kJ/(kg K), C_{p2} = 0.655 kJ/(kg K), $q_1 = 9.5\cdot10^{-4}$ K^{-1}, $q_2 = 11.4\cdot10^{-6}$ K^{-1}, $\rho_2 = 5.21\cdot10^3$ kg/m³, φ = 0.1, T = 300 K, we obtain $A^*T\varphi = 0.026$.

The simplest methods of selecting the value of $B^*$is the equalisation of the derivative $\frac{\partial c_{ST}}{\partial\varphi}$ to the tangent of the angle of inclination of the tangent to the curve $c(\varphi)$ at the point $\varphi = 0$, plotted on the basis of the experimental data. At the same time, this method is most accurate because the expressions (3.5) and (3.9) were obtained for low concentrations of the disperse phase. From equations (3.6), (3.11) we obtain

$$\frac{\partial c_{ST}}{\partial\varphi} = \frac{\partial}{\partial\varphi}\left\{c_1\rho_1^{0.5}\left(1+B^*\varphi\right)\left[\rho_1 + (\rho_2-\rho_1)\varphi\right]^{-0.5}\right\}$$

$$= c_1\rho_1^{0.5}\left\{\begin{matrix} B^*\left[\rho_1 + (\rho_2-\rho_1)\varphi\right]^{-0.5} - 0.5\left(1+B^*\varphi\right) \\ \left[\rho_1 + (\rho_2-\rho_1)\varphi\right]^{-1.5}\times(\rho_2-\rho_1) \end{matrix}\right\}.$$

Specifically,

$$\left.\frac{\partial c_{ST}}{\partial\varphi}\right|_{\varphi=0} = -c_1\left[B^* - 0.5\left(\frac{\rho_2}{\rho_1} - 1\right)\right]. \tag{3.12}$$

The values obtained for six specimens of the magnetic fluid of the magnetite–kerosene–oleic acid type with different concentration of the solid phase, investigated in [64], were: $B^* = 0.875$, $\alpha = 0.75$, $\varepsilon = 1.27$.

At the same value of ε, the equation (3.10) has the form ($\varphi << 1$):

$$c_{ST} = c_1\left\{1 - 0.5\left[\left(\frac{\rho_2}{\rho_1} - 1\right) - 2.27\right]\varphi\right\} \quad (3.13)$$

According to the data in [64] $\rho_1 = 791$ kg/m^3, $\rho_2 = 5210$ kg/m^3 so that for the magnetic fluids of this type we can write the following equation:

$$c_{ST} \cong c_1(1 - 1.6\varphi).$$

Changes in the structure of the magnetic fluids result in changes in the speed of sound as a result of changes of the density of the fluid and its compressibility. In the magnetic fluid, as in the microheterogeneous medium, there is another operating mechanism – internal heat exchange. As a result of simple transformations, the equation (3.10) can be presented in the following form:

$$c_{ST} = c_1\frac{1 - A^*T\varphi\left[1 + (\rho_2/\rho_1 - 1)\varphi\right]^{-1}\left[1 - (1+\varepsilon)\varphi\right]^{-1}}{\left[1 + (\rho_2/\rho_1 - 1)\varphi\right]^{0.5}\left[1 - (1+\varepsilon)\varphi\right]^{0.5}}.$$

The expression $[1 + (\rho_2/\rho_1 - 1)\varphi]^{-0.5}$ is determined by the increase of density. The expression $[1 - (1 + \varepsilon)\varphi]^{-0.5}$ is associated with the reduction of the volume fraction of the component with high compressibility. The last pair of the square brackets contains the expression characterising the effect of internal heat exchange and the speed of sound. This factor, like the first one, reduces the speed of sound with increasing φ. At $\varphi<<1$, the increase of the speed of sound in the magnetic fluid can be described by the sum:

$$\Delta c_{ST} = \Delta c_\rho + \Delta c_\beta + \Delta c_T$$

in which the individual components represent the contribution of each of the above-mentioned factors:

$$\Delta c_\rho = -\frac{c_1}{2}\left(\frac{\rho_2}{\rho_1} - 1\right)\varphi; \Delta c_\beta = \frac{c_1}{2}(1+\varepsilon)\varphi; \Delta c_T = -c_1 A^* T\varphi.$$

For example, at $\varphi = 0.05$ we obtain $\Delta c_\rho = -184$ m/s, $\Delta c_\beta = 75$ m/s and $\Delta c_T = -17$ m/s.

The thermodynamic parameters of the aggregates in the magnetic fluid of the third type are slightly different from the parameters of the surrounding medium and this could be reflected in the elastic properties of the system [51, 99].

3.2. Method for measuring the speed of sound

The speed of propagation of the sound waves in the magnetic fluid can be measured by any currently available method of ultrasonic determination of this acoustic parameter, with the exception of the methods designed for the optical transparency of the medium.

In a number of studies, the speed of sound in the non-magnetised magnetic fluid was measured using a sound interferometer whose operating principle is based on the recording of standing waves formed in the fluid between an emitting quartz and a reflector in displacement of the latter [196, 197]. Knowing the distance covered by the reflector, and the total number of the standing waves, situated along this distance, we can calculate the speed of sound. The error of measurement by this method usually does not exceed ±0.1%.

This method is well-known and we will not discuss it because no changes have been made in the method. It should only be mentioned that in the case of magnetic fluids, the methods of ultrasonic interferometry, as any other method realised in the mode of continuous oscillations, should be used with a certain degree of care. The appearance in the medium of scattering centres whose role in the magnetic fluid, placed in the magnetic field, can be played by aggregates of magnetic particles, in the continuous radiation mode is accompanied by the formation of a complicated interference pattern causing difficulties in the interpretation of the data. In this connection, it is important to consider the results of the 'anomalously' large variation of the speed of sound in the magnetic fluid placed in the magnetic field, obtained by D.Y. Chung and W.E. Isler which are now regarded as erroneous and irreproducible by the majority of investigators [119]. They used the phase method, based on comparing phases of two continuous packets of coherent sinusoidal oscillations one of which passes through the investigated medium and the other one through a calibrated delay line. Having the phase difference, corresponding to the delay of the acoustic signal in the fluid, we can determine the speed of sound for the given length of the column of the fluid and frequency.

Regardless of the fact that as regards the spectral frequency of the experiment the method of non-damping oscillations has an obvious

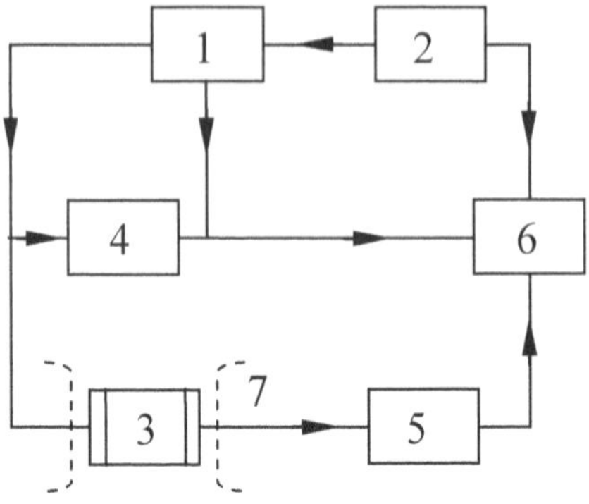

Fig. 3.1. The flow diagram of pulsed equipment.

advantage in comparison with the pulsed methods, the speed of sound in the magnetised magnetic fluid can be measured efficiently by the pulsed method and, in particular, the pulsed method with a fixed base. The constancy of the acoustic base in the course of measurements prevents the failure of the resultant structure of the magnetic circuits which is unavoidable during movement of the reflector. Usually, for a fixed measurement base it is possible to greatly reduce the volume of the investigated fluid. This is very important in cases in which the amount of the investigated fluid is small [308].

In addition to the interference method, the speed of sound in the magnetic fluid can be measured by the pulsed method [196, 197] which has been used successfully in the investigation of conventional fluid media. The flow diagram of measuring equipment is shown in Fig. 3.1. The generator 2 starts the gated sweep of the oscilloscope 6 and the two-channel pulse generator 1.

The right-angled pulse of the first channel of the generator 1 with the duration of 0.5 μs is supplied to the emitting piezoelement (X-section quartz plate, 2 MHz). The ultrasound pulse, containing 18–20 oscillations, after passage through the investigated medium is converted by the receiving piezotransducer to an electrical pulse which, after amplification by the wide band amplifier 5, is supplied to the X-input of the oscilloscope. The generator 1 can be used to delay the pulse of the second channel in relation to the pulse of the first channel, and the delay time is determined with high accuracy of ±0.05% using the frequency meter 4; the delay pulse of the second channel is supplied to the cathode of the electron beam tube of the oscilloscope and is used as a moving time mark.

In measurements of the speed of sound, the mark is combined with the first maximum of the high-frequency component of the pulse observed on the screen of the oscilloscope and subsequently is combined with the first maximum of the first of the series of reflected pulses. When using this method, the speed of propagation of sound in the investigated material is determined by the relationship

$$c = 2L/(t - t_0)$$

where L is the length of the acoustic path; t is the delay time of the reflected pulses; t_0 is the delay time of the first pulse, passed through the fluid.

The measuring cell is produced from a non-magnetic material (brass, bronze, copper) in order to avoid distorting the geometry of the magnetic field. The ends of the support cylinder, to which the piezoelements are compressed, are carefully machined and ground to ensure the parallel form. The distance between them was measured with an IZV–2 comparator with an accuracy of up to 0.001 mm. The measuring cell, containing the investigated fluid, is placed between the poles of the electromagnet 7. The measuring cell can be rotated around the axis, normal to the lines of the strength of the field.

This pulsed method is used for measuring simple and organic fluids [269]. The absolute error of the absolute measurements of the speed of ultrasound is determined by a set of partial errors which include the error, typical of the given methods, and the temperature measurement error. Since the time mark is not combined with the maximum of the video pulse and is combined with the first high-frequency maximum, in the given method, the variation of the curvature of the leading front of the video signal as a result of attenuation of the higher harmonics of the acoustic signal is not important. The effect of 'spreading' of the pulse is not large. Actually, the width of the spectrum of the pulse in this case is ~100 kHz, the lower spectral component of the pulsed spectrum $v_l = 0.975\ v_0$, the upper frequency of the spectrum $v_u = 1.025\ v_0$. If it is assumed that the absorption coefficient is proportional to the square of frequency, the ratio of the adsorption coefficient at the upper α_u and lower α_l frequencies can be represented by the following equation

$$\alpha_u/\alpha_l = v_u^2 / v_l^2 = 1.1.$$

The experimental error in the determination of the speed of ultrasound can be expressed using the equation

$$\frac{\Delta c}{c}=\sqrt{\left(\frac{\Delta L}{L}\right)^2+\left(\frac{2\Delta t}{t-t_0}\right)^2+\left(\frac{1}{c}\frac{\partial c}{\partial T}.\Delta T\right)^2}.$$

Consequently, we determine $\Delta c/c \leq 0.15\%$.

The experimental verification of the accuracy of calculating the error and setting equipment was carried out by measuring the speed of sound in a reference fluid – distilled water at different temperatures and by the comparison of the results with the calculated results obtained using the well-known equation proposed by M. Greenspan and C. Tschiegg, which is the most accurate interpolation equation:

$$c_G=1402.736+5.03358T_c-0.0579506T_c^2+3.31636\cdot 10^{-4}T_c^3 \\ -1.45262\cdot 10^{-6}T_c^4+3.0449\cdot 10^{-9}T_c^5.$$

The difference between the values of the speed, obtained by the experiments and calculations, does not exceed 1÷2 m/s which is in complete agreement with the calculated value of the error.

3.3. Results of measurement of the speed of sound in non-magnetised magnetic fluids

The speed of propagation of the sound waves in the magnetic fluid in the absence of the magnetic fields was measured in [7, 64, 100–104]. In [7, 64] and [102] the authors use the pulsed method in which the frequency of filling of the pulses had discrete values in the range 0.7 ÷ 45 MHz. The measurement error was 0.5% in [7] and 2% in [102]. The measurements show that the speed of sound in the magnetic fluid is lower than the speed of sound in the pure fluid-carrier. This relationship is characteristic both of the magnetic fluids based on kerosene [7, 64] and for water-based magnetic fluids.

The magnetic fluid, investigated in [64], contains kerosene as a dispersed medium and magnetite particles as the magnetic component. Oleic acid is used as the stabiliser. Five specimens with different concentration of the solid phase were obtained by diluting with kerosene. In addition to this, kerosene was one of the investigated specimens. The speed of ultrasound was measured by the pulsed method at a pulse filling frequency of 2 MHz. The density of the liquid was measured with a pycnometer. The measurements were taken at a temperature of 14.5°C. Figure 3.2 shows the results of experimental investigation of the speed of propagation of the

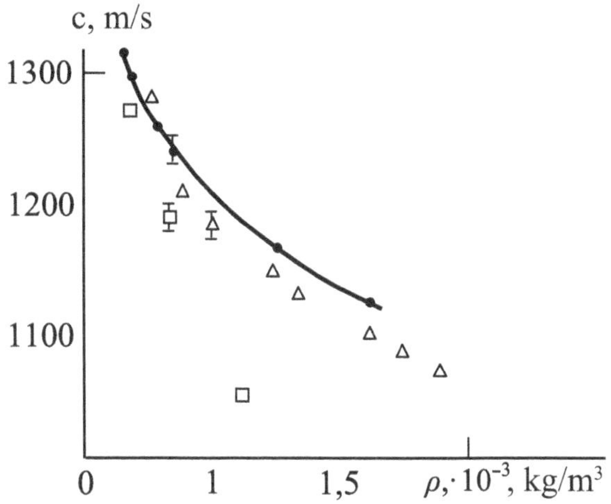

Fig. 3.2. Dependence of the speed of sound for the magnetic fluid on the basis of kerosene: □ – [102]; • – [64]; Δ – [106].

Table 3.1

Specimen No.	φ, %	$\rho \cdot 10^{-3}$, kg/m³	c, m/s	c_{SS}, m/s	c_{ST}, m/s	$-\frac{\partial c}{\partial \varphi}$, m/s·%	$-\frac{\partial c_{SS}}{\partial \varphi}$, m/s·%	$-\frac{\partial c_{ST}}{\partial \varphi}$, m/s·%
1	0	0.791	1318	1318	1318	25.5	25	25
2	0.72	0.823	1299	1300	1300	25	23	23
3	1.86	0.87	1273	1275	1275	18	21	21
4	4.77	1.002	1234	1223	1223	13	15	15
5	10.6	1.259	1166	1157	1156	10	8	9
6	16.6	1.525	1119	1127	1120	6	3.8	

sound waves in the magnetic fluids based on kerosene obtained by different authors. The temperature at which the individual results were obtained slightly differed. The data obtained in [102, 106] were recorded at 20°C. The frequencies of the sound oscillations differed slightly, 2 ÷ 3.17 MHz. The errors of measurements in [102] and [106] were equal to 30 and 5 m/s, respectively. The displacement of the experimental points, obtained in [106] downwards by ~17 m/s in relation to the experimental curve, constructed on the basis of the data published in [64], is satisfactorily explained by differences in the temperature by 5.5 K.

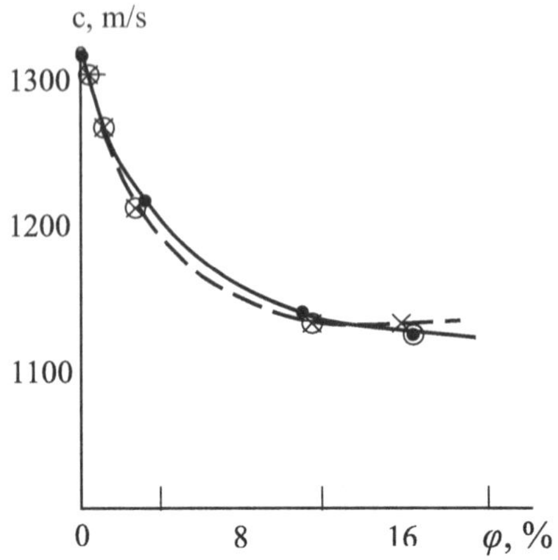

Fig. 3.3. The curves of the dependences $c(\varphi)$, $c_{SS}(\varphi)$ and $c_{ST}(\varphi)$:• – c; ∘– c_{SS}; × – c_{ST}.

Table 3.1 shows the results of measurements of the density ρ and the speed of sound c and also the appropriate values of φ, c_{SS} and c_{ST}, calculated from the equations (3.6), (3.5) and (3.10). The table also gives the values of $-\partial c/\partial\varphi$, obtained from the slope of the tangents to the curve of the experimental dependence $c(\varphi)$ and to the curves of the dependences c_{ST} (φ) and c_{SS} (φ). It was assumed that $\rho_2 = 5210$ kg/m^3, ε = 0.75.

Figure 3.3 shows the curves of the dependences $c(\varphi)$, c_{SS} (φ) and c_{ST} (φ). In the investigated concentration range $\dfrac{\partial c}{\partial \varphi} < 0$. The increase of the concentration is accompanied by a monotonic increase of $\dfrac{\partial c}{\partial \varphi}$, i.e. $\dfrac{\partial^2 c}{\partial \varphi^2} > 0$.

At concentrations lower than 2%, the differences between the experimental and calculated data do not exceed the measurement error (<2 m/s). At higher concentrations, the differences c_{SS} slightly increase. The large quantitative difference between the parameters $\dfrac{\partial c_{ST}}{\partial \varphi}, \dfrac{\partial c_{SS}}{\partial \varphi}$, on the one hand, and the parameter $\dfrac{\partial c}{\partial \varphi}$, on the other hand, for sample No. 6 is not accidental. In addition, a further increase of concentration may lead to an even larger difference in these parameters and also c, c_{SS} and c_{ST}. The method for taking into account the internal heat exchange in the magnetic fluid is justified only for the diluted dispersed system. At the same time, the quantitative correspondence between the parameters c, $\dfrac{\partial c}{\partial \varphi}$ and

c_{SS}, c_{ST}, $\frac{\partial c_{ST}}{\partial \varphi}$, $\frac{\partial c_{SS}}{\partial \varphi}$, in the region of not too high concentrations is the consequence of the physically substantiated representations forming the base of the additive model [8, 65].

3.4. Temperature dependence of the speed of sound

The initial experimental investigations of the dependence of the speed of sound on temperature in the magnetic fluids were carried out in [7]. The results show that the temperature dependence of the speed of sound in the magnetic fluid is identical to the dependence in a pure dispersion liquid. However, this study does not contain the results for the magnetic fluid with different values of φ so that no conclusions can be made regarding the tendency for the variation of the temperature coefficient with concentration.

The experimental investigations of the temperature dependence of the speed of propagation of the sound waves in the non-magnetised magnetic fluid based on magnetite and kerosene in [65] are carried out using an ultrasonic interferometer at a frequency of 2.7 MHz. The error of measurement of the speed was ±1.5 m/s. The thermostat made it possible to record the temperature with the accuracy of 0.1°C in the range 20–85°C. The density of the investigated specimens at different temperatures was measured with a pycnometer. The numerical density values of the investigated specimens at two temperatures (20 and 80°C) are in Table 3.2.

Figure 3.4 shows the graphs of the dependences $c(T_c)$ and $c_{SS}(T_c)$. The curves 1, 3 and 5 (circles – experiments) were obtained respectively for kerosene, MF-1 and MF-2, the curves 2 and 4 – $c_{SS}(T_c)$ for MF-1 and MF-2.

Table 3.2

Specimen	T_C, °C	ρ, kg/m³	φ, %	c m/s	$\Delta\rho/\Delta T$, kg/(m³·K)	$\Delta c_{SS}/\Delta T$, m/(s·K)	$\Delta c/\Delta T$, m/(s·K)
Kerosene	20	800	0	1318	−0.7	–	−3.7
	80	758	0	1095			
MF-1	20	977	4	1230	−0.8	−3.5	−3.3
	80	929	3.8	1034			
MF-2	20	1183	8.7	1260	−0.84	−3.4	−3.0
	80	1132	8.4	976			

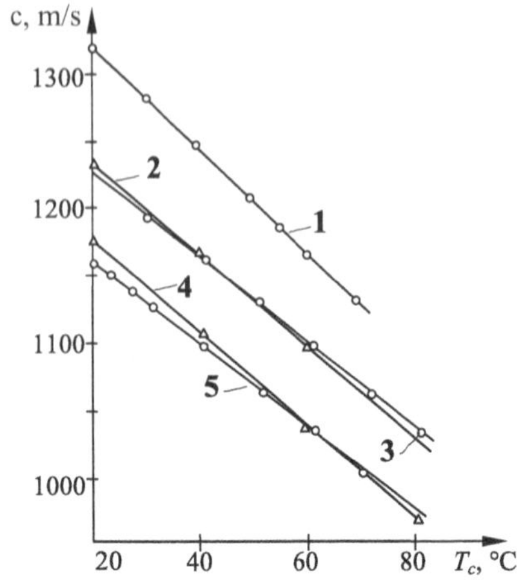

Fig. 3.4. Graph of the dependences $c(T_c)$ and $c_{SS}(T_c)$.

It is important to note the reduction of the speed of sound in the magnetic fluid with increasing temperature which, as is well-known, is typical of a great majority of the fluids; there is a tendency for the reduction of the absolute value of the temperature coefficient of the speed with the concentration of the magnetic fluid for kerosene.

Since the graph of the dependence $\rho(T_c)$ in the investigated temperature range is a straight line, to obtain any intermediate value of density, it is necessary to use the value of the temperature coefficient of density $\frac{\Delta\rho}{\Delta T}$, presented in Table 3.2.

The MF-1 and MF-2 differ from each other by the concentration of the solid phase. The data for the dependence $c(T_c)$ and also the parameters $\Delta c/\Delta T$, $\Delta c_{SS}/\Delta T$, obtained on the basis of the linear interpolation of the dependences $c(T_c)$ and $c_{SS}(T_c)$, are also presented in the same table.

The data are in agreement with the results published in [7, 102]. The tendency for the reduction of the absolute value of the temperature coefficient of the speed with the concentration of the magnetic fluid for kerosene was determined in the experiments [65]. Later, identical results were obtained in [107].

In addition to the previously mentioned comparison of the values of the speed of propagation of sound, it is also interesting to compare the temperature coefficients $\Delta c/\Delta T$, determined in experiments and calculated on the basis of the additive model $c_{SS}(T_c)$. This detailed comparison of the theory and experiment makes it possible to draw a more justified conclusion on the physical adequacy of the theoretical model.

The expression for the temperature coefficient of the speed has the following form

$$\frac{\partial c_{SS}}{\partial T} = \frac{c_{SS}}{c_1} \cdot \frac{\partial c_1}{\partial T} + \frac{c_1}{2\rho_1} \cdot \frac{\partial \rho_1}{\partial T} - \frac{c_{SS}}{2\rho} \cdot \frac{\partial \rho}{\partial T} + \frac{c_{SS}(1+\alpha)}{2(1-\varphi-\alpha\varphi)} \cdot \frac{\partial \varphi}{\partial T}. \quad (3.14)$$

Tables 3.1 and 3.2 shows that the values of c_{SS} and c_{SS} (T_c) are in satisfactory quantitative agreement with the experiment. Calculations of the temperature coefficient from equation (3.14) gives the correct sign but the numerical value is slightly lower than the absolute value. The main contribution is provided by the first term of the sum – 3.15 m/(s·K), and the smallest contribution by the last term – 0.07 m/(s·K), i.e., the factor of the temperature dependence of the concentration has almost no effect on the nature of changes of the elastic properties of the magnetic fluid with the variation of the temperature of the liquid. The second and third terms of the sum, determined by the temperature dependence of the density of kerosene and magnetic colloid, are equal to respectively 0.49 and 0.4 m/(s·K). Taking into account the internal heat exchange, i.e., calculations of $\partial c_{ST}/\partial T$, adds several terms to the equation (3.14) but does not cause any large change in the numerical value of the temperature coefficient of the speed of sound.

3.5. Temperature dependence of adiabatic compressibility

Prior to the publication of the study [108] there was no information on the adiabatic compressibility and the wave resistance in relation to the temperature of the magnetic fluid. To obtain these parameters, it was necessary to measure (in addition to the speed) the density of the fluid at different temperatures. At the same time, the results are very useful for examining the structure of the magnetic fluid and determination of the mechanisms of formation of the elastic properties of the disperse systems of this type.

In [108] the authors investigated the adiabatic compressibility β_S and the wave resistance Z of the magnetic fluid in relation to temperature. The experiments were carried out on specimens of a magnetic fluid in which the disperse phase was magnetite, and the dispersion medium one of the fluids: transformer oil, PES-5 polyethyl siloxane or water. The speed of sound was measured by the pulsed method at a filling frequency of the pulses of 4 MHz. To maintain and regulate the temperature in the range from 20 to 90°C, the authors used a thermostat and the temperature was recorded with a

Table 3.3

No.	ρ, kg/m³	φ, %	c, m/s	$Z\cdot 10^{-3}$, kg/(m²s)	$-\Delta Z/\Delta T \cdot 10^{-3}$, kg/(m² s K)	$\Delta\rho/\rho \cdot \Delta T$, K^{-1}	$-\Delta c/\Delta T$, m/(s K)	β_{SS} 10^{10}, Pa^{-1}	$\beta_{ST} \cdot 10^{10}$, Pa^{-1}	$\beta \cdot 10^{10}$, Pa^{-1}
1	899	0	1450	1304	4.0	−7.34	3.5	–	–	5.29
2	1113	4.96	1328	1478	4.4	−6.65	3.2	4.94	5.03	5.09
3	1277	8.77	1270	1622	4.8	−6.42	3.0	4.66	4.82	4.86
4	1422	12.13	1224	1741	5.0	−6.40	2.8	4.42	4.65	4.69
5	992	0	1301	1291	3.64	−6.75	2.88	–	–	5.96
6	1010	0.42	1295	1307	3.72	−6.73	2.87	5.92	5.93	5.91
7	1042	1.19	1279	1333	3.72	−6.53	2.82	5.85	5.87	5.87
8	1095	2.44	1258	1377	3.84	−6.48	2.78	5.74	5.78	5.77

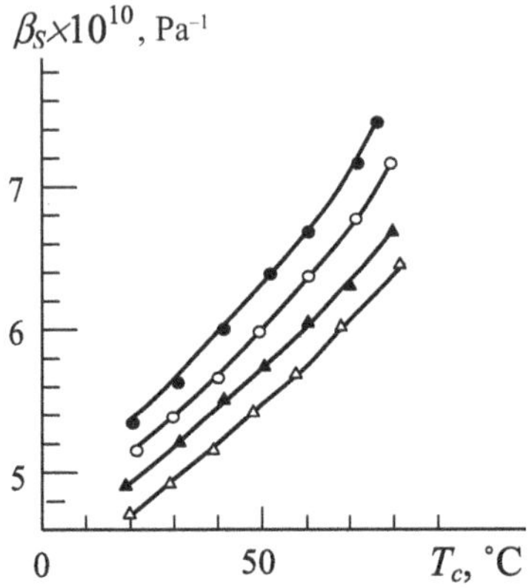

Fig. 3.5. Dependence of adiabatic compressibility on temperature for the specimens of the magnetic fluid 1–4 based on transformer oil: ● – 1; ○ – 2; ▲ – 3; Δ – 4. The numbers correspond to the number of the specimens.

mercury thermometer with the 0.1°C scale divisions. The specimens of the magnetic fluids based on transformer oil with the concentration φ: 0; 4.96; 8.77 and 12.13% were given the numbers 1, 2, 3 and 4, and the number 5, 6, 7 8 were given to the specimens based on PES-5 with the concentration φ: 0; 0.42; 1.19 and 2.44%.

Table 3.3 gives the following parameters of the specimens Nos. 1–8: ρ, φ, c, Z at T = 20°C and also the temperature coefficients $\Delta Z/\Delta T$, $\rho^{-1}\,\Delta\rho/\Delta T$, $\Delta c/\Delta T$ because the experimental dependences $\rho(T)$, $c(T)$ and $Z(T)$ for the given specimens are linear.

Figure 3.5 shows the dependence $\beta_S(T)$ for the specimens No. 1–4, and Fig. 3.6 for the specimens 5 and 8.

Figure 3.7 shows the dependence of the adiabatic compressibility on the temperature for the specimens of the magnetic fluids based on water. The following correspondence exists between the numbers

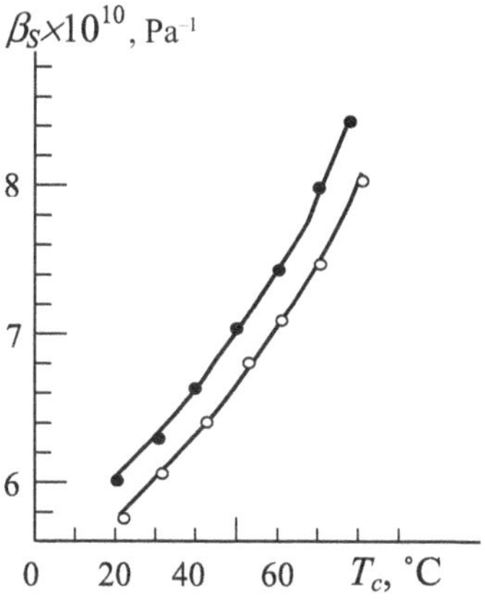

Fig. 3.6. Dependence of the adiabatic compressibility on temperature for samples of the magnetic fluid Nos. 5 and 8 based on PES-5: ● – 5; ○ – 8. The numbers correspond to the numbers of the sample.

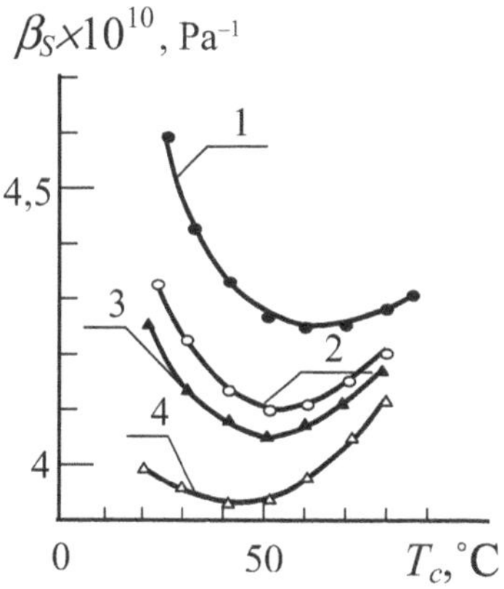

Fig. 3.7. Temperature dependence of adiabatic compressibility for the specimens of the magnetic fluids based on water with the concentration φ: 0 –●–1; 1.66–○–2; 2.71–▲–3; 6, 12–Δ–4.

and the concentrations of the specimens: 1 – φ = 0%; 2 – φ = 1.66%; 3 – φ = 2.71%; 4 – φ = 6.12%.

The dependences β_S (T_c) and Z (T_c) for the magnetic fluids and the carrier fluid are identical. At a fixed temperature the increase of φ is accompanied by a reduction of compressibility and increase of the wave resistance. This fact is explained by the increase of the volume fraction of the solid phase. In the additive model, the adiabatic compressibility with the interphase heat exchange taken into account can be presented in the form of a linear function of φ:

$$\beta_{ST} = (1 - \varphi - \alpha\varphi)\beta_{S1} + \alpha\varphi\beta_{S\alpha} + \rho_2 C_{p2} T \left(\frac{q_2}{\rho_2 C_{p2}} - \frac{q_1}{\rho_1 C_{p1}} \right)^2 \varphi.$$

The sum of the first two terms is the adiabatic compressibility of the system without taking the heat exchange between the phases

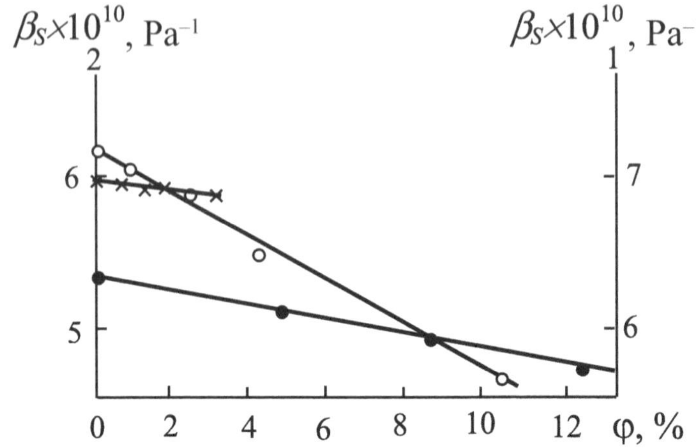

Fig. 3.8. Dependence of the but it compressibility β_S on the concentration of the magnetic fluid φ: × - magnetic fluids based on PES-5; •- magnetic fluids based on transformer oil; o - magnetic fluids based on kerosene.

into account, and the last term is the addition, determined by internal heat exchange.

Circles in Fig. 3.8 show the dependence $\beta_S(\varphi)$ for the magnetic fluids based on kerosene (axis 1), the solid circles refer to the magnetic fluids based on transformer oil (axis 2), and the crosses to the magnetic fluids based on PES-5 (axis 2) according to the data in Table 3.3.

Table 3.3 gives the values of β_{SS} and β_{ST}, calculated assuming that for the specimen Nos. 2, 3, 4 C_{p1} = 1710 J/(kg K), $q_1 = 6.7 \cdot 10^{-4}$ K^{-1}, $\beta_{S\alpha} = 0.8\ \beta_{S1}$ for the specimens Nos. 6, 7, 8 C_{p1} = 1586 J/(kg K), $q_1 = 6.7 \cdot 10^{-4}$ K^{-1}, $\beta_{S\alpha} = 0.7\ \beta_{S1}$, and that $\alpha = 1.75$; $T = 293$ K, 2 ρ_2= 5210 kg/m^3, $q_2 = 11.4 \cdot 10^{-6}$ K^{-1}, C_{p2} = 655 J/(kg K).

It is interesting to obtain information on the displacement of the temperature minimum of the dependence of adiabatic compressibility (or the temperature maximum of the speed of sound) of the water-based magnetic fluid with the concentration of the solid phase using the additive model. This is important because the study [102] did not solve this problem unambiguously. The adiabatic compressibility of the magnetic fluids based on water can be described by the equation:

$$\beta_{ST} = (1 - \varphi - \alpha\varphi)\beta_{S1} + \alpha\varphi\beta_{S\alpha}.$$

In this equation, the term associated with the internal heat exchange is ignored. This is justified due to the small value of the coefficient of volume expansion of the carrier. The minimum compressibility condition has the form:

$$\frac{\partial\beta_{ST}}{\partial T} = (1-\varphi-\alpha\varphi)\frac{\partial\beta_{S1}}{\partial T} + \alpha\varphi\frac{\partial\beta_{S\alpha}}{\partial T} = 0.$$

For water $\frac{\partial\beta_S}{\partial T} < 0$, for all other fluids $\frac{\partial\beta_S}{\partial T} > 0$. If it is assumed, as suggested by the additive model, than the surface active agent, adsorbed in the shells, is in the liquid state, characterised by the 'normal' temperature dependence of compressibility, then $\frac{\partial\beta_{ST}}{\partial T} = 0$ already at some vlue $\frac{\partial\beta_{S1}}{\partial T} < 0$, i.e., in the temperature range in which β_1, continues to decrease with increasing T. Consequently, the value β_{ST} passes through a minimum, and the speed of sound c through a maximum at a lower temperature than in the water free from the dispersed phase [109]. This conclusion was also confirmed in [110], and the results of this experiment were described previously and shown in Fig. 3.7. Thus, the prediction of the additive model has been confirmed by experiments.

If the internal heat exchange is taken into account, the additive model becomes very similar to the actual process and is physically more substantiated. Therefore, it is possible to evaluate the adiabatic compressibility of the shell of the stabiliser β_{sa}. Equating the derivative $(\partial\beta_S/\partial\varphi)_m$, obtained from the above equation for β_{ST}, to the experimental value $(\Delta\beta_S/\Delta\varphi)$, we can estimate the adiabatic compressibility of the material of the stabiliser substance

$$\beta_{sa} = \left[\beta_{s1}(1+a) + \Delta\beta_s/\Delta\varphi - \rho_2 C_{p2} T\left(\frac{q_2}{\rho_2 C_{p2}} - \frac{q_1}{\rho_1 C_{p1}}\right)^2\right]\alpha^{-1}.$$

Such an estimate was made for the first time in [207] for the magnetic fluids based on kerosene and assuming that the ratio of the volume of the stabiliser to the volume of the solid particle $a = 1.75$ (this value of a is obtained at the radius of the solid particle of $5\cdot10^{-9}$ m and the thickness of the shell of the stabiliser $2\cdot10^{-9}$ m, corresponding to the generally accepted assumptions [35, 260]), $\rho_1 = 791$ kg/m^3, $\rho_2 = 5210$ kg/m^3, $C_{p2} = 0.7$ kJ/kg K, $C_{p1} =$ 2 kJ/kg K, $q_1 = 9.5\cdot10^{-4}$ $q_2 = 11.4\cdot10^{-6}$ K^{-1}, $T = 287.5$ K, $\Delta\beta_s/\Delta\varphi = -1.4\cdot10^{-9}$ Pa^{-1}, at $\Delta\varphi = 0.1$. The value $\beta_{sa} = 1.3\cdot10^{-10}$ Pa^{-1} was obtained. Thus, $\beta_{sa} = 0.2\ \beta_{S1}$, since at $T = 287.5$ K, $\beta_{s1} = 7.28\cdot10^{-10}$ Pa^{-1}.

The first term in the equation for β_{sa} is positive and provides the main contribution equal to 1 in relative units. The second and third terms have the negative sign and their contribution equals 0.7 and 0.187. Therefore, the numerical value β_{sa} depends not only on the selection of the parameter a but also on the width of the range $\Delta\varphi$ for determining $\Delta\beta_s/\Delta\varphi$.

Thus, if we consider the initial section of the curve $\beta_s(\varphi)$, which is of greatest interest from the viewpoint of the theory of microheterogeneous media, developed for the diluted disperse media, it can be seen that the value $\left|\frac{\Delta\beta_s}{\Delta\varphi}\right|$ slightly decreases, and $\beta_{sa} = 0.3\ \beta_{s1}$, and the scatter in the values $\left|\frac{\Delta\beta_s}{\Delta\varphi}\right|$ and the relative calculation error also increase.

Consequently, if the assumptions of the ratio of the volumes of the shell and the solid particle are accurate, the compressibility of the shell should be lower than the compressibility of the carrier fluid in the given specimen of the magnetic fluid. Possibly, this is supported by the forces of intermolecular interaction of the particles of the stabiliser and the ferroparticle.

The same procedure was used to obtain the values of the adiabatic compressibility of the stabilising shells for the MF specimens based on PES-5 organic silicon and transformer oil by L.I. Roslyakova [101]. The value of the parameter a was estimated taking into account the ratio of the lengths of the molecules of the stabiliser and the diameter of the ferroparticle, obtained on the basis of magnetic granulometric measurements.

These data show that the adiabatic compressibility of the molecules of the stabiliser of the investigated MF specimens are in satisfactory agreement and slightly different from those of the dispersed media (they are lower).

It was also noted that in the temperature range from 20°C to 80°C, the ratio of the compressibility of the molecules of the stabiliser and the dispersed medium remains constant.

It should be mentioned that we considered the problem of the elastic properties of the magnetic colloids without taking into account the possible presence in the carrier fluid of a certain number of molecules of a non-adsorbed stabiliser, for example, the molecules of oleic acid in kerosene. Unfortunately, we could not determine by experiments the concentration of the material of the stabiliser, dissolved in the carrier fluid.

However, the literature contains reports on the effect of the dissolved surface active agent on the elastic properties of the carrier fluid. For example, the experimental material for the effect of the oleic acid on the elastic properties of the carrier fluid was presented in [7]. The addition to kerosene of the oleic acid in the amount corresponding to its content in the magnetic fluid results in the 'upward' displacement of the linear dependence $c(T_c)$ by the value Δc = 12 m/s, whereas in the magnetic fluid, the linear dependence $c(T_c)$ is displaced 'downwards' by the value $\Delta c = -140$ m/s.

Identical data were published in a later study [107]. This means that the presence of oleic acid in kerosene increases the elasticity of the system.

The simplest calculations show that the compressibility of the component of oleic acid in the solution in the given example is $\beta_s = 6.53 \cdot 10^{-10}$ Pa^{-1} which is only 1.1 times smaller than the compressibility of the carrier fluid and ~5 times greater than the compressibility of material in the shell. In this case, the difference between the compressibility of the material in the shell and in the dissolved state is large but there are other examples where this difference is not large.

Possibly, special features of the elastic properties of the actual magnetic fluids play a significant role in this case. At the same time, it should be remembered that in this case we are concerned with a very approximate and unique evaluation of the compressibility of the stabilising shell, and its approximate nature is associated not only with the error of determination of $\Delta\beta_s/\Delta\varphi$, which is equal to approximately 10%, but also with the conventional nature of the selection of parameter a.

3.6. The non-linearity parameter

The propagation of acoustic waves of a finite amplitude in the medium is accompanied by non-linear effects [197], which include the change in the shape of the wave, i.e., the variation of the time dependence of the wave parameters, formation of combination tones, self-focusing of the wave, sound pressure, acoustic flows, and others.

The non-linear properties of the medium are characterised by the non-linearity parameter Γ_n. This parameter Γ_n can be calculated using the data for the dependence of the speed on sound pressure [113, 114]

$$\Gamma_n = \left(\rho^2 \frac{\partial^2 p}{\partial \rho^2}\right)_0 \Bigg/ \left(\rho \frac{\partial p}{\partial \rho}\right)_0 = \rho_0 \left(\frac{\partial^2 p}{\partial \rho^2}\right)_0 \Bigg/ \left(\frac{\partial p}{\partial \rho}\right)_0 .$$

From equation (3.7) at T = const we obtain:

$$\left(\frac{\partial c}{\partial P}\right)_T = \frac{c}{c_1}\left(\frac{\partial c_1}{\partial p}\right)_T$$

$$+\frac{c^3}{\rho_1 c_1^4}\left\{\left[1-\frac{\rho_2}{2\rho_1}+\frac{\alpha}{2}-\varphi-\alpha\varphi+\frac{\rho_2}{\rho_1}\varphi(1+\alpha)\right](1-\varphi)+\frac{\rho_2}{\rho_1}[1-\varphi-\alpha\varphi]\right\}\varphi.$$

Using the numerical values c_1 = 1318 m/s, ρ_1 = 791 kg/m³, ρ_2 = 5210 kg/m³, α = 1.75, we obtain

$$\left(\frac{\partial c}{\partial p}\right)_T = \frac{c}{1318}\left(\frac{\partial c_1}{\partial p}\right)_T + 42\cdot 10^{-17} c^3\left(1.37\varphi + 5.95\varphi^2 - 9.79\varphi^3\right). \quad (3.15)$$

The contribution of the second term in the right-hand part of the equation (3.15) is 0.2% at φ = 0.72% and monotonically increases, reaching ~5% at $\varphi = 16.6\%, \frac{\partial c_1}{\partial p} = 0.5\cdot 10^{-5}$ m/(s Pa) [111].

Table 3.4 gives the dependence of the relative variation of the baric coefficient of the speed on the concentration for a kerosene-based colloid.

It can be seen that the additive model predicts the reduction of the baric coefficient of the speed with the concentration of the solid phase (specimens Nos. 1–6).

This conclusion is in agreement with the results obtained in the experiments by S.P. Dmitriev and V.V. Sokolov [112].

Taking these studies into account, the value $\left(\frac{\partial c}{\partial p}\right)_T \Big/ \left(\frac{\partial c_1}{\partial p}\right)_T$ for the specimen No. 7 was calculated

The method of the experimental determination of parameter Γ_n is based on the approximate relationship [114]:

$$\Gamma_n \cong 2\rho_0 c_0 \left(\partial c / \partial p\right)_T .$$

The results of calculation of the relative variation of the non-linearity parameter (Γ_n/Γ_{n1}) are presented in Table 3.4.

For specimen No. 7, the calculations were carried out using the data from [112], with corrections of the value of density ρ_0 and the

Table 3.4.

No.	φ, %	ρ, kg/m^3	c, m/s	$\left(\frac{\partial c}{\partial p}\right)_T \Big/ \left(\frac{\partial c_1}{\partial p}\right)_T$	Γ_n/Γ_{nl}	β_Γ
1	0	791	1318	1.00	1.0	1.0
2	0.72	823	1299	0.99	1.01	–
3	0.86	873	1277	0.97	1.04	1.007
4	4.77	1000	1234	0.95	1.11	1.03
5	10.6	1260	1166	0.90	1.26	1.13
6	16.6	1525	1119	0.89	1.47	
7	20.6	1720	1117	0.83	1.51	

speed of sound c for the temperature varying in the range from 30°C to 14.5°C.

In the investigated range of variation of φ the value of Γ_n increases almost 1.5 times.

The experiments described in [115] were carried out to examine the non-linear acoustic parameter of the magnetic fluids based on magnetite and kerosene. The data published there relate to the temperature of 40°C.

Table 3.4 gives the values of the relative variation of the parameter Γ_n (i.e. β_Γ), obtained from the average experimental curve of the dependence $\Gamma_n(\varphi)$.

3.7. Dispersion of the speed of sound in an unlimited magnetic fluid

The relative movement of the particles in the carrier fluid results in the dispersion of the speed of sound [93, 94].

The expression for the speed of sound in the heterogeneous medium in which the linear dimensions of the disperse particles are small in comparison with the wavelength, the particles are absolutely rigid and do not interact with each other, was obtained in [67]:

$$\frac{1}{c} = \frac{1}{c_{or}}\left[1 - b_1 \frac{\psi_\vartheta\sqrt{\psi_\vartheta}\,(1 + b_2\sqrt{\psi_\vartheta})}{\left(1+\sqrt{\psi_\vartheta}\right)^2 + \psi_\vartheta\left(1 + b_2\sqrt{\psi_\vartheta}\right)^2}\right],$$

where c_{or} is the speed of sound in the dispersed medium at low frequencies; $b_1 \equiv 2/9\cdot\varphi(\gamma_0 - 1)^2$; $b_2 \equiv 2/9\cdot(1 + 2\gamma_0)$; $\gamma_0 = \rho_2/\rho_1$; $\psi_\vartheta = \omega\rho_1 R^2/2\eta_{s1}$; R is the particle radius.

Since $c = c_{or} + \Delta c$, then

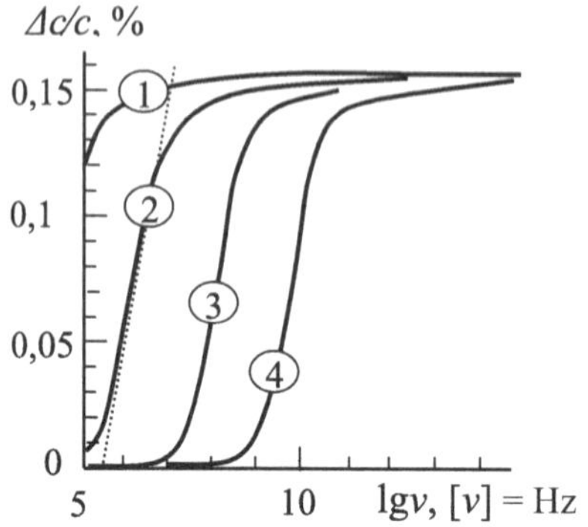

Fig. 3.9. Dispersion of the speed of sound a dispersed medium.

$$\frac{\Delta c}{c} = b_1 \frac{\psi_\vartheta \sqrt{\psi_\vartheta}(1 + b_2\sqrt{\psi_\vartheta})}{\left(1 + \sqrt{\psi_\vartheta}\right)^2 + \psi_\vartheta\left(1 + b_2\sqrt{\psi_\vartheta}\right)^2}. \tag{3.16}$$

For the specimens of the magnetic fluid based on magnetite and kerosene with the parameters $\gamma_0 = 3$, $b_1 = 0.245$, $b_2 = 1.558$ and $R = 7\cdot10^{-9}$m, $R = 7\cdot10^{-8}$ m, $R = 7\cdot10^{-7}$m, $R = 7\cdot10^{-6}$ m, the equation (3.16) was used to plot the dependence $\Delta c/c$ (log ν) (Fig. 3.9). The variation of the speed of sound on the curve 4, relating to the magnetic fluid of the first type, starts to be evident in the hypersonic frequency range. The 'step' on the curve 2 forms in the frequency range 10^6–10^7 Hz. Consequently, the dispersion of the speed of sound, determined by the relative movement of the particles, can be detected in the dispersed system with the particles of the micron dimensions, i.e., in the ferrosuspension. To ensure that $\nu_k = 10^7$ Hz, we should have $R \approx 1.5$ μm. The maximum dispersion of the speed of sound in these systems should have the order of 10–20 m/s [116]. The particles of the micron dimensions can be present also in the magnetic fluid as a result of the aggregation process. However, the density of the aggregates is lower than the density of the solid ferroparticles because the composition of the aggregates contains, in addition to the solid ferromagnetics, also a stabiliser.

If the dependence $\Delta c/c(\log \nu)$ in the dispersion range is approximated by the section of the straight line (indicated by the dotted line in Fig. 3.9), then at $c = 1200$ m/s

$$\Delta c / c = K_d \log\left(\nu / 1.26\cdot10^5\right),\ K_d = 0.0649.$$

Experimental investigations of the dispersion of the speed of sound in the non-magnetised magnetic fluid were started in [7]. In two specimens of the magnetic fluid, based on kerosene and vacuum oil, in the frequency range 50 kHz–1.2 MHz there were no changes in the speed of sound outside the range of the measurement error of

0.6%. A.N. Vinogradov et al [117] reported on the absence of the dispersion of the speed of sound in the magnetic fluid on the basis of the dodecane and magnetite ferroparticles, stabilised with oleic acid, in the frequency range 12–132 MHz.

There is only one study [17] reporting on the observation of the dispersion of the speed of sound in the magnetic fluids based on kerosene and magnetite in the frequency range 3–50 MHz. According to the data in [17], the speed of sound increases by ~20 m/s, i.e. $\Delta c/c \approx 0.017$.

3.8. Effect of the magnetic field on the speed of sound

The first attempt to determined by experiments the dependence of the speed of sound in the magnetic fluid on the strength of the magnetic field was made in [7]. According to the data [7], the dependence $c(H)$ is not evident in the magnetic fluid based on kerosene and vacuum oil.

The field dependence of the speed of sound in the magnetic fluid was also investigated by D. Chung and W. Isler [10, 14, 119]. In a water-based magnetic fluid, the authors detected the strong dependence of the speed on the magnitude and direction (in relation to vector **k**) of the strength of the magnetic field [10].

The sample, investigated in [10] was in the form of a commercial magnetic fluid; the magnetic properties of the fluid were described in [120]. The volume concentration of the magnetic phase was 2.84%, saturation magnetisation 16 kA/m. Measurements of the speed of ultrasound were taken by the phase method at a frequency of 2.44 MHz. The authors noted that at the strength of the magnetic field of $H = 6$ kA/m and 12 kA/m, the speed (c) changed by at least ~10%. At $H = 20$ kA/m in the transverse magnetic field the value c reached 30% and then decreased to 0 at $H = 60$ kA/m. A further increase of the strength of the magnetic field resulted in a repeated increase of $\Delta c/c$, which at $H = 80$ kA/m reached ~50%. In the longitudinal field, the shape of the curve $c(H)$ was different, but in this case $\Delta c/c$ reached extremely high values (~30%).

In another study by the same authors [14], performed on the same object and using the same procedure, measurements were taken of the 'response' time. The 'response' time is the time during which the equilibrium value of the phase of the ultrasound signal, transmitted through the magnetic fluid, is established. The magnetic field is generated in a 'jump' to a specific value and this is followed by the rotation of the magnetic field in relation to the wave vector in 'jumps' of 10°. The response time reaches 12–24 s which, according to the authors, indicates the formation of aggregates with the diameter of 1–2 μm in the magnetised liquid (see also [121]).

From the purely physical viewpoint, the dependence $c(H)$, determined in [10], is unlikely to be valid because it is difficult to find the reason for more than doubling of the compressibility of the liquid.

In [11] the speed of sound was measured by the pulsed method at a frequency of 2–4 MHz and a temperature of 20°C. A magnetic fluid sample based on kerosene and magnetite was used. The saturation magnetisation of the magnetic fluid was 48 kA/m. The speed of ultrasound was measured without applying the magnetic field to the fluid and in the magnetic field with the strength changing smoothly in the range 0–100 kA/m. The speed of ultrasound in the absence of the magnetic field was 1135 m/s. In the magnetic field with the collinear and perpendicular distribution of the wave vector and the vector of the strength of the magnetic field the values of the speed were equal to respectively $c_{||}$ = 1139 m/s, $c_{\perp}$ = 1137 m/s. The repeatability of the resultant absolute values of the speed of sound in the magnetic fluid under the effect of the magnetic field is ensured only if the condition of efficient preliminary mixing and demagnetisation of the magnetic fluid is fulfilled.

Figure 3.10 shows the dependence for the given value of the variation of the speed of sound in the MF-1 $\Delta c/\Delta c_{max}$ in relation to the strength of the magnetic field $H_{||}$ in three series of consecutive measurements.

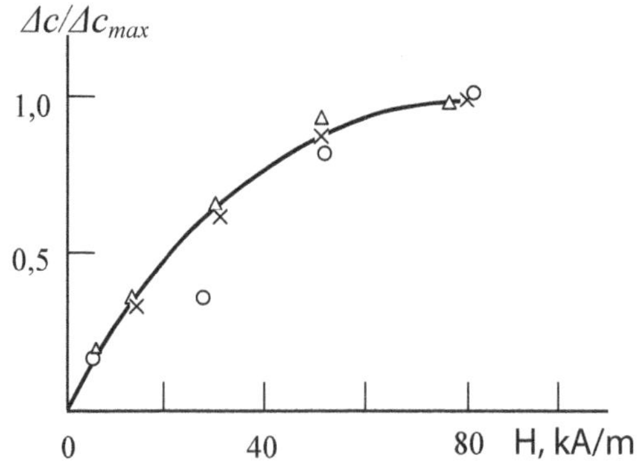

Fig. 3.10. Dependence of the reduced value of the variation of the speed of sound in the magnetic fluid on the strength of the magnetic field: ○ – the values obtained in the first series of the measurement; Δ – in the second series; × – in the third series.

In [166, 122] the authors investigated the field dependence of the speed of sound in two specially prepared specimens of the magnetic colloid MF-1 and MF-2.

The MF-1 specimen in [122] was prepared by dilution of a stable colloid with the density $\rho = 1.24 \cdot 10^3$ kg/m^3 of a mixture of kerosene with oleic acid at a ratio of 5/2 at room temperature. The density of the specimen was 10^3 kg/m^3.

The MF-2 specimen [116] relates according to the accepted classification to the third type of the magnetic fluid because it is characterised by the formation of a deposit after holding for 8–10 hours in the magnetic field with the strength of 400 kA/m and several days of holding, and also by the presence of aggregates of the micron in the absence of the magnetic field, observed in an optical microscope. At the start of the acoustic measurements, the density of the samples was $0.945 \cdot 10^3$ kg/m^3.

The average size of the aggregates in the MF-1 was ~2 μm, and in the MF-2 ~4.5 μm. The optical studies were carried out using a magnetising field, directed along the layer of the fluid with a thickness of 20–30 μm, enclosed between two covering sheets. The strength of the magnetic field was varied in steps of 1.5 kA/m in the range from 0 to 30 kA/m.

The experimental results show that during magnetisation the MF-1 sample shows the process of elongation of the aggregates in the field, and in the MF-2 sample the process of elongation of the aggregates slows down at $H \approx 20$ kA/m. It was assumed that as a result of the magnetic interaction of the ferroparticles in the MF-2 aggregates, the molecules of the stabiliser are displaced from the region of contact of the particles, accompanied by the formation of a denser packing

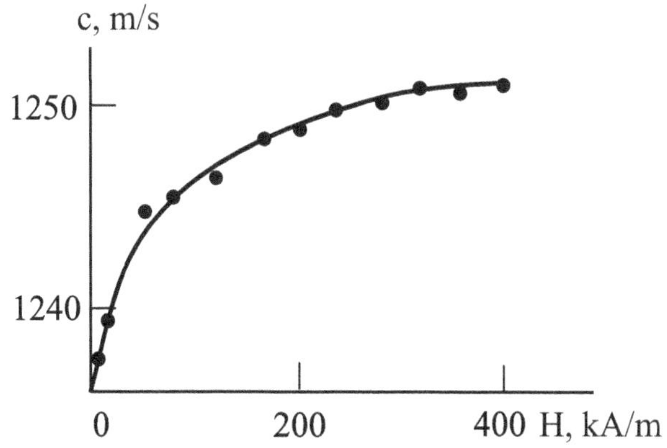

Fig. 3.11. Dependence $c(H)$ in MF-1 at $\mathbf{H} \| \mathbf{k}$.

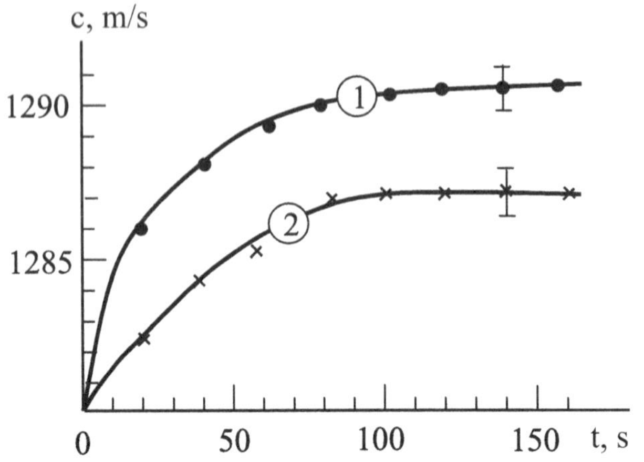

Fig. 3.12. Dependence of the speed of sound on the holding time of the MF-2 specimen in the magnetic field: 1 – **H**||**c**; 2) – **H**⊥**c**.

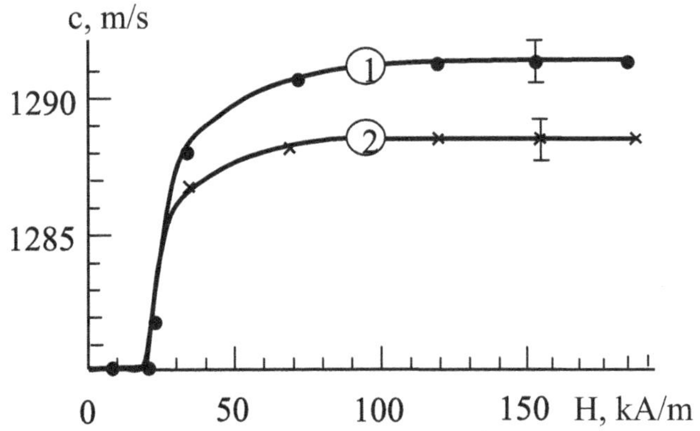

Fig. 3.13. Dependence of the speed of sound on the strength of the magnetic field in the MF-2 sample: 1 – **H**||**c**; 2) **H**⊥**c**.

of the ferromagnetic particles and, consequently, a reduction of the size of the stretched aggregates. This assumption is in agreement with the fact that when the field is switched off, the aggregates in MF-2 were more stable than in MF-1. Spherical droplets–aggregates in the MF-2 sample were on average 2–2.5 times greater than prior to the application of the magnetic field.

Figure 3.11 shows the results of measurement of the field dependence of the speed of propagation of ultrasound in the MF-1 specimens at **H**||**k**. In the measurement error range (±1 m/s), the $c(H)$ curve, with the origin at zero, monotonically increases and reaches saturation in the magnetic field $H \approx 350$ kA/m. In this case, $\Delta c \approx 12$ m/s, whereas the measurements taken in the initial colloid in the same equipment show that the speed of ultrasound is independent of H.

Figure 3.12 shows the dependence of the speed of ultrasound on time $c(t)$ at $H = 200$ kA/m for the MF-2 sample; similar experiments were carried out in a magnetic field with the strength of 40, 80, 120 and 160 kA/m. The rearrangement time of the structure in the specimen is 1.5–3 min, and the duration of the process decreases with the increase of the strength of the magnetic field.

Figure 3.13 shows the dependence $c(H)$ for the MF-2 sample with the value of H changing in the range 0–200 kA/m. Curve 1 was obtained for the longitudinal magnetic field, curve 2 for the transverse field. The delay time of the pulse, transmitted through the investigated medium, was measured at the moment of arrest of the received signal in relation to the time mark on the screen of the oscilloscope. The speed varies starting from a specific threshold value of the strength of the magnetic field ~20 kA/m, and at $H \approx 40$ kA/m the anisotropy of the speed of sound becomes evident. In the field $H \approx 80$ kA/m the dependence $c(H)$ becomes saturated. The speed of sound in the longitudinal magnetic field is ~3m/s greater than the speed of sound in the transverse direction.

Multiple magnetisation of the MF-2 sample in the field with the strength of ~250 kA/m results in an increase of the speed of sound by 3 m/s. At the same time, the increase of the speed at the application of the magnetic field decreases by approximately the same value. Evidently, this fact is the consequence of the residual phenomena in the structure formation in magnetisation of the magnetic fluid which have been reported, in particular, in [45]. In [122] it was also noted that not all liquids belonging in the group of the third type of the magnetic fluid show such large changes in the speed of sound in the magnetic field. This fact is explained by the absence of characteristic properties of the aggregates – sufficiently high density and capacity for changes of the dimensions and shape in the magnetic field.

3.9. Mechanisms of the field dependence of the speed of sound

We consider the currently known physical mechanisms of the dependence of the speed of propagation of the sound waves in magnetic fluids on the strength of the magnetic field: magnetic-hydrodynamic factor, magnetostatic pressure, magneto-calorific effect, the dynamic factor, magnetic diffusion.

Magnetohydrodynamic factor. The most accurate information on the $c(H)$ dependence in the case of the magnetic fluids of the first type is obtained by algebraic solution of the magnetohydrodynamic system of equations, proposed by I.E. Tarapov [2, 123, 124]. Here, we are interested in the speed of sound in the magnetised fluid medium, and not in the attenuation of the sound. Therefore, in the system of the equations published in [124] we ignore the terms containing viscosity and heat conductivity:

$$\partial\rho / \partial t + \operatorname{div}\rho\dot{\mathbf{u}} = 0; \tag{3.17}$$

$$\rho T \frac{d}{dt}\left(S + \frac{\mu_0}{\rho}\int_0^H \left(\frac{\partial M}{\partial T}\right)_{\rho,H} dH \right) = 0; \tag{3.18}$$

$$\rho\frac{d\dot{\mathbf{u}}}{dt} + \nabla\left[p + \mu_0 \int_0^H \left\{ M - \rho\left(\frac{\partial M}{\partial T}\right)_{T,H} \right\} dH \right] - \mu_0 M \nabla H = 0; \tag{3.19}$$

$$\operatorname{div}\mathbf{B} = 0. \tag{3.20}$$

$$\operatorname{rot}\mathbf{H} = 0. \tag{3.21}$$

$$\mathbf{B} = \mu_0\left(\mathbf{H} + M(\rho, T, H)\mathbf{H}/H\right). \tag{3.22}$$

The algebraic solution of the system of equations (3.17)–(3.22), is relatively complicated, even for the case of flat waves. The equation for the speed of sound, obtained using this procedure, contains more than 10 parameters, many of which are often repeated in various combinations with each other and this complicates the physical interpretation of the calculation results [125]. Evidently, better results upon be obtained using the approach based on determining the corrections for the speed of sound, defined by the effect of magnetohydrodynamic factors.

In [12] the authors obtained the correction for the speed of sound in magnetisation of the magnetic fluid in an isothermal wave, and attention was also given to the relaxation of magnetisation of the magnetic fluid which is not taken into account by the system of equations (3.19)–(3.22). The correction is associated with the modulation of the magnetic field which is the result of the ponderomotive interaction of the magnetised medium with an inhomogeneous (as a result of the propagation of sound in the liquid) magnetic field. The formal reason for the appearance of this correction is the term $\mu_0 M\nabla H$ in the equation (3.19). Actually, it can be assumed that $(\partial M/\partial \rho)_T = M_0/\rho$ and, consequently, this equation is reduced to the form

$$\rho(\partial \dot{\mathbf{u}} / \partial t) + \nabla p^* = 0,$$

where $p^* \equiv p - \mu_0 M_0 H$.

From this it follows that the elasticity modulus E_* increases appropriately in comparison with $E\left(E = \rho\left(\frac{\partial p}{\partial \rho}\right)_S\right)$.

The correction, obtained by this procedure, has a positive sign and, as shown in section 4.1 (The effect of the relaxation of magnetisation) its order of magnitude is 10^{-3} m/s.

The correction for the speed of sound, determined by the magnetocalorific effect, forms as a result of the appearance of the second term in the round brackets of the equation (3.18). At $(\partial M/\partial T)_{\rho,H} = 0$, i.e., in the absence of the magnetocalorific effect, this equation converts to the adiabaticity condition $\delta S = 0$. In the case of a flat sinusoidal wave, propagating in the transverse magnetic field, the magnetocalorific effect is not realised because of the absence of oscillations of the strength of the magnetic field [3, 12]. For the magnetic field longitudinal in relation to the sound waves, equation (3.18) gives

$$\delta S = -\frac{u_0}{\rho}\left(\frac{\partial M}{\partial T}\right)_{\rho,H} \delta H = \mu_0 M_T \delta M / \rho, \tag{3.23}$$

since $\int\limits_0^{H+\delta H} M_T dH - \int\limits_0^{H} M_T dH = \int\limits_H^{H+\delta H} M_T dH = M_T \delta H.$

After substitution of the expression for dM, forming the real part of (2.10) at $\omega\tau << 1$, into equation (3.23) we obtain

$$T\delta S = \frac{\mu_0 M_T (nM_n + \gamma_* M_T) T}{\rho(1 + M_H)} \cdot \frac{\partial u}{\partial x}.$$

The equation of energy conservation has the following form:

$$\left[C_V - \frac{\mu_0 M_T (nM_n + \gamma_* M_T) T}{\rho\gamma_*(1 + M_H)}\right]\delta T - \frac{qT}{\beta_T \rho^2}\delta\rho = 0.$$

We denote $C_{VH} \equiv -\frac{\mu_0 M_T (nM_n + \gamma_* M_T) T}{\rho\gamma_*(1 + M_H)}$, so that

$$\hat{\gamma} = \frac{C_V + C_{VH} + \Delta}{C_V + C_{VH}},$$

where $\hat{\gamma}$ is the analogue of Poisson's ratio [63], $\Delta \equiv C_p - C_V$.

The correction for the speed is obtained from the equation [63]:

$$c^2 = \hat{\gamma}\frac{E_T}{\rho} = \frac{c_0^2 C_V \hat{\gamma}}{C_p} \cong c_0^2\left[1 - C_{VH}\left(\frac{1}{C_V} - \frac{1}{C_p}\right)\right].$$

At $M_0 = 2.5 \cdot 10^4$ A/m, $M_T = -0.3 \cdot 10^2$ A/m·K, the numerical value $C_{VH} \sim 10^{-3}$J/kg·K and the correction for the speed is $\sim 1.5 \cdot 10^{-4}$ m/s.

In [13] the authors presented the linear hydrodynamic theory of the magnetic fluid situated in strong external magnetic fields $\sim 10^6$ A/m. The equations were solved for the sound waves with a small amplitude propagating under an angle in relation to the direction of the external field. Special attention is given to the perturbation of the magnetisation of the fluid by the sound waves, and the special feature of the study is the inclusion in the system of the equations of the equation of motion of the 'director' – the unit vector the direction of which coincides with the direction of the magnetic chain of the ferroparticles. The possibility of the rotation of the magnetic chains around the equilibrium orientation, defined by the direction of the magnetic field, is taken into account. The results show that the relative increase of the speed is always positive and its order of magnitude is approximately 10^{-5}.

Magnetostatic pressure and magnetocalorific effect. According to [57], the electromagnetic part of the differentials of the thermodynamic potentials of the magnetic material in the magnetic field is supplemented by the term $V\,(\mathbf{B} \cdot d\mathbf{H})$, where V is the volume of the magnetic. Therefore, the enthalpy differentials and the Gibbs differentials have the following form:

$$d\tilde{H} = TdS + Vdp - V(\mathbf{B}.d\mathbf{H}); \tag{3.24}$$

$$d\tilde{G} = -SdT + Vdp - V\left(\mathbf{B} \cdot d\mathbf{H}\right). \tag{3.25}$$

From this is follows that $V = (\partial\tilde{H}/\partial p)_{S,H}$ or $V = (\partial\,\tilde{G}/\partial p)_{T,H}$.

In the absence of the field

$$d\tilde{H}_0 = TdS + Vdp$$

and, therefore, $d\tilde{G}_0 = -SdT + Vdp$.

Denoting the volume of the magnetic at $H = 0$ by V_0, we obtain

$$V - V_0 = -\left(\frac{\partial}{\partial p}\int_0^{H_0} V\mathbf{B}\cdot d\mathbf{H}\right)_{S,H}, \tag{3.26}$$

$$V - V_0 = -\left(\frac{\partial}{\partial p}\int_0^{H_0} V\mathbf{B}\cdot d\mathbf{H}\right)_{T,H}. \tag{3.27}$$

If the process of application of the field is adiabatic, we use the equation (3.26), and if it isothermal we use the equation (3.27).

Let the magnetic fluid have the form of a plane parallel layer. We examine separately the case of the longitudinal and transverse (in relation to the layer of the magnetic fluids) fields. The linear dependence $\mathbf{B} = \mu\mu_0\mathbf{H}$ is satisfied. In the longitudinal field, the equation (3.27) gives

$$\frac{V - V_0}{V} = \frac{\beta_T\mu\mu_0 H_0^2}{2} - \frac{\mu_0 H_0^2}{2}\left(\frac{\partial\mu}{\partial p}\right)_T. \tag{3.28}$$

Since $\frac{\partial\mu}{\partial p} = \frac{\partial\mu}{\partial\rho}\cdot\frac{\partial\rho}{\partial p}$, then

$$p(\rho,T) = \frac{\mu_0 H_0^2}{2} + \frac{\mu_0 H_0^2}{2}\left[\rho\left(\frac{\partial\mu}{\partial\rho}\right)_T - \mu + 1\right].$$

The equation, which determines the density of the liquid, has the form

$$p(\rho,T) - p_{atm} = \frac{\mu_0 H_0^2}{2}\left[\rho\left(\frac{\partial\mu}{\partial\rho}\right)_T - (\mu - 1)\right], \tag{3.29}$$

which is the partial case of the expression, obtained using the stress tensor [57]:

$$p(\rho,T) - p_{atm} = \frac{\mu_0\rho H_i^2}{2}\left(\frac{\partial\mu}{\partial\rho}\right)_T - \frac{\mu_0(\mu-1)}{2}\left(\mu H_n^2 + H_\tau^2\right), \tag{3.30}$$

where H_i is the strength of the field inside the magnetic; H_n and H_τ are the normal and tangential (in relation to the surface of the magnetic) component of the field.

In the Langevin approximation at $\xi \ll 1$ we have:

$$\frac{\partial\mu}{\partial\rho} = \frac{\partial\mu}{\partial n}\cdot\frac{\partial n}{\partial\rho} = \frac{\partial\mu}{\partial n}\cdot\frac{n}{\rho} = \frac{\mu_0 n m_*^2}{3k_0 T\rho}.$$

Since in this case $\mu - 1 = \chi_0 = \frac{\mu_0 n m_*^2}{3k_0 T}$, in equation (3.29) the expression in the square brackets converts to 0. Consequently, the density of the magnetic fluid in isothermal magnetisation of the fluid in the longitudinal magnetic field remains unchanged.

In the magnetic field transverse to the magnetic fluid layer, on the basis of equation (3.30) we obtain

$$p(\rho, T) - p_{atm} = -\frac{\mu_0 M H_n}{2} - \mu_0 M^2.$$

At the 'moderate' values $M = 30$ kA/m, $H_n = 10^5$ A/m, we obtain $p(\rho, T) - p_{atm} = -3$ kPa. In such static tensile loading of the liquid, the speed of sound in the fluid decreases by $\sim 1.5 \cdot 10^{-2}$ m/s.

A higher static pressure in the magnetic fluid can be obtained by placing it in a heterogeneous magnetic field. In this case, the static equilibrium condition has the following form

$$\nabla p = \mu_0 M \nabla H + \rho g.$$

Ignoring the hydrostatic pressure, in the approximation of the linear form of the equation of the magnetic state we obtain

$$\delta p_M = \mu_0 \chi \left(H_{\max}^2 - H_{\min}^2\right)/2.$$

Let it be that H changes along the column of the fluid from 0 to $H_{\max}$ and, consequently, at $H_{\max} = 10^6$ A/m and $\chi = 0.5$, we have $\delta p_M \approx 3 \cdot 10^5$ Pa.

Consequently, in the direction ∇H the mean speed along the length of the specimen increases by ~ 1 m/s.

We estimate the role of the magnetocalorific effect in the magnetic fluid. The adiabatic magnetisation process will be considered. From equation (3.25) be obtain

$$T - T_0 = -\frac{\mu_0 T H_0 M}{2\rho C_p}\left(q - \frac{1}{T}\right). \quad (3.31)$$

At $q \ll T^{-1}$ we have

$$T - T_0 \approx -\frac{\mu_0 H_0 M}{2\rho C_p}. \quad (3.32)$$

The temperature increment in the fluid in magnetisation in the transverse field is calculated from the equation [1]:

$$T - T_0 = -\frac{\mu_0 T H_0^2 \chi\left(\mu q - T^{-1}\right)}{2\rho C_p \mu^2}. \quad (3.33)$$

Taking into account that $\mu g \ll T^{-1}$, we obtain

$$T - T_0 \approx \frac{\mu_0 H_0^2 \chi}{2\rho C_p \mu^2} = \frac{\mu_0 H_\perp M}{2\rho C_p}. \tag{3.34}$$

Comparing (3.32) and (3.34), it may be seen that the temperature increments of the fluid is a result of its magnetisation in the longitudinal and transverse (in relation to the magnetic fluid layer) magnetic fields are similar. As regards the order of magnitude at $H = 10^5$ A/m, $M = 30$ kA/m, $C_p = 2$ kJ/(kg·K) we have $T_0 \approx 5 \cdot 10^{-4}$ K. After demagnetisation of the liquid, the temperature of the liquid decreases by the same value. The corresponding increase of the speed of sound $|\Delta c| = \alpha_c \cdot \Delta T = 3.5 \cdot 5 \cdot 10^{-4} \approx 2 \cdot 10^{-3}$ m/s.

The dynamic factor. The anisotropy of the speed of sound. The dynamic theory of propagation of sound in the magnetised magnetic fluids, which takes into account the relative movement of the particles of the dispersed medium, was investigated for the first time in [18, 126, 127–130]. This theory is more general in relation to the available theory [67, 93, 94] because the particles of the dispersed phase are regarded as ellipsoids, and the orientation of the axes of the ellipsoids is given by the magnetic field.

The theory [126, 128] takes into account the possibility of changes of the volume and shape of the particles of the dispersed phase during magnetisation, i.e., the structural defects which develop during magnetisation of actual magnetic fluids. The aggregates are simulated by the ellipsoids of rotation whose dimensions, the ratio of the half-axes and their orientation can change in relation to the value of the applied magnetic field. We use the system of the equations describing the magnetic fluids as a dispersed medium with the aggregates of the dispersed magnetic nanoparticles. The system of the equations consists of the equations of continuity and motion for the dispersed liquid and the dispersed phase of the aggregates, the Maxwell equations and the equations of state. The equations of motion take into account the exchange of the pulses. The orientation effect of the magnetic field and the scatter of the aggregates with respect to the orientations as a result of thermal motion are taken into account. The variation of the dimensions and shape of the aggregates resultd in changes in the interaction force between the phases, determined by the Stokes and Basse effects and the attached mass. This in turn determines the degree of slipping of the particles of the dispersed phase in relation to the liquid matrix during the propagation of the sound waves and predetermines the anisotropy of the elastic properties.

The estimates, made in [127] for the droplet-shaped aggregates, show that the specific features of the acoustic properties of the magnetic fluids are associated mainly with the dynamic and, to a lesser extent, thermodynamic effects. In [126, 128], the internal heat exchange processes are not taken into account and, as in [127, 129], low-concentration magnetic fluids are investigated. The dependence of the speed of ultrasound on the dimensions and form of the aggregates, frequency, the direction and the strength of the magnetic field is determined. A method for determination of the dimensions and concentration of the aggregates on the basis of the analysis of experimental data for the propagation of ultrasound in the magnetic fluid is proposed.

The equation for the speed of sound in the magnetised magnetic fluid has the following form [126]

$$c = c_e\sqrt{\left(W_*^2 + U_*^2\right)/\left(U_*S_* - W_*Q_*\right)},$$

where c_e is the equilibrium speed ($\omega \to 0$); W_*, U_*, Q_*, S_* are the values which depend on the number of parameters – the frequency of oscillations, the size and density of the particles, the density of the dispersed medium, the viscosity of the medium, the parameters of the ellipsoidal form of the aggregates, the angle $\hat{\varphi}$, formed between the vectors **H** and **k**, and the concentration of the solid phase.

At $\omega \to 0$ and $\omega \to \infty$, the speed of propagation of sound is equal to the equilibrium speed c_e and the 'frozen' speed c_∞, respectively:

$$c_e = c_1\sqrt{\rho_1/(1-\varphi)\rho};\ c_\infty = c_e\sqrt{1+m'^2q'}. \tag{3.35}$$

$$m'^2 = \varphi(1-\varphi)(\rho_{\alpha 0} - \rho_{f0})^2/\rho_1\cdot\rho_2;$$

$$q' = \frac{1 + k_\tau\left[\left(\lambda_{//}/N_*\right)\sin^2\hat{\varphi} + N_*^2\lambda_\perp\cos^2\hat{\varphi}\right]/9}{1 + k_\tau\left[\left(\lambda_{//}/N_*\right) + N_*^2\lambda_\perp\right]/9 + \left(k_\tau/9\right)^2 N_*\lambda_{//}\lambda_\perp},$$

where N_* is the ellipsoidal form parameter; $\lambda_{||}$, $\lambda_\perp$ are the correction coefficients; $k_\tau = 9/2\cdot\rho/(1-\varphi)\ \rho_{\alpha 0}$; $\rho_{\alpha 0}$ and ρ_{f0} are is the mean density of the aggregates and the dispersed medium.

The first of the equations (3.35) corresponds as regards the accuracy to the equation (3.5) obtained from the additive model under the condition $\gamma' = 1$ – the equality of the compressibilities of the dispersed medium and the shell which does not take into account the internal heat exchange. The value c_∞ is always higher than c_e, as is also observed in the dispersed phase, consisting of spherical particles (see equation (3.16)), but in this case the first of the equations (3.35) indicates the appearance of anisotropy with its axis represented by the direction of the magnetic field. The theory shows that the inequality $c_e \le c \le c_1$ is always satisfied, and the maximum relative variation of the speed of sound $(c - c_e)/c_e = 4.9\%$ for the characteristic values of the parameters of the medium accepted in [126].

If there are no large aggregates without the field, and their number in the field is sufficiently large, then the variation of the speed of sound in the case **k**||**H** ($\Delta c_\parallel$) and **k** ⊥ **H** ($\Delta c_\perp$) are linked, according to [127], by the following relationship

$$\frac{\Delta c_\parallel}{\Delta c_\perp} \cong 1 + \frac{1 - b_* / a_*}{\left(1 + \dfrac{\rho_{f\ell} - \rho_{10}}{\rho_{10}}\right) + \dfrac{b_*}{2a_*}} < 1 + \frac{\rho_{10}}{\rho_{f\ell}}, \tag{3.36}$$

where ρ_{10} is the density of the magnetic fluid; $\rho_{f\ell}$ is the density of the aggregates (floccules); a_* and b_* are the major and minor half axes of the ellipsoids which gives the upper estimate of 10% for the anisotropy of the increase of the speed expressed in the relative units.

If the number of large aggregates without the field is still large and the variation of the speed of sound is associated only with the change of the shape of the aggregates, then

$$\Delta c_{//} / \Delta c_\perp \approx -\left(1 - b_* / a_*\right) / \left(1 + b_* / a_*\right). \tag{3.37}$$

Thus, the dynamic theory permits also the increment of the speed of sound with the opposite sign in the magnetised magnetic fluid in two mutually orthogonal directions – collinear and perpendicular to the magnetic field.

The results of measurements of the field dependence of the speed of ultrasound in [11] are interpreted by A.I. Lipkin [131, 135] on the basis of the dynamic theory. It is shown that the increase of the speed in the direction of the field $\Delta c_\parallel$ is greater than the increase of the speed in the direction transverse to the direction of the field $\Delta c_\perp$:

$$c_\parallel = c_0 \left[1 + 0.5\varphi_{f\ell} \frac{\left(\rho_{f\ell} - \rho_{11}\right)^2}{\rho_{f\ell} \cdot \rho_{11}} \right],$$

$$c_{\perp} = c_0 \left[1 + 0.5\varphi_{f\ell} \frac{\left(\rho_{f\ell} - \rho_{11}\right)^2}{\rho_{11}\left(\rho_{f\ell} + \rho_{11}\right)} \right]$$

and

$$\Delta c_{\parallel} = \Delta c_{\perp} = 0.5\varphi_{f\ell} c_0 \left(\rho_{f\ell} - \rho_{11}\right)^2 / \left[\rho_{f\ell}\left(\rho_{11} + \rho_{f\ell}\right)\right],$$

where φ_{fl} is the volume concentration of the floccules; ρ_{11} is the density of the fluid, surrounding the floccules.

Assuming that φ_{fl} = 0.04 and ρ_{fl} = 2000 kg/m^3, ρ_{11} = 1000 kg/m^3 $\Delta c_{\parallel} - \Delta c_{\perp} \approx 4$ m/s in accordance with the experimental results published in [11].

Taking into account the assumptions of the dynamic theory, A.I. Lipkin [131] proposed the method of modulation of the speed of sound in the colloid based on the forced deformation of microdroplet aggregates under the effect of the alternating magnetic field.

However, there are also other approaches to explaining the anisotropy of the speed of sound based on the specific physical effects. For example, V.V. Sokolov and V.V. Tolmachev [136] use the concept of 'frozen-in magnetisation'.

Magnetic diffusion. Magnetisation of the actual magnetic fluids and ferroparticles may be accompanied by the rearrangement of the structure disrupting the microscopic homogeneity of the system. These disruptions of homogeneity may be caused by magnetic diffusion of ferromagnetic particles.

The possible increase of the speed of sound in a dispersed system with relatively large particles ~1 μm, for which the condition $k_0T/\mu_0 m_* Gh << 1$ is satisfied, will be estimated. The theory of the process of magnetic diffusion is discussed in [132].

A cube with the size h will be hypothetically defined in the fluid. The equilibrium particle concentration in the absence of the field is n_e. In the heterogeneous magnetic field, each particles is influenced on average by the force $\mathbf{F}_1 = \mu_0 m_* L\mathbf{G}$. Here $\mathbf{G}$ is collinear with the axis Z and perpendicular to one of the faces of the cube, and its value increases linearly along the length h from G_1 to G_2. In the quasi-stationary mode, the particles carry out directional movement along $\mathbf{G}$ with the constant speed ϑ whose value is determined from the following relationship:

$$\vartheta = \mu_0 M_{SO} V_0 L(\xi) G / 6\pi\eta_S R.$$

During the time Δt ΔN_1 particles arrive additionally through the left phase of the cube, and through the right face ΔN_2 particles arrive at the same time. In this case, $\Delta N_1 = n_e\vartheta h^2\Delta t$ and $\Delta N_1 = n_e\vartheta_2 h^2\Delta t$, i.e., the value n differs only slightly from n_e within the limits of the investigated cube. Consequently, in the specified cube the number of particles decreases by ΔN and this loss is in the first approximation uniformly distributed in the volume of the cube. Consequently, the concentration increment is

$$\Delta\varphi = -2\mu_0 n_e M_{SO} V_0 L(\xi) R^2 \Delta t / 9\eta_S h.$$

Consequently

$$\Delta c = -2\left(\frac{\partial c}{\partial\varphi}\right)\mu_0\varphi M_{SO}\left(G_2 - G_1\right)L(\xi)R^2\Delta t / 9\eta_S h. \qquad (3.38)$$

At $\varphi<<1$ on the basis of (3.12) ($B^* = 0.875$) we obtain

$$\Delta c = -2c_1\left[0.875 - 0.5\left(\frac{\rho_2}{\rho_1} - 1\right)\right]\frac{\mu_0\varphi M_{SO} L(\xi)\left(G_2 - G_1\right)R^2\Delta t}{9\eta_S h}. \qquad (3.39)$$

The speed of sound in the specified volume increases. If, for example, $\partial c/\partial\varphi = -2500$ m/s, $L(\xi) = 1$, $R = 1$ μm, $\Delta t = 300$ s, $\Delta G = 10^7$ A/m^2, $\varphi = 0.1$, $\eta_S = 1.3$ kg/(m·s), $M_{SO} = 4.77\cdot10^5$ A/m, $h = 10^{-2}$ m, then $\Delta c \approx 7.5$ m/s, Δc increases 4in proportion to R^2 and ΔG. At high concentrations $\partial c/\partial\varphi \approx 0$ and magnetic diffusion has only a slight effect on the value of c.

The dispersed systems may also show gravitational diffusion (baric diffusion). The ratio the magnetic field, acting on the particle, to the gravitational force β' does not depend on the particle dimensions: $\beta' = \mu_0 M_{SO} G/(\rho_2 - \rho_1)G$. At $\beta' > 1$, the magnetic force is greater the gravitational force, and at $\beta' < 1$ the gravitational force is greater than the magnetic force. The character of the magnetic effect greatly differs from that of the gravitational effect due to the fact that it can be used to produce the mass transfer of the magnetic material in the finite volume, for example, in an acoustic cuvette, of variable intensity and direction and, in turn, this makes it possible to produce the required geometry of the distribution of the concentration and the speed of sound in the volume and control the cross-section and direction of the sound beams.

The redistribution of the concentration of the dispersed phase in the volume determines the appearance of the gradient of the speed of sound and, consequently, the refraction of sound beams

[133, 134]. The quantitative estimate of the refraction can be made using the results of the geometrical (beam) theory used in the case of small changes of the heterogeneity parameter $a_H = G_c/c$ (G_c is the gradient of the speed of sound), i.e., at $k/a_H \gg 1$ or $\omega/G_c \gg 1$. In the above example, the increment of the speed of sound in the section $h = 1$ cm is of the order of 10 m/s and, consequently, $G_c \approx 10^3$ s^{-1}. Therefore, at the frequency 2 MHz $\omega/G_c \approx 10^4$ and the above equality is satisfied. In the case of a constant gradient of the speed of sound, the trajectory of the beam is the circle with the radius $R_r = c_s / G_c \cos\widehat{\varphi}_0$, where $\widehat{\varphi}_0$ is the angle of exit of the beam from the source formed between the normal to the gradient of the speed and the wave vector, and c_s is the speed in the vicinity of the source [134]. For the beam, originating from the source of the normal to the gradient of the speed, $\widehat{\varphi}_0 = 0$ and $R_r = c_s/G_c$. If $c = 1200$ m/s, $G_c = 750$ s^{-1}, then $R_r = 1.6$ m. The angle of rotation of the wave vector at the distance S_a from the source, counted around the arc of the circle, is $\Delta\widehat{\varphi}_0 = \Delta L \cdot G_S / c$ Therefore, at $\Delta L = 3$ cm $\Delta\widehat{\varphi}_0 \approx 1°$.

In [305], the authors noted the increase of the speed of sound in water-based magnetic fluids amounting to ~18 m/s approximately 20 min after activation of the magnetic field with the strength of 85 kA/m. The investigated specimen evidently belonged to the third type of magnetic fluid.

In conclusion, we compare the contribution of the individual physical mechanisms to the increment of the speed of sound Δc in the process of magnetisation of the magnetic fluid:

1. The mechanism of heating/cooling in magnetisation demagnetisation of the magnetic fluid in accordance with the equation

$$\Delta T = -\frac{\mu_0 T}{C_p}\left[\frac{\partial V}{\partial T}\int_0^{H_0} M \cdot dH\right]_p$$

gives $\Delta c \approx 2 \cdot 10^{-3}$ m/s.

2. The baric mechanism – as a result of the 'jump' of the normal component of the field at the interface of the medium in accordance with the expression

$$p(\rho, T) - p_{\text{atm}} = \frac{\mu_0 \rho H^2}{2}\left(\frac{\partial\mu}{\partial\rho}\right) - \frac{\mu_0(\mu - 1)}{2}\left(\mu H_n^2 - H_\tau^2\right)$$

gives $\Delta c \approx 1.5 \cdot 10^{-2}$ m/s.

3. The baric mechanism – as a result of the heterogeneity of the magnetic field, according to the expression $\Delta\rho = 0.5\mu_0\chi\left(H_{\max}^2 - H_{\min}^2\right)$ leads to the estimate $\Delta c \approx 1$ m/s.

4. The magnetohydrodynamic mechanism – as a result of the induced inhomogeneity of the magnetic field $\Delta c = \mu_0 M^2/2\rho c^2 \approx 10^{-3} \div 10^{-2}$ m/s.

5. The magnetohydrodynamic mechanism – as a result of the magnetic–calorific effect gives the estimate $\Delta c \approx 10^{-4}$ m/s.

6. The magnetic–diffusion mechanism, determined by the equation $\mathbf{F} = \mu_0 m_* L(\xi)\mathbf{G}$, gives for the ferrosuspension $\Delta c \leq 10$ m/s.

7. The dynamic mechanism, determined by the relative movement of the phases of the dispersed system, gives $\Delta c \approx 10 \div 50$ m/s.

The presented values of Δc should be regarded as the upper estimates obtained on the basis of simulation representations.

Thus, the controlling contribution to the increase of the speed of sound in magnetisation of the magnetic fluid is provided by the effects of the relative movement of the phases of the dispersed system, taken into account by the dynamic theory.

4

Absorption and scattering of sound

4.1. The mechanisms of absorption of sound waves

We examine the most probable physical mechanisms of the absorption of sound waves in a magnetic fluid: the Stokes mechanism of absorption of ultrasound, the mechanism of relative movement of the phases, the mechanism of internal heat exchange, magnetohydrodynamic processes, relaxation of magnetisation, dynamic processes.

This problem is of obvious interest not only for science but also for practice, for example, in ultrasonic flaw inspection the acoustic contact media are represented in most cases by the magnetic fluids, transmitting not only longitudinal [107, 137] but also transverse ultrasonic waves [138].

The main mechanisms, which determine the largest contribution to the dissipation of sound energy in the magnetic fluid are determined by their unique structure: the ferroparticles (FP) are dispersed in the fluid–carrier. These particles are coated with a layer of the stabiliser and exist separately from each other (magnetic fluids of the first and second type) or merge in the absence of the magnetic field into aggregates (magnetic fluids of the third type).

The most stable fluids, i.e., the magnetic fluids of the first, belong in the group of microheterogeneous media in which the propagation of sound depends strongly on the course of the 'non-local' relaxation processes [93] which can be investigated by the acoustic methods of studying the dispersed media, developed in [93–96, 156–158].

The special features of passage of ultrasound through the fluid magnetising media are determined by a set of the physical mechanisms of the losses and conversion of the elastic energy which include the viscous losses, scattering on the disperse phase, internal heat exchange, magnetohydrodynamic processes, the magnetisation relaxation phenomenon, refraction, dispersion and non-linear effects.

The Stokes mechanism of absorption of ultrasound waves. This mechanism is determined by the presence of shear viscosity, formed in any fluid. It is therefore convenient to discuss separately the results of investigations of the shear viscosity of the magnetic colloids.

Outside the magnetic fields, the magnetic fluid has the form of a conventional colloidal solution whose viscosity depends on the content of the dispersed phase [35, 139, 140]. In the colloidal solutions with a low particle concentration, this increase is described by the Einstein equation [141]:

$$\eta / \eta_0 = 1 + 2.5\varphi_g,$$

where φ_g is the hydrodynamic concentration of the particles of the dispersed phase which includes, in addition to the volume fraction of the solid phase φ, also the volume fraction of the shielding shells; η and η_0 are the dynamic viscosities of the colloid and of the fluid-base.

The viscosity of the concentrated suspensions is described by the dependence derived by Vand [142]:

$$\eta / \eta_0 = \exp[(2.5\varphi_g + 2.7\varphi_g^2) / (1 - 0.609\varphi_g)].$$

There are also other expressions for describing the concentration dependence of the viscosity of colloidal solutions [35, 143, 151].

V.M. Buzmakov and A.F. Pshenichnikov [143] published an equation for the relative viscosity of the solution of the 'magnetite in kerosene' type

$$\eta / \eta_0 = 1 + 4.4\varphi_g + 28.5\varphi_g^2$$

obtained by approximating the experimental data for the temperatures of 25, 40 and 60°C.

The most probable reason for the high viscosity of the ferrocolloids is the content of the aggregates in the fluid. At a low shear rate, the behaviour of the aggregated magnetic fluid becomes non-Newtonian and its deviation from the Newtonian behaviour becomes greater with the reduction of the shear rate and the increase of the volume fraction of the solid phase.

In the magnetic fluids, in addition to the hydrodynamic interactions, there is also the magnetic interaction of the particles influencing the relative movement of the particles and, therefore, the viscosity of the magnetic fluid also depends on the intensity of this interaction. In a stable magnetic fluid, the magnetic interaction can be ignored. In this case, the viscosity of the magnetic fluid is determined by the hydrodynamic concentration of the particles $\varphi_g = p\varphi$ and corresponds to the relationships, obtained for the suspensions of non-magnetic particles. Here p is the coefficient which does not depend on the concentration of the solid phase.

The viscosity of the actual magnetic fluids may also depend on the prior history of the specimen [144, 145] (i.e. on the preliminary external effects, such as stirring and magnetisation) and the shear rate.

Aggregation of the magnetic fluid is accompanied by the increase of the effective hydrodynamic concentration φ_g. In addition to this, the large structures may penetrate throughout the entire volume of the magnetic fluid and can be inhibited by its boundaries. Both mechanisms increase the viscosity and lead to the non-linear dependence of the viscous stresses on the strain rate [144-148].

The viscosity and rheological behaviour of the magnetic fluids is influenced by the variation of temperature. Above all, the temperature determines the viscosity of the base of the magnetic fluids and surface-active agents and, in addition to this, the variation of temperature influences the contribution of rotational diffusion to the viscosity and the process of aggregation of the particles in the fluid. Consequently, the temperature dependences of the viscosity of the magnetic fluid and of the base differ. This difference increases with increasing concentration of the magnetic phase in the magnetic fluid and with increasing temperature. Viscometric experiments [149, 150] confirm the large difference between the temperature dependence of the effective viscosity of the magnetic fluid in comparison with that for the fluid of the base. The effective coefficient of viscosity of the magnetic fluid based on kerosene is described quite accurately by the Andrade equation:

$$\eta_{ef} = P \cdot \exp(N / RT),$$

where R is the gas constant; T is temperature; P and N are some constants which depend on the concentration of the ferrophase and the width of the temperature range.

M.M. Maiorov [152] published the results of investigations of the viscosity of colloidal solutions of ferromagnetics for two samples. One of them was the magnetic fluid based on magnetite, the other one – the colloidal solution of cobalt ferrite. The carrier fluid was kerosene. The volume fraction of the particles was 0.24 and 0.19, respectively. The measured anisotropy of viscosity corresponds to the suspension of the ellipsoids with the ratio of the half-axes of 1.1–1.5. The dependence of viscosity on the shear rate was stronger in the field perpendicular to the flow speed of the fluid. The increase of the viscosity of the magnetic fluid for the magnetite specimen did not exceed 5–6%. The results of detailed determination of viscosity, the relaxation time of the magnetic moment and the measurements of magnetisation show that the colloidal particles are represented by the aggregates of the single-domain particles.

Yu.D. Varlamov and A.B. Kaplun [144, 153] investigated the viscosity of magnetite magnetic fluids in which the carrier medium was kerosene and also water. The rheological properties were studied by the method of vibrational viscometry. In [144] special attention was given to the process of preparation of the specimens of the magnetic fluid. The results of gradual centrifuging of the 'real' fluids and of the methods of preparation of the specimens of the magnetic fluid with different concentrations of the dense phase were analysed. The following conclusions were made: in the magnetic field, the centrifuged magnetic fluids show reverse formation of the aggregates and their properties do not depend on various external effects such as agitation and magnetic treatment. In contrast to the data published in [152], the maximum values of the viscosity of the magnetic colloids η_∞ with the concentration of the solid phase of ~6% in the magnetic field of H = 58 kA/m, published in [144], were 27% higher than the viscosity of the same fluids in the absence of the field for the centrifuged samples and by 62% for the non-centrifuged samples.

The viscosity of the magnetic fluids based on silicone oil in strong magnetic fields was studied by A.B. Kaplun and Yu.D. Varlamov also by the vibrational method. In the absence of the magnetic field, the viscosity of the magnetic fluid and silicone oil is described by the exponential temperature dependences. This shows indirectly that the magnetic fluid is Newtonian. To classify the nature of the flow of the magnetic fluids, the investigations were carried out at different shear rates. Similar studies were performed for the magnetic fluids in the magnetic field. No deviations from the Newtonian nature of the flow were found for the selected parameters of the vibrational system.

Investigations of the effective viscosity of the magnetic fluids in the homogeneous external magnetic field [154] show that in the case of the collinear orientation of the hydrodynamic flow and the strength of the magnetic field, no variation of viscosity with increasing strength of the magnetic fields was detected in the experimental error range. In the experiments with the mutual orthogonal orientation of the flow speed and the field, the viscosity increase monotonically with increasing strength of the magnetic field. The maximum increase of viscosity was 28% of the initial value.

In the shear flow, the solid particle is affected by the moment of forces causing the particle to rotate. The magnetic field orients the magnetic moment of the particle and in the presence of a bond between the moment of the particle and the particle complicates the free rotation of the particle. This results in local gradients of the speed of the fluid–base in the vicinity of the particles and increases the effective viscosity of the magnetic fluid [199]. The saturation of the so-called 'rotational' viscosity starts when the strong field leads to a rigid orientation of the particle. For the magnetic fluids with the spherical particles, the additional internal friction in the strong fields ($H \gg k_0T/\mu_0m_*$) is determined by the relationship [34, 58]:

$$\Delta\eta_H = 1.5\varphi_g \cdot \eta_0 \cdot \sin^2\beta,$$

where $\Delta\eta_H$ is the variation of the viscosity coefficient of the magnetic fluid in the magnetic field; η_0 is the viscosity coefficient at H=0; φ_g is the hydrodynamic concentration of the magnetic fluid; β is the angle between **H** and the angular velocity of the magnetic particle.

The calculations of viscosity for the particles in the shape of the ellipsoids of rotation, when the field complicates the flow of a symmetric (vortex-free) flow around the particles, were described in [155].

The formation of rotational viscosity as a result of the orientation effect of the magnetic field and its influence on the effective viscosity of the magnetic fluid in principle should be independent of the shear rate in the fluid, i.e., it should be Newtonian. However, the experimental data show that with increasing strength of the field, the flow of the fluid deviates from the Newtonian flow and another mechanism of the effect of the field on the rheology of the magnetic fluid is added to the orientation effect of the field. This mechanism is associated with the formation and movement of aggregates of the particles in the magnetic fluid which depend on the strength of the field and the particle concentration. In particular, this mechanism is typical of the ferrosuspensions, and also some unstable (non-centrifuged) magnetic fluids [151].

Thus, the rheological properties of the magnetic fluid greatly differ. They depend on the composition, the method of production and subsequent purification of the magnetic colloid, and on its 'magnetic prior history'.

We take into account the 'excess' (Einsteinian) viscosity of the dispersed system, using the expression [35, 51]:

$$\frac{\eta_s - \eta_{s1}}{\eta_s} = 2.5\left(1+\frac{\delta}{R}\right)^3 - \left(\frac{2.5\varphi_{cr}-1}{\varphi_{cr}^2}\right)\left(1+\frac{\delta}{R}\right)^6 \varphi, \qquad (4.1)$$

where η_S is the viscosity of the magnetic fluid; η_{S1} is the viscosity of the fluid–carrier; $\varphi_{cr} = 0.745$; δ is the thickness of the stabilising shell; R – is the particle radius.

The additional adsorption (in relation to the pure dispersed medium) per single wavelength and unit concentration is estimated ignoring the relaxation of the shear viscosity and using the following equation:

$$\frac{\Delta\alpha_1\lambda}{\varphi} = \frac{2\pi^2 v(\eta_S - \eta_{S1})}{3\varphi\rho c^2},$$

where λ is the wavelength of sound; v is the frequency of oscillations.

Assuming ν = 25 MHz, φ = 0.1, ρ= 1230 kg/m^3, c = 1135 m/s, η_{S1} = $0.13\cdot10^{-2}$ kg/m·s, and η_S = 0.37 10^{-2} kg/m·s [15], we obtain $\frac{\Delta\alpha_1\lambda}{\varphi} = 10^{-2}$.

The mechanism of relative motion of the phases. Differences in the density of the components of the dispersed system results in relative movement of the components during propagation of the sound waves. As a result of the viscosity of the fluid–carrier, the movement of the particles in relation to the medium is accompanied by friction which tries to equalise the velocities of the medium and the particle. The process of momentum exchange between different particles of the medium takes place with a delay in relation to the sound waves and this results in additional adsorption of sound [67, 68, 156, 157]. The following equation was derived for the additional adsorption, determined by this process, in the studies [68, 67]:

$$\frac{\Delta\alpha_2\lambda}{\varphi} = \frac{4\pi(\gamma_0-1)^2\Psi_V(1+\sqrt{\Psi_V})}{9\left(1+\sqrt{\Psi_V}\right)^2 + \Psi_V\left(1+b_2\sqrt{\Psi_V}\right)^2}, \qquad (4.2)$$

where $\gamma_0 \equiv \frac{\rho_2}{\rho_1}, \Psi_V \equiv \frac{\pi R^2\rho_1 v}{\eta_{S_1}}$.

The dependence of additional absorption on the radius of the particle is shown graphically in Fig. 4.1.

The calculation of the additional absorption in the conventional emulsions is carried out using the equation [156, 158]

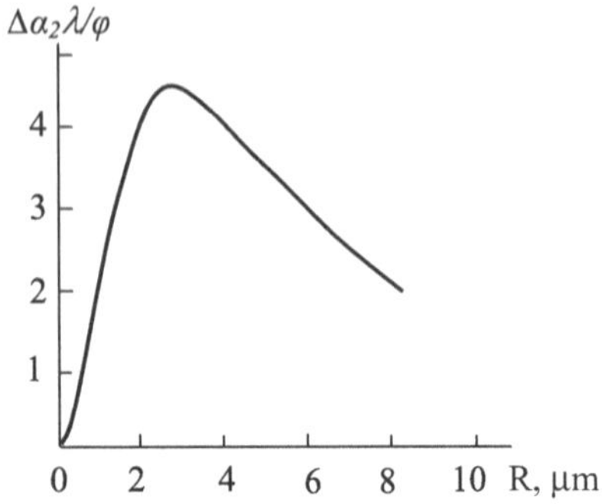

Fig. 4.1. Dependence of the additional absorption, determined by the relative movement of the particles and the fluid–carrier, on the radius of the dispersed particles *R*.

$$\frac{\Delta\alpha_2\lambda}{\varphi} = 9\pi\frac{(\rho_1 - \rho_2)^2}{(\rho_1 + 2\rho_2)^2\sqrt{\psi_V}}$$

which is a partial case of expression (4.2) at $\Psi_V \gg 1$; it is not suitable for the investigated media, and the values of $\frac{\Delta\alpha_2\lambda}{\varphi}$, calculated from this equation, are of the order of 10^3–10^4 and, consequently, on the basis of these values it could be concluded that the sound can not propagate in such a medium. The reason why this equation is not suitable for the magnetic fluids is that for these fluids the value Ψ_V is in the range 10^{-6}–10^{-4}. When $\Psi_V \ll 1$, equation (4.2) transforms to the following form:

$$\frac{\Delta\alpha_2\lambda}{\varphi} = \frac{4\pi}{9}(\gamma_0 - 1)^2\Psi_V. \tag{4.3}$$

As regards the magnetic fluid, in equation (4.3) it is necessary to carry out transition from *R* to *R* + δ and at the centre, replace φ by $\varphi\left(1+\frac{\delta}{R}\right)^3$ and $\gamma_0 = \frac{\rho_2}{\rho_1}$ by $\gamma_0' = \frac{\gamma_0 - 1}{(1+\delta/R)^3} + 1$. Taking this into account, equation (4.3) has the following form:

$$\frac{\Delta\alpha_2\lambda}{\varphi} = \frac{4\pi^2(\gamma_0 - 1)^2 R^2\rho_1 v}{9\eta_{S_1}(1+\delta/R)}.$$

Assuming that ρ_2 = 5210 kg/m³ (the dispersed medium is magnetite) and taking into account the previously introduced numerical values of the quantities, included in this expression, we obtain $\frac{\Delta\alpha_2\lambda}{\varphi} \approx 3.7\cdot10^{-2}$.

The additional damping, associated with the scattering of sound on the particles $\frac{\Delta\alpha_3\lambda}{\varphi}$, is investigated in section 4.2.

The mechanism of internal heat exchange. The heat exchange between the adjacent layers in the region bordering with the ferroparticle takes place with a delay as a result of the low heat conductivity of the fluid and this is also the reason for the thermal adsorption of the sound in the magnetic fluids [9].

For $R\sqrt{\pi v\rho_2 C_{p_2}/\chi_2} << 1$ (for the magnetic fluids of the first and second kind and the frequency of 25 MHz, this parameter is equal to $\sim 10^{-2}$) and low concentration, the thermal adsorption is specified by the equation [94]

$$\frac{\Delta\alpha_4\lambda}{\varphi} = \frac{2\pi^2}{3\chi_2} Tc_1^2\rho\rho_2^2 C_{p_2}^2 R^2 v\left(\frac{\chi_2}{\chi_1}+\frac{1}{5}\right)\left(\frac{q_2}{\rho_2 C_2}-\frac{q_1}{\rho_1 C_{p_1}}\right)^2, \tag{4.4}$$

where χ is the heat conductivity coefficient; q is the thermal expansion coefficient.

We accept the following numerical values of the quantities included in the equation: C_{p1} = 2 kJ/(kg·K), C_{p2} = 0.655 kJ/(kg·K), χ_1 = 0.12 W/(m·K) [2], χ_2= 5.9 W/m/(kg·K), $q_1 = 9.5\cdot 10^{-4}$ K^{-1}, $q_2 = 11.4\cdot 10^{-6}$ K^{-1} [159]. Since $\chi_2 >> \chi_1$, $q_1 >> q_2$, the equation (4.4) can be simplified:

$$\frac{\Delta\alpha_4\lambda}{\varphi} = \frac{2\pi^2}{3\chi_1} Tc_1^2\rho\rho_2^2 C_{p_2}^2 R^2 v\frac{q_1^2}{\rho_1^2 C_{p_1}^2}.$$

Assuming that T = 300 K, ρ = 1230 kg/m^3, we obtain $\frac{\Delta\alpha_4\lambda}{\varphi} = 8.8\cdot 10^{-2}$.

The ratio of the losses, determined by the heat exchange and the relative movement of the phases of the dispersed system is:

$$\frac{\Delta\alpha_4\lambda}{\varphi}\Big/\frac{\Delta\alpha_2\lambda}{\varphi} = \frac{3Tc_1^2\rho\rho_2 C_{p_2}^2 vq_1^2\eta_{S_1}(1+\delta/R)}{2\chi_1\rho_1^3 C_{p_1}^2(\gamma-1)^2}$$

This relationship depends only slightly on the ratio of the ferroparticles, is independent of the heat conductivity of the ferroparticles, and its numerical value for the investigated system is ~2.4.

Within the framework of the model of the non-local relaxation processes, the complete theoretical attenuation is described by the sum [9, 76]:

$$\left(\frac{\Delta\alpha\lambda}{\varphi}\right)_{th} = \frac{\Delta\alpha_1\lambda}{\varphi} + \frac{\Delta\alpha_2\lambda}{\varphi} + \frac{\Delta\alpha_3\lambda}{\varphi} + \frac{\Delta\alpha_4\lambda}{\varphi}$$

In this case, $\left(\frac{\Delta\alpha\lambda}{\varphi}\right)_{th} = 0.135$. It should be noted that the term, determined by the structural relaxation [160], is not taken into account, and the value of this term is close to $\kappa \sim \frac{\Delta\alpha_1\lambda}{\varphi}$.

From equation (4.2) we obtain easily the relationship for calculating the most informative parameter $\Delta\alpha/\nu^2$:

$$\Delta\alpha/\nu^2 = c_* \cdot f(\nu)$$

where $c_* = \frac{2\pi^2 R^2 \rho_1 b_1}{c_1 \eta_{S_1}}, f(\nu) = \frac{1+\sqrt{\Psi_\nu}}{\left(1+\sqrt{\Psi_\nu}\right)^2 + \Psi_\nu\left(1+b_2\sqrt{\Psi_\nu}\right)^2}$.

The above relationship and also equation (3.16) can be used to obtain the dispersion curves of additional absorption and dispersion of the speed in the entire frequency range. We compare the dispersion of the acoustic parameters of the magnetic fluids of the first and third type. Taking into account the typical values of the physical parameters of the components of the magnetic fluids of the first type [51, 161]: ρ_1 = 760 kg/m^3, η_{S_1} = $0.7\cdot10^{-2}$ Pa·s, ρ_2 = 5240 kg/m^3, δ/R = 0.4, we obtain that in the range 1–1000 MHz the value $\Delta\alpha/\nu^2$ decreases only slightly – from 0.226 to 0.197 m^{-1} Hz^{-2}, and the increment of the speed is in the range from $1.8\cdot10^{-7}$ to $5.64\cdot10^{-3}$ %. Consequently, the mechanism of relative movement of the phases in the magnetic fluid of the first type does not cause any dispersion of the acoustic parameters. As regards the magnetic fluids of the third kind, it will be assumed that the particles of the dispersed phase with the same dimensions and the radius R = 0.7 μm consist of ferroparticles with the maximum packing density. The concentration of the disperse phase is φ_δ= $3.43\cdot10^{-2}$. The density of the floccules under the condition of filling of the cavities between the particles by the fluid-carrier is $\rho_{f\ell}$ = 1880 kg/m^3.

Figure 4.2 shows the dispersion curves of adsorption and speed (the curves 1 and 2, respectively). In this case, the 'slipping' mechanism of the particles determines the relatively strong dispersion of the parameter $\Delta\alpha/\nu^2$, and the lower absolute value of the dispersion of the speed ≤1%.

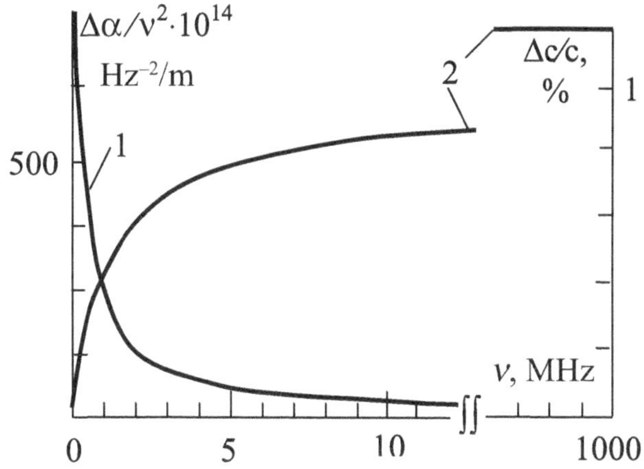

Fig. 4.2. Dispersion of the speed and adsorption in the aggregated magnetic fluid.

In the aggregated magnetic fluids, the particles of the dispersed phase and the fluid–carrier can greatly differ from each other in the thermodynamic parameters and only slightly in density. For such a system, the parameter $R\sqrt{\pi\nu\rho_2 C_{p2}/2\chi_2} = C_2 >> 1$. It is assumed that $\chi_2 = 1$ W·m^{-1}K^{-1}, i.e., it is an order magnitude smaller than in the case of magnetite, and an order magnitude higher than in the case of kerosene, $C_{p2} = 10^3$ J/kg·k, $\rho_2 = 1880$ kg/m^3, $\nu = 25$ MHz, consequently $C_2 \cong 10$. In this case, the thermal adsorption is defined by the equation [156]:

$$\frac{\Delta\alpha_4}{\nu^2} = \frac{3\varphi_{f\ell} T\rho_1 c_1}{\sqrt{2/\pi}\cdot\nu^{3/2} R}\left(\frac{q_2}{\rho_2 C_{p2}} - \frac{q_1}{\rho_1 C_{p1}}\right)^2 \cdot \frac{\sqrt{\chi_1\chi_2\rho_1\rho_2 C_{p1} C_{p2}}}{\left(\chi_1\rho_1 C_{p1}\right)^{1/2} + \left(\chi_2\rho_2 C_{p2}\right)^{1/2}}.$$

If $q_2 = 0.25\cdot 10^{-3}$ K^{-1}, $q_1 = 10^{-3}$ K^{-1}, then $\Delta a_4/\nu^2 = 4\cdot 10^{-14}$ m^{-1} Hz^{-2}. At lower frequencies, for example $\nu = 2.5$ kHz, for the same dispersed system the parameter $C_2 = 0.1$ and the additional adsorption is calculated from equation (4.4) which shows that that there is no dispersion of the parameter $\Delta a_4/\nu^2$, and the numerical value of the parameter is of the order of $1\cdot 10^{-12}$ m^{-1} Hz^{-2}. Thus, the aggregated magnetic fluids may be characterised by the overlapping of the dispersion ranges of the dynamic and thermal mechanisms in the frequency range ≥1 MHz.

The conclusion of the controlling contribution to the additional absorption of the non-magnetised magnetic fluids of the mechanisms, associated with the interphase heat exchange and the relative movement, was also made by I.S. Kol'tsova in [162]. In addition to this, on the basis of the 'dimensional–frequency plane', proposed by the author of [162], the distribution of the dispersion regions along the frequency axis was indicated.

Magnetohydrodynamic processes. Initially, the theoretical investigations of the special features of the absorption of sound waves in the magnetised magnetic fluids were carried out in the framework of the classic magnetohydrodynamics and mechanics of solid media. The system of equations for the wave process in the magnetic fluid was published by B.M. Berkovskii and V.G. Bashtov [163], and in a more complete form – taking into account the magnetocalorific effect – by I.E. Tarapov [2, 124]. The system of the acoustic equations for the continuous electrically conducting medium, which can be magnetized non-uniformly and isotropically, has the following form [2]:

$$\frac{\partial \rho}{\partial t} + \operatorname{div} \rho \mathbf{v} = 0; \tag{4.5}$$

$$\rho T \frac{d}{dt}\left(S + \frac{\mu_0}{\rho}\int_0^H \left(\frac{\partial M}{\partial T}\right)_{\rho,H} dH \right) = \tau_{i,k}\frac{\partial v_i}{\partial x_k} + \operatorname{div}(\chi \nabla T) + (\operatorname{rot}\mathbf{H})^2/\sigma; \tag{4.6}$$

$$\begin{gathered} \rho\frac{\partial \mathbf{v}}{\partial t} = -\nabla\left\{ p + \mu_0 \int_0^H \left[M - \rho\left(\frac{\partial M}{\partial \rho}\right)_{T,H}\right] dH \right\} \\ + [\operatorname{rot}\mathbf{H}\times\mathbf{B}] + \mu_0 M \nabla H + \eta_s \Delta v + (\eta_v + \eta_s/3)\nabla \operatorname{div}\mathbf{v}; \end{gathered} \tag{4.7}$$

$$\frac{\partial \mathbf{B}}{\partial t} = \operatorname{rot}[\mathbf{v}\times\mathbf{B}] - \operatorname{rot}\operatorname{rot}\mathbf{H}/\mu_0\sigma; \tag{4.8}$$

$$\operatorname{div}\mathbf{B} = 0, \tag{4.9}$$

where τ_{ik} is the tensor of viscous stresses; η_s, η_v are the constant coefficients of the shear and bulk viscosity; σ is the specific electrical conductivity of the medium.

The remaining notations have the same meaning as previously. It is assumed that $\mathbf{B} = \mu_0 (\mathbf{H} + \mathbf{M})$, and $\mathbf{M}||\mathbf{H}$, and the equation of state of the medium is given in the form

$$p = f(\rho, T). \tag{4.10}$$

If it is assumed that $\sigma \to \infty$, $M \approx 0$ and the dissipative processes are ignored, we obtain the system of equations well-known in the theory of waves in plasma [57], indicating the possibility of propagation in such a medium of several waves with different speeds – Alfven, slow and fast magnetic sound waves. The solution of the system of equations, published in [2], also indicates the existence of these waves which is the consequence of the assumption on the electrical conductivity of the fluid magnetising media.

For the non-conducting magnetic fluids we may obtain $\sigma \to 0$. If the role of the bias currents is not significant, then rot $H = 0$. As a result of this and also taking into account that $M - \rho\left(\frac{\partial M}{\partial \rho}\right)_{T,H} = 0$, the equations (4.6)–(4.8) have the following form:

$$\rho T \frac{d}{dt}\left(S + \mu_0 \int_0^H \left(\frac{\partial M}{\partial T}\right)_{\rho,H} dH / \rho \right) = \tau_{i,k} \frac{\partial v_i}{\partial x_k} + div(\chi \nabla T); \quad (4.11)$$

$$\rho \frac{\partial \mathbf{v}}{\partial t} = -\nabla\left(p - \mu_0 M_0 H\right) + \eta_s \Delta \mathbf{v} + \left(\eta_\upsilon + \eta_s / 3\right) \nabla \mathrm{div}\, \mathbf{v}; \quad (4.12)$$

$$\mathrm{rot}\ \mathbf{H} = 0. \quad (4.13)$$

Taking into account also the equations (4.5), (4.9) and (4.10), we obtain a new system of acoustic equations suitable for describing the non-conducting magnetising media.

In the absence of dissipation of the energy determined by the viscous and thermal mechanisms, this system of equations describes non-damping sound waves. This fact can be explained quite easily physically taking into account the fact that the magnetic–calorific and ponderomotive effects, introduced in the equations (4.11) and (4.12), are assumed to be instantaneous, i.e., following without a delay in relation to the elastic oscillations in the sound waves. For this reason, the elastic modulus and heat capacity of the medium may be actual values, and the sound waves – non-damping. Therefore, it may be assumed that the attenuation of sound waves in the continuous instantaneously remagnetized magnetic fluid should be determined by the viscosity and heat conductivity as in the conventional fluids [63].

The exact analytical solution of the system of equations, taking into account viscosity and his conductivity, is extremely cumbersome [125]. In order to make a number of qualitative inclusions, we use the method of successive approximations in which the 'zero' approximation is represented by the non-magnetic medium. The magnetic–calorific effect is formally manifested in the system of acoustic equations for the non-conducting magnetic fluids by the additional term in the heat conductivity equation:

$$\rho T \frac{d}{dt}\left[\mu_0 \rho^{-1} \int_0^H (\partial M / \partial T)_{\rho,H}\, dH \right].$$

If the viscosity is ignored, the equation of heat conductivity in the case of flat waves has the following form:

$$T \frac{d}{dt}\left(S + \mu_0 \rho^{-1} \int_0^H \left(\frac{\partial M}{\partial T} \right)_{\rho,H} dH \right) = \frac{\chi}{\rho} \frac{\partial^2 T}{\partial x^2}.$$

It is assumed that in the zero approximation, the wave is sinusoidal. Consequently, the heat conductivity equation has the form:

$$T\delta\left(S + \mu_0 \rho^{-1} \int_0^H \left(\frac{\partial M}{\partial T} \right)_{\rho,H} dH \right) = \frac{i\omega\chi}{\rho c^2} \delta T. \tag{4.14}$$

In turn, the energy conservation equation, with (4.14) taken into account, acquires the following form:

$$i\omega\chi c^{-2} \rho^{-1} \delta T + T\delta\left[\mu_0 \rho^{-1} \int_0^H \left(\frac{\partial M}{\partial T} \right)_{\rho,H} dH \right] = C_V \delta T - qTc^2 \rho^{-1} \gamma^{-1} \delta\rho. \tag{4.15}$$

In the case of instantaneous magnetic reversal we have

$$\delta M = \gamma_*^{-1} \left[nM_n + \gamma_* M_T \right] \delta T,$$

and consequently

$$\left(\frac{\partial M}{\partial T} \right)_{\rho,H} = \gamma_*^{-1} \left[nM_n + \gamma_* M_T \right]$$

and

$$\begin{aligned} \delta\left[\mu_0 \rho^{-1} \int_0^H \left(\frac{\partial M}{\partial T} \right)_{\rho,H} dH \right] &= \mu_0 \rho^{-1} \gamma_*^{-1} \left[nM_n + \gamma_* M_T \right] \delta H - \\ -\mu_0 \gamma_*^{-1} \rho^{-2} \left[nM_n + \gamma_* M_T \right] H \cdot \delta\rho & \end{aligned} \tag{4.16}$$

For evaluation, we consider the case $\delta H = -\delta M$ which is observed at $\mathbf{H} || \mathbf{k}$. Substituting (4.16) into (4.15), we obtain:

$$\begin{aligned} &\left[C_V - \mu_0 T (M_0 + \gamma_* M_T)^2 \rho^{-1} \gamma_*^{-2} - i\omega\chi c^{-2} \rho^{-1} \right] \cdot \delta T = \\ &= \left[qc^2 T (\rho\gamma)^{-1} + \mu_0 \gamma_*^{-1} \rho^{-2} (M_0 + \gamma_* M_T) H \right] \cdot \delta\rho \end{aligned} \tag{4.17}$$

In the square brackets of the left and right parts of the equation there is a single additional term and each of these terms is at least four orders of magnitude smaller than the appropriate actual term.

Therefore, in the final analysis, we obtain the conventional algebraic equation of the adiabate:

$$(C_V - i\omega\chi c^{-2}\rho^{-1})\delta T = qc^2 T\rho^{-1}\gamma^{-1}\delta\rho$$

which leads to the expressions for damping and dispersion, determined by the heat conductivity of the medium [63, 164].

Effect of the relaxation of magnetisation. The periodic variation of the magnetisation of the magnetic fluids, in which a flat sound wave propagates in the direction of the vector of the strength of the magnetic field, results in the modulation of the field. The periodic inhomogeneity of the magnetic field and the finiteness of the relaxation time of magnetisation can in principle result in additional absorption of sound. This circumstance was stressed in the study by B.I. Pirozhkov and M.I. Shliomis [12] in which appropriate calculations were carried out for the low-concentration magnetic fluids without taking into account the oscillations of temperature in the sound waves.

This theory will be generalised for the case of a magnetic fluid with any concentration in propagation of an isoentropic wave in it [3].

In a heterogeneous magnetic field, the unit volume of the magnetised fluid is subjected to the effect of the ponderomotive force $f = \mu_0 M \frac{\partial H}{\partial x}$. This force results in the formation of additional complex elasticity $\tilde{E}_M$. In this case $(\mathbf{H}||\mathbf{k})$

$$\tilde{E}_M \frac{\partial^2 u}{\partial x^2} = -\mu_0 M_0 \frac{\partial M}{\partial x}. \tag{4.18}$$

The expression for δM will be substituted to (4.18) and $\tilde{E}_M$ will be expressed from the following equation:

$$\tilde{E}_M = \frac{\mu_0 M_0^2 (nM_n / M_0 + \gamma_* M_T / M_0 + i\omega\tau_1)}{1 + M_H + i\omega\tau_1}$$

and we introduce the notations

$$\tau \equiv \tau_1 / (1 + M_H)$$

and

$$A_\beta \equiv (nM_n / M_0 + \gamma_* M_T / M_0) / (1 + M_H)$$

Taking into account the notations, the real part of the complex elasticity E_M and the additional viscosity η_M have the form:

$$E_M = \frac{\mu_0 M_0^2 (A_\beta + \omega^2\tau^2)}{1+\omega^2\tau^2};$$

$$\eta_M = \frac{\mu_0 M_0^2 (1 - A_\beta)\tau}{1+\omega^2\tau^2}$$

The speed of propagation c_M and the coefficient of absorption of sound α_M in the magnetised magnetic fluid can be calculated from the equations:

$$c_M = \left[c^2 + \mu_0 M_0^2 (A_\beta + \omega^2\tau^2) / \rho(1+\omega^2\tau^2)\right]^{1/2} \tag{4.19}$$

$$\alpha_M = \mu_0 M_0^2 (1 - A_\beta)\omega^2\tau / 2\rho c^3 (1+\omega^2\tau^2) \tag{4.20}$$

In the range of 'low' frequencies ($\omega\tau \ll 1$)

$$c_{M0} = (c^2 + \mu_0 M_0^2 A_\beta / \rho)^{1/2} \tag{4.21}$$

At 'high frequencies ($\omega\tau \gg 1$)

$$c_{M\infty} = (c^2 + \mu_0 M_0^2 / \rho)^{1/2} \tag{4.22}$$

Using the expressions (4.21) and (4.22), we represent the relation (4.20) in the form

$$\alpha_M = (c_{M\infty}^2 - c_{M0}^2)\omega^2\tau / 2c^3 (1+\omega^2\tau^2) \tag{4.23}$$

The values of $c_{M\infty}$ and α_M in this form were found in [12], but the values of c_{M0} and τ were determined using different expressions. Complete agreement is achieved if it is assumed that $\delta T = 0$ and we transfer to low concentrations at which the Langevin equation can be used. Estimates will now be obtained.

Let it be that $\Delta c_M \equiv c_{M\infty} - c$, from equation (4.22) we obtain $\Delta c_M = \frac{\mu_0 M_0^2}{2\rho c}$. To estimate Δc_M, we use the values from [64]: $c = 1119$ m/s, $\rho = 1525$ kg/m^3, $M_0 = 52$ kA/m, consequently $\Delta c_M = 10^{-3}$ m/s. The value α_M is determined from the equation (4.20), assuming that $\partial\rho/\partial T = -0.8$ kg/(m$^3\cdot$K) [65], $\partial M/\partial T = -0.3\cdot10^2$ A/(m$\cdot$K) [35], $\tau = 10^{-5}$ s [60], $\partial M/\partial H = 0.5$, $nM_n/M_0 = 1$, $C_p = 2100$ J/kg$\cdot$K, $T = 293$ K, $\nu = 25$ MHz – the case of 'high' frequencies. Consequently, $\alpha_M \approx 5.6\cdot10^{-5}$ m^{-1}. However, if $\tau = 3.5\cdot10^{-9}$ s, then $\omega\tau \approx 1$ and $\alpha_M = 2.4\cdot10^{-2}$ m^{-1}. At the same time, the experimental value of the coefficient of absorption of ultrasound in the magnetic fluid, prepared on the basis of kerosene and magnetite, is of the order of 10^2 m^{-1}.

The role of dynamic processes. In the displacement of the conductor in the magnetic field, as a result of the effect on the Lorenz force on the free electrical charges, a potential difference e forms at the ends of the conductor. This process is similar to the Hall effect and, therefore, can be analysed using the same procedure [57, 165]. In the transverse magnetic field the effect will be maximum. Assuming that the ferroparticles has the form of the cube with the side $\tilde{d}$, with the direction of displacement of the cube coinciding with the normal to one of its faces, under the condition of dynamic equilibrium of the fluxes of the electrical charges we obtain $e/\tilde{d} = \dot{u}B$. The formation of alternating EMF at the ends of the conductor ensures the passage of alternating current through the conductor and this is accompanied by the generation of Lenz–Joule heat. The amount of heat, generated in 1 s in a single particle is $Q_{T1} = e_m^2 \cdot \tilde{d} / 2\rho_e$, where e_m is the amplitude value of the EMF, ρ_e is the specific electrical resistance of the material of the ferroparticle. The amount of heat generated in 1 s in the plane-parallel layer of the disperse medium whose surface area is 1 m^2, and the thickness dx, is

$$dQ_T = \dot{u}_m B^2 \varphi dx / 2\rho_e$$

These losses determine the reduction of the intensity of the sound waves:

$$dJ = -\dot{u}_m B^2 \varphi dx / 2\rho_e$$

Since $J = 0.5z\dot{u}_m^2$, then $\alpha_{dis} = B^2\varphi/\rho_e z$, where z is the wave resistance of the magnetic fluid.

If the ferromagnetic phase is represented by iron, for which $\rho_e = 10^{-7}$ ohm·m, then at $H_0 = 10^6$ A/m, $\varphi = 0.1$, $z = 1.44 \cdot 10^6$ kg/m^2·s, we obtain $\alpha_{dis} \approx 1$ m^{-1}. When using the magnetite as the ferromagnetic phase with $\rho_e = 5 \cdot 10^{-5}$ Ω·m, the losses will be 2–3 orders of magnitude smaller.

The magnetic field is characterised not only by the change of the mean volume of the aggregates but also of their shape and the distribution. Therefore, depending on the level of development of the structure (ellipsoids of filament-like aggregates, randomly distributed in the fluid–carrier, the redistribution of the concentration of the magnetic particles in the volume of the fluid, formation of the structure from filament-like aggregates characterised by a more or less distinctive long-range order), it is possible to obtain the high anisotropy of the coefficient of absorption (attenuation) of the ultrasound in the magnetised magnetic fluid whose axis is the direction of the magnetic field.

The theory of dissipation of sound energy as a result of the relative movement of the aggregates of the ellipsoidal shape in the fluid–carrier was developed by V.V. Gogosov *et al* [18, 126, 128]. For the additional (in comparison with the disperse medium) attenuation of sound with the inequalities $\xi >> 1$, $\omega\tau_r << 1$ fulfilled, where $\tau_r = (2/9)d^2(1-\varphi)\rho_1\rho_2/\eta_{s1}\rho$, we obtain the following expression

$$\frac{\Delta\alpha\lambda}{\varphi} = \pi(1-\varphi)\frac{(\rho_2-\rho_1)}{\rho_1\rho_2}\omega\tau_r\frac{L_{011}}{L_{011}L_{022}-L_{012}^2} \tag{4.24}$$

where the coefficients L_{011}, L_{022}, L_{012} depend on the angle $\tilde{\varphi}$ = **H^k**, the parameter $N_* \equi l/d$, and the correction coefficients $K_{\|}(d/\ell)$, $K_\perp$ (d/ℓ) using the equations:

$$L_{011} = \frac{K_{\|}}{N_*}\cos^2\widehat{\varphi} + K_\perp \sin^2\widehat{\varphi}$$

$$L_{011}L_{022} - L_{012}^2 = \frac{K_{\|}K_\perp}{N_*}$$

Equation (4.24) predicts in particular the anisotropy of the additional absorption of sound. The ratio of the additional absorption at **H**||**k** and **H**⊥**k** is:

$$\left(\frac{\Delta\alpha\lambda}{\varphi}\right)_{\|} \Bigg/ \left(\frac{\Delta\alpha\lambda}{\varphi}\right)_{\perp} = \frac{(L_{011})_{\|}}{(L_{011})_{\perp}} = \frac{K_{\|}}{N_* K_\perp}$$

When the ratio between the thickness and length of the chain d/ℓ changes from 0.990 to 10^{-7}, the ratio of the additional absorption decreases from 0.998 to 0.531. At low concentrations, the expression (4.24) differs from (4.23) which was derived for the 'conventional' suspensions by the fractional multiplier $L_{022}/(L_{011}L_{022}-L_{012}^2)$, reflecting the condition of preferential orientation of the aggregates in the field. In the absence of aggregation in the magnetic field (the magnetic fluid of the first type), the elastic movement of the phases of the dispersed system does not lead to the anisotropy of the absorption coefficient. If it is assumed that paired merger of the ferroparticles has taken place ($\ell=2d$), equation (4.24) shows that as a result of the changes of the coefficients L_{011}, L_{022} and L_{012}, the additional absorption increases in the transverse direction 1.67 times and in the collinear direction 1.43 times.

For the small and moderate magnetic fields, i.e., at $\xi \leq 1$, the thermal Brownian motion of the aggregates becomes important, and the coefficients of this motion L_{011}, L_{022} and L_{012} depend on ξ by the equations:

$$L_{011} = \cos^2\hat{\varphi}\left[\frac{K_{\|}}{N_*}\cdot\left(1-\frac{2\operatorname{cth}\xi}{\xi}+\frac{2}{\xi^2}\right)+K_{\perp}\cdot\left(\frac{2\operatorname{cth}\xi}{\xi}-\frac{2}{\xi^2}\right)\right]+$$
$$+\sin^2\hat{\varphi}\left[\frac{K_{\|}}{N_*}\cdot\left(\frac{\operatorname{cth}\xi}{\xi}-\frac{1}{\xi^2}\right)+K_{\perp}\cdot\left(1-\frac{\operatorname{cth}\xi}{\xi}+\frac{1}{\xi^2}\right)\right] \tag{4.25}$$

$$L_{022} = \sin^2\hat{\varphi}\left[\frac{K_{\|}}{N_*}\cdot\left(1-\frac{2\operatorname{cth}\xi}{\xi}+\frac{2}{\xi^2}\right)+K_{\perp}\cdot\left(\frac{2\operatorname{cth}\xi}{\xi}-\frac{2}{\xi^2}\right)\right]+$$
$$+\cos^2\hat{\varphi}\left[\frac{K_{\|}}{N_*}\cdot\left(\frac{\operatorname{cth}\xi}{\xi}-\frac{1}{\xi^2}\right)+K_{\perp}\cdot\left(1-\frac{\operatorname{cth}\xi}{\xi}+\frac{1}{\xi^2}\right)\right] \tag{4.26}$$

$$L_{012} = \left(\frac{K_{\|}}{N_*}-K_{\perp}\right)\left(1-\frac{3\operatorname{cth}\xi}{\xi}+\frac{3}{\xi^2}\right)\cos\hat{\varphi}\sin\hat{\varphi} \tag{4.27}$$

According to the numerical value, the additional absorption the sound in the magnetised magnetic fluid, determined by the kinetics of ellipsoidal aggregates, is considerably more intensive than the absorption caused by the relaxation perturbation of magnetisation and the magnetic field. It should be remembered, however, that the theory was developed for the low-concentration magnetic fluids in which the particles of the disperse phase do not interact hydrodynamically with each other.

The problems of the dynamic theory of absorption of sound in the magnetised magnetic fluids were also discussed in the theoretical study by Taketomi Susami [166]. Attention was given to the 'cluster model' of the fluid according to which the ferroparticles or aggregates formed by the merge into chains with the given linear density ρ_1 and are oriented preferentially in the field. Two dynamic mechanisms of dissipation of energy are investigated: the rotational movement of the clusters around the direction **H**, and the translational oscillations of the clusters in the fluid–carrier.

4.2. Acoustic scattering

The stationary particles, dispersed in the medium in which the sound wave propagates, become sources of secondary waves – scattering waves [71, 280]. The scattering waves carry with them part of the energy of the main beam, i.e., they reduce the intensity of the main beam. For the fraction of the energy, scattered by a single absolutely rigid particle whose dimensions are small in comparison with the length of the flat sound wave, H. Lamb derived the following equation [251]:

$$\alpha_\lambda = (7/9)\pi R^2 (kR)^4$$

Thus, the fraction of the scattered energy is proportional to the radius of the particle to the sixth degree. In the condition of acoustic non-interaction of the particles in the unit volume of the disperse systems the scattered energy is:

$$\alpha_s = (7/9)n\pi R^2 (kR)^4 = (7/12)\varphi k^4 R^3 = 909\varphi R^3 / \lambda^4$$

If it is assumed that λ=1 mm, R = 5 nm, φ = 0.1, we obtain $\alpha_S \approx 10^{-11}$ which already in the framework of the assumptions made in this case indicates that this process is not important with respect to the weakening of the soundwave.

The additional attenuation, associated with the scattering of sound on the particles $\frac{\Delta\alpha_3\lambda}{\varphi}$, will be estimated using the equation obtained taking into account the finite compressibility of the dispersed medium and the dispersed phase, derived by I.A. Ratinskaya [156]:

$$\frac{\Delta\alpha_3\lambda}{\varphi} = \frac{8\pi^4 (\nu R / c_1)^3 (\rho_2 c_2^2 - \rho_1 c_1^2)^2}{3(\rho_2 c_2^2)^2}$$

It will be assumed that the adiabatic compressibility and the density of the stabilising shells and of the fluid–carrier do not differ from each other, and that $\rho_2 c_2^2 \gg \rho_1 c_1^2$ and, consequently, we obtain

$$\frac{\Delta\alpha_3\lambda}{\varphi} = \frac{8\pi^4 (vR / c_1)^3}{3}$$

The numerical value $\frac{\Delta\alpha_3\lambda}{\varphi}$ is insignificantly small, ~$3.4 \cdot 10^{-10}$.

The validity of these estimates can be doubted owing to the fact that the accepted model assumes the stationary solid particles in relation to the fluid–carrier, whereas the nanoparticles of the disperse phase carry out random thermal motion. Evidently, the criterion of suitability of the model in this case may be the small size of the path travelled by the particle as a result of Brownian motion during the oscillation period in comparison with the wavelength of the sound.

The mean square of the distance over which the particle travels from the initial point during the time t [219] is:

$$< r^2 > = 6Dt$$

which shows that the mean distance, travelled by the particle during the time period t is proportional to the square root of this time. The diffusion coefficient D can be calculated from the mobility of the suspended particles b:

$$D = k_0 Tb$$

For the spherical particles (radius R) the mobility is $b = (6\pi\eta R)^{-1}$. Therefore

$$< r^2 > = k_0 Tt(\pi\eta R)^{-1}$$

If it is assumed that T = 300 K, t = 10^{-6} s (the period of the ultrasound wave), η = $1.38 \cdot 10^{-3}$ Pa·s, R = $5 \cdot 10^{-9}$ m, then $<r> \approx 5.5 \cdot 10^{-9}$ m, since $\lambda \approx 1$ mm.

The order of the time τ during which the particle suspended in the fluid rotates around its axis as a result of Brownian motion through the 'large' angle [219]:

$$\tau \approx \eta R^3 (k_0 T)^{-1} \approx 4 \cdot 10^{-8} \text{s}$$

Thus, in the ultrasound frequency range the condition of stationarity of the disperse magnetic particles in the carrier fluid is fulfilled only in the case of translational movement. This is sufficient for ensuring that the given model is suitable for the given application.

The process of aggregation of the ferroparticles may result in a large increase of the scattering of ultrasound energy because the value $\frac{\Delta\alpha_3\lambda}{\varphi} \sim R^3$. Actually, assuming that R = 2.3 μm and $\rho_{f\ell} c_{f\ell}^2 / \rho_1 c_1^2 \geq 10$, we obtain $\frac{\Delta\alpha_3\lambda}{\varphi} = 0.022$.

The rearrangement of the structure, formed by the ferroparticles, in magnetisation of the fluid of the third type may greatly change the damping of the sound beam in the normal propagation (in relation to the system of filament-like aggregates) of the flat sound wave as a result of acoustic scattering.

It will be assumed that as a result of joining of the spherical aggregates and their stretching in the magnetic field, the structure of cylinders elongated along the field forms in the fluid–carrier. The length of the cylinders coincides with the width of the sound beam, and the radius equals R_c. We consider the limiting case of scattering of the waves by the surface of an absolutely rigid cylinder [72]. For the waves with the long wavelength ($R_c/\lambda \ll 1$), the ratio of the scattered power per unit length of the cylinder to the intensity of the incident wave is expressed by the equation

$$Q_p \approx \frac{3}{4} k^3 R_c^4$$

The reduction of intensity J in passage of the wave through the fluid layer with a thickness dx is $JQ_p n_c dx$, where $n_c = \varphi/V_c$ is the concentration of the cylindrical aggregates. Therefore,

$$\alpha\lambda / \varphi = 6\pi^2 \nu^2 R_c^2 / c^2$$

assuming that $R_c = 2.3$ μm, $\nu = 25$ MHz, $c = 1200$ m/s, we obtain $\alpha\lambda/\varphi \approx 0.15$, i.e., additional attenuation of the sound beam has the order of the absorption coefficient in the non-magnetised magnetic fluid [100].

The sound pressure at large distances from the surface of the cylinder decreases as a result of cylindrical propagation with the distance as $1/\sqrt{r}$, whereas in scattering by the sphere it decreases as $1/r$ [72]. It may be assumed that in the case of the ordered distribution of the magnetic change in the fluid–carrier, when the distances between them are comparable with the length of the sound wave, diffraction phenomena similar to Bragg diffraction of x-ray radiation on a crystal lattice will appear.

The formation of the ordered structure – the hexagonal lattice of the magnetic chains in the magnetised magnetic fluid – was observed using an optical microscope [167]. In [168] it was attempted to take into account the dissipation of sound energy, determined by the magnetostriction rotation of the vector of spontaneous magnetisation in the ferroparticle. Unfortunately, comparison with the experimental data was not carried out.

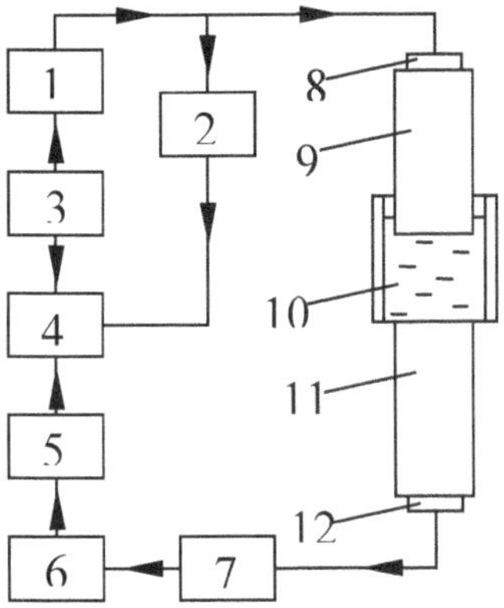

Fig. 3. Flow diagram of equipment for absolute measurements of absorption in the non-magnetised magnetic fluid.

4.3. The method for measuring the absorption coefficient

As in the case of non-magnetic fluids, absolute measurements of the coefficient of absorption of ultrasound in the non-magnetised magnetic fluid can be taken by the pulsed method of the variable base [172, 196, 197], because the non-magnetised fluid is a macroscopically homogeneous medium.

The flow diagram of the experimental equipment is shown in Fig. 4.3. The master pulse generator 3 operates in the auto-modulation mode with the pulse repetition frequency of ~4 kHz. The main shortcoming of the generator with auto-modulation is the instability of repetition frequency. However, this shortcoming is not of any considerable importance because the oscilloscope 4 operates in the standby mode and is activated by the start pulses, produced by the master generator. The RF pulses, generated by the master generator, travel to the output cascade 1 acting as a resonance amplifier. The frequency of filling of the RF pulses can be changed smoothly in the range of 2 to 50 MHz. The voltage amplitude is regulated in the range 0.5–200 V. The application of the high-frequency transformer connection at the output of the generator makes it possible to connect the piezotransducer with low and large electrical capacitance. The duration of the RF pulses τ_u is ~50/ν, i.e., the pulse contains approximately 50 complete high-frequency oscillations. One of the advantages of the self-regulation regime of generation is that the generated pulses are bell-shaped and, consequently, are characterised by a narrow frequency spectrum in comparison with the rectangular pulses. To control the stability of the operation of the generator, the RF pulses are transferred, using the high-frequency transformers 2, to the input of the free channel of the two-beam oscilloscope 4. The emitting piezoelement 8 transforms the high-frequency voltage into the ultrasound signal. The latter propagates through the acoustic delay line 9, the investigated magnetic fluid and the acoustic delay line 11 to the reception piezoelement 12, which transforms the sound pressure, acting on it, to a radio pulse. The RF pulse travels to the input of the receiver of the super-heterogeneous type 7, which has a smooth calibrating attenuator in the IFA channel. The intermediate frequency is 1.9 MHz, the transmission band 300 kHz. Further, the RF pulses are detected by the detector 6, amplified by the video amplifier 5, and travel to the input of the first channel of the oscilloscope. As a result of detection of the received pulses, the reception quality of the image on the screen of the oscilloscope is high. To reduce the oscillations of the mains voltage, it is recommended to use a ferroresonance stabiliser. To eliminate the effect of remanent magnetisation on the results of measurements of the absorption coefficient in the magnetic fluid, all the elements of the acoustic cell, including the acoustic delay lines 9 and 11, are produced from a non-magnetic material (duralumin, brass, plexiglass).

It is well-known that the absorption of ultrasound in macro-heterogeneous media is characterised by the exponential laws of the variation of the amplitude of sound pressure or the intensity of the sound. This assumption is valid for the case of flat waves when there are no changes in intensity as a result of the divergence of the beam of the ultrasound waves. Therefore, in the case of the magnetic fluid, the criterion of the minimum divergence of the beam in the near field $D_a/\lambda > 20$, remains valid. Here D_a is the diameter of the active surface of the emitter [177].

The specific errors, typical of the pulse method, include the errors, caused by the large length of the spectrum of the pulse in the frequency range [97, 196]. The spectral components reach high values in the frequency range $\nu_2 - \nu_1 = \Delta\nu_c$, $\Delta\nu_c = \tau_u^{-1} = \nu/50$. To obtain the non-distorted results, the transmission band of the entire measuring circuit – transducers, amplifier – must not be smaller than the spectrum of the signal. For example, at $\nu = 50$ MHz, $\Delta\nu_c = 200$ kHz. The width of the transmission band of the receiver reaches 300 kHz, i.e., it is greater than ν_c. The width of the frequency range in which the piezotransducers operate efficiently is determined from the expression $\Delta\nu_n = \nu/Q$, where Q is the mechanical quality factor of the transducer loaded to the investigated medium. Taking into account that $Q \leq 10$, we obtain $\Delta\nu_n = 1$ MHz.

It is necessary to fulfil the requirement of the smallness of the amplitude of sound determined by the existence of non-linear defects. In the case of the magnetic fluid this requirement becomes even more stringent because the superposition of powerful ultrasound fields may be accompanied by the failure of the shell of the stabiliser, coalescence of ferroparticles and delamination of the dispersed medium.

If the above requirements are fulfilled, the absorption coefficient of ultrasound is calculated from the equation

$$\alpha = (L_2 - L_1)^{-1} \ln U_1 / U_2$$

where L_2–L_1 is the change of the distance between the emitting and receiving quartz, and U_1/U_2 is the ratio of the amplitudes of the voltage of the signals received at the input of the receiver prior to and after displacement of the piezovibrator. The value $\ln U_1/U_2$ is determined by modifying the level of the second received pulse to the level of the first pulse using a measuring attenuator. The total error of measurement of the absorption coefficient is determined by a large number of partial errors. They include: the error of calibration of the smooth attenuator, the error of measuring the displacement distance, the error, determined by the instability of mains voltage, the level of RF interference, and also the errors associated with diffraction phenomena and non-linear effects in the acoustic fields.

It is quite difficult to calculate with high reliability the measurement error α. However, this error can be evaluated by experiments, by examining the repeatability of the measurement results and determining α in the so-called reference fluids which include distilled water. The estimates show that the relative error of the absolute measurements of the absorption coefficient of ultrasound is 5–7%, if we use the method of determining from the angle of inclination of the straight-line $\ln U_1/U_2 = \alpha(x_2 - x_1)$, plotted on the basis of the results of 7–10 measurements in a single movement of the piezoreceiver.

In investigating the dependence of the absorption coefficient of ultrasound in the magnetic fluid on the strength of the external magnetic field, the application of the variable base method is associated with certain difficulties of technical nature. The specimen of the magnetic fluid, placed between the poles of the magnet, is subjected to the effect of forces, determined by the demagnetising factor, even if the degree of homogeneity of the external field is high.

These forces results in large changes of the geometry of the free surface affecting the projector zone. Therefore, measures should be taken to ensure the leak tightness of the measuring cuvette which in turn imposes restrictions on the selection of the design of the device for the displacement of the transducer. Another reason why the displacement of the transducer in the magnetic fluid is undesirable is that this may be accompanied by an unavoidable mechanical failure of the structure of the magnetic chains, i.e., the 'natural' process of structure formation in the magnetic fluid will be distorted.

Therefore, special attention is given to the possibility of using the 'constant base' method. In this method, two piezoelectric quartz sheets of X-cut (emitter and receiver) are placed coaxially at the fixed distance from each other L_0, and the measurement of the increment of the absorption coefficient $\Delta\alpha$ is reduced to the determination of the relative amplitude of the video pulse on the screen of the oscilloscope $\beta_a = U_1/U_2$ [172, 196]. The application of this method for the magnetic fluids will now be justified.

It is assumed that an electrical pulse with the amplitude U_m arrives at the emitting transducer. Consequently, as a result of the reverse piezoeffect, the active surface is displaced by $u_0 = u_s Q$, where u_s is the static displacement of the surface, Q is the quality factor of the half-wave vibrator. The static displacement is proportional to the amplitude of electrical voltage $u_s = C_1 U_m$, and the proportionality coefficient C_1 depends on the value of the piezocoefficient and elasticity moduli of the piezoelement. The quality factor is calculated from the expression

$$Q = \pi\rho c / 4\rho_f c_f$$

where ρc and $\rho_f c_f$ are the wave resistances of the material of the transducer and the fluid.

The amplitude of sound pressure at the surface of the emitter is

$$p_0 = \rho c u_0 \omega_r$$

Here ω_r is the resonance frequency of the sheet. Taking into account these relationships

$$p_0 = \pi C_1 \omega_r (\rho c)^2 U_m / 4\rho_f c_f$$

After passage of the sound pulse through the investigated medium, its amplitude becomes equal to

$$p = p_0 e^{-\alpha_0 L}$$

The amplitude of electrical pulses, obtained on the faces of the receiving piezotransducer and amplified by the wide-band amplifier, has the following value:

$$U_{m1} = C_2 q_n d \cdot p$$

where C_2 is the piezoelectric pressure modulus; d is the thickness of the sheet; q_n is the gain factor.

If the fluid is replaced by another fluid, with the wave resistance $\rho'_f c'_f$, the pulse amplitude has the new value U_{m2}. Therefore, the following equation can be written

$$\frac{U_{m_1}}{U_{m_2}} = \frac{\rho'_f c'_f}{\rho_f c_f} \exp L(\alpha - \alpha_0)$$

or

$$\alpha - \alpha_0 = L^{-1} \ln(U_{m1} \rho_f c_f / U_{m2} \rho'_f c'_f)$$

In the magnetic fluid of the first and second type, the speed of sound remains constant with the accuracy to 0.1% with the variation of both the longitudinal and transverse fields up to 500 kA/m. In some samples of the magnetic fluid of the third type the speed of sound increases, but no more than by 1%. As regards the shear viscosity, the changes of this parameter in the magnetic field are insignificant, as already mentioned. Assuming that the wave resistance of the magnetic fluid in the magnetic field does not change, the above equation can be written in the new form

$$\Delta\alpha = L_0^{-1} \ln(U_{m1} / U_{m2})$$

The absolute error of measurement of the increment of the absorption coefficient by this method is 2.5 m^{-1} at $L_0 = 2$ cm.

The applicability of this procedure to the solution of the given task can be regarded as limited to the case of not too large changes in the bulk and shear viscosity in the process of magnetisation of the magnetic fluid.

4.4. Discussion of the experimental results Non-magnetised fluid

The initial experimental data on the absorption of ultrasound in the the non-magnetised magnetic fluid were obtained in [15, 70, 100, 107].

In [15] samples of a magnetic fluid based on magnetite and kerosene were studied. Examination in the optical microscope with a magnification of ×600 of a layer of the magnetic fluid with a thickness of ~30 μm, enclosed between two covering glass sheets, did not show the presence of aggregates of the micron size in the fluid.

The absorption coefficient was measured using the pulsed method with the variable base. The speed of ultrasound in the investigated samples was measured with a pulsed interferometer at a frequency of 25 MHz. The measurements of density ρ and static shear viscosity η_{S0} were taken by the conventional method using a pycnometer and an Ostwald viscometer.

The parameter characterising the magnetic properties was the saturation magnetisation M_S, which was measured by the induction method. All the measurements were taken at a temperature of 24°C. In calculating φ it was assumed that $\rho_2 = 5210$ kg/m³, $\rho_1 = 800$ kg/m³.

The results of measurements of a/ν^2 at frequencies of 15 and 25 MHz and also of the values of the volume concentration of the solid phase φ, determined from the density of the magnetic fluid, are presented in Table 4.1.

Figure 4.4 shows the results of measurements of additional absorption, obtained in [15, 100, 106 and 162] for different concentrations φ. The results of the study [106], obtained for a frequency of 26, 57 MHz, were corrected for a frequency of 25 MHz. Since the difference value $\Delta\alpha=\alpha-\alpha_0$, was calculated, the absolute and relative errors slightly increased, to 0.35 cm^{-1} and 12%, respectively.

At φ~0, the dependence was almost linear, and this claim is in agreement to a certain extent with the conclusions made in [104] in which an empirical equation was proposed for describing this dependence $\Delta\alpha/\alpha_1 = A'\varphi + B'\varphi^2 + C'\varphi^3$; $A' = 34$, $B' = 216$, $C' = 28$

Table 4.1

No.	$\rho\cdot 10^{-3}$, kg/m³	φ, %	M_S, kA/m	$\eta_{S0}\cdot 10^2$, kg/m·s	c, m/s	$\alpha/\nu^2\cdot 10^{14}$, s²/m 15 MHz	25 MHz
1	0.80	0	–	0.13	1280	18	14
2	0.86	1.37	6.3	0.14	1220	35	30
3	0.97	3.85	14.6	0.17	1175	52	47
4	1.09	6.35	23.0	0.23	1150	63	58
5	1.14	7.94	27.9	0.28	1145	71	60
6	1.23	9.75	36.9	0.37	1135	77	69

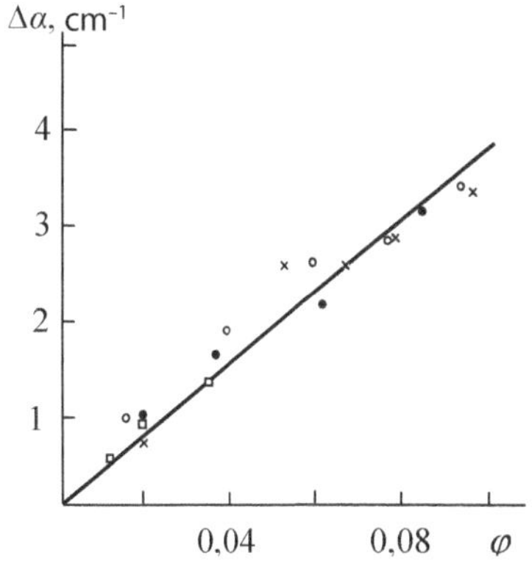

Fig. 4.4. Dependence of the additional absorption of ultrasound on the concentration of the magnetic fluid based on kerosene: •– [100],○– [15], □– [162], × – [106].

for the magnetic fluid based on kerosene. This formula generalises the results of the experiments carried out at frequencies of 3–21 MHz. The data, shown in Fig. 4.4, are approximated by the straight-line $\Delta\alpha/\alpha_1 = 37.5\varphi$.

The absorption coefficient depends on the shear η_s and volume viscosity η_v of the fluid:

$$\alpha = \frac{\omega^2}{2\rho c^3}\left(\frac{4}{3}\eta_s + \eta_v\right)$$

Consequently, the expression for the calculation of the volume viscosity has the form

$$\eta_v = 2\alpha\rho c^3\omega^{-2} - \frac{4}{3}\eta_s \tag{4.28}$$

In turn, η_v and η_s depend on the frequency of the sound wave. In the simplest case of a single relaxation mechanism, this dependence has the form

$$\eta_s = \eta_{so} / (1+\omega^2\tau_s^2)\ \eta_v = \eta_{vo} / (1+\omega^2\tau_v^2) \tag{4.29}$$

where η_{v0} and η_{s0} are the static values of the shear and volume viscosities; τ_s and τ_v is the relaxation time of the shear and volume viscosity, respectively.

The Stokes theory of absorption of sound in matter ignores the volume viscosity and, consequently, the expression for the absorption coefficient has the form

$$\alpha_s = \frac{2}{3}\omega^2\eta_{so} / \rho c^3 \tag{4.30}$$

Table 4.2

Parameter	Specimen No.					
	1	2	3	4	5	6
$\eta_v 10^2$, kg/ms	0.98	1.6	3.5	4.5	4.9	5.8
η_v/η_{s0}	7.5	11	21	19	17	16

The difference $\Delta\alpha = \alpha - \alpha_s$ is the so-called super Stokes absorption. Table 4.2 gives the values of volume viscosity η_v, calculated from the equation derived from (4.28) by replacing η_s by η_{s0}. This was carried out using the values of $\frac{\alpha}{\nu^2}$ for the frequency of $\nu = 25$ MHz, taken from Table 4.1.

Taking into account the possible relaxation of shear viscosity, the values should be regarded only as a bottom estimate. The same table gives the values of the ratio η_v / η_{s0} for each of the six investigated samples.

The equation (4.30) was used to calculate the value α_s. The graph of the dependence $\alpha_s(\varphi)$ is shown in Fig. 4.5. In the investigated concentration range of the magnetic fluid, the absorption increases almost four times. Approximately the same increase was also obtained for the shear viscosity η_{s0}. However, the increase of $\frac{\alpha}{\nu^2}$ is not associated with the increase of η_{s0}. This is shown by comparing η_v and η_{s0}. Volume viscosity is almost an order of magnitude higher than shear viscosity, and its changes with concentration determine practically the change of α.

Figure 4.6 shows the dependence of 'super Stokes' absorption $\Delta\alpha$ on the concentration of the magnetic fluid φ. The numerical values of the super Stokes absorption, taken at different concentrations, are many times greater than as α_s.

The angle of the straight dependence $\Delta\alpha(\varphi)$, shown in Fig. 4.4, is used to determine $(\Delta\alpha\lambda/\varphi)_{ex} = 0.17 \pm 0.02$. The contribution of each

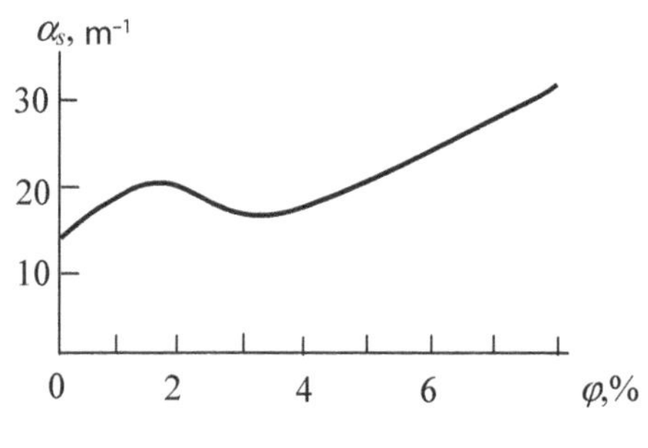

Fig. 4.5. Dependence of the Stokes absorption coefficient on the concentration of the magnetic fluid.

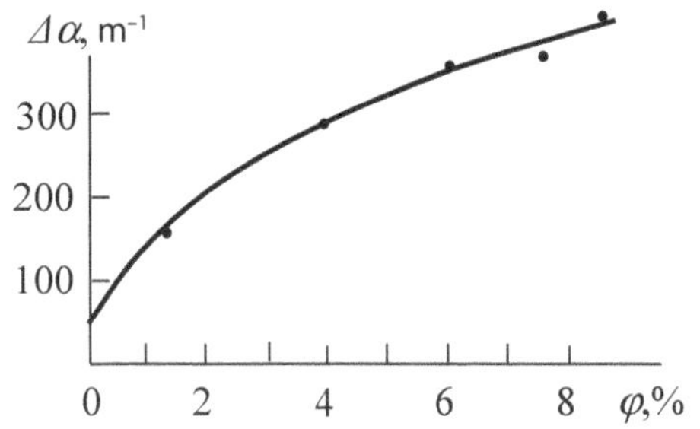

Fig. 4.6. Dependence of 'super Stokes' absorption on the concentration of the magnetic fluid.

of these mechanisms of dissipation of sound energy is evaluated as follows:

- Einstein viscosity: $\Delta\alpha_1\ \lambda/\varphi$ = 5.5% from $(\Delta\alpha\lambda/\varphi)_{ex}$,
- relative movement of the phases: $\Delta\alpha_2\ \lambda/\varphi$ = 21%,
- interphase heat exchange: $\Delta\alpha_4\ \lambda/\varphi$ = 50%.

The total theoretical damping is $(\Delta\alpha\lambda/\varphi)_{th}$ = 77%.

Taking into account the evaluation nature of these calculations, we can note the satisfactory agreement between the theoretical and experimental results. Therefore, it may be assumed that the observed additional absorption of sound in the magnetic fluid is determined mainly by the internal heat exchange and viscous friction.

In [169], the authors presented the results of processing the experimental data for the speed of propagation and the absorption coefficient of sound waves in the magnetic fluid based on dodecane, obtained in the frequency range 12–2000 MHz. The theoretical model, used in the analysis of the field dependence of the elastic properties of the magnetic fluid, is based on the concept of 'sliding of magnetic fluid aggregates in relation to the fluid– carrier'.

It is interesting to examine the results obtained by I.S. Kol'tsova [162] for the frequency range from $2{\cdot}10^5$ to $3{\cdot}10^7$ Hz. The analysis results confirm the conclusion on the controlling contribution of the microheterogeneity mechanisms. A.N. Vinogradov, V.V. Golosov, *et al* [117] observed the dispersion of the value $\frac{\alpha}{\nu^2}$ in a magnetic fluid on the basis of dodecane with magnetite particles, stabilised with oleic acid, in the frequency range 12–132 MHz.

At the same time, the application of the conclusions of the classic acoustics of microheterogeneous media to the systems with the disperse nanoparticles requires additional verification because of a number of reasons. Firstly, the theory does not take into account the presence of thermal chaos in the movement of the particles of the disperse phase and, secondly, the theory does not take into account the fractal nature of the surface of the nanoparticles on the level of interatomic spacing which is especially important for evaluating the internal heat exchange mechanism.

4.5. Discussion of the experimental results The magnetised fluid

The short-term effect of the magnetic field. The duration of the 'short-term' effect of the magnetic field and the magnetic fluid is the duration of the 'conventional' experiments with the measurement of the absorption coefficient of ultrasound in the fluid of ~1 hour. The 'long-term' effect of the field is the effect of the field for 24 hours or longer. The introduction of this terminology is associated with the determination of qualitatively different results of observations of the variation of the amplitude and the form of the ultrasound pulses, transmitted through the layer of the magnetised magnetic fluid, with large differences in the duration of the effect of the magnetic field.

Observations of the change of the amplitude of the ultrasound pulses in the passage of the pulses through the magnetised magnetic fluid were carried out on a large number of samples with different types of the magnetic fluid [16, 125, 200, 305, 309]. They included the sample of the magnetic fluid, prepared on the basis of magnetite and water with stabilisation with sodium oleate [16]. Observations using the optical microscope showed that the investigated sample does not have any microdroplet aggregates. The speed of sound remains constant with the accuracy of 1 m/s when the magnetic field with the strength of up to 400 kA/m is applied. The application of the magnetic field of up to ~400 kA/m with different orientation in relation to the wave vector, the amplitude of the ultrasound pulses with the filling frequency of 4 MHz, passed through the 2 cm layer of the fluid, decreased by less than 0.2 of the initial value, i.e., $\Delta\alpha \leq 10\ \mathrm{m}^{-1}$.

The solid line in Fig. 4.7 shows the dependence of the relative variation of the amplitude of ultrasound pulses, passed through the magnetic fluid, on the strength of the magnetic field H at $\mathbf{H}||\mathbf{k}$. In this case, the sample of the magnetic fluid was prepared on the basis of magnetite and kerosene. The concentration of the solid phase was 19.5%. According to the external features (absence of the deposit on the bottom of the vessel after holding for many days, the absence

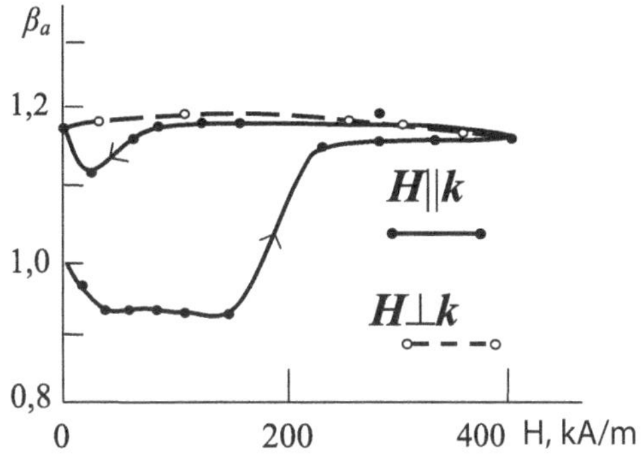

Fig. 4.7. The field dependence of the relative amplitude of the ultrasound pulse.

of aggregates of the micron size and larger, visible with the naked eye during decanting) the given sample is stable. The measurements were taken at a frequency of ultrasound oscillations of 3 MHz at a temperature of 20°C.

If in the first magnetisation cycle, shown by the solid line in Fig. 4.7, the change of the signal amplitude was relatively small, after three magnetisation cycle, the signal amplitude stabilised on the same level.

In subsequent rotation of the cuvette through 90° there were no changes in the signal amplitude in the range of variation of *H* up to 440 kA/m. These results are indicated by the dashed line in Fig. 4.7. The fact that the changes in the absorption of ultrasound in magnetisation of the given sample are small is obviously associated with its microscopic homogeneity and aggregate stability. The characteristic circumstance, which was also observed in many other specimens, was the 'smoothing' of the curves $\beta_a(H)$ after several magnetisation cycles.

Under the multiple (more than 10 cycles) effect of the magnetic field up to ~500 kA/m, both at $\mathbf{H}\|\mathbf{k}$, and at $\mathbf{H}\perp\mathbf{k}$, the signal amplitude was not restored to its initial value. Usually, the amplitude in the specimen of the magnetic fluid, subjected to magnetic 'training', is smaller and, consequently, the absorption is greater than prior to placing in the field.

To measure the increment of the absorption coefficient of ultrasound in the magnetic fluid in magnetisation in the heterogeneous magnetic field, experiments were carried out [16] based on the method of the single fixed distance. The length of the acoustic path in the magnetic fluid was 2 cm. The filling frequency of the pulses was 3 MHz. The source of the magnetic field was an electromagnet. The cuvette was placed either in the section of the homogeneous field or in the section of the heterogeneous magnetic field situated above the axial line of the poles of the terminals. The section of the heterogeneous field is characterised by the gradient of the strength $G = 23H_{mid}$ kA/m^2. Here H_{mid} is the numerical value of the strength of the magnetic field at the point situated in the middle of the acoustic path. The measurements of the amplitude of the pulses and the determination of the volume concentration of the magnetic fluid were carried out at a temperature of 25°C.

Three specimens of the magnetic fluid, prepared on the basis of kerosene and magnetite, were investigated. Two of the specimens (with the volume concentration of the solid phase $\varphi = 3.9\%$ and $\varphi = 15\%$) were stable. After holding for many days and placing in the saturated magnetic field, the specimens did not contain any deposit, the density and saturation magnetisation did not change, and there were no aggregates of micron dimensions.

Figure 4.8 shows the results of measurement of the relative amplitude in a stable magnetic fluid ($\varphi = 3.9\%$) in homogeneous (circles) and heterogeneous (squares) magnetic fields at different values of H_{mid}. The symbols without crosshatching refer to the results of the direct path, the crosshatched symbols are the results of the reverse path. The amplitude of the ultrasound pulse, passed through the stable magnetic fluid, slightly changed when the value of H_{mid} change to the values of ~120 kA/m in the homogeneous and heterogeneous magnetic fields (both with increasing and decreasing H_{mid}). The maximum increment of the absorption coefficient was 4 m^{-1}. The same results were obtained for the stable specimen with $\varphi = 15\%$.

The passage of ultrasound through the unstable magnetic fluid, place in the homogeneous (Fig. 4.9) or heterogeneous magnetic field (Fig. 4.10) is accompanied by a large change of the pulse amplitude.

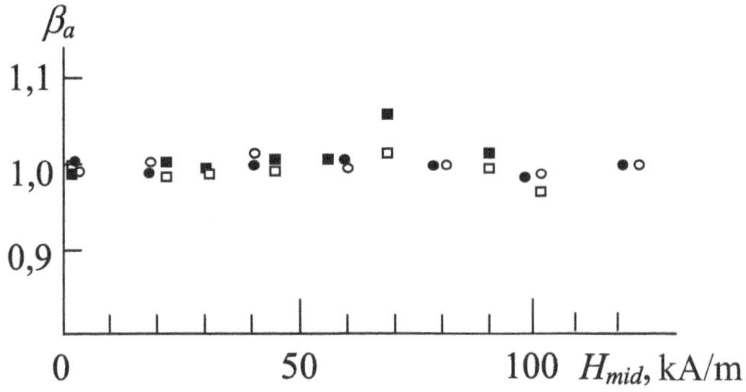

Fig. 4.8. Dependence $\beta_a(H_{mid})$ for the stable magnetic fluid.

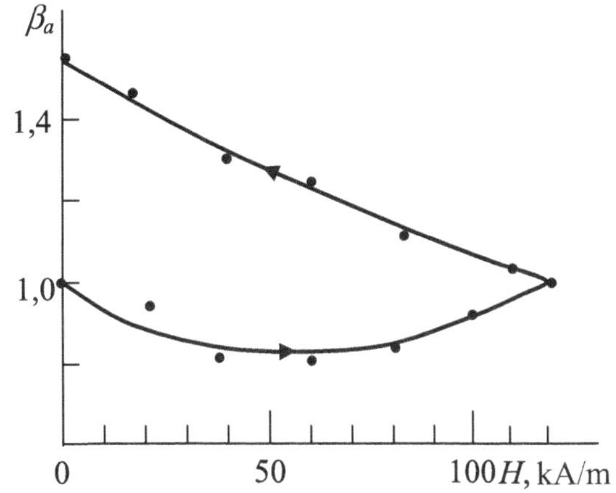

Fig. 4.9. Dependence $\beta_a(H)$ for the unstable magnetic fluid in the homogeneous magnetic field.

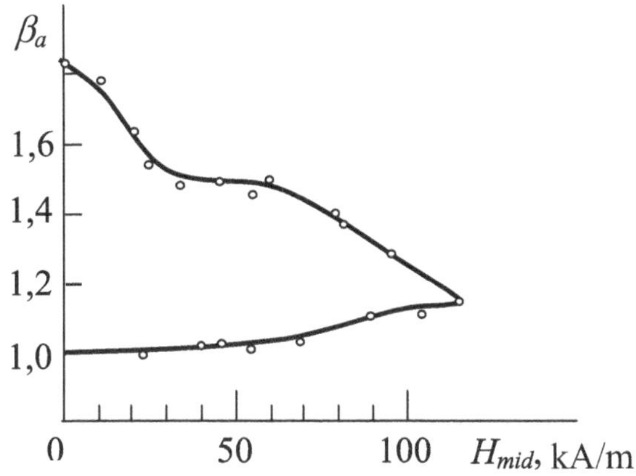

Fig. 4.10. Dependence $\beta_a(H_{mid})$ for the unstable magnetic fluid in the heterogeneous magnetic field.

The dependence $\beta_a(H_{mid})$ is non-monotonic, and the branches of the direct and reversed parts do not coincide with each other, and when the magnetic field is switched off the amplitude does not restore to the initial value. In the repeated experiment (without thorough stirring), the resultant dependence $\beta_a(H_{mid})$ greatly differs from the initial dependence. On the curve of the dependence $\beta_a(H_{mid})$, shown in Fig. 4.10, the divergence of the branches of the direct and reversed paths is stronger than in Fig. 4.9, although in both cases the fluid was thoroughly mixed prior to the start of the measurements.

The increase of the signal amplitude in the homogeneous field is evidently associated with sedimentation and the corresponding reduction of the concentration of the dispersed phase. When the fluid is placed in the inhomogeneous magnetic field, sedimentation is accompanied by magnetic diffusion [16, 132, 170] which increases the intensity of sucking away of particles from the projector zone. Finally, taking into account the instability of the system, the curves (Figs. 4.9, 4.10) should be regarded as the illustration of the discussed phenomenon.

The refraction of the sound beam, caused by the directional redistribution of the concentration of the magnetic phase in the volume of the magnetic fluid and described in section 3.9 (*Magnetodiffusion. Control of the cross section and direction of sound beams*) may also reduce the amplitude of the received signal.

Under the effect of refraction the flat ultrasound wave falls on the surface of the piezoelectric sheet–receiver under the angle $\Delta\hat{\varphi}$:

$$\Delta\hat{\varphi} = \frac{\Delta L \cdot G_c}{c} = \frac{\Delta c}{c}$$

The wave falling on the square piezosheet with the side Δh causes oscillations of the sheet with a certain distribution of the phases on the surface. For example, the first minimum of the amplitude of the signal, received by the resonant sheet, is detected at $\Delta\hat{\varphi}_{\min}$ at which the two halves of the sheet oscillate in the opposite phases. With a further increase of $\Delta\hat{\varphi}$ the phases of the oscillations are redistributed on the sheet surface and this is accompanied by pulsations of the amplitude of the received signal. Calculations are carried out by the method of the Frenel zones well known in the theory of light diffraction. The first minimum of the amplitude of voltage corresponds to the arrival of oscillations from the second Frenel zone on the surface of the receiving piezosheet, i.e.

$$\frac{\Delta h \cdot \sin\Delta\hat{\varphi}}{\lambda / 2} = 2 \text{ or } \Delta h \cdot \sin\Delta\hat{\varphi} = \lambda$$

At $\lambda/\Delta h << 1$, $\Delta\hat{\varphi} \approx \lambda/\Delta h$. Thus

$$\frac{\Delta c}{c} = \frac{\lambda}{\Delta h} = \frac{0.253}{20} = 1.25 \cdot 10^{-2}$$

The increase of the speed of sound of the same magnitude as a result of the redistribution of the concentration of the solid phase is possible so that the investigated mechanism of the reduction of the amplitude of the received signal can be included in the group of probable mechanisms.

The independence of the amplitude of the transmitted acoustic signal on the strength of the field, detected in some specimens of the magnetic fluid, in particular, in the specimen with $\varphi = 3.9\%$, confirms one of the main conclusions of the dynamic theory of additional absorption of sound in the magnetised magnetic fluids [18, 126, 128], according to which $\Delta\alpha = 0$ in the absence of aggregation of the ferroparticles in the magnetic field.

The above considerations regarding the origin of the physical nature of the reversible changes of the absorption coefficient of the magnetic fluid under the effect of magnetic training are confirmed by comparing the results of acoustic and electron microscopic studies [171].

In the dilution of a magnetic fluid with a mixture of kerosene and oleic acid at a ratio of 5:2 the resultant dispersed system becomes susceptible to the aggregation of magnetic particles with the formation of microdroplet aggregates. The passage of the ultrasound pulse through the magnetic fluids, diluted in this manner, and placed in the magnetic field, has a number of special features. These will be investigated on an example of one such sample.

The sample was produced by diluting a magnetic fluid based on magnetite and kerosene with stabilisation with oleic acid, with a density of $1.66 \cdot 10^3$ kg/m^3, and a mixture of kerosene with oleic acid with the given ratio. The density and volume concentration of the diluted specimen was $1.2 \cdot 10^3$ kg/m^3 and 9.1%. The specimen was poured into a cuvette immediately after preparation, the cuvette was placed between the poles of the electromagnet and the strength of the field was then smoothly increased.

Figures 4.11–4.14 show oscillograms for several values of H in the process of magnetisation and demagnetisation of the magnetic fluid at $\mathbf{H} \perp \mathbf{k}$. The upper oscillogram contains the probing pulse used for counting the delay time of the acoustic signal and is also used as an indicator of the stability of generated oscillations with respect to amplitude.

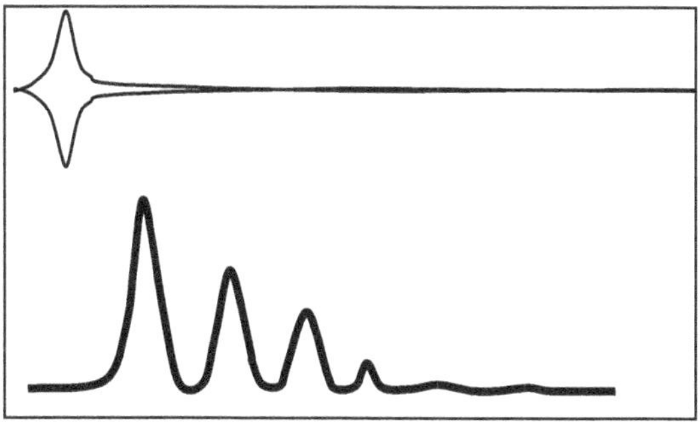

Fig. 4.11.

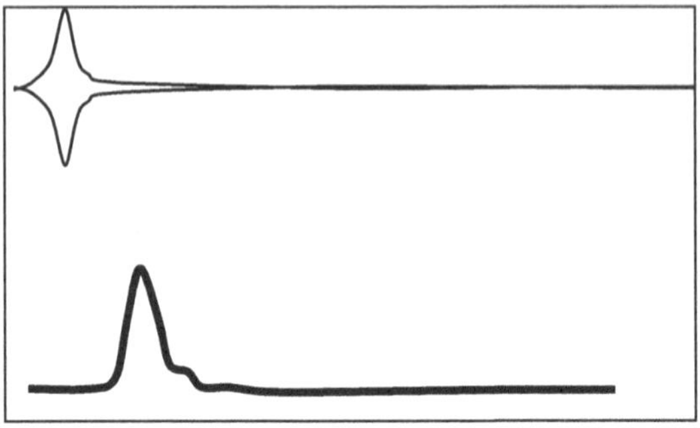

Fig. 4.12.

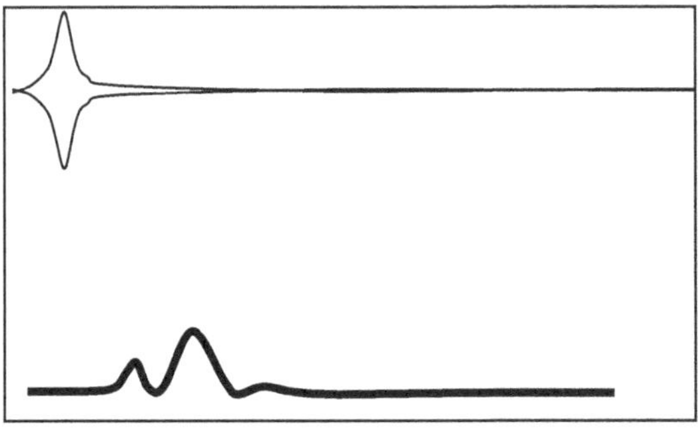

Fig. 4.13.

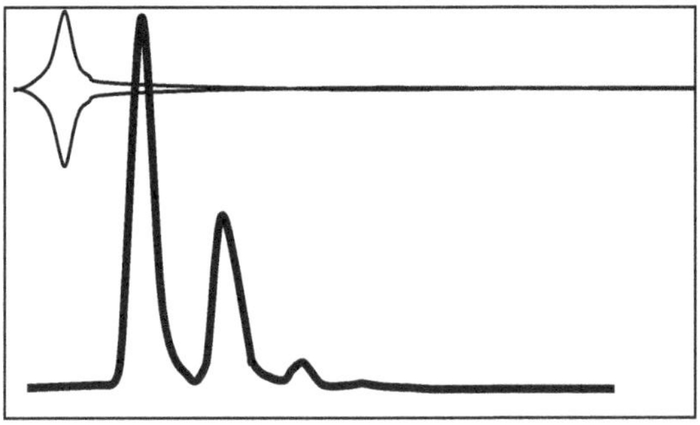

Fig. 4.14.

The lower oscillogram (Fig. 4.11) is a group of four pulses each of which passes $(2m+1)$ times the acoustic path between the emitter and the receiver (m = 0, 1, 2, 3) at H = 0.

The application of the magnetic field with the strength of 12 kA/m results in a large reduction of the amplitude of the 'first' pulse and in almost complete disappearance of the other pulses (Fig. 4.12). The typical feature of the systems is that the reduction of the amplitude is non-monotonic, in 'jumps', similar to interference of two pulses with sinusoidal filling, if the delay time of one of them changes. Further increase of the strength to $H = 240$ kA/m results in a qualitative change of the observed pattern – there is a 'dip' between the pulses whose delay time shortens in the course of magnetisation (Fig. 4.13). In the magnetic field with $H = 320$ kA/m both pulses are almost completely equal in amplitude. The maximum value in the investigated range of the strength of the field is $H = 440$ kA/m.

As the direct course of the experiment is completed (increase of the strength of the magnetic field), the rotation of the cuvette through 90° when reaching $\boldsymbol{H}||\boldsymbol{k}$ leads to the almost complete restoration of the initial oscillogram, shown in Fig. 4.11. The oscillograms, produced with the reduction of the strength to 160 kA/m, repeat in the general features their previous form. However, with a further reduction of the strength the 'first' pulse becomes considerably stronger. When the electromagnet is switched off, the amplitude of the 'first' and subsequent pulses is ~1.5 times greater than the initial value (Fig. 4.14). Similar features of the passage of the sound pulses through the magnetised magnetic fluid have been detected many times in the specimens of the magnetic fluids, classified as the third type.

As regards the origin of the effects observed in the last experiment, it may be assumed that there is a long-range order in the structure of filament-like aggregates, formed in the magnetic fluid in the magnetic field, and the diffraction of the ultrasound beam on it.

The speed of propagation and the absorption coefficient of sound waves in the magnetic fluid were also studied in [172, 252, 309]. In [172] the equations for the speed of propagation and the damping coefficient of sound were determined from hydrodynamic equations.

In a theoretical study [252], the system of equations derived, as stated by the authors, on the basis of the molecular–kinetic theory, was used to find frequency-dependent expressions for the speed and absorption coefficient, reflecting the force effect on the ferroparticle from the side of the heterogeneous magnetic field. The resultant very cumbersome mathematical expressions for these parameters are regarded as general expressions for the speed and absorption coefficient of the acoustic waves in the magnetic fluid taking into account the rearrangement of the structure of the fluid in a wide range of density, concentration, temperature, the frequency of oscillations and the external inhomogeneous magnetic field. The given model of the magnetic fluid does not take into account the rotation of the ferroparticles around their axes, the presence of the aggregates produced from the ferroparticles, and the duration of Néel relaxation is assumed to be equal to 0. Numerical calculations were carried out of the frequently dependence of the rate and coefficient of absorption of sound in the magnetic fluid, prepared on the basis of kerosene and Fe_3O_4 particles, at the values of the gradient of the strength of the field of 10 A/m^2 and 10^3 A/m^2 in the frequency range 10^5–10^{13} Hz. According to the data presented in the graphical form, the speed of sound monotonically increases with frequency, and the curvature of increase of the curve $c(\nu)$, obtained for the strength gradient of 10^3 A/m^2, is considerably greater than for the gradient 10 A/m^2. The frequency dependence a/ν^2 is observed at a frequency of $\sim 10^7$ Hz and monotonically decreases with frequency and at $v \sim 10^{12}$ Hz tends to 0. For the homogenous magnetic field and the given dependences the theory predicts the zero effect.

Unfortunately, in [252] the reasons for the insensitivity of the model to the dimensional effects were not discussed. Theory predicts the monotonic variation of the acoustic parameters with a frequency, although in the frequency range 10^5–10^{13} Hz the length of the sound wave λ changes from $\sim 10^{-2}$ to $\sim 10^{-10}$ m. With the decrease of λ its value becomes gradually comparable with the dimensions of the dispersed nanoparticles, the molecules, with the distances between them and the dimensions of the atoms of the substance. Conditions form in which the assumption of the continuity of the medium cannot be used and it is necessary to explain it on the basis of the very principle of the investigated wave process.

Long-term effect of the magnetic field. The authors of [173, 174] investigated the relative amplitude and the form of the ultrasound pulse, transmitted through the acoustic cell with the magnetic fluid, under the long-term effect of the magnetic field.

The filling frequency and the repetition frequency of the ultrasound pulses were equal to respectively 6 MHz and 4 kHz. The video pulses, obtained from the output of the receiving device, were transported to the input of the oscilloscope. For subsequent computer processing, the observed oscillograms were recorded with a digital video camera. The source of the magnetic field was a permanent magnet, positioned on a rotating platform. The acoustic cell had the form of a parallelepiped with the base of 2 × 2 cm and a height of 7 cm.

The investigated specimens were represented by magnetic colloids prepared on the basis of kerosene. The dispersed phase in them was magnetite Fe_3O_4, and oleic acid is used as a stabiliser. Table 4.3 shows the main physical parameters of the specimens of the magnetic fluids, used in the experiments.

Experiments were carried out at a temperature of 290±1 K. The strength of the field in the gap between the poles of the magnet was 122 kA/m, the initial angle between the vector of the strength of the magnetic field **H** and the wave vector **k** was $\vartheta = 90°$.

The results of the first part of the experiments (prior to the change of the form of the video pulse) are presented in Fig. 4.15. Triangles represent the results of measurements of the relative amplitude of the video pulse U/U_{max} and the squares – the increment $\Delta\alpha$ of the absorption coefficient. The experimental results showed a large reduction of the amplitude of the ultrasound pulse during the time $t \approx 30$–35 h for the MF-1, and for the MF-2 this time was ~60 hours, and the non-monotonic form of the angular dependence of which in general features is in agreement with the data obtained by B.E. Kuzin and V.V. Sokolov [175].

The additional absorption of ultrasound is associated with the changes in the structure of the magnetic fluid. The effect of the

Table 4.3

Specimen	ρ, kg/m³	φ, %	M_s, kA/m
MF-1	1350	12.8	54±1
MF-2	1300	11.7	50±1

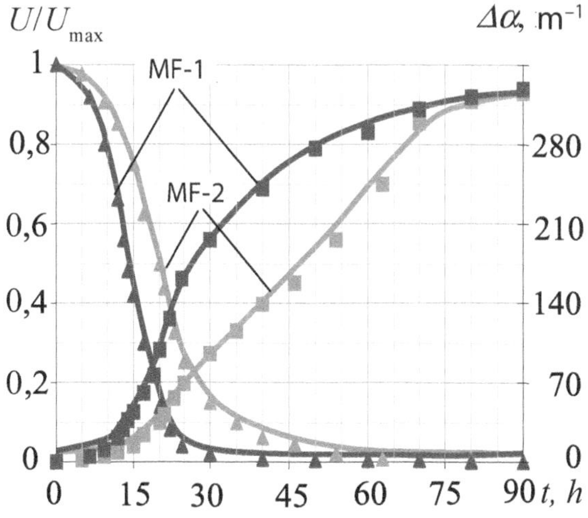

Fig. 4.15. Dependence of the relative amplitude of the video pulse U/U_{max} and the increment of the absorption coefficient $\Delta\alpha$ on time t.

magnetic field results in the formation of chain aggregates of ferroparticles. In passage through the dispersed system the sound wave is partially absorbed and scattered on the aggregates. The fact that $\Delta\alpha$ approaches the equilibrium value over a period of several days indicates the relatively lower rate of the process of structure formation. After $t \approx 50$ hours (MF-1) and $t \approx 150$ hours (MF-2) from the start of the experiment with additional amplification of the signal in the receiver, ensured by the regulation of the input attenuator for the specific angle ϑ, the form of the video pulse greatly changes. In the previously published studies the effect of this type was not reported.

Figure 4.16 shows the oscillograms of the ultrasound pulse, transmitted through the MF-2, for the given values of ϑ. The Y axis is divided in 0.5 V/division and X in 5 μs/division. The oscillograms, obtained for the MF-1, are identical.

At the angles of 130°–150° the central part of the video pulse shows a 'dip' which is again detected at the angles of 205°–225°, but in the former case the 'dip' travels from right to left, and in the latter case in the reverse direction.

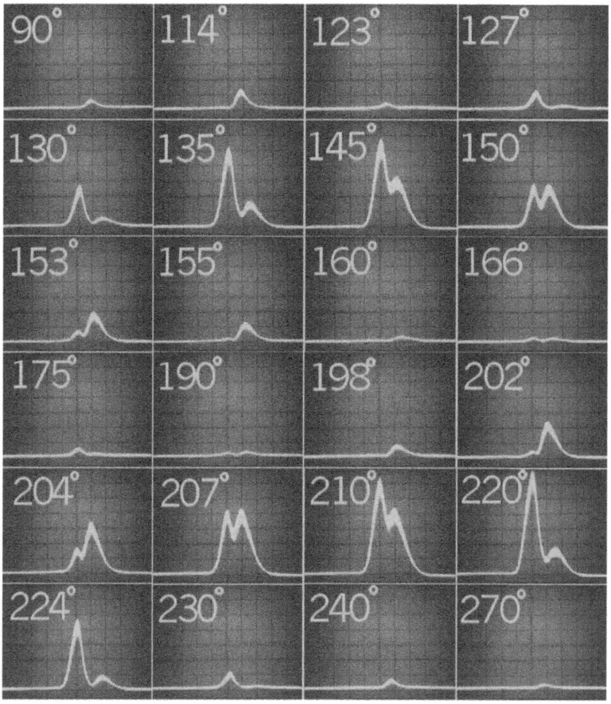

Fig. 4.16. Oscillograms of the ultrasound pulse.

When the cuvette is removed from the magnetic field, the 'dip' disappears and the signal amplitude increases, although it does not reach the initial level; multiple (up to 20–30 times) rotation of the magnetic field results in the restoration of the form similar to the initial form, and in a slight increase of the amplitude of the video pulse, and after several days the 'dip' appears again: the restoration of the initial form of the amplitude of the ultrasound pulse is achieved by thorough mechanical mixing of the magnetic fluid. It may be assumed that the change of the shape of the ultrasound pulse, detected in the experiments, is the consequence of the self-modulation of the wave during propagation in the medium with non-linearity and dispersion [173, 176]. As a result of the dipole-dipole interaction, the structure formation processes take place in a specific sequence, and the structures formed from the largest particles are the first to appear. They have high magnetic moments. The aggregates, consisting of small particles, are less stable and in the rotation of the magnetic field they easily fracture. In this case, the dispersion may be associated with the appearance in the magnetic colloid of the aggregates, consisting of the fine fraction of the ferroparticles and characterised by the resonance properties in the megahertz frequency range in the magnetic field. The expected resonance mechanism is associated with the forced rotation oscillations of the magnetic chain around the direction of the external magnetic field $\mathbf{H}_0$ (Fig. 4.17). The chain is affected by the rotational moment $\mathbf{M}_r = [\mathbf{p}_m \times \mathbf{B}]$ from the side of the magnetic field, and its value is:

$$M_r = -\mu_0 M_s V H_0 \cdot \sin \vartheta'$$

where $M_s V = p_m$ is the magnetic moment of the chain; V is the volume of the chain.

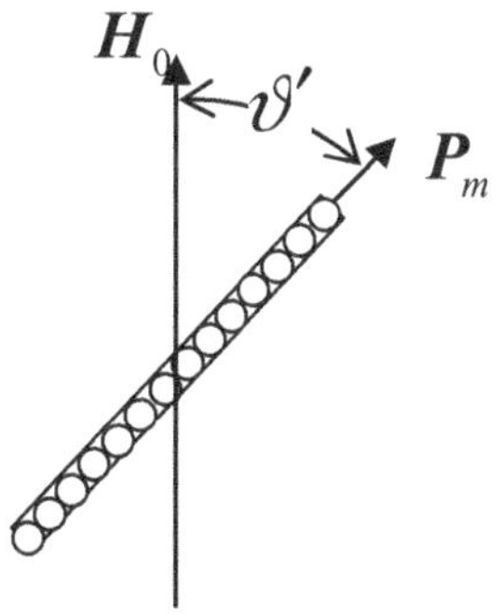

Fig. 4.17. The model of rotational oscillations.

The differential equation of the free non-damping oscillations in the approximation of small angles has the form ϑ':

$$\ddot{\vartheta}' + \frac{\mu_0 M_S V H_0}{J}\vartheta' = 0$$

where J is the moment of inertia of the chain in relation to the centre of rotation.

The resonance frequency of the oscillatory system res is v_{res}:

$$\nu_{\text{res}} = \frac{\sqrt{12\mu_0}}{2\pi N_p d_p} \cdot \sqrt{\frac{M_s H_0}{\rho}} \tag{4.31}$$

where N_p is the number of the particles in the chain; d_p is the diameter of the particle with the shell of the surfactant.

The role of the driving force is played by the orientation mechanism proposed by Ya.I. Frenkel [69], acting on the ellipsoidal particles in the ultrasound wave. Assuming that H_0 = 100 kA/m; $\bar{\rho} \approx 3 \cdot 10^3$ kg/m³; $\varphi_m/\varphi_s < 0.6$; $M_s = 0.2\ M'_s$ (taking the shell into account), $M'_s = 4.71 \cdot 10^5$ A/m; d_p = 15 nm, from equation (4.31) we obtain $\nu = 15 \cdot 10^7/N_p$, i.e. if the number of the particles is $N_p = 12$, then for $\nu \approx 6$ MHz, the length of the chain is $\ell = 15 \cdot 12 = 180$ nm.

However, the small chains should take part in thermal Brownian motion and this is not taken into account in the calculations.

As an alternative explanation of the observed modulation of the ultrasound pulse, it is interesting to examine the previously mentioned assumption of the mechanism of formation of diffraction beams on the system of the filament-like aggregates.

4.6. Some special features of the passage of ultrasound through a ferrosuspension

The magnetic properties of the concentrated ferrosuspensions were investigated in sufficient detail [151, 245]. These investigations resulted in the assumption of the presence of the stable structure in the ferrosuspensions, subjected to preliminary magnetisation and of the specific rearrangement of the structure under cyclic remagnetisation. However, the results of measurements of the magnetic parameters can be regarded only as indirect data supporting this assumption.

The direct confirmation of the presence of the structure are the results of optical, x-ray and acoustic investigations. The restrictions, associated with the application of the optical and x-ray methods of measurement for the studies of ferrosuspensions, are determined by their almost complete opacity to the light, and by the relatively large size of the particles dispersed in them whose linear dimensions are considerably greater than the wavelength of x-ray radiation. The acoustic methods are the most effective methods in this case because they do not require optical transparency and are sufficiently sensitive to different structural changes [63].

A special feature of the structures, formed in the ferrosuspensions during magnetisation, is their inability to resist mechanical mixing. This is indicated by, in particular, the fact that mechanical mixing of the ferrosuspensions completely eliminates remanent magnetisation [245]. For this reason, the method of the 'constant measurement base' was used in [246] and the method is described in detail in section 4.3.

The flow diagram of the equipment, designed for measuring the amplitude of the transmitted ultrasound pulse in relation to the strength of the magnetising field, is shown in Fig. 4.18. The RF pulses are generated by the pulse generator 1 and travel to the emitting piezoelement 2. The ultrasound signals propagate along the cylindrical delay line made of duralumin 3 and through the investigated medium – ferrosuspension 4, embedded in a special cell. The cell is placed in a magnetising solenoid 5. The signal then travels through the receiving delay line 6 to the piezoelement 7 where it is converted into the RF pulse. After amplification by the receiver 8 and detection by the device 9, the high-frequency RF pulse travels to the input of the oscilloscope 10, working in the regime of external synchronisation from the generator 1. The filling frequency and the pulse repetition frequency of the RF pulses are equal to approximately 2 MHz and 4 kHz.

The system was used for measurements of the relative variation of the amplitude of the ultrasound signal, transmitted through the

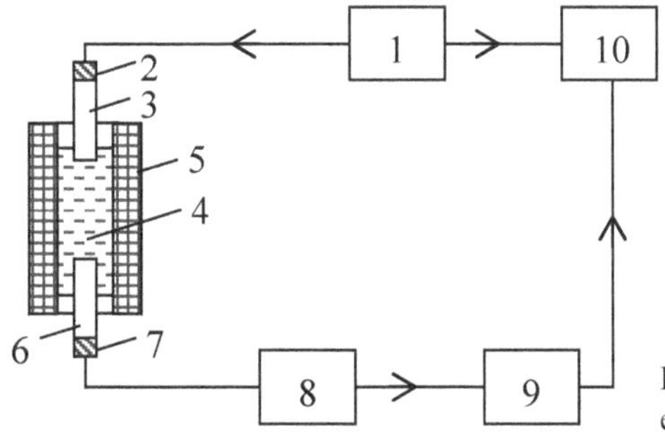

Fig. 4.18. The flow diagram of equipment.

ferrosuspension, in relation to the strength of both the direct and alternating magnetic fields at a constant thickness of the layer of the ferrosuspension d = 7 mm.

The ferrosuspensions were prepared by thorough mixing of the ferromagnetic powder with castor oil which was used as the fluid-carrier owing to the fact that, firstly, it is sufficiently viscous and, secondly, homogeneous as regards its chemical composition. The volume concentration of the investigated suspensions was 30%. Measurements were taken at room temperature, T_c = 20°C.

Figure 4.19 shows the dependences of the relative variation of the amplitude $\beta = A/A_0$ of the ultrasound pulse, transmitted through the suspension, on the strength of the constant homogeneous magnetic field H. The reference value of the amplitude was the initial value of the signal amplitude (prior to the application of the magnetic field) A_0.

Curve 3 shows the dependence for the suspension prepared using the F-600 ferrite powder (the mean size of the solid particles was ~3÷5 μm). Curve 2 relates to the suspension, prepared on the basis of the F-2000 ferrite powder with the particle size not greater than 63 μm. Both curves are characterised by a slope whose curvature decreases with increase of the strength of the magnetising field and, starting at $H \geq 16$ kA/m, the value of β for these curves remains almost constant and equals 0.07 and 0.19 for the suspensions prepared using the F-600 and F-2000 ferrite powders, respectively. When the magnetic field is applied, the amplitude of the transmitted signal does not change.

The dependence (the curves 2 and 3) is not unexpected because the relative change of the amplitude of the oscillations at constant acoustic resistance of the investigated medium is determined by the absorption coefficient α which is assumed to be proportional to the

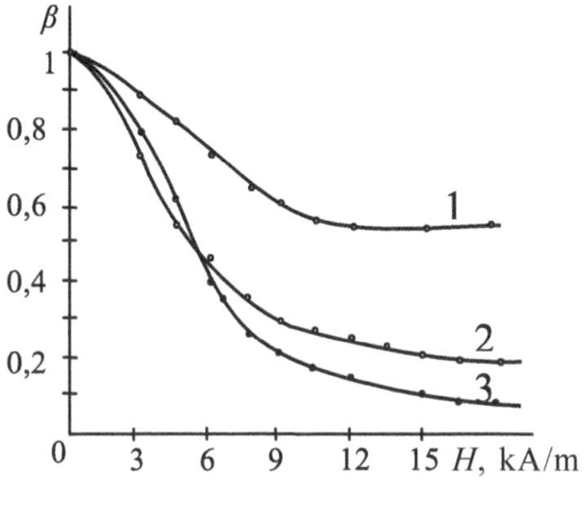

Fig. 4.19. Dependence β(H) for the suspensions of powders: 1 – F-600 with quartz; 2 – F-2000; 3 – F-600.

shear viscosity of the fluid. It is well-known that the shear viscosity of the concentrated ferrosuspensions depends strongly on the strength of the magnetising field (strictly speaking, the ferrosuspension is not a Newtonian fluid and, therefore, the assumption of the usual shear viscosity in this case can be used only in a first approximation). The curve 1 relates to the ferrosuspension in which the solid component contains, in addition to the F-600 powder, a non-magnetic quartz powder at a ratio of 1:1. In contrast to the curves 2 and 3, the slope of the curve 1 is flatter which is associated evidently with the small volume fraction of the magnetic component. The completed rotation of the particles in the formation of the structure in this case are obtained in the presence of smaller fields.

Figure 4.20 shows the dependence of β on the amplitude of the strength of the magnetic field H_0, changing at a frequency of 50 Hz, for the suspensions, produced on the basis of the F-2000 (curve 1) and F-600 (curve 2) powders. It may be noted that in this case the amplitude of the signal decreases with the increase of H_0, and the curves 1 and 2 in Fig. 4.20 diverge with increasing H_0, like the curves 2 and 3 in Fig. 4.19. Comparison of the dependences, shown in Fig. 4.19 and 4.20, indicates, firstly, that the application of the constant magnetic field results in a more rapid reduction of the amplitude of the transmitted signal than when using the alternating magnetic field and, secondly, the slope of the curves 1 and 2 (Fig. 4.20) in the investigated range of the strength of the field changes less markedly than in the curves 2 and 3 in Fig. 4.19. Therefore, it may be assumed that the rate of structure formation is higher when using the static magnetic fields and that there is a characteristic duration of this process which in this case is longer than 0.02 s.

Figure 4.21 shows the relative variation of the amplitude β of the signal, passed through the ferrosuspension, produced on the basis

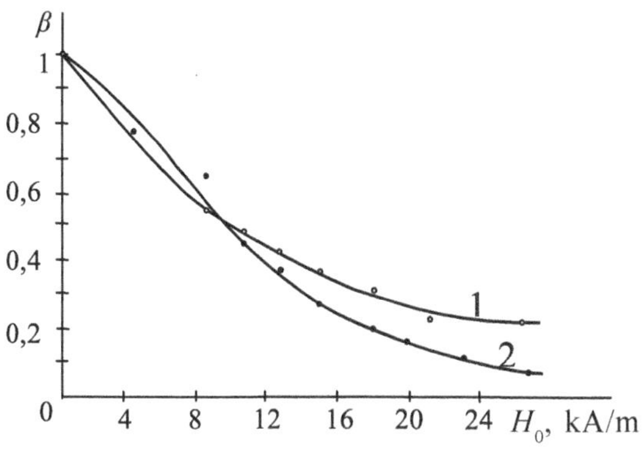

Fig. 4.20. Dependence $\beta(H_0)$ for powder suspensions: 1 – F-2000; 2 – F-600.

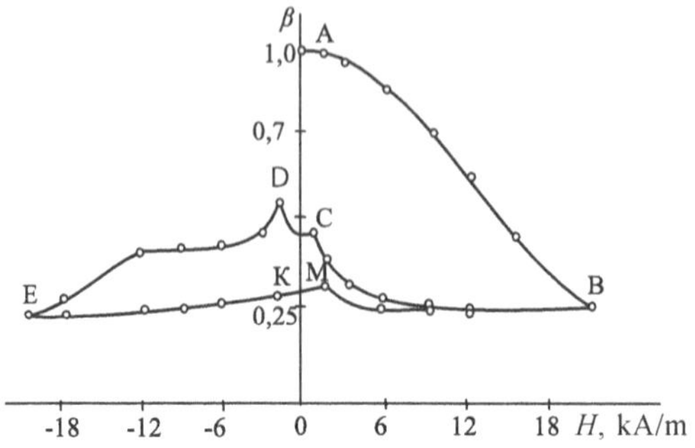

Fig. 4.21. Dependence β(*H*) in cyclic magnetisation reversal.

of the powder of the concentrate of magnetite of the fourth stage of wet magnetic separation, in relation to the strength of the constant magnetic field. Measurements were taken in cyclic magnetisation reversal of the specimen in the quasi-static mode.

In initial magnetisation (the section of the curve *AB*), the pulse amplitude monotonically decreases. The reduction of the strength of the magnetic field with the initial direction results in a small increase of the signal amplitude in weak fields (section *BC*). When the direction of the magnetic field changes to the opposite direction, the curve shows an extremum (section *DE*).

Thus, the curve of the dependence β(*H*) for the iron-ore suspension under cyclic magnetisation reversal indicates structure formation in the stage of initial magnetisation and the rearrangement of the structure when the direction of the field changes.

4.7. Optimisation of the acoustic parameters of magnetic fluids and ferrosuspensions

The application of the fluid dispersion media with the dispersed magnetic nanoparticles for the construction of emitters and receivers of sound oscillations has significant advantages in comparison with solid state materials.

Several possibilities of optimisation of the acoustic parameters of the magnetic fluids and ferrosuspensions will now be discussed [250].

To avoid non-uniformity in the frequency characteristic of the transducer (from the resonance, characteristic of the solid state transducers), it is sufficient to use the unique special feature of the magnetic fluids – the presence of the wave resistance, with the value similar to the wave resistance of the fluid media, for example, water. It is therefore interesting to derive relationships for the calculation of the wave resistance of the magnetic fluid and the concentration of the fluid with the given value of the wave resistance, based on the additive model of the formation of compressibility of the magnetic fluid.

For the wave resistance of the magnetic fluid, the additive model leads to the equation:

$$\rho c = \rho_1 c_1 \left\{ \frac{\varphi\left(\frac{\rho_2}{\rho_1} - 1\right) + 1}{1 - \left[1 + (1 - \gamma')a\right]\varphi} \right\}^{0.5} \tag{4.32}$$

where ρ_1 and ρ_2 are the densities of the fluid–carrier and the particles of the disperse phase, respectively; φ is the concentration of the solid phase; α and γ' are the relative volume and the relative compressibility of the stabiliser.

The equation (4.32) can be transformed to the following form:

$$\rho c = \left\{ \rho_1 c_1 \frac{(\rho_2 - \rho_1)\rho / \rho_1}{\rho_2 - \rho - a(1 - \gamma')(\rho - \rho_1)} \right\}^{0.5} \tag{4.33}$$

The concentration of the magnetic fluid with the given value of the wave resistance can be derived from the relationship:

$$\varphi_{00} = \frac{\left[(\rho c)_0^2 / (\rho_1 c_1)^2\right] - 1}{(\rho_2 / \rho_1 - 1) + (\rho c)_0^2 \left[1 + (1 - \gamma')a\right] / (\rho_1 c_1)^2} \tag{4.34}$$

For example, for a magnetic fluid, prepared on the basis of kerosene, for which $\rho_1 c_1 = 1.043 \cdot 10^6$ kg/m²s, with the wave resistance equal to the wave resistance of distilled water $(\rho c)_0 = 1.444 \cdot 10^6$ kg/m²s, the equation (4.34) gives $\varphi_{00} = 9.23\%$.

Table 4.4 shows the values of the wave resistance determined in the experiments, and the values of $(\rho c)_T$, calculated from the equation (4.33).

The results show that the determined value of φ_{00} is close to the corresponding value φ obtained in the experiments, and fits in the investigated concentration range.

The non-linear properties of transducers are very important for hydroacoustics. The additive model predicts the increase of the non-linearity of the parameter of the magnetic fluid with concentration. It is important to discuss the possible dependence of the non-linear parameter of the magnetic fluid on the degree of these magnetisation. It is well known, for example, that the non-linear parameter of the

Table 4.4

No.	M_s, kA/m	φ, %	$\rho c \cdot 10^{-6}$, kg/m²·s	$(\rho c)_T \cdot 10^{-6}$, kg/m²·s
1	0	0	1.043	1.043
2	2.3	0.72	1.069	1.070
3	5.4	1.9	1.111	1.113
4	15	4.8	1.236	1.225
5	36	10.6	1.468	1.457
6	52	16.6	1.707	1.719

solid ferromagnetics in magnetisation can be 1–2 orders of magnitude higher than their value as a result of the purely elastic non-linearity. The origin of this phenomenon is associated with the variation of the elastic modulus of the material under the effect of the magnetic field, i.e., with the value ΔE – the effect which reaches several tens of percent for the solid ferromagnetics. However, the elastic parameters of the magnetic fluid are almost completely independent of the magnetisation of the fluid. Actually, equation (4.22) leads to:

$$E_M / E = \mu_0 M_0^2 / \rho c^2$$

Assuming that M_0 = 80 kA/m, ρ = 1700 kg/m³, c = 1080 m/s, we obtain $E_M/E = 4 \cdot 10^{-6}$. Therefore, it is not justified to expect any large change of the non-linear parameter of the magnetic colloid during magnetisation.

The technology of production of ferrosuspensions with the given acoustic properties includes the task of production of the ferrosuspension in which the total contribution of shear viscosity to the absorption of sound will be minimal. This task is physically justified because the additional absorption $\Delta\alpha_2 \lambda/\varphi$ decreases, and the absorption in the fluid–carrier $\Delta\alpha_1 \lambda/\varphi$ increases with increase of its shear viscosity. The equation for shear viscosity (4.1) can be presented in the form:

$$\eta = \eta_{s1} \left\{ 1 - 2.5\left(1 + \frac{\delta}{R}\right)^3 \varphi + \frac{2.5\varphi_{cr} - 1}{\varphi_{cr}^2}\left(1 + \frac{\delta}{R}\right)^6 \varphi^2 \right\}^{-1}$$

The part of the absorption of sound, determined by viscosity, can be expressed as follows:

$$\frac{\alpha_\eta \lambda}{\varphi} = \frac{\pi\omega\eta_{s1}}{3\varphi\rho c^2 A_\eta} + \frac{2\pi\omega(\rho_2/\rho_1 - 1)^2 (R+\delta)^2 \rho_1}{9(1+\delta/R)^3 \eta_{s1}}$$

where A_η is the notation of the expression in the braces.

The dependence $\alpha_\eta\lambda/\varphi(\eta_{s1})$ has an extremum, and at some value $\eta_{s1} = \eta_{EXT}$ the contribution of $\alpha_\eta\lambda/\varphi$ to the total resistance will be minimal. Investigating the extremum, we find

$$\eta_{EXT} = \left[\frac{6\varphi\rho c^2 A_r (\rho_2/\rho_1 - 1)\rho}{(1+\delta/R)}\right]^{0.5} R \qquad (4.35)$$

Substituting the numerical values of the quantities into equation (4.35) we obtain $A_\eta = 0.43$ and $\eta_{EXT} = 3.3 \cdot 10^{-3}$ Pa·s. Equation (4.35) shows that the magnetic fluids with large particles will have higher values of η_{EXT}.

The field dependence of the amplitude of the ultrasound pulse, transmitted through the layer of the ferrosuspensions, does not show any large qualitative differences. This cannot be said for the magnetic fluid. The physical nature of this diversity is associated with the special features of structural rearrangement in a specific sample of the magnetic fluid. The magnetic fluid is characterised by the formation of solid or microdroplet aggregates with different densities of packing and spatial structure.

The Appendix gives the physical parameters of some fluids and solids. These data can be used for approximate calculations the acoustic parameters of the designed magnetic fluid and ferrosuspensions.

5

Ponderomotive mechanism of electromagnetic excitation of sound

5.1. Magnetoacoustic effect in the kilohertz frequency range

The ponderomotive mechanism of electromagnetic excitation of elastic oscillations in the magnetic fluid was initially seem to be the only possible mechanism. Formally, this conclusion was made on the basis of the main equation of the quasi-equilibrium hydrodynamics of magnetic fluids in which the irreversible magnetisation processes can be ignored and $\mathbf{M} \uparrow\uparrow \mathbf{H}$ [58]:

$$\rho \frac{d\mathbf{v}}{dt} \quad \nabla p + \eta \Delta \mathbf{v} + \mu_0 M \nabla H$$

where the strength of the magnetic field is determined by the magnetostatic equations: div $\mathbf{B} = 0$; rot $\mathbf{H} = 0$.

There are only a small number of studies concerned with the examination of the physical nature of the mechanism of generation of elastic oscillations of the magnetic fluid, situated in the alternating magnetic field with the frequency of oscillations of several tens of kilohertz [211, 212, 213]. We have attempted to investigate this process.

Figure 5.1 shows the flow diagram of experimental equipment for the investigation of the magnetoacoustic effect in the magnetic fluid in the range of low ultrasound frequencies [213].

The L-shaped glass tube 1 is filled with the magnetic fluid to be investigated 2. The lower horizontal nozzle is situated between the poles of the laboratory electromagnet 3, producing the constant homogeneous field. The exciting inductance coil 4 is inserted coaxially on the nozzle. The coil and the glass tube are separated by

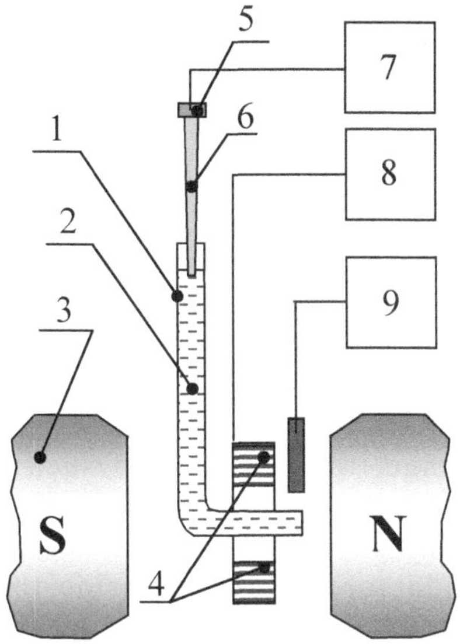

Fig. 5.1. Schematic of the experimental equipment.

an air gap of ~2 mm. The presence of the standing elastic wave in the magnetic fluid–cylindrical tube system is fixed using the piezoelectric sheet 5, located on the end of the metallic bar – waveguide 6. The oscilloscope 7 is designed for recording the alternating electrical voltage, taken from the piezoelement 8 – the generator of alternating voltage. 9 is the device for measuring the magnetic induction.

The experiments were carried out using the magnetic fluid based on kerosene with the density of ρ = 1300 kg/m^3 and the saturation magnetisation M_s = 51 kA/m.

Figure 5.2 shows the dependence of the relative amplitude β_e of the excited oscillations on the strength of the constant component of the magnetic field H_0, for the frequency of 20 kHz. The crosshatched circles are the values obtained as a result of increasing the strength of the magnetic field, the open circles – with the reduction of the

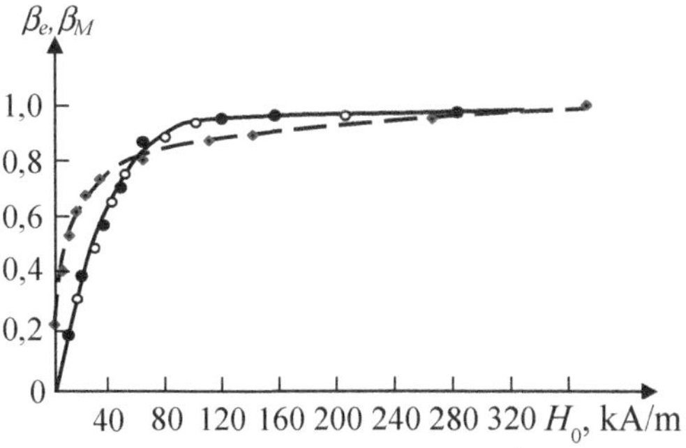

Fig. 5.2. Graphs of the $\beta_e(H)$ and $\beta_M(h)$ dependences.

strength of the magnetic field. The same figure shows the results of measurements of $\beta_M = M/M_{max}$ (the crosshatched rhombs, the dashed line) at $M_{max} = 49$ kA/m. In the range of technical saturation both curves are qualitatively identical – they reach saturation with increasing strength of the field.

As indicated by the equation $f_p = \mu_0 MV\nabla H$, the ponderomotive mechanism determines the effect of the driving force, proportional to the magnetisation of the fluid. This is also observed in these experiments.

The ponderomotive mechanism is also characterised by the linear dependence of the amplitude of the generated sound on the amplitude of the alternating magnetic field.

In the investigated frequency range, the presence of such a dependence is confirmed in the experiments, regarding the equipment shown schematically in Fig. 5.3. The generator 1 is connected to the exciting inductance coil (inductor) 2, placed inside the circular magnet 3. The displacement of the magnet is fixed by the cathetometer 4. The magnetic fluid 5, filling the cylindrical shell 6, is characterised by the formation of a standing wave whose presence is recorded using the piezoelectric sheet 7, positioned on the end of the waveguide 8. The oscilloscope 9 is designed for studying and measuring the electrical signal.

If the source of the alternating magnetic field is the inductance coil (inductor), the amplitude of the strength of the magnetic field H_m, according to the Biot–Savart–Laplace law, is directly proportional to the amplitude of the current intensity I in the inductor. The same dependence on the current intensity I is obtained for ∇H_m and the ponderomotive force f_p.

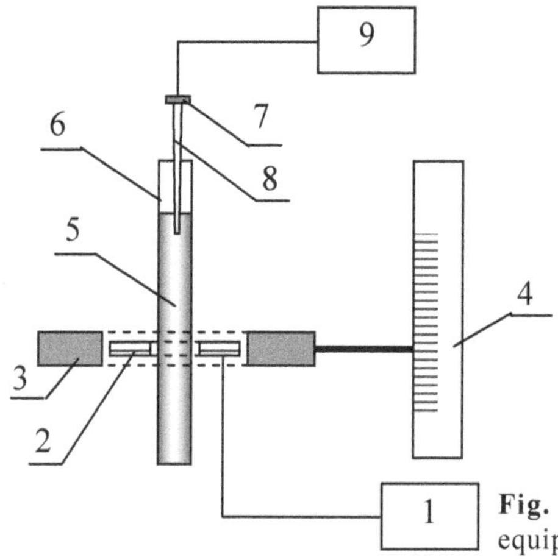

Fig. 5.3. Flow diagram of experimental equipment.

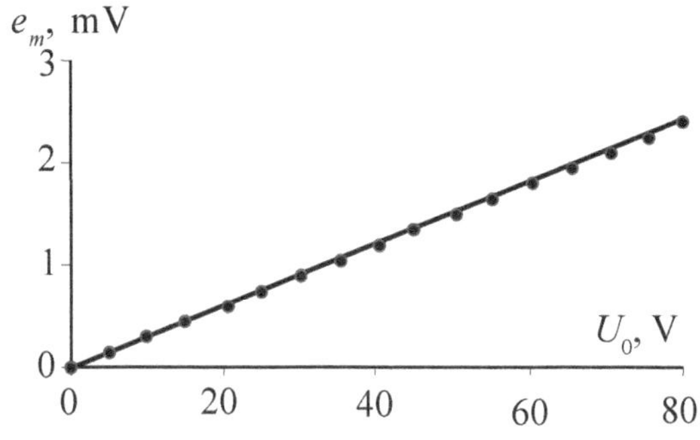

Fig. 5.4. Dependence $e_m(U_0)$.

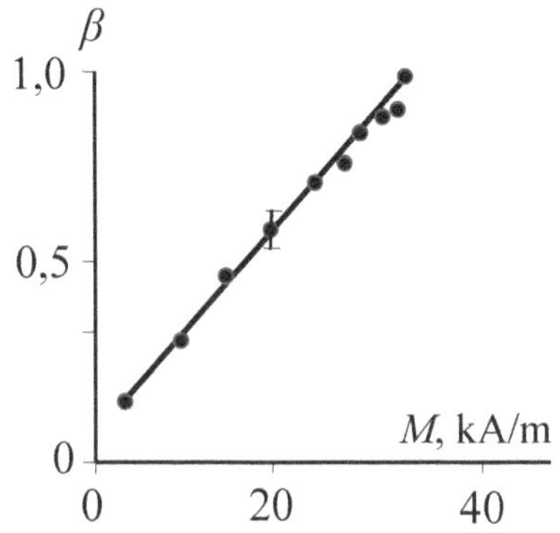

Fig. 5.5. Dependence $\beta(M)$.

Figure 5.4 shows the results of measurement of the amplitude of the oscillograms e_m for different values of the amplitude of alternating voltage U_0 and the approximating straight line.

The linear form of the dependence of the relative amplitude of the oscillations on the magnetisation of the magnetic fluid $\beta(M)$ is confirmed by the graph shown in Fig. 5.5. The experimental data were obtained at a frequency of 24.2 kHz [212].

Thus, the assumption of the dominating role of the ponderomotive mechanism of electromagnetic excitation of the sound oscillations in the magnetic fluid, used frequently in various theoretical models, does not contradict the experimental data, obtained in the kilohertz frequency range. In the study of the specific cases of electromagnetic excitation of elastic oscillations in the magnetic fluid we will use the concept of the ponderomotive mechanism.

5.2. Cylindrical magnetic fluid resonator

The effect of the electromagnetic field on the magnetic fluid may lead to the formation of different types of oscillations in the fluid: elastic, surface and oscillations of the form. A special position is occupied by the elastic – sound and ultrasound oscillations, with obvious scientific interest. The conversion effect of this type will be referred to as the magnetoacoustic effect (MAE).

In the applied aspect, the magnetic fluid in the framework of the investigated problem is regarded as the material used for the conversion of the energy of the electromagnetic field to the energy of elastic oscillations. The conversion devices have a number of advantages in comparison with conventional solid-state magnetostriction piezoelectric transducers. These advantages may be described as follows: the working body of the transducer has a lower density and speed of sound in comparison with the solid state transducer and this reduces by more than an order of magnitude the mass of the emitter for the same resonance frequency; the surface of the magnetic fluid is capable of acquiring any geometry, specified by the shape of the container; the resonance frequency and the directional diagram can be smoothly changed; the equality of the wave resistances of the magnetic fluid and seawater predetermines the possibility of producing a wide band source of sound oscillations.

The first attempts to solve this problem were based on the application of coarse ferrosuspensions [177]. However, as a result of rapid delamination of these systems and the very rapid damping of the elastic oscillations in them, the transducers of this type have found no application. These shortcomings are no longer observed in the transducers in which the active elements are the magnetic fluids. The problems of electromagnetic excitation of the acoustic oscillations in the magnetic fluid were investigated for the first time in the theoretical study by B. Cary and F. Fenlon [1]. The active element of the transducer which they studied was a cylindrical disc – tablet. The case of an infinite plane-parallel layer, with the external magnetic field directed along the normal to the layer, was investigated. Thermodynamic transformations show that a pressure gradient, determined by the 'jump' of the strength of the magnetic field, forms at the boundaries of the layer. It is assumed that only the ponderomotive mechanism may operate in the range of technical saturation, resulting in the hope to produce a source of oscillations competing with the conventional magnetostriction and piezoelectric transducers in the frequency range 100–150 kHz in which the non-conducting magnetic fluids are characterised by small losses due to eddy currents and magnetic reversal.

The first results of experimental studies of the special features of functioning of magnetic fluid transducers were obtained by A.R. Baev and P.P. Prokhorenko at frequencies of 16–26.7 kHz [178].

In cases in which the magnetic fluid fills a cavity with a specific geometry – resonator, we can use the simplest source of the magnetic field whose main purpose is the formation of the alternating component of the field. Resonance excitation of the oscillations is achieved by selecting the frequency of variation of the driving force.

This problem was solved in the studies [22, 23] in which attention was given to the cylindrical model of the magnetic fluid emitter (MFE). In the simplest case, this model can be produced by immersing a straight conductor, through which the alternating current flows, in a magnetic fluid.

In the theoretical aspect, the advantage of this model is that in its analytical examination it is not necessary to use any empirical or semi-empirical equation which determines the geometry and the time dependence of the magnetic field. The investigated magnetic field is the field of a straight infinite conductor with current whose geometry, determined by the Biot–Savart–Laplace law, is well-known. The principle of the method is investigated using a model shown in Fig. 5.6.

This method of excitation of resonance oscillations in the magnetic fluid is based on the application of a heterogeneous magnetic field containing the component which changes in time according to the harmonic law. The magnetic fluid fills the space between the coaxial infinitely long cylindrical surfaces with the radii r_0 and r_1. The proposed model uses a non-viscous, homogeneous fluid with no heat conductivity, and the special feature of magnetisation of this fluid will be discussed below.

The cylinders, restricting the fluid, are regarded as non-magnetic, electrically non-conducting and absolutely rigid. The assumption of the electrical and magnetic 'neutrality' of the restricting cylinders makes it possible to ignore electromagnetic induction in them and all the resultant consequences.

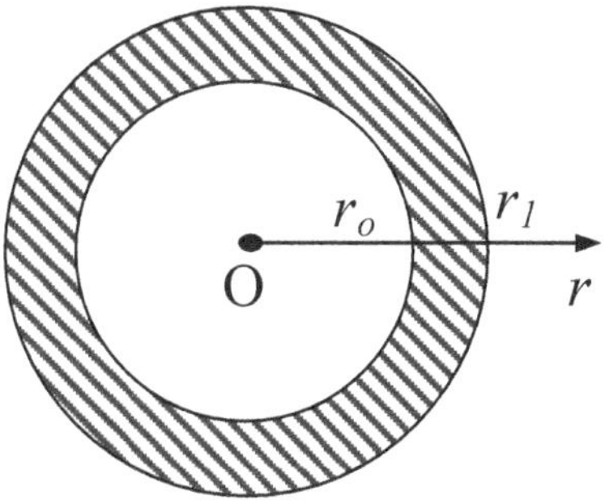

Fig. 5.6. The cylindrical model of the magnetic fluid emitter.

In the investigated cylindrical coordinate system, the axis Z coincides with the axis of the cylinders and is directed behind the drawing. The wave equation, describing the displacement of the fluid particles from the equilibrium position in the cylindrical coordinates, has the form

$$\frac{\partial^2 u}{\partial t^2} = c^2\left(\frac{1}{r}\cdot\frac{\partial}{\partial r}r\frac{\partial u}{\partial r}\right) + F_H, \tag{5.1}$$

where $u(r, t)$ is the displacement of the particles from the equilibrium position; r is the coordinate; t is time; c is the speed of propagation of sound in the fluid which is a real value, because there are no dissipative processes, determined by the viscosity or heat conductivity of the medium; F_H is the driving force which in the investigated model is associated with the heterogeneity of the magnetic field, generated by the axial current in the medium with no electrical conductivity.

The right-hand part of equation (5.1) is the ratio of the elasticity force, acting on the elementary volume of the fluid dV, to the mass of the fluid enclosed in the volume. When the heterogeneous magnetic field is applied to the volume dV of the magnetic fluid with no electrical conductivity, the ponderomotive force $d\mathbf{F}$ starts to operate and its intensity and direction can be determined on the basis of the quasi-static hydrodynamics of the isotropic magnetic fluid [34, 179] using the equation:

$$\mathbf{F} = \mu_0 M \nabla H / \rho$$

The driving force in the investigated model strongly depends on the type of function $\mathbf{M}(\mathbf{H})$, i.e., on the specific form of the equation of the magnetic state which in turn should satisfy the Maxwell equations div $(\mathbf{H} + \mathbf{M}) = 0$ and rot $\mathbf{H} = 0$.

This requirement is in agreement with, in particular, the linear equation

$$M = \chi H,$$

where χ is the magnetic susceptibility, and the equation of magnetic saturation

$$\mathbf{M} = \mathbf{M}_s = \text{const}.$$

In the first case, the mass unit is subjected to the effect of the additional force $\mathbf{F}_M$ equal to

$$\mathbf{F}'_M = \mu_0 \chi \nabla \mathbf{H}^2 / 2\rho \tag{5.2}$$

In the second case

$$\mathbf{F}''_M = \mu_0 \mathbf{M}_s \nabla \mathbf{H} / \rho \tag{5.3}$$

To ensure the cylindrical symmetry of the investigated problem, we use the magnetic field, generated by an infinite conductor with the current, situated along the axis OZ. The dependence of current I on time is given in the form

$$I = I_0 + I_m \cos \omega t, \tag{5.4}$$

where I_0 and I_m is the constant component of current and the amplitude of the variable component of the current, respectively; ω is the circular frequency of alternating current.

In the quasi-static approximation, on the basis of the Biot–Savart-Laplace law, the following equation is obtained for **H**

$$\mathbf{H} = (2\pi r)^{-1}(I_0 + I_m \cdot \cos \omega t)\mathbf{e}_\varphi, \tag{5.5}$$

where $\mathbf{e}_\varphi$ is the unit vector.

Substituting (5.5) to (5.2 and (5.3), we determine the projection of the force $\mathbf{F}'_M$ and $\mathbf{F}''_M$ on the direction of the unit vector

$$F'_M = -\mu_0 \chi I^2_m (1 + \cos 2\omega t) / 8\pi^2 r^3 \rho; \tag{5.6}$$

$$F''_M = -\mu_0 M_s (I_0 + I_m \cos \omega t) / 2\pi \rho r^2 \tag{5.7}$$

When deriving the equation (5.7) it is assumed that the amplitude of the alternating component of current is considerably smaller than the direct component, i.e. $I_m << I_0$, and when deriving (5.6) $I_0 = 0$. The equations (5.6) and (5.7) show that in the investigated magnetic field, each cylindrical element of the fluid of the unit class is affected by the force containing the station the component F_{M0} and the non-stationary component F_{M1}:

$$F'_{M0} = -\frac{\mu_0 \chi I^2_m}{8\pi^2 r^3 \rho}; \tag{5.8}$$

$$F''_{M0} = -\frac{\mu_0 M_s I_0}{2\pi r^2 \rho}; \tag{5.9}$$

$$F'_{M1} = -\frac{\mu_0 \chi I^2_m \cos 2\omega t}{8\pi^2 r^3 \rho}; \tag{5.10}$$

$$F''_{M1} = -\frac{\mu_0 M_s I_m \cos \omega t}{2\pi r^2 \rho}; \tag{5.11}$$

The effect of the stationary force on the fluid at the fixed cylindrical boundaries can result only in some distribution of the static pressure. If the absolute value of this pressure is not too high, it has no significant effect on the nature of oscillatory movement, and the components F'_{M0} and F''_{M0} can be excluded from further considerations. The components F'_{M1} and F''_{M1} should be regarded as the driving force, and the minus sign should be removed. This is equivalent to the change of the initial phase by π.

Further, to simplify the calculations, we consider the case of 'thin' cylindrical layers for which the following inequality is fulfilled

$$h / r << 1, \tag{5.12}$$

where $h \equiv r_1 - r_0$ is the thickness of the layer.

Taking into account the monotonic form of the dependences F'_{M1} and F''_{M1} on r and the smallness of the investigated range of variation of r, the equations (5.8) and (5.11) will be replaced by their mean values in the range $r_1 - r_0$:

$$F'_{M1} = \frac{\mu_0 \chi I_m^2 (r_1\ r_0) \cos 2\omega t}{16\pi^2 \rho r_0^2 \rho_1^2}; \tag{5.13}$$

$$F''_{M1} = \frac{\mu_0 \mathrm{M}_s I_m \cos \omega t}{2\pi \rho r_0 r_1}. \tag{5.14}$$

As a result of averaging (5.10) and (5.11), the differential equation (5.1) can be solved using special functions [180].

Substituting successively (5.13) and (5.14) to the right-hand part of (5.1), gives to differential equations with the general form

$$\frac{\partial^2 u}{\partial t^2} = c^2 \left(r^{-1} \frac{\partial}{\partial r} r \cdot \frac{\partial u}{\partial r} \right) + A_v \cdot \cos \alpha_v t. \tag{5.15}$$

To solve the cases 1 and 2, in equation (5.15) it should be accepted that

$$A_v = \frac{\mu_0 \chi I_m^2 (r_1 + r_0)}{(4\pi r_0 r_1)^2 \rho} \text{ and } \alpha_v = 2\omega$$

or

$$A_v = \frac{\mu_0 M_s I_m}{2\pi \rho r_0 r_1} \text{ and } \alpha_v = \omega$$

The solution of the differential equation (15) is obtained by dividing the variables r and t. For this, $u\ (r, t)$ is written in the form

$$u = U(r) \cdot \cos \alpha_v t. \tag{5.16}$$

After substituting (5.16) to (5.15) and carrying out the simplest transformations, we obtain the new differential equation

$$\frac{\partial^2 R}{\partial r^2}+\frac{1}{r}\frac{\partial R}{\partial r}+k^2 R=-\frac{A_v}{c^2}, \tag{5.17}$$

where $k = k_1 = 2\omega/c$ for the case 1 and $k = k^2 = \omega/c$ for the case 2.

The differential equation (5.17) is a heterogeneous equation, and the solution of the appropriate homogeneous equation is well-known – it is expressed by the Bessel functions of the zeroth order [180]:

$$R^* = a_0 J_0(kr) + b_0 N_0(kr), \tag{5.18}$$

where J_0 is the Bessel function of the zeroth order; N_0 is the Neumann function of the zeroth order; a_0 and b_0 are the arbitrary constant values.

To obtain the general solution of the heterogeneous equation (5.17), the constant value A_v/c^2k^2 should be added to the solution (5.18)

$$R^* = a_0 J_0(kr) + b_0 N_0(kr) = -A_v / c^2 k^2 \tag{5.19}$$

As shown below, the highest value of the length of the resonant sound wave is $2s$. Consequently, on the basis of (5.12) $kr >> 1$ and, consequently, the Bessel functions can be replaced by asymptotic expressions [180]

$$J_0(kr) \approx \sqrt{\frac{2}{\pi kr}}\cdot\cos\left(kr-\frac{\pi}{4}\right); \tag{5.20}$$

$$N_0(kr) \approx \sqrt{\frac{2}{\pi kr}}\cdot\sin\left(kr-\frac{\pi}{4}\right), \tag{5.21}$$

and taking these equations into account, the solution (5.19) has the form

$$R^*(r) = \alpha_{01}\frac{\cos kr}{\sqrt{kr}} + b_{01}\frac{\sin kr}{\sqrt{kr}} - \frac{A_v}{c^2 k^2} \tag{5.22}$$

The values of the constants a_{01} and b_{01} are determined from the boundary conditions $R^*(r_0) = 0$ and $R^*(r_1) = 0$, since the restricting cylinders are assumed to be absolutely rigid

$$a_{01}\frac{\cos kr_0}{\sqrt{kr_0}} + b_{01}\frac{\sin kr_0}{\sqrt{kr_0}} - \frac{A_v}{c^2 k^2} = 0; \tag{5.23}$$

$$a_{01}\frac{\cos kr_1}{\sqrt{kr_1}} + b_{01}\frac{\sin kr_1}{\sqrt{kr_1}} - \frac{A_v}{c^2 k^2} = 0. \tag{5.24}$$

From the system of the equations (5.23) and (5.24) we obtain

$$a_{01} = \frac{A_v}{c^2k^2}\frac{\sqrt{kr_0}\sin kr_1 - \sqrt{kr_1}\, sunkr_0}{\sin k(r_1 - r_0)}; \tag{5.25}$$

$$b_{01} = \frac{A_v}{c^2k^2}\frac{\sqrt{kr_1}\cos kr_0 - \sqrt{kr_0}\cos kr_1}{\sin k(r_1 - r_0)}. \tag{5.26}$$

Substituting a_{01} and b_{01} into (5.22) and the resultant equation into (5.16), we obtain the solution of the differential equation (5.15) for the case 1 and case 2:

$$u_1(r,t) = \frac{\mu_0 \chi I_m^2 (r_1 + r_0)}{\rho(8\pi\omega r_0 r_1)^2}\left[\frac{\sqrt{r_1}\sin k_1(r - r_0) + \sqrt{r_0}\sin k_1(r_1 - r)}{\sqrt{r}\cdot \sin k_1 h} - 1\right]\cos 2\omega t \tag{5.27}$$

$$u_2(r,t) = \frac{\mu_0 M_s I_m}{2\pi\rho\omega^2 r_0 r_1}\left[\frac{\sqrt{r_1}\sin k_2(r - r_0) + \sqrt{r_0}\sin k_2(r_1 - r)}{\sqrt{r}\cdot \sin k_2 h} - 1\right]\cos \omega t. \tag{5.28}$$

Analysis of the equations (5.27) and (5.28), which describe the movement of the fluid particles, shows that the particles in the conditions of this model carry out the radial harmonic oscillations with a circular frequency 2ω in the case 1 and ω in the case 2. The amplitude of the oscillations depends on the magnetic parameters of the magnetic fluid – magnetic susceptibility χ in the case 1 and saturation magnetisation M_S in the case 2. If $\chi = 0$ or $M_S = 0$, which holds for conventional non-magnetic fluids, no oscillations form. The magnetic liquids from the vast number of different liquid media, disregarding liquid metals, are characterised by high electrical conductivity and have the unique property of transformation of the energy of electromagnetic oscillations to the energy of elastic mechanical oscillations.

It may easily be seen that at $\omega = \omega_m = \pi cm/2h$ (in case 1) and $\omega = \omega_m = \pi cm/h$ (in case 2), where $m = 1, 2, 3...$, the nominator of the expression in the square brackets of the equations (5.27) and (5.28) converts to 0, and the amplitude of the oscillations has an infinitely large value. Consequently, at the frequencies of alternating current $\omega = \omega_m$ the oscillations become resonant. The formation of the infinitely large amplitude of the oscillations at resonance is the consequence of the assumption made regarding the absence of energy dissipation in the medium and the absence of sound radiation into the restricting cylinders.

A specific example be considered. Let it be that $h = 5$ mm at $r_0 = 50$ mm. Since at resonance $h = m\lambda/2$, the wavelength of the main resonance frequency will be equal to 10 mm. Accepting for the magnetic fluid $c = 1300$ m/s, we obtain the value of the main resonance frequency $\nu = c/\lambda = 130$ kHz; this value is obtained in the cases 1 and 2 by using the alternating current with a frequency of 130 and 65 kHz.

Thus, the proposed method can in principle be used for the direct excitation of the resonance ultrasound oscillations in the magnetic fluid. The magnetic fluid emitter with the same geometry of the magnetic field was investigated by P. Dubbelday [181]. He reported that the cylindrical model uses most efficiently the properties of the liquid magnetic material, and concluded that it can also be used in the sound frequency range 100–3000 Hz. As in [23], both fields – constant and exciting – have the azimuthal geometry and depend only on the radial distance, and the fields are generated by the currents, flowing through the conductors, wound in the azimuthal direction.

5.3. The flat magnetic fluid source of sound oscillations

V.G. Bashtovoi and M.S. Krakov investigated theoretically the model of the magnetic fluid emitter represented by a resonator in the form of a right-angled parallelepiped, filled with the magnetic fluid [182, 183]. It was assumed that the dissipation of acoustic energy is caused only by the viscous properties of the magnetic fluid and that one of the walls of the parallelepiped is a membrane with zero elasticity. These assumptions were used to derive equations for the three oscillatory velocities of the particles of the fluid, directed along the faces of the parallelepiped, in the general case and at resonance. The resultant relationships were subsequently used for numerical estimation of the amplitude of the physical quantities in the resonant mode of operation of the magnetic fluid emitter in the absence of load and in radiation into air. If the linear dimensions of the right angle resonate have the order of 10^{-2} m, then at $c = 1200$ m/s the resonance frequency is $\nu = 90$ kHz, i.e., the magnetic fluid emitter in the form of the right-angle parallelepiped is designed mainly for the low frequency region of the ultrasound range 16–90 kHz [184].

The magnetic fluid source of elastic oscillations in the megahertz frequency range was investigated theoretically by V.M. Polunin [24]. Special features of the operation of the magnetic fluid emitter,

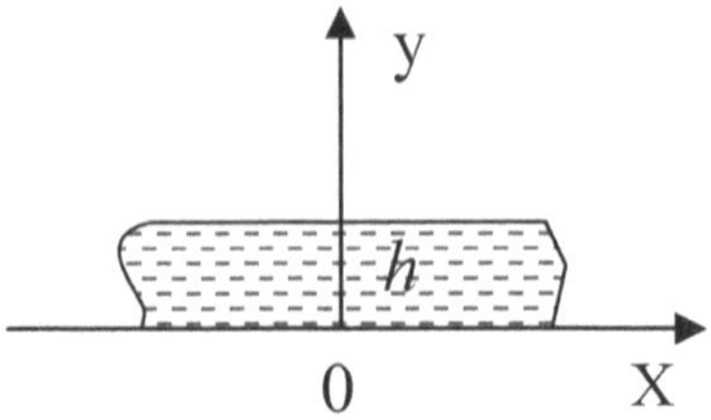

Fig 5.7. Calculation model.

simulated by the infinite plane-parallel layer of the magnetic fluid, were investigated (Fig. 5.7). The homogeneous, viscous magnetic fluid, with no electrical conductivity, was situated on the surface of the solid body made of the absolutely rigid, non-magnetic material with no electrical conductivity and had the form of a plane-parallel layer with a thickness h. The plane XOZ of the coordinate system coincided with the lower surface of the liquid layer. The unit mass of the fluid along the axis y is subjected to the effect of the driving force

$$F = f_0 \cdot \cos \omega t, \tag{5.29}$$

where ω is the circular frequency, and the amplitude of the force f_0 is independent of the coordinates.

This equation for the driving force can be derived, for example, assuming that it is determined by the ponderomotive interaction of the fluid with the magnetic field:

$$H_x = H_0 + H_m(y) \cdot \cos \omega t, H_y = 0 \text{ and } H_z = 0, \tag{5.30}$$

in which H_0 is independent of the coordinates, ∇H_m is the constant vector, $H_m \ll H_0$. Consequently

$$f_0 = \mu_0 M \,|\nabla H_m|/\rho. \tag{5.31}$$

The differential equation of oscillatory motion for this case has the following form

$$\frac{\partial^2 u}{\partial t^2} = c^2 \frac{\partial^2 u}{\partial y^2} + \frac{\eta}{\rho} \cdot \frac{\partial^3 u}{\partial y^2 \partial t} + f_0 \cdot \cos \omega t, \tag{5.32}$$

where $\eta = \eta_v + 4\eta_s/3$ is the total viscosity of the fluid (η_s and η_v is the shear and volume viscosity of the fluid).

The boundary conditions for the stationary lower surface of the fluid layer and its free upper surface have the following form:

$$u\Big|_{y=0} = 0 \text{ and } \frac{\partial u}{\partial y}\Big|_{y=h} = 0.$$

The system of the flat standing waves formed in the fluid:

$$u = \sum_{m=1}^{\infty} u_{2m-1} \cdot \sin\frac{(2m-1)\pi y}{2h} \cdot \cos(\omega t + \psi_{2m-1}). \quad (5.33)$$

Here u_{2m-1} is the amplitude of displacement of the fluid particles in the antinode of the $2m$–1th harmonics, and Ψ_{2m-1} is the phase shift between these displacements and the driving force.

The parameters of oscillatory motion u_{2m-1} and Ψ_{2m-1} should be determined. For this purpose, (5.33) is substituted into (5.32) and after transformations we obtain:

$$\left[\begin{array}{l}\sum_{m-1}^{\infty}(\omega_{2m-1}^{2} - \omega^{2})u_{2m-1} \cdot \sin\frac{(2m-1)\pi y}{2h} \cdot \cos\psi_{2m-1} \\ -\frac{\omega\eta}{\rho c^{2}}\sum_{m=1}^{\infty}(\omega_{2m-1}^{2}u_{2m-1} \times \cdot \sin\frac{(2m-1)\pi y}{2h} \cdot \sin\psi_{2m-1} - f_{0}\end{array}\right]$$
$$\cdot \cos\omega t - \left\{\begin{array}{l}\sum_{m=1}^{\infty} u_{2m-1} \cdot \sin\frac{(2m-1)\pi y}{2h} \times \\ \left[\left[(\omega_{2m-1}^{2})\sin\psi_{2m-1} - \frac{\omega\eta}{\rho c^{2}}\omega_{2m-1}^{2} \cdot \cos\psi_{2m-1}\right]\right]\end{array}\right\} \quad (5.34)$$
$$\cdot \sin\omega t = 0,$$

where $\omega_{2m-1} \equiv (2m-1)\pi c / 2h.$

The equality (5.34) should be fulfilled at any moment of time t and at any point y of the investigated material. Therefore

$$(\omega_{2m-1}^{2} - \omega^{2})\sin\psi_{2m-1} - \frac{\omega\eta}{\rho c^{2}}\omega_{2m-1}^{2}\cos\psi_{2m-1} = 0,$$

from which $$\operatorname{tg}\psi_{2m-1} = \frac{\omega\eta\omega_{2m-1}^{2}}{\left(\omega^{2} - \omega_{2m-1}^{2}\right)\cdot \rho c^{2}}. \quad (5.35)$$

In addition, we have

$$\sum_{m=1}^{\infty}\left(\omega_{2m-1}^{2} - \omega^{2}\right)u_{2m-1} \cdot \sin\frac{(2m-1)\pi y}{2h} \cdot \cos\psi_{2m-1} - \frac{\omega\eta}{\rho c^{2}}$$
$$\sum_{m=1}^{\infty}\omega_{2m-1}^{2}\, u_{2m-1} \times \sin\frac{(2m-1)\pi y}{2h} \cdot \sin\psi_{2m-1} - f_{0} = 0$$

Multiplying the last term by the trigonometric series

$$\sum_{m=1}^{\infty}\frac{4\sin\left[(2m-1)\pi y/2h\right]}{\pi(2m-1)},$$

diverging to 1 in the period $0 < y < 2h$ [185], we obtain

$$\sum_{m=1}^{\infty}\left\{u_{2m-1}\cdot\left[\left(\omega_{2m-1}^2-\omega^2\right)\cos\psi_{2m-1}+\frac{\omega\eta\omega_{2m-1}^2}{\rho c^2}\sin\psi_{2m-1}\right]-\frac{4f_0}{\pi(2m-1)}\right\}\cdot$$

$$\sin\frac{(2m-1)\pi y}{2h}=0.$$

The expression in the braces converts to 0. Replacing in this equation the functions cos Ψ_{2m-1} and sin Ψ_{2m-1} by tg Ψ_{2m-1} using the trigonometric equality, and using the expression (5.35), we obtain

$$u_{2m-1}=\frac{4f_0}{\pi(2m-1)\sqrt{\left(\omega_{2m-1}^2-\omega^2\right)^2+\left(\frac{\omega\eta\omega_{2m-1}^2}{\rho c^2}\right)^2}} \tag{5.36}$$

Substitution of (5.36) to (5.33) gives

$$u=\sum_{m=1}^{\infty}\frac{4f_0}{\pi(2m-1)\sqrt{\left(\omega_{2m-1}^2-\omega^2\right)^2+\left(\frac{\omega\eta\omega_{2m-1}^2}{\rho c^2}\right)^2}}\times$$

$$\sin\frac{(2m-1)\pi y}{2h}\cdot\cos\left(\omega t+\psi_{2m-1}\right). \tag{5.37}$$

The following expression for the resonance frequency can be written for each harmonics ω_{2m-1}^r:

$$\omega_{2m-1}^r=\omega_{2m-1}\sqrt{1-\frac{(\eta\omega_{2m-1})^2}{2(\rho c^2)^2}}.$$

As a result of the smallness of the second term in the radicand

$$\omega_{2m-1}^r\approx\omega_{2m-1}=\pi(2m-1)c/2h. \tag{5.38}$$

The amplitude of the resonance oscillations u_{2m-1}^r at not too high values of *m* is several orders of magnitude greater than the amplitude of the adjacent harmonics u_{2m+1} and u_{2m-3}. Actually, using (5.36) we obtain

$$u_{2m-3}/u_{2m-1}^2=\left[c_u\ell_1^2(1-\ell_1^2)^2+\ell_1^6\right]^{-0.5}$$

and

$$u_{2m+1} / u^r_{2m-1} = \left[c_u \ell_2^2 (\ell_2^2 - 1)^2 + \ell_2^6 \right]^{-0.5}$$

where

$$C_u \equiv \rho^2 c^4 / \eta^2 \omega^2_{2m-1}; \ell_1 \equiv (2m-3)/(2m-1); \ell_2 \equiv (2m+3)/(2m-1).$$

The value η can be determined from the results of measurement of the coefficient of absorption of ultrasound α [63] connected with η by the dependence $\alpha = \omega^2\eta/2\rho c^3$. Using the results obtained in the previous chapter, it is assumed that $c = 1200$ m/s, $\alpha = 200$ m^{-1} at $\nu = 25$ MHz. Consequently, $m = 2 : C_u \approx 10^5$, $\ell_1 = 1/3$, $\ell_2 = 7/3$, $u_1 u_3^r \approx 10^{-2}$, $u_5 u_3^r \approx 1.3 \cdot 10^{-4}$.

The assumption of the smallness of the second term in the radicand (5.38) is consistent with the numerical value of C_u.

For frequencies close to the resonant frequency ω^r_{2m-1}, instead of the series (5.37), we can write

$$u = \frac{4 f_0 \sin \frac{(2m-1)\pi y}{2h} \cos(\omega t + \psi_{2m-1})}{\pi (2m-1) \sqrt{\left(\omega^2_{2m-1} - \omega^2\right)^2 + \left(\frac{\omega \eta \omega^2_{2m-1}}{\rho c^2}\right)^2}}. \quad (5.39)$$

At $\omega = \omega^r_{2m-1}$:

$$u = \frac{4 f_0 \rho c^2 \sin \frac{(2m-1)\pi y}{2h} \cos(\omega t + (2m-1)\pi/2)}{\pi (2m-1) \eta \omega^3_{2m-1}}. \quad (5.40)$$

If there is only one ponderomotive mechanism of excitation of the oscillations, then at $\omega \approx \omega_{2m-1}$

$$u = \frac{4 u_0 M G c^2 \sin \frac{(2m-1)\pi y}{2h} \cos(\omega t + \psi_{2m-1})}{\pi (2m-1) \rho \sqrt{\left(\omega^2_{2m-1} - \omega^2\right)^2 + \left(\frac{\omega \eta \omega^2_{2m-1}}{\rho c^2}\right)^2}}.$$

At $\omega = \omega_{2m-1}$

$$u = \frac{4 u_0 M G c^2 \cdot \sin \frac{(2m-1)\pi y}{2h} \cdot \cos(\omega t + \pi/2)}{\pi (2m-1) \omega^3 \eta}$$

The form of the equations (5.35) and (5.39) indicates the analogy between the oscillatory motion, carried out by the fluid particles, and the mechanical system with concentrated parameters, which can be confirmed by introducing the notations $\delta'_{2m-1} \equiv \eta\omega^2_{2m-1} / 2\rho c^2$ and regarding this quantity as the analogue of the coefficient of damping of the mechanical system.

Consequently, the logarithmic coefficient of damping Θ'_{2m-1} and the quality factor Q'_{2m-1} of the transducer taking into account only the internal losses, can be described by [186]:

$$\Theta'_{2m-1} = \pi\eta\omega_{2m-1} / \rho c^2, \tag{5.41}$$

$$\Theta'_{2m-1} = \rho c^2 / \eta\omega_{2m-1}. \tag{5.42}$$

For the above mentioned numerical values of α and c we obtain at $m = 1$: $\Theta'_1 = 9.5 \cdot 10^{-3}$ and $Q'_1 = 330$.

These calculations clarify why the authors of [21] obtained a high value of the quality factor of the magnetic fluid transducer. In the experiment conditions, there is almost no emission of sound into the surrounding medium.

The acoustic–mechanical analogy can be used to calculate the amplitude of oscillations of the particles at resonance u^r_{2m-1} taking into account both the internal losses and the losses due to emission.

It is well-known [186] that

$$u^r_{2m-1} = u^s_{2m-1}\, Q_{2m-1}, \tag{5.43}$$

where u^s_{2m-1} is the effective value of the static displacement.

Consequently

$$u^s_{2m-1} = \frac{4f_0 \sin[(2m-1)\pi y / 2h]}{\pi(2m-1)\omega^2_{2m-1}}.$$

In order to find Q_{2m-1}, we use the property of additivity of the energy losses in the period ΔW : $\Delta W = \Delta W_i + \Delta W_e$, where ΔW_i and ΔW_e are the internal losses and the losses due to emission in the period. If W is the mechanical energy of the system at time t, then its logarithmic attenuation coefficient is $\Theta = -0.5|\Delta W|/W$. Consequently

$$\Theta_{2m-1} = \Theta'_{2m-1} + 0.5|\Delta W|/W. \tag{5.44}$$

If the inequality $\rho_c\rho_c >> \rho c$ is fulfilled (wave resistance of the solid is considerably higher than the wave resistance of the fluid), the pressure antinode forms at the interface [186] and a flat ultrasound wave with the intensity

$$J = p^2 / 2\rho_c c_c,$$

passes through the interface. Here p is the amplitude of the pressure in the antinode of the standing wave.

Ignoring the energy emited into the air, we obtain

$$\left|\Delta W_i\right| = \pi p^2 s / \rho_c c_c \omega \tag{5.45}$$

where s is the area of the active surface of the liquid layer.

The total energy of the liquid layer

$$W = (2m-1)\pi p^2 s / 8\rho c \tag{5.46}$$

Substituting (5.46) and (5.45) to (5.44) we obtain

$$\Theta_{2m-1} = \Theta'_{2m-1} + 4\rho c / (2m-1)\rho_c c_c \tag{5.47}$$

and since $Q_{2m-1} = \pi / \Theta_{2m-1}$

$$Q_{2m-1} = \frac{Q'_{2m-1}(2m-1)\pi\rho_c c_c / 4\rho_c}{Q'_{2m-1} + (2m-1)\rho_c c_c / 4\rho c} \tag{5.48}$$

Let α and c have the previous values, and $\rho = 1200$ kg/m^3, $\rho_c c_c = 1.33 \cdot 10^5$ kg/s·m^2, then at $m \leq 3$ the inequality $Q'_{2m-1} \gg (2m-1)$ $\rho_c c_c/4\rho c$ is fulfilled and, consequently, we can write:

$$\Theta_{2m-1} = 4\rho c / (2m-1)\rho_c \rho_c \text{ and } Q_{2m-1} = (2m-1)\pi\rho_c \rho_c / 4\rho c$$

Therefore, for $m = 1$ we obtain $Q_1 = 7.8$, which in almost complete agreement with the quality factor of quartz emitting into water. It is quite natural that for and absolutely rigid medium the equation (5.48) gives $Q_{2m-1} = Q'_{2m-1}$.

The quality factor of the flat magnetic fluid emitter in emission into the solid medium is determined by the ratio of the acoustic resistance of this medium to the acoustic resistance of the transducer – magnetic fluid, whereas the solid vibrator, emitting into the fluid, is characterised by the reversed relationship.

At ‘low’ frequencies of variation of the driving force, when $\omega \ll \omega_{2m-1}$ the ‘thin’ layer of the magnetic fluid oscillate. The amplitude of the oscillations of the particles in the ‘thin’ layer of the fluid can be evaluated using the effective value of static displacement u_{0ef}:

$$u_{0ef} = \frac{4 f_0 \sin(\pi y / 2h)}{\pi \omega_1^2} \tag{5.49}$$

However, the equation (3.49) does not take into account the contribution of the harmonics with the numbers higher than 1. More accurate is the expression obtained from (5.37) taking into account that $\omega \ll \omega_1$ and the series $\sum_{m=1}^{\infty} \cdot \frac{\sin[(2m-1)\pi y / 2h]}{(2m-1)^3}$ converges in the range $0 \le y \le 2$ to the function $(\pi^3 y/16h)(1 - y/2h)$.

This expression has the form

$$u_{0ef} = \pi^2 f_0 y(1 - y/2h)/4\omega_1^2 h \qquad (5.50)$$

5.4. Resonance excitation of sound in an unlimited magnetic fluid

Theoretical investigations of the excitation of sound by a travelling magnetic field in the unlimited volume of the fluid were carried out by V.G Bashtovoi, B.M. Berkovskii and M.S. Krakov [19, 20, 187].

In the fluid, being magnetised and situated in a heterogeneous magnetic field, the pressure is greater in the areas in which the magnetic field is stronger. In the stationary fluid the distribution of pressure is determined by the equation $\nabla p = \mu_0 M \nabla H$. If the strength of the field is the periodic function of the spatial coordinates, then the periodic distribution of pressure forms in the fluid. If, at the same time, the field periodically changes with time, the pressure becomes a periodic function of both the spatial and time coordinates. In the compressed fluid, this leads to the compression periodic in space and time which is in fact nothing else but forced sound waves.

The travelling magnetic field excites the travelling sound wave, and the stationary field (proportional to cos kx cos t) – the standing sound wave. It should be expected that if the characteristics of the driving force (then the role of the driving force in the investigated case is played by the ponderomotive force $\mu_0 M \nabla H$) k and ω coincide with the appropriate characteristics of the free sound waves, their excitation by the magnetic field periodic in space and in time will be resonant.

In fact, the frequency and the phase of the free sound wave at every point coincide with the frequency and phase of the driving force creating favourable conditions for the supply of energy to the system from the source of the external magnetic field. Mathematical description of the investigated phenomenon is carried out in the simplest geometry of an infinite plane-parallel layer with thickness ℓ. The fluid fills the layer, and the travelling field is generated by induction coils, situated at the boundaries of the layer. The axis

OY is perpendicular to the layer, as shown in Fig. 5.7. The special configuration of the magnetic field is proposed

$$B_x = \frac{B_0 shky}{shk\ell}\cos(kx-\omega t), B_y = \frac{B_0 chky}{shk\ell}\sin(kx-\omega t)$$

$$H_x = \mu_0^{-1}B_x, H_y = \mu_0^{-1}B_y \tag{5.51}$$

However, if the magnetic moment of the layer $\mathbf{M} \neq 0$, the configuration of the field is determined from the system of equations

$$\Delta\psi = \operatorname{div}\mathbf{M},\ \Delta\mathbf{A} = -\mu_0 \operatorname{rot}\mathbf{M} \tag{5.52}$$

where $\mathbf{H} = -\nabla\psi$ and $\mathbf{B} = \operatorname{rot}\mathbf{A}$. The system is solved very simply in two cases. If $\mathbf{M}$ = const, the travelling field coincides with (5.51) with the only difference being that we have a constant addition in this case. For the linear dependence $\mathbf{M} = \chi\mathbf{H}$:

$$B_x = \frac{B_0 shky}{shk\ell}\cos(kx-\omega t),\ B_y = \frac{B_0 chky}{shk\ell}\sin(kx-\omega t), \tag{5.53}$$

$$H_x = \frac{B_x}{\mu_0(1+\chi)}, H_y = \frac{B_y}{\mu_0(1+\chi)}, M_x = \frac{\chi B_x}{\mu_0(1+\chi)}, M_y = \frac{\chi B_y}{\mu_0(1+\chi)}$$

Let the speed of displacement of the fluid particles be v; ρ' and p' are the deviations from the equilibrium values of the density ρ and pressure p, in the linear approximation they are linked by the relationship $p' = \frac{\partial p}{\partial \rho}\rho' = c_f^2\rho'$, where c_f is the speed of sound in the magnetic fluid.

The fluid is assumed to be non-conducting and the processes of viscous friction and heat conductivity are ignored. The magnetic equation of state is assumed to be linear: $M = \chi H$. The process is assumed to be adiabatic.

Therefore, the system of ferrohydrodynamic equations has the form:

$$\rho\frac{\partial\mathbf{v}}{\partial t} = -c_f^2\nabla\rho' + \frac{1}{2}\mu_0\chi\nabla H^2,\ \frac{\partial\rho'}{\partial t} + \rho\operatorname{div}\mathbf{v} = 0 \tag{5.54}$$

From the system of equations we obtain in particular the expression for the oscillatory speed:

$$v_x = \frac{(A_H/\rho)\cdot\omega/k}{(\omega/k)^2 - c_f^2}\cos 2(kx-\omega t),\ \ v_y = 0 \tag{5.55}$$

where $A_H = \mu_0\chi \mathrm{H}_a^2/4; H_a = B_0/\mu_0(1+\chi)sh\,k\ell$.

Thus, the travelling magnetic field excites longitudinal sound waves in the infinite layer of the magnetised fluid. The frequency of the excited sound is twice the frequency of the field. The condition of occurrence of resonance is the coincidence of the speed of the free sound waves and the speed of the travelling field.

Let the travelling magnetic field have the constant component H^*, directed along the axis OY. Consequently

$$H^2 = H^{*2} + 2H^* H_a chky \sin(kx - \omega t) + H_a^2 [sh^2 ky + \sin^2 (kx - \omega t)] \quad (5.56)$$

$$v_x = \frac{(A_H / \rho) \cdot \omega / k}{(\omega / k)^2 - c_f^2} \cos 2(k x - \omega t) - \frac{4A_H}{\rho} \frac{H^*}{H_a} \frac{k}{\omega} chky \sin (kx - \omega t) \quad (5.57)$$

$$v_y = \frac{4A_H}{\rho} \frac{H^*}{H_a} \frac{k}{\omega} shky \cos (kx - \omega t) \quad (5.58)$$

The equation for v_y (5.58) shows that in the presence of the constant component in the travelling field not only longitudinal but also transverse oscillations can form in the ideal fluid. Taking into account the viscous forces leads to the system of equations for **v** and ρ' in the form

$$\rho \frac{\partial \mathbf{v}}{\partial t} = -c_f^2 \nabla \rho' + \frac{1}{2} \mu_0 \chi \nabla H^2 + \eta \nabla \mathbf{v} + \left(\eta_v + \frac{\eta_s}{3} \right) \nabla (\nabla \mathbf{v}) \frac{\partial \rho'}{\partial t} + \rho \, div \, \mathbf{v} = 0 \quad (5.59)$$

If again $H^2 = H_a^2 [sh^2 ky + \sin^2 (kx - \omega t)]$, the longitudinal component of the speed wave is determined from the equation:

$$v_x = v_{xa} \cos [2(kx - \omega t) + \varphi_{vx}]$$

where φ_{vx} is the difference of the phases between the travelling magnetic field and the excited sound wave.

In this case

$$v_{xa} = \frac{(A_H / \rho) \cdot \omega / k}{\left\{ \left[c_f^2 - (\omega / k)^2 \right] + 4\omega^2 b^2 \right\}^{1/2}}, \quad \mathrm{tg}\, \varphi_{vx} = \frac{2b\omega}{c_f^2 - (\omega / k)^2} \quad (5.60)$$

where $A_H = \mu_0 \chi H_a^2 / 4, b = [\eta_v + (4/3)\eta_s] / \rho$.

As indicated by equation (5.60), when the speed of the travelling magnetic field becomes similar to the speed of sound, the amplitude of the forced sound oscillations in the fluid starts to increase to a limited extent – resonance forms in the system with the dissipation of elastic energy.

A very characteristic feature is the fact that to excite the sound in the magnetic fluid it is not necessary to have a solid wall carrying out oscillations. It is sufficient to ensure that the travelling magnetic field has the appropriate parameters of the wave vector **k** and the circular frequency of oscillations ω.

5.5. Oscillations of the form of the magnetic fluid droplet

Evidently, the problem of the physical mechanism of electromagnetic excitation of the oscillations of the magnetic fluid active element, whose volume remains unchanged, was discussed for the first time in the study by V.I. Drozdova, Yu.N. Skibin and V.V. Chekanov [231] in which theoretical and experimental investigations were carried out into low-frequency (2–3 Hz) axisymmetric oscillations of the spherical droplet of the magnetic fluid suspended in the non-magnetic liquid medium in the magnetic field. The theoretical model of elasticity of the oscillatory system, proposed by the above authors, takes into account the capillary forces and the forces of the ponderomotive effect of the magnetic field.

The magnetic fluid droplet, suspended in the non-magnetic liquid medium, has an additional degrees of freedom, associated with the deformation of the shape of the magnetised droplet. The oscillations are accompanied by the perturbation of the internal magnetic fields. When the droplet is deformed (elongated) to a greater extent than in the equilibrium position, the strength of the internal field increases as a result of the reduction of the demagnetising factor and this should result in even greater deformation. In the droplet, deformed to a lesser degree than in the equilibrium position, the internal field becomes weaker as a result of the increase of the intensity of the demagnetising field which should result in a further reduction of the degree of deformation. In contrast to the capillary forces, which always act in the direction of restoration of the spherical form, the ponderomotive force of the magnetic field act in the opposite direction. As a result, the elasticity of the oscillatory system, determined in the absence of the magnetic field by the forces of surface tension of the fluid, decreases when the field is activated, and the axial symmetry, directed along the magnetic field, appears in deformation.

At $H = 0$, the frequencies of the natural oscillations of the magnetic fluid droplet are:

$$\omega_0\big|_{H=0} = \sqrt{\frac{\sigma_0 \ell(\ell-1)(\ell+2)(\ell+1)}{R^3[\rho(\ell+1)+\rho_1\ell]}}$$

where σ_0 is the surface tension coefficient; ρ and ρ_1 are the density of the magnetic fluid and the non-magnetic liquid medium, respectively; $\ell = 1, 2, 3, \ldots$.

At $H \neq 0$ the frequency of the natural oscillations of the magnetic fluid droplet is determined from the following expression:

$$\omega_0 = \sqrt{\omega_0^2\big|_{H=0} - \frac{\mu_0(\mu_i - \mu_e)H^2\ell(\ell+1)}{R^2[1+(\mu_i/\mu_e - 1)N_x][\rho(\ell+1)+\rho_1\ell]}}$$

where μ_i and μ_e are the magnetic permittivities of the substance inside the droplet of the magnetic fluid and outside the droplet; N_x is the demagnetising factor.

The investigations were carried out on the droplets of the magnetic fluid, immersed in an aqueous solution of calcium chloride with the density of the magnetic fluid. The droplets were transferred from the equilibrium position using a homogeneous magnetic field, generated by a system of Helmholtz coils. Under the effect of the field the droplets became ellipsoidal. After switching off the external magnetic field, the droplets carried out attenuating oscillations. The period of the free oscillations of the magnetic fluid droplet with a radius of 2.65 mm in the homogeneous magnetic field H = 1.12 kA/m was 0.24 s, whereas in the absence of the field it was 0.215 s.

In the study [31], published by Yu.K. Bratukhin and A.V. Lebedev, it was shown that the presence of viscosity results in a reduction of the resonant frequency of the oscillations of the droplet determined by the increase of the 'effective mass', i.e., the appearance of the attached mass.

5.6. Oscillations of the magnetic fluid chain

The magnetic field, modulated in space, can be used to produce the 'fluid chain' (FC) system [229]. Here, the FC have the form of fluid droplets, and the connecting elements are elastic gas cavities.

Figure 5.8 shows the model of such a system in the form of a discontinuous fluid column enclosed in an absolutely rigid cylinder shell with a constant cross-section S, b is the length of the fluid member, k is the thickness of the gas cavity, d is the identity period, is the number of the link.

The elastic properties of such a system were investigated in [229]. It was assumed that only the longitudinal sound wave (zero mode) propagates in the chain, the viscous friction and heat exchange purposes are not present, and the fluid is incompressible.

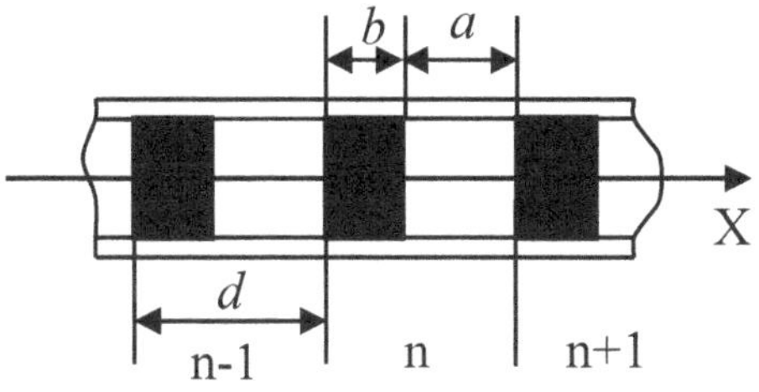

Fig. 5.8. Model of the magnetic fluid chain.

In the absence of heat exchange in the system, the process of propagation of the elastic waves is adiabatic. The displacement of the liquid members from the equilibrium position are governed by the harmonic law, and their values are denoted as U_{n-1}, U_n, U_{n+1}.

The excess pressure, applied to the n-th fluid member, is determined from the equation:

$$\Delta p_n = p_0 + \delta p_n - p_0 - \delta p_{n+1} = \delta p_n - \delta p_{n+1}$$

where p_0 is the equilibrium pressure in the gas cavity; δp_n and δp_{n+1} is the pressure on the fluid droplet from the left and right.

The force, accelerating the n-th member is:

$$F_n = S \cdot \Delta p_n$$

Taking into account the adiabatic nature of the process, we can write the Poisson equation of state for the gas cavity:

$$p(Sa)^\gamma = \text{const},$$

where γ is the Poisson coefficient.

Consequently,

$$p = \frac{\text{const}}{a^\gamma} \text{ and } \delta p = -\frac{p_0\gamma}{a} \cdot \delta a$$

Therefore, for the left and right gas cavities in relation to the n-th member, we have

$$\delta p_n = -\frac{p_0\gamma}{a} \cdot (U_n - U_{n-1}), \delta p_{n+1} = -\frac{p_0\gamma}{a} \cdot (U_{n-1} - U_n)$$

Consequently,

$$\Delta p_n = -\frac{p_0\gamma}{a}(U_n - U_{n-1} - U_{n+1} + U_n) = \frac{p_0\gamma}{a}(U_{n-1} + U_{n+1} - 2U_n)$$

Denoting $\frac{p_0\gamma S}{a} \equiv k_g$ – the coefficient of the quasi-elastic force and consequently

$$F_n = k_g(U_{n-1} + U_{n+1} - 2U_n)$$

In this case, in projection on the axis X, the second Newton law is written in the following form:

$$\rho_f Sb \frac{d^2U_n}{dt^2} = k_g(U_{n-1} + U_{n+1} - 2U_n) \tag{5.61}$$

The differential equation (5.61) coincides with the well-known equation of motion of the one-dimensional chain of atoms [63].

The soundwave, propagating along the axis X, is flat and, therefore, the displacement of the members is governed by the equation

$$U = U_0 \cdot \cos(\omega \cdot t - k_w x)$$

where k_w is the wave number; ω is the circular frequency.

Let it be that at $X = 0$ there is a term characterised by the zero delay in phase; consequently, the second link ($n = 1$) has the phase delay q, the third link ($n = 2$) – $2q,\ldots$, n-th – nq.

Therefore, for the displacement U_n from equilibrium of the n-th fluid link, we can write

$$U_n = U_0 \cdot \cos(\omega t - nq) \tag{5.62}$$

Thus, $k_w x = nq$ at $x = nd$ from which $k_w d = q$ and $\lambda = \dfrac{2\pi d}{q}$.

Substituting (5.62) into equation (5.61) and shortening by U_n, we obtain

$$-\rho_f \cdot S \cdot b \cdot \omega^2 = 2 \cdot k_g(\cos q - 1) = -4 \cdot k_g \cdot \sin^2 \frac{q}{2}$$

or

$$\omega = \Omega' \cdot \sin \frac{q}{2} \quad \text{where} \quad \Omega' \equiv 2\sqrt{\frac{k_g}{\rho_f S.b}}$$

If the wavelength is considerably greater than the identity period $\lambda \gg d$, then φ_0 is small and, consequently

$$\omega = \Omega' \frac{\pi d}{\lambda}, \text{ or for the frequency } \nu_0 = \Omega' \frac{d}{2\lambda} \tag{5.63}$$

Denoting the speed of propagation of the perturbation by v, the obtain

$$\mathbf{v}_0 = \lambda \nu_0 = \sqrt{\frac{k_g}{p_f Sb}} d \tag{5.64}$$

Assuming that $b \gg a$, we write

$$v_0 \sqrt{\frac{p_0\gamma}{\rho_f}\frac{d}{a}} \tag{5.65}$$

Using the well-known equation for the speed of sound in the gases $c_g = \sqrt{\frac{p_0\gamma}{\rho_g}}$, the equations (5.64) and (5.65) it can be written in following form:

$$\mathbf{v}_0 = c_g\sqrt{\frac{\rho_g}{\rho_f ab}}d, \quad \mathbf{v}_0 = c_g\sqrt{\frac{\rho_g}{\rho_f}\frac{d}{a}}.$$

Taking into account the values of ρ_g, ρ_f, c_g, and assuming, for example, that $d = 2.5$ cm, $a = 1$ mm, we obtain:

$$\mathbf{v}_0 = 0.18c_g.$$

In a more general case, the length of the soundwave, propagating along the linear liquid chain, can be both considerably higher and comparable with d. Regarding the differential equation (5.61) as the equation of the harmonic oscillators, the equation for the spectrum of frequency can be presented in the following form:

$$\omega_k = \sqrt{\frac{k_g}{\rho_f Sb}} \cdot 2\sin\frac{k_w d}{2}$$

All possible oscillations can be determined, neglecting the wave numbers k_w from the range:

$$-\frac{\pi}{d} < k_w \le \frac{\pi}{d}$$

For the small values of k ($\lambda \gg d$), and taking into account that $\sin\frac{k_w d}{2} \approx \frac{k_w d}{2} = \frac{\pi d}{2}$, we obtain

$$\omega_0 = \sqrt{\frac{k_g}{\rho_f Sb}}\pi d \quad \text{and} \quad \mathbf{v}_0 = \sqrt{\frac{k_g}{\rho_f Sb}}\frac{d}{\lambda}$$

Correspondingly, $\mathbf{v}_0 = \sqrt{\frac{k_g}{\rho_f Sb}}d$ which coincides with equation (5.64)

At high values of k, the speed of the wave does not remain constant. Thus, accepting $k_w d = \frac{\pi}{3}$, $k_w d = \frac{\pi}{2}$ and $k_w d = \pi$, we obtain

$$\mathrm{v}=\frac{3}{\pi}\sqrt{\frac{k_g}{\rho_f Sb}}d;\ \mathrm{v}=\frac{2\sqrt{2}}{\pi}\sqrt{\frac{k_g}{\rho_f Sb}}d \text{ and } \mathrm{v}_\infty=\nu_\infty\lambda=\frac{2}{\pi}\sqrt{\frac{k_g}{\rho_f Sb}}d$$

The FC system (fluid chain) should be characterised by the dispersion of the speed of sound. The speed of the longitudinal wave decreases with the reduction of the wavelength and at $\lambda = 2d$ becomes equal to $\mathrm{v}_0 = \frac{\pi v_\infty}{2}$.

5.7. Magnetic fluid chain with the elasticity of the ponderomotive type

The elastic properties of the ferrosuspension system are influenced also by the magnetoelastic component, determined by the interaction of the magnetic fluid with a source of the magnetic field [230]. The displacement of the phase boundary U is caused by the parallel effect of both these elasticity factors, and the coefficient of the quasi-elastic force of the system k is therefore equal to the sum of the coefficients of the magnetoelastic and gas cavity: $k = k_p + k_g$.

Figure 5.9 shows the model of a chain with magnetic stabilisation of the magnetic fluid links. The chain of the magnetic fluid bridges is enclosed in the absolutely rigid cylinder shell with the constant cross-section S; b is the length of the fluid link; k is the thickness of the gas cavity, d is the identity period.

The magnetic field is produced by a system of permanent magnets spaced from each other by the distance d.

It is assumed only the longitudinal soundwave (zero mode) propagates in the chain, and the processes of viscous friction and heat exchange are ignored, and the fluid is assumed to be incompressible.

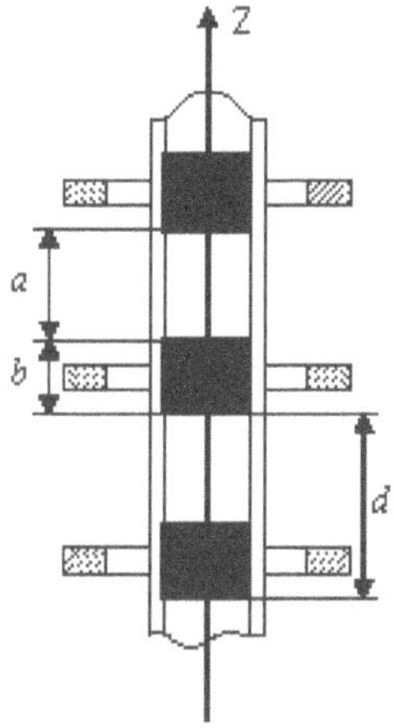

Fig. 5.9. The model of the chain with magnetic stabilisation.

In this case, to the forces, acting on the droplet with the number n on the side of air cavities

$$\left(S\frac{p_0\gamma}{a}(U_{n+1}+U_{n-1}-2U_n) \right)$$

we add the ponderomotive force, acting on the droplet from the side of the magnetic field during its displacement from the equilibrium position.

Under the condition of coincidence of the equilibrium position with the plane of symmetry of the magnet, this force is equal to

$$f_m = -2\mu_0 SM_z\left(\frac{\partial H_z}{\partial z}+\frac{\partial M_z}{\partial z}\right)U_n$$

As a result, the second Newton law for this droplet has the following form:

$$\rho_f Sb\frac{d^2U_n}{dt^2} = S\frac{p_0\gamma}{a}(U_{n+1}+U_{n-1}-2U_n)-2\mu_0 SM_z\left(\frac{\partial H_z}{\partial z}+\frac{\partial M_z}{\partial z}\right)U_n$$

Introducing the notations $\chi' = \rho_g c^2/\rho_f ab$, and also

$$\omega_m = \sqrt{\frac{2\mu_0 M_z}{\rho_f b}\left(\frac{\partial H_z}{\partial z}+\frac{\partial M_z}{\partial z}\right)}$$

for the cyclic frequency of oscillations of the droplet under the effect of the ponderomotive forces, we can write the following equation, describing the propagation of the waves in the chain:

$$\frac{d^2U_n}{dt^2}+\omega_m^2U_n = \chi'(U_{n+1}+U_{n-1}-2U_n) \tag{5.66}$$

Equation (5.66) has the form of the standard equation of the connected interacting oscillators [222]. It is well known that its solution has the form of the travelling wave:

$$U_n = A\cdot\exp i(\omega t - nk_w d)$$

and the frequency and the wave number are linked by the dispersion equation:

$$\omega^2 = \omega_m^2 + 4\chi'\sin^2\frac{k_w d}{2} \tag{5.67}$$

The analysis of the equation (5.67) shows that in the investigated system with excitation in the system of the waves with frequency ω the waves which propagate are only those whose wavelength is in the 'transparency range' [222]:

$$\omega_m \le \omega \le \sqrt{\omega_m^2 + 4\chi'} \tag{5.68}$$

This magnetic fluid chain operates as a band filter of sound oscillations. Only the perturbations with the frequencies from the 'window' (5.16) propagate over a sufficiently large distance, and others exponentially attenuate with increasing distance from the source. The effect of the oscillatory system is equivalent to the LC band filter.

The elastic properties of the gas and magnetic subsystems are compared on the basis of the parameter ψ:

$$\psi \equiv \frac{4\chi'}{\omega_m^2}$$

Assuming that the magnetic susceptibility of the magnetic fluid is equal to unity, we obtain

$$\psi = \frac{\rho_g c^2}{a\mu_0 MG} \tag{5.69}$$

where ρ_g is the density of the gas; c is the speed of sound in the gas; M is the magnetisation of the fluid; G is the gradient of the strength of the magnetic field; μ_0 is the magnetic constant.

Taking into account our data obtained for the field of the annular magnet, at M = 20 kA/m, $G = 4.5 \cdot 10^6$ A/m^2, ρ_g = 1.29 kg/m^3, e = 340 m/s, a = 0.1 m, and using equation (3.89), we obtain: $\Psi \approx 13$.

When the value of n increases by an order of magnitude, the contributions of the magnetic and gas elasticity become similar to each other. However, if $\Psi \gg 1$, the role of the magnetic elasticity is small, and the magnetic field only maintains the form of the magnetic fluid droplet.

5.8. The mechanism of formation of sound oscillations in an air resonator

The mechanism of ponderomotive excitation of the oscillations is related only indirectly to the case investigated here. Figure 5.8 shows the model of excitation of sound in a resonator. The effect of the ponderomotive force of the heterogeneous field of the circular magnet results in the formation of a magnetic fluid bridge (MF-bridge). When the circular magnet 1 is lifted in the hermetically sealed air cavity 2, a pressure gradient forms in relation to the pressure of air in

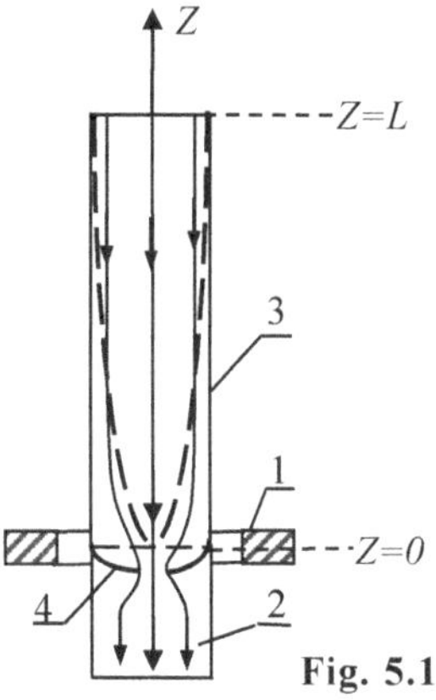

Fig. 5.10. The model of excitation of sound in an air resonator.

the resonator 3. Under the effect of the excess pressure the magnetic fluid bridge 4 breaks and air travels into the orifice resulting in a jump-like increase of pressure.

If the bridge at the moment of closure is stationary, the effect of the aerodynamic impact of the airflow on the obstacle in the upper of the tube results in the formation of a system of sound waves, described by the following expression [71] in accordance with the Rayleigh model

$$\delta p = -\frac{4u_0\rho_g c}{\pi}\sum_{n=1}^{n=\infty}\frac{\cos k_n z}{2n-1}\sin\omega_n t$$

where k_n is the wave number of the n-th harmonics; q_0 is the speed of the airflow in the pipe at the moment of closure of the bridge.

For the main harmonics (quarter-wave tube) we obtain

$$\delta p_1 = -\frac{4}{\pi}u_0\rho_g c\cdot\cos\frac{\pi z}{2L}\sin\frac{\pi c}{2L}t$$

where L is the length of the open section of the tube.

The bridge ($z = 0$) is subjected to the effect of the following pressure

$$\delta p_1 = -\frac{4}{\pi}u_0\rho_g c\sin\frac{\pi c}{2L}t \tag{5.70}$$

The ratio of the oscillatory parameters of the bridge–resonator oscillatory system determines the nature of oscillations of the membrane. In particular, wobbling may occur with a frequency equal exactly to the difference of the frequencies of the initial oscillations, i.e., detuning [225]: $\omega = |\omega_1 - \omega_2|$.

These considerations do not contradict the experimental results obtained on a tube 51 cm long with a diameter 1.35 cm using a colloid with a high concentration of the ferroparticles in the magnetic fluid 2. The oscillograms show distinctive beats at the main frequency ≈170 Hz when the magnetic head approaches the bottom of the tube ($h_0 \approx 1 \div 0.5$ cm). Beats form twice: initially as a result of the fact that the frequency of the sound oscillations of the air column is exceeded and, subsequently, since the frequency of the oscillations of the bridge is exceeded; in the intermediate period, the oscillogram has the form of the 'classic' curve of the damping oscillations. In the displacement of the head in the reverse direction, beats form again but in this case the detuning of the frequencies takes place in the reverse sequence.

The ponderomotive mechanism of electromagnetic excitation of sound operates in all cases in which a heterogeneous magnetic field is superposed on the magnetic fluid, but is it the only and main mechanism in other situations? This question will be answered in the following chapter.

6

Magnetoacoustic effect in the megahertz frequency range

6.1. Experimental equipment for investigating the magnetoelastic effect in the megahertz frequency range

Experimental studies of the special features of the electromagnetic excitation of elastic oscillations in the magnetic fluid in the megahertz frequency range were carried out for the first time in the Laboratory of Magnetoacoustic measurements of the Department of Physics of the Southwest State University (Kursk) [25, 26, 189–193]. These investigations were carried out because of the fact that the experimental data on the excitation of the ultrasonic oscillations in the magnetic fluid at frequencies of 16–26.7 kHz [21, 194) where not sufficient for understanding the physical mechanisms of transformation of the oscillations of the magnetic fluid, in particular, there were no data for verifying the theory of the flat magnetic fluid emitter (MFE) [24].

The excitation of the oscillations in the magnetic fluid in the megahertz frequency range is associated with certain technical difficulties. To obtain an echo signals sufficient for measurements of the amplitude, it is necessary to construct a powerful pulse generator of high-frequency EMF and, the same time, a sensitive receiver–amplifier with high frequencies is required for receiving the useful signal. In this case, the excitation of the oscillations should be carried out at the frequency, resonant for the magnetic fluid transducer. In the investigated frequency range, the resonant thickness of the plane-parallel layer of the magnetic fluid, deposited on the solid surface, equals tenths of a millimetre.

However, there are also advantages in the selected frequency range, i.e., as a result of the application of the pulsed method, the reliability of received information is higher because interference phenomena are eliminated.

The method for examining the acoustic pulses, separated using a delay line from the probing pulse, has been known for a long time [195, 196, 197], and is widely used in measurements of the coefficient of absorption of ultrasound [70]. It was not necessary to work using the principal scheme of this method and, at the same time, it was necessary to solve a number of technical problems associated with specific features of the magnetic fluid. They include: the deposition and removal of the fluid from the substrate, situated in the gap between the poles of the electromagnet; the variation and inspection of the thickness of the layer of the magnetic fluid; removal of the possibility of formation of the acoustic contact between the investigated specimen in the induction coil as a result of 'needle formation' on the surface of the magnetic fluid, placed in the magnetic field with the component normal to the surface; prevention of contamination of the induction coil by the magnetic fluid resulting in the distortion of the field in the gap of the induction coil.

Equipment consists of four functional units, shown in Fig. 6.1. Unit 1 is the source of the alternating magnetic field generating powerful RF pulses, supplied to the induction coil. The functional unit 2 is an acoustic cell designed for placing the investigated specimen and separating the useful acoustic signal. Unit 3 is designed for amplification and indication of the received signal. Unit 4 has the function of the source of the direct magnetic field.

The most important elements of these functional units and their relationship are shown in the flow diagram (Fig. 6.2). The functional unit 1 consists of radio units 8, 9, 10. The units 9 and 10 form the amplifier of power and the master generator. The electronic unit 8 is the resonance amplifier of power. The current pulses travel to the induction coil 1, produced from two inductance coils, connected in

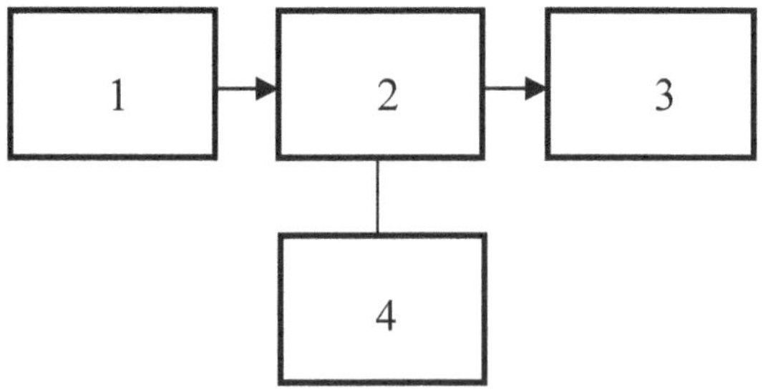

Fig. 6.1. Functional diagram of experimental equipment.

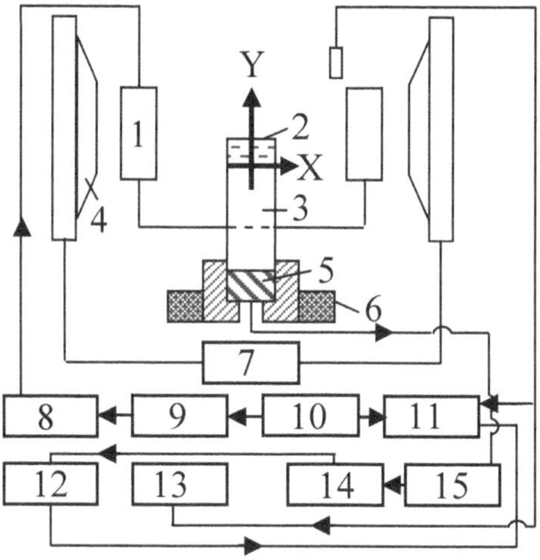

Fig. 6.2. Flow diagram of experimental equipment for investigating electromagnetic excitation of elastic oscillations in the magnetic fluid.

series. The axes of the inductance coils and the pole terminals of the electromagnet are parallel. The operation of the functional unit is inspected using an induction sensor with the pulsed voltmeter 13 connected to it and the two-channel oscilloscope 11. The master generator 10 carries out synchronisation of the start-up of the oscilloscope. The RF pulses, produced by the generator, have the filling frequency of 2–4 MHz and the duration ~25 μs, the repetition frequency is ~4 kHz.

The functional unit 2 includes the glass sheet 3. The end of the sheet is coated with the layer of magnetic fluid 2, with a thickness of 0.15 mm. The sheet is used not only for placing the magnetic fluid and ensuring that the fluid has the form of a flat layer, but has also the function of the acoustic delay line. The opposite ends of the sheet are parallel to each other. The area of the surface of the end for different sheets is (3–7)·30 mm^2. To prevent the interference of the direct and reflected (from the opposite end) sound pulses, the length of the sheets L_n is selected from the condition $L_n > 50c_{g\ell}/\nu$, where $c_{g\ell}$ is the speed of ultrasound n the glass. At $c_{g\ell} = 6000$ m/s, $\nu = 2$ MHz, $L_n = 15$ cm.

The RF pulses from the receiving piezoelement travel to the functional unit 3, containing the amplifier 15, the detector 14, the amplifier of video pulses 12, and the oscilloscope 11. To increase the gain factor and reduce the noise level, the IFA frequency band is reduced to 2 MHz ± 50 kHz. The conditions, ensuring the operating regime of the functional unit 3 in its dynamic range are fulfilled in this case.

The main elements of the functional unit 4 are the laboratory electromagnet 4 and stabilisation device 7.

At the selected geometry and orientation of the inductance coils, it may be assumed that the strength of the magnetic field depends only on x and y (Fig. 6.2):

$$H_x = H_0 + H_{mx}(x, y) \cdot \cos \omega t,\ H_y = H_{my}(x, y) \cdot \cos \omega t \text{ and } H_z = 0,$$

where H_{mx} and H_{my} of the amplitude values of the x-th and y-th components of the field.

For the investigated planar case, the Maxwell equations div **H** = 0 and rot **H** = 0 have the following form

$$\frac{\partial H_y}{\partial x} - \frac{\partial H_x}{\partial y} = 0 \quad \text{and} \quad \frac{\partial H_x}{\partial x} - \frac{\partial H_y}{\partial y} = 0$$

Symmetry considerations show that in the plane XOZ $H_y = 0$ and, therefore, $\frac{\partial H_y}{\partial x} = 0$, and from the first Maxwell equations at $y = 0$ we obtain $\frac{\partial H_{mx}}{\partial y} = 0$. From the second equation for the XOZ plane $\frac{\partial H_{mx}}{\partial x} = 0$.

The error of measurement of the amplitude of the video pulse is determined by the error of the measuring device – oscilloscope– which equals 5–6% for the relative measurements.

Since even small changes in the thickness of the layer of the magnetic fluid (~10%) lead, as shown later, to an almost tenfold change of the oscillation amplitude, to stabilise the thickness of the magnetic fluid layer, the upper surface of the layer is covered with a thin (~0.5 mm) glass sheet.

A cooling chamber, blown with liquid nitrogen vapours, was constructed for investigating the magnetoacoustic effect at low temperatures. Nitrogen is evaporated from a Dewar vessel using an electric heater. The cooling system can be used for measurements from room temperature to –150°C. The temperature of the cooled body is recorded using a differential thermocouple and a potentiometer with the error smaller than 3 K.

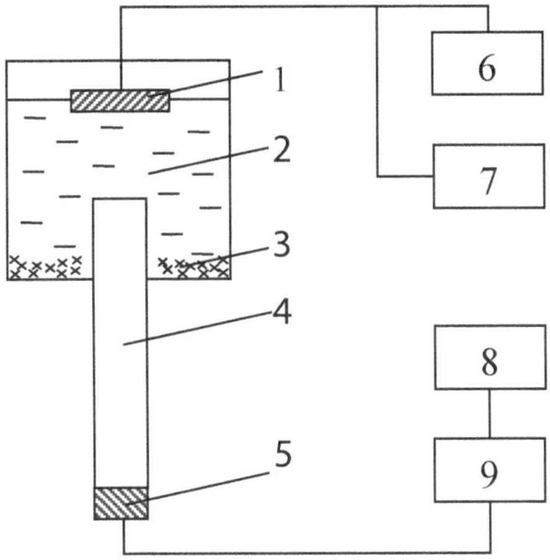

Fig. 6.3. The flow diagram of equipment for absolute measurements of the amplitude of the magnetoacoustic effect.

6.2. The method of absolute measurements of the oscillation amplitude

As a result of the small resonance thickness of the layer of the magnetic fluid, the measurement of the oscillation amplitude of these frequencies cannot be taken using the microphone immersed in the fluid. Therefore, absolute measurements of the amplitude of the elastic oscillations excited by the electromagnetic field in the magnetic fluid are taken by the method of substituting the investigated source of oscillations by a *X*-cut quartz sheet excited by the alternating electrical voltage with the same pulse shape and filling frequency.

The flow diagram of the equipment for determining the absolute value of the amplitude of the excited oscillations is shown in Fig. 6.3. The generator of RF pulses 6 is placed on the quartz plate 1 with the resonance frequency of 2 MHz. One surface of the piezosheet is immersed in water 2, the other surface is free.

After passage through the layer of water and the glass sheet 4, the sound pulse travels to the piezoelement 5 where it is transformed to the RF pulse. The RF pulse is amplified by the receiver 9 and then travels to the oscilloscope 8.

The amplitude of the RF pulse, generated by the generator 6, is controlled with the pulsed voltmeter 7.

Since this method of absolute measurements of the oscillation amplitude is based on the principle of substituting one source of oscillations by another – reference, the elements 4, 5, 8, 9 are transferred into this equipment from the equipment for electromagnetic excitation of the oscillations of the magnetic fluid, and the mode of amplification of the elements 8 and 9 is maintained on the same level.

To eliminate the effect of the signals, deflected from the bottom of the cuvette, the bottom is covered with the layer of glass wool 3. To prevent the formation of the system of standing waves in the gap between the quartz and steel sheets, the thickness of the water layer is selected greater than the length of the sound pulse in water.

The excitation voltage of quartz U, selected in the measurements, is the voltage which produces on the screen of the oscilloscope the same amplitude of the video pulse as from the magnetic fluid emitter, excited by the alternating magnetic fields with the strength H_m. If the intensities of the sound waves, transmitted into the glass sheet from the reference source and from the magnetic fluid emitter, are equal, the static displacement of the surface of the investigated emitter u_{st} can be described by the following equation:

$$u_{st} = 4z_q U d_q/(z_w + z_g),$$

where z_w, z_q, z_g are the wave resistances of water, quartz and the glass, respectively; d_q is the piezomodulus of quartz.

Taking into account that $z_w = 149\cdot10^4$ kg/m^2 s, $z_q = 153\cdot10^5$ kg/m^{2}s, $z_q = 133\cdot10^5$ kg/m^2 s, $d_q = 2.3\cdot10^{-12}$ m/V, and that at $H_m = 1.5$ kA/m $U = 1$ V, we obtain that $u_{st} = 0.01$ nm. Taking into account the expression for the quality factor, derived in section 5.3, from which at $m = 1$ $Q_1 = \pi\rho_c c_c/4\rho c = 7/8$ we obtain the value of the amplitude of the oscillations of the surface of the flat magnetic fluid emitter $u_m \approx 0.1$ nm.

The scatter of the data for the voltage of excitation of quartz reaches ~30% as a result of the detuning of the emitting quartz–glass sheet system and, therefore, the results should be regarded only as an experimental evaluation of the absolute value of the amplitude of the oscillations of the magnetic fluid emitter.

Another method of evaluating the amplitude of the oscillations, excited in the magnetic fluid emitter, used in our experiments, consists of the ‘immediate’ substitution of the magnetic fluid by a quartz sheet [24]. The piezosheet is pressed to the end surface of the glass sheet. Castor oil is used as the acoustic contact medium. In this case, at the equality of the intensities of the transmitted sound we obtain:

$$u_{st} = 2z_q d_q U / \sqrt{z_{ko} z_g}$$

where z_{ko} is the wave resistance of the castor oil.

If we use the simplified measurement method, we obtain the following value for the amplitude of static displacement at $z_{ko} = 127 \cdot 10^4$ kg/m²s and the experimentally determined value $U =$ 0.5 V:

$$u_{st} = 0.008 \text{ nm}$$

In this method of substitution, the scatter of the data reaches 50–100% as a result of the differences in the strength of compressing the quartz sheet to the end surface of the glass sheet.

Thus, the order of magnitude of the amplitude of displacement of the free surface of the plane-parallel layer of the magnetic fluid at $H_m = 1.5$ kA/m equals 0.1 nm.

6.3. Measurement results

Resonance curve

Figure 6.4 shows the resonance curve of the magnetic fluid emitter–dependence of the relative amplitude of the elastic oscillations β_α on the thickness of the fluid layer *h* [192]. The specimen of the magnetic fluid, investigated in this work, was in the form of a colloidal solution of magnetite particles in kerosene, stabilised with oleic acid. Saturation magnetisation of this fluid was 52 kA/m. Its elastic properties were investigated in our study [64] (Table 3.1, specimen No. 6).

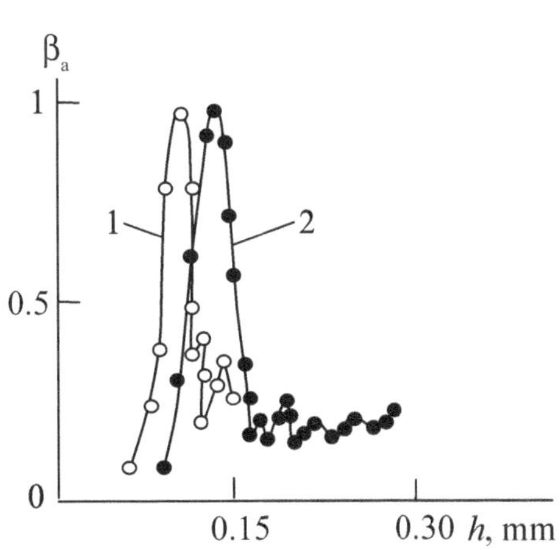

Fig. 6.4. Resonance curve – dependence of the relative amplitude of excited oscillations β_a on the thickness of the layer of the magnetic fluid *h*.

Measurements were taken using glass sheets – waveguides in which the active end surface was 4 × 20 and 7 × 20 mm² in size. The increase of the width of the end surface of the glass sheet resulted in an increase of the radius of curvature of the upper (free in the given experiments) surface of the liquid layer, determined by the surface tension of the magnetic fluid. Since direct measurements of the thickness of the layer of the magnetic fluid (~0.15 mm) could not be taken, the indirect method of determination of h from the equation

$$h = m_f / \rho S$$

was used. Here m_f is the mass of the fluid, deposited on the sheet; S is the area of the end surface.

The thickness of the layer was changed by taking specific portions of the fluid. The mass of the removed portions was determined by weighing (on an analytical balance) a wire bar, used for sampling of the fluid. The error of measurement of the thickness of the layer of the magnetic fluid was determined from the following equation:

$$\frac{\Delta h}{h} = \frac{\Delta m_f}{m_f} + \frac{\Delta \rho}{\rho} + \frac{\Delta x_1}{x_1} + \frac{\Delta x_2}{x_2}$$

where x_1 and x_2 are the linear dimensions of the end of the sheet.

The last two components provide the main contribution, ~0.3%. Therefore, it may be assumed that $\frac{\Delta h}{h} \sim 0.5\%$. The open circles in Fig. 6.3 shows the results of measurements obtained for the 4 × 20 mm² glass sheets, the crosshatched circles – for the sheet with a cross-section of 7 × 20 mm².

The experimental results show that the magnetic fluid emitter, emitting into a solid, is characterised by distinctive resonance. The thickness of the resonant layer was equal to 0.1 mm for the sheet with the smaller area of the end, and 0.12 mm for the sheet with a large area of the end. At the same time, the calculated value of the resonant thickness using equation (5.38) at $m = 1$ was 0.14 mm.

The difference between the calculated and experimental values is situated outside the range of the measurement error h and, evidently, can be explained as follows.

Under the effect of the surface tension forces the edges of the fluid layer are slightly rounded, as shown in Fig. 6.5. Consequently, the value h, obtained from equation (5.38), is smaller than the maximum thickness of the fluid layer h_m in its central part. This

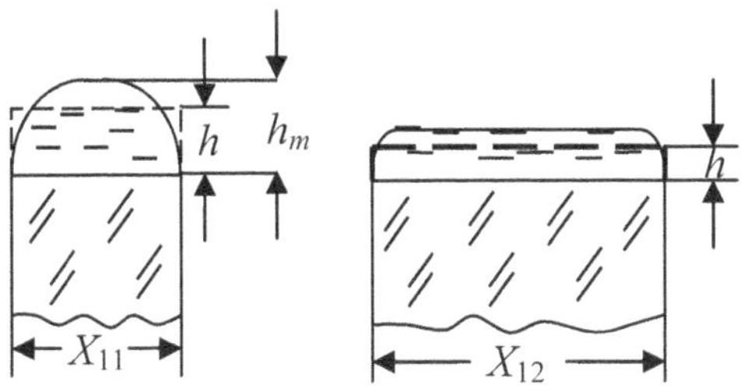

Fig. 6.5. The shape of the surface of the fluid layer for different thickness of the glass sheet x_{11}, x_{12}: $x_{12} > x_{11}$.

difference will increase with the increase in the thickness of the sheet.

Evidently, the resonant oscillations at the main frequency form at $h_m = c/4\nu$, i.e., when the central part of the liquid layer reaches the resonant thickness, consequently, at $h < h_m$.

This assumption is also confirmed by the fact that the value of the resonant thickness, obtained for the thicker sheet, was closer to the calculated value.

The presence of satellites – secondary maxima on the resonance curves – can be the consequence of some non-monochromatic nature of the driving force and distortion of the surface of the layer.

The effect of the magnetic field and temperature

The dependence of the amplitude of forced oscillations on the gradient of the alternating magnetic field was investigated in [26, 189, 190]. Using the dependences H_{mx} (0, y), obtained in the experiments, it was possible to determine the distribution of the gradient of the strength of the magnetic field on the y axis.

Figure 6.6 shows the relative values of the gradient β_G (open circles). The reference value was the maximum gradient, i.e.

$$\beta_G = (\Delta H_{mx} / \Delta y) / (\Delta H_{mx} / \Delta y)_{\max}$$

The active element of the magnetic fluid emitter was a specimen of the magnetic fluid with M_S = 52 kA/m. The dark circles in Fig. 6.6*a* and 6.6*b* show the results of measurement of the relative amplitude β_a for the inductance coils situated at the distance L_d = 7 mm and L_d = 10 mm from each other.

With increase of the distance from the point $y = 0$ along the *OY* axis at $H_m(0,0)$ = const, there are changes not only in the gradient of the field but also in H_{mx}.

This leads to an indeterminacy in the interpretation of the results of measurements of β_a. To remove this indeterminacy, in the displacement of the glass sheet with a specimen it is necessary to

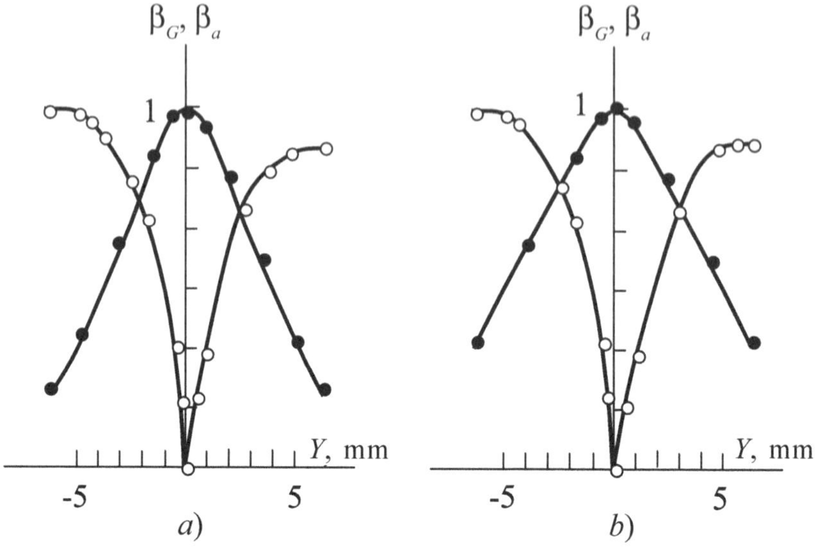

Fig. 6.6. Relative variation of the gradient of the field C (open circles) and the amplitude of the oscillations β_α (dark circles) at L_d = 7 mm (a) and L_d = 10 mm (b).

adjust the amplitude of the alternating magnetic field to ensure the constant value of H_m (0, y).

Similar experiments were also carried out in the case of the magnetic field transverse to the fluid layer. The experimental results show that the variation of the gradient of the alternating magnetic field from 10^3 to 10^5 A/m^2 at a constant amplitude of the field $H_m = 10^3$ A/m in the experimental condition does not lead to any changes in the amplitude of the excited elastic oscillations of the magnetic fluid.

In the experiments carried out to determine the field dependence of the amplitude of the excited elastic oscillations in the magnetic fluid, both surfaces of the magnetic fluid layer were in contact with the solid surface: the lower – with the end surface of the glass sheet, the upper – with the surface of the plexiglass sheet. The width of the gap, formed between the sheets according to the estimates based on the measurement of the mass of the introduced fluid, was ~0.15 mm. When the coating sheet was used, both surfaces of the magnetic fluid layer were parallel. Measurements were taken at a temperature of 25°C.

Experiments were carried out using magnetic fluids based on kerosene and instrument oil MVP, prepared using the procedures described in [39, 40, 198] and [48], respectively. The concentration of the solid phase of the specimens in the given sequence was 9.4% and 13.5%. In addition to this, further three specimens were produced using kerosene by diluting the initial sample. Measurements of the magnetisation of the investigated specimens of the magnetic fluid in the strength range 0–500 kA/m were taken.

Figure 6.7 shows the dependence of the magnetisation of the investigated fluids M on the strength of the magnetic field.

Curve 1 relates to the magnetic fluid based on MVP oil, and the curves 2, 3, 4 and 5 refer to the fluids based on kerosene with the initial concentration of 0.094, 0.077, 0.062 and 0.048.

Figure 6.8 shows the dependence of the relative amplitude of the ultrasound oscillations β_a excited in the magnetic fluid on the strength of the magnetising field H_0 at H_m = 1.6 kA/m. In calculations, the reference value was the maximum amplitude of the oscillations.

Figure 6.9 shows the graphs of the dependences of the relative amplitude of the excited oscillations in the investigated specimens of the magnetic fluid on their magnetisation $\beta_a(M)$. The values of β_a and M were obtained on the basis of the curve of the dependences $\beta_a(H_0)$ and $M\ (H_0)$ for the same values of H_0 [191].

Figure 6.10 shows the dependences of the relative amplitude of the excited oscillations in the investigated specimens of the magnetic

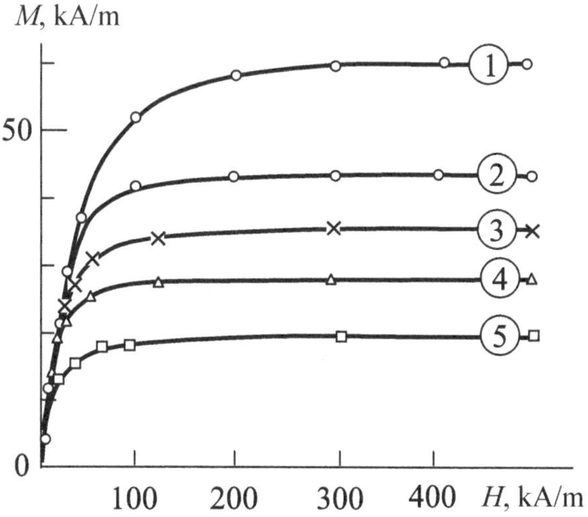

Fig. 6.7. Dependence $M\ (H_0)$ for magnetic fluids based on MVP oil (curve 1) and kerosene (curves 2–5) with the concentration of the magnetic phase of 0.094, 0.077, 0.062 and 0.048, respectively.

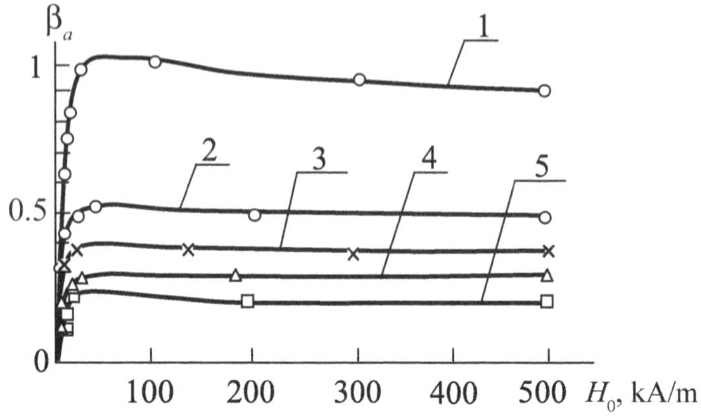

Fig. 6.8. Dependence β_α (H_0) at H_m = 1.6 kA/m for the fluids 1–5.

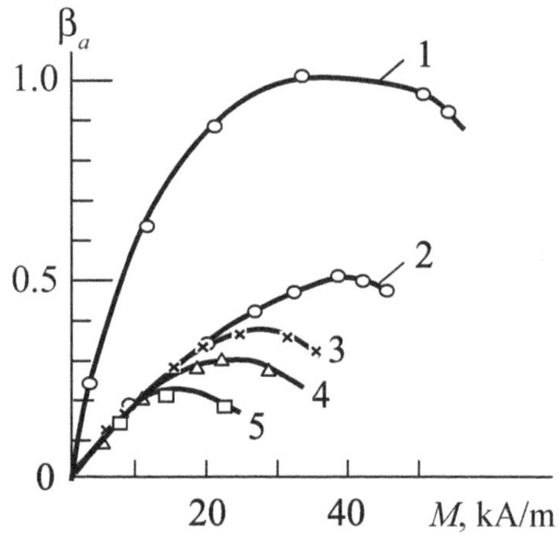

Fig. 6.9. Dependence β_α (M) for the fluids 1–5.

fluid on the amplitude of the alternating magnetic field at H_0 = 25 kA/m for the magnetic fluids 1–5.

Figure 6.11 shows the dependence of the relative amplitude of the elastic oscillations, excited in the fluid 2, on the amplitude of the alternating magnetic field at different strengths of the magnetising field H_0 = 15, 9 and 6 kA/m (the curves 1, 2 and 3, respectively). These relationships are non-linear and show a tendency for the increase of the derivative $\partial\beta_a/\partial H_m$ in the investigated range of the amplitudes of the alternating magnetic field.

Experimental studies of the electromagnetic excitation of the elastic oscillations in the frozen and hardened dispersed system of the magnetic particles were carried out by N.M. Ignatenko *et al* [201].

Figure 6.12 shows the data for a specimen of a magnetic fluid based on magnetite and kerosene with the saturation magnetisation

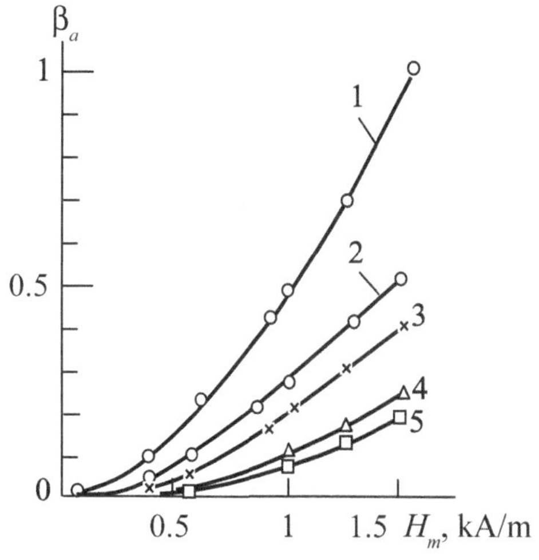

Fig. 6.10. Dependence $\beta_\alpha(H_m)$ at H_0 = 25 kA/m for the fluids 1–5.

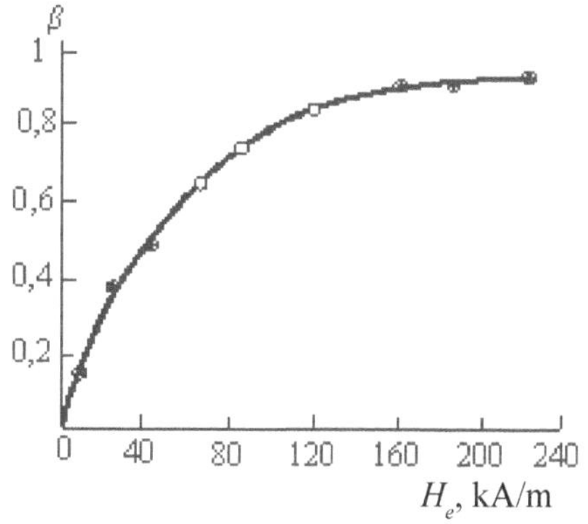

Fig. 6.11. Dependence $\beta_\alpha(H_m)$ for the fluid 2 at H_0 : 1 – 15, 2 – 9, 3 – 6 kA/m.

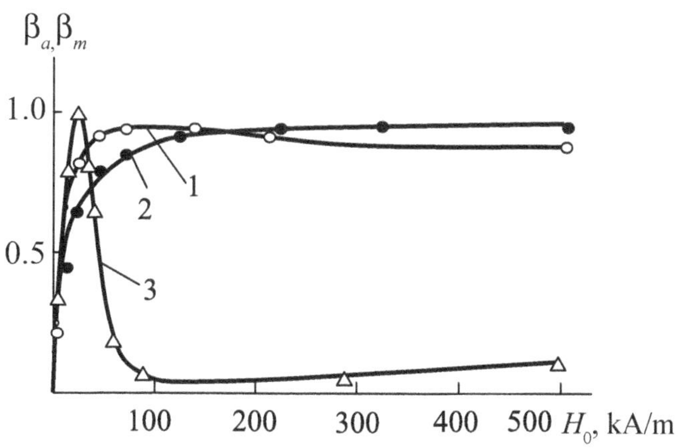

Fig. 6.12. Dependence β_a (H_0) – curve 1 – and the dependence β_a (H_0) – curve 2 at 24°C. Dependence β_a (H_0) – curve 3 at –110°C.

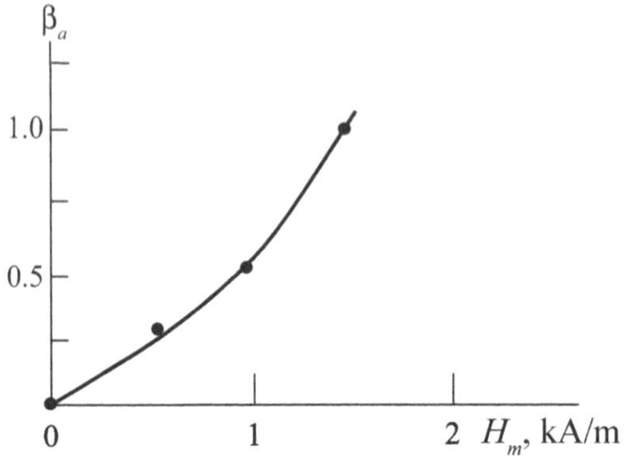

Fig. 6.13. Dependence $\beta_a(H_m)$ for a frozen sample of the magnetic fluid at $T_c = -110°C$.

of 43 kA/m. Here the graph shows the dependences of the relative amplitude of the excited elastic oscillations β_a and the relative magnetisation β_m of the investigated specimen of the magnetic fluid on the strength of the magnetising field H_0. The open circles show the values of the relative amplitude of the oscillations in the magnetic fluid at 24°C, the triangles at –110°C. The full circles show the values of the magnetisation of the magnetic fluid at 24°C, expressed in relative units. Curve 3, relating to the magnetic fluid at –110°C (frozen specimen), is characterised by a sharp maximum and a subsequent rapid fall of the dependence $\beta_a(H_0)$ in the range of variation of H_0 from 0 to ~100 kA/m. The increase of H_0 from 100 kA/m to 500 kA/m is accompanied by a small increase of β_a.

It should be noted that the transition, observed when the temperature varies from the dependence β_a (H_0), corresponding to the curve 1, to the dependence, described by the curve 3, takes place quite rapidly in the temperature range from –90°C to –110°C. The region of solidification of the magnetic fluid is also characterised by the increase of the amplitude of the maximum echo signal by more than an order of magnitude. Figure 6.13 shows the dependence of the relative amplitude of the excited oscillations on the amplitude of the alternating magnetic field H_m in the investigated specimen at $T_c = -110°C$, $H_0 = 40$ kA/m. The dependence of the amplitude of the excited oscillations on the amplitude of the alternating magnetic field H_m for the frozen dispersed system of the ferroparticles, as for the conventional magnetic fluids, is non-linear.

6.4. Failure of the ponderomotive mechanism

The resonance curves, shown in Fig. 6.4, can be used to calculate the quality factor of the magnetic fluid emitter on the basis of the well-known equation [202]

$$Q = \omega^r / 2\Delta\omega^r$$

where ω^r is the resonance frequency; $2\Delta\omega^r$ is the width of the frequency band on the level of 0.7 of the resonance curve.

At $m = 1$ we obtain

$$Q = h_{max} h_{min} / \left[h^r (h_{max} - h_{min}) \right] \approx h^r / (h_{max} - h_{min})$$

Using the curve 2, we determine $Q \approx 7$ which is close to the results obtained by theoretical considerations using equation (5.48) for the quality factor of the flat magnetic fluid emitter [24].

We determine the static displacement of the free surface of the fluid u_S under the effect of the exclusively ponderomotive mechanism. This will be carried out using equation (5.49). In the investigated case $\omega_1 = \pi c / 2h$, $f_0 = \mu_0 M \nabla H_m \rho^{-1}$. Consequently

$$u_s = 16\mu_0 M \nabla H_m h^2 / (\pi^3 \rho c^2)$$

In the experiments [203] the maximum value of G was equal to 10^5 A/m^2, M = 50 kA/m, $h = 0.15 \cdot 10^{-3}$ m, ρ = 1200 kg/m^3, c = 1200 m/s and, consequently, we obtain $u_S \approx 4 \cdot 10^{-14}$ m, which is three orders of magnitude smaller than the experimental value.

The given experimental and calculated data explain why the variation of the gradient of the field from 10^3 to 10^5 A/m^2 (at the constant value of the amplitude of the alternating component H_m) did not result in any increase of the amplitude of the elastic oscillations excited in the magnetic fluid and why there is no correlation between the dependences $\beta_a(y)$ and $\beta_G(y)$, shown in Fig. 6.5. Evidently, the ponderomotive mechanism is not capable of playing the role of the main driving force in the given experimental conditions.

The resultant dependence between the relative amplitude of the oscillations of the particles of the magnetic fluid and its magnetisation (Fig. 6.8), and the dependence of the amplitude of the sound oscillations on the amplitude of the exciting field (Figs. 6.10 and 6.11) also do not agree with the assumption on the controlling role of the ponderomotive mechanism because these relationships, as indicated by the equations (5.31) and (5.37), should be linear.

Therefore, in the experimental conditions concerned with the excitation of the oscillations in the megahertz frequency range the ponderomotive mechanism provides only a small contribution to the generation of oscillations, including the range of technical saturation, where, according to the assumptions made in [1], its intensity is maximum and has no competition. The results of experimental investigation of the magnetoacoustic effect in magnetic colloids in the megahertz frequency range have been taken into account in the development of theoretical models [43, 206–208].

6.5. The mechanism of linear magnetostriction

Two mechanisms, leading to deformation, may operate in the ferromagnetic materials subjected to the electromagnetic effect. They are the magnetoelastic (magnetostriction) effect and the effect associated with the interaction of induced eddy currents with the magnetising field [204, 205]. Since the specific resistivity of the magnetite and the liquid–carrier is large, the contribution of the second mechanism is relatively small. Therefore, the probable mechanisms of magnetoacoustic effect in the magnetic fluid should also include the magnetostriction of the ferroparticles.

In the 'conventional' thick ferromagnetics, magnetostriction shows a distinctive extreme during magnetisation and rapidly decreases already in the magnetic fields with comparatively low strength as a result of the completion of the processes of displacement of the boundaries between the domains and the rotation of the magnetic moments of the domains with respect to the field [41]. The identical phenomenon is also detected in the hardening of the liquid–carrier, although in this case, only the rotation of $\mathbf{m}_*$ can take place, as indicated by the experimental results obtained in the investigation of the magnetoacoustic effect of low temperatures and in the hardened mixture of the ferroparticles with the non-magnetic substance [201] and, in particular, the curve 3 of the field dependence of the relative amplitude of the excited oscillations, shown in Fig. 6.12.

Thus, if the section of the curve $\beta_a(H)$ with the distinctive maximum corresponds to the magnetoelastic effect, the right slowly rising section of this curve evidently corresponds to the interaction of the eddy currents and the magnetising field.

The fact that the dependences $\beta_a(H)$ for the liquid system and for the frozen system (Figs. 6.10, 6.11, 6.13) are non-linear and similar to each other, possibly indicates a degree of agreement in the physical nature of the mechanisms of excitation of the oscillations.

At the same time, the large qualitative difference in the field dependences of the relative amplitude of the oscillations in the liquid (curve 1) and frozen (curve 3) dispersed systems of the ferroparticles

(Fig. 6.12) should be attributed especially to the possibility of the thermal Brownian motion of the ferroparticles in the liquid matrix and to the absence of such a possibility in the solidified system.

In [209, 210] A.A. Rodionov *et al* proposed to include the magnetostriction of the ferroparticles in the number of the probable mechanisms of electromagnetic excitation of the elastic oscillations in the magnetic fluid, and carried out the preliminary evaluation of the efficiency of this mechanism in the given experimental conditions. In [209] attention was given to the linear magnetostriction of the dispersed ferroparticles assuming that the duration of rotational diffusion of the magnetite particles is greater than the period of the ultrasound wave, and that the angle θ_0 between the direction of the magnetising field and the equilibrium direction of the 'mean' dipole moment of the particle depends on the Langevin function according to the equation θ_0 = arc cos $L(\xi)$.

The following equation was obtained for the single-domain and single crystal particle in the approximation of two magnetostriction constants:

$$\Lambda(\alpha_{*1},\alpha_{*2},\alpha_{*3}) = \Lambda_{100} + 3(\Lambda_{100} - \Lambda_{100})(\alpha_{*1}^2\alpha_{*2}^2 + \alpha_{*2}^2\alpha_{*3}^2 + \alpha_{*3}^2\alpha_{*1}^2) \quad (6.1)$$

where α_{*1}, α_{*2}, α_{*3} are the directional cosines of the vector $\mathbf{m}_*$ in relation to the directions [103] of the magnetite crystal. (At 20°C the direction [103] is the direction of the axis of 'easy' magnetisation in magnetite). Therefore, α_{*1} = cos θ_0, α_{*2} = sin θ_0, α_{*3} = 0. When $\mathbf{m}_*$ in the particle deviates from the 'easy' direction, deformation $\Delta\Lambda = 3 = (\Lambda_{111} - \Lambda_{100})\sin^2 2\delta\theta / 4$ takes place in the single crystal. Since $\delta\theta \approx 0$, then

$$\Delta\Lambda = 3(\Lambda_{111} - \Lambda_{100})(\delta\theta)^2 \quad (6.2)$$

If the magnetic field changes in accordance with the law $H = H_0 + H_m \sin \omega t$, then the value of the angle θ_1 is established in the first half of the period of variation of H, and in the second half θ_2. According to the assumption regarding the orientation of the mean magnetic dipole, the angles of deviation of the magnetisation vector from the 'easy' direction will be equal to θ_0–θ_1 and θ_2–θ_0. In this case, equation (6.2) has the following form:

$$\Delta\Lambda = 3(\Lambda_{111} - \Lambda_{100})[\arccos L(H_0) - \arccos L(H_0 - H_m)]^2 \quad (6.3)$$

At $H_m \ll H_0$ equation (6.3) has the form

$$\Delta\Lambda = 3(\Lambda_{111} - \Lambda_{100})(\xi^{-1} - \xi \cdot \mathrm{sh}^{-2}\xi)^2 H_m^2 / (1 - L^2(\xi))H_0^2 \quad (6.4)$$

If $H_0 = 0$, then $\theta_0 = \pi/2$ and equation (6.3) is transformed to the form

$$\Delta\Lambda = 3(\Lambda_{111} - \Lambda_{100})(\mu_0 m_* / 3k_0 T)^2 H_m^2 \qquad (6.5)$$

The equations (6.4) and (6.5) show that there is a quadratic dependence between the deformation of the particles and the amplitude of the alternating magnetic field. This result is in qualitative agreement with the experimental data. At $H_m = 1$ kA/m, $H_0 = 20$ kA/m and the values $\Lambda_{111} = 77.6 \cdot 10^{-6}$, $\Lambda_{100} = -19.5 \cdot 10^{-6}$ [41], characteristic of magnetite, the equation (6.5) yields $\Delta\Lambda \approx 10^{-7}$. At dense packing of the cubic magnetite particles, separated from each other by the double monomolecular layer of the stabiliser, the deformation along the chain of the particles is $0.7 \cdot 10^{-7}$ which coincides in the order of magnitude with the experimental value.

However, as noted in [209], the resultant dependence $\Delta\Lambda$ (H_0) is characterised by a rapid decrease in the range of high values of H_0. From equation (6.4) we obtain $\Delta\Lambda \sim H_0^2$ which contradicts the experimental values. The magnetoacoustic effect is also detected in the magnetising field with the strength of technical saturation, where the well-known magnetostriction mechanism of generation of the oscillations by solid multi-domain ferromagnetics does not operate.

If the models of the dense crystal packing of the particles is not assumed to be applicable only in the case of the magnetic fluids with the maximum concentration, the model of linear magnetostriction, used also in a later study by A.A. Rodionov *et al* [210] for explaining the effect of electromagnetic excitation of the elastic oscillations in the magnetic fluid, requires clarification. The mechanism of appearance of the directional component of the speed of displacement of the particles of the fluid at linear oscillations that are small in comparison with the wavelength of the emitters whose volume remains constant and which carry out random thermal motion, is doubtful.

6.6. The mechanism of bulk magnetostriction

The considerations of this type stimulate search for the solution of the given problem on the basis of the model of bulk magnetostriction. The latter is acceptable from the viewpoint of the possibilities of excitation of elastic oscillations in the fluid because evidently the superposition of the coherent oscillations of the volume of the individual particles should lead to oscillations of the volume of the dispersed system.

The theoretical model of the excitation of the elastic oscillations of the magnetic fluid, based on volume magnetostriction of the dispersed ferroparticles, was proposed by A.O. Tsebers *et al* [206].

In constructing the physical model of excitation of the ultrasound oscillations of the magnetic fluid as a result of bulk magnetostriction of uniaxial ferroparticles, suspended in the fluid, attention is given to the previously described dependence of the amplitude of ultrasound oscillations, excited by the alternating field H_m, on the strength of the constant magnetising field H_0 characterised by a number of typical features, such as the more rapid (in comparison with the increase of magnetisation) initial increase of the amplitude of the oscillations with the field, and its slower reduction with the increase of the strength of the field in the region close to saturation magnetisation. The latter feature is characteristic especially of suspensions in the liquid state. The dried magnetite powder or frozen magnetic fluid is characterised by the rapid decrease of the amplitude of the ultrasound oscillations with the increase of the strength of the magnetising field in the saturation range.

When explaining the above special features of the excitation of the ultrasound oscillations of the magnetic fluid by bulk magnetostriction of the dispersed ferroparticles, theoretical studies were carried out of the effect of the rotational Brownian motion of the particles in the suspension on the dependence of the amplitude of the ultrasound oscillations on the strength of the magnetising field. Taking into account the highly probable nature of this model, we will discuss in detail the article [206].

For bulk magnetostriction ω_k, determined by the variation of the direction of magnetisation in relation to the axis of magnetic anisotropy of the uniaxial particle, it can be assumed [211]

$$\omega_k = \lambda(1-(en)^2)$$

where e is the unit vector along the direction of the magnetisation vector; n is the unit vector along the direction of the axis of magnetic anisotropy.

As a result of the thermal motion of the magnetic moment in the ferroparticles and of the particle itself, the intensity of emission of sound by the suspension is determined by the statistically mean magnetostriction with respect to the ensemble expressed in the form

$$\omega_k = \int P(\mathbf{e},\mathbf{n},t)(\mathbf{en})^2 d^2\mathbf{e} d^2\mathbf{n} \tag{6.6}$$

here $P(\mathbf{e}, \mathbf{n}, t)$ is the combined function of the distribution of the magnetic moment and the anisotropy axis. Consequently, it may be seen that the thermodynamically equilibrium bulk magnetostriction of the ferrosuspensions does not depend on the strength of the magnetic field.

The bulk magnetostriction in the thermodynamically non-equilibrium state can be described on the basis of the kinetic equation of combined Brownian motion of the ferroparticle and its magnetic moment. In the case in which the Néel time of the thermal fluctuations of the magnetic moment τ_N is considerably smaller than the characteristic time of rotational Brownian movement of the ferroparticle τ_B and the period of the alternating field, this phenomenon can be described in the approximation of the locally equilibrium state in which the magnetic moment of the ferroparticles at the given direction of the anisotropy axis is distributed in accordance with the law $(\xi = \mu_0 m * H / k_0 T, \sigma * = K_a V_f / k_0 T)$

$$P(\mathbf{e},\mathbf{n}) = exp(\xi(\mathbf{eh}) + \sigma * (\mathbf{en})^2) / \int d^2\mathbf{e}\, exp(\xi(\mathbf{eh}) + \sigma * (\mathbf{en})^2)$$

In such a case, for the function of the density of probability of the combined distribution of the magnetic moment and the axis of magnetic anisotropy we have $P(\mathbf{e}, \mathbf{n}) = P(e; n)\, P(\mathbf{n})$, where the kinetic equation holds for the function of distribution of the anisotropy axees $(\mathbf{K}_n = \mathbf{n} \times \partial/\partial\mathbf{n})$:

$$\partial P(\mathbf{n})/\partial t = \alpha^{-1} K_n (K_n F_m(\mathbf{n}) P(\mathbf{n})) + k_0 T \alpha^{-1} K_n^2 P(\mathbf{n}); \qquad (6.7)$$

here $F_m(\mathbf{n}) = -k_0 T \ln \int \exp(\xi(\mathbf{eh}) + \sigma_*(\mathbf{en})^2) d^2 e$ is the free energy of the magnetic state of the ferroparticles.

In the thermodynamic equilibrium state

$$P(\mathbf{n}) = P_0(\mathbf{n}) = Q^{-1} \exp(-F_m(\mathbf{n}) / k_0 T)$$

and in the alternating magnetic field $\mathbf{H}_m$,when the non-equilibrium distribution $P(\mathbf{n})$ forms as a result of the finite speed of the relaxation of the anisotropy axes, it is in the form $<n_i n_k>_0 = \int d^2\, \mathbf{n} P_0(\mathbf{n}) n_i n_k$; $(\lambda_{ik} \sim H_m)$

$$P(\mathbf{n}) = P_0(\mathbf{n})(1 + \lambda_{ik}(n_i n_k - < n_i n_k >_0)) \qquad (6.8)$$

Substituting (6.8) into (6.7), multiplying by $n_l n_m$ and integrating with respect to n, with the accuracy to the terms of the first order with respect to the amplitude of the alternating field H_m we obtain the relationship $(< n_l n_m > = \int d^2 \boldsymbol{n} P(\boldsymbol{n}) n_l n_m)$

$$\frac{\partial}{\partial t} < n_l n_m > = -\frac{k_0 T}{\alpha}\left(2\lambda_{ik} < n_k n_m >_0 + 2\lambda_{mk} < n_k n_l >_0 - 4 < n_l n_m n_i n_k >_0 \lambda_{ik}\right) \tag{6.9}$$

The relationship between λ_{lm} and $<n_l n_m>$ follows from expansion (3.77) $(\xi_m = \mu_0 m_* H_m / k_0 T)$

$$< n_l n_m > = < n_l n_m >_0 + d < n_l n_m >_0 / d\xi \xi_m + \lambda_{ik}(< n_l n_m n_i n_k >_0 - < n_l n_m >_0 < n_i n_k >_0) \tag{6.10}$$

In this case, the bulk magnetostriction is proportional

$$\omega_k = \int P(\mathbf{e},\mathbf{n}) P_0(\mathbf{n}) e_i e_m n_l n_m \lambda_{ik} (n_i n_k - < n_i n_k >_0) d^2\mathbf{e} d^2\mathbf{n}$$

so that, using the expression for the moments ($\mathbf{h} = \mathbf{H}_0/H_0$) we obtain

$$<n_i n_k>_0 = A L_2 h_i h_k + (B + AL/\xi)\delta_{ik},$$
$$<n_i n_k n_l n_m>_0 = (A_1 + 2B_1 L/\xi + C_1 L_2/\xi^2)(\delta_{ik}\delta_{lm} + \delta_{il}\delta_{km} + \delta_{im}\delta_{kl}) + (B_1 L_2 + C_1 L_3/\xi)(h_i h_k \delta_{lm} + h_i h_l \delta_{km} + h_i h_m \delta_{kl} + h_k h_l \delta_{im} + h_k h_m \delta_{il} + h_l h_m \delta_{ik}) + C_1 L_4 h_i h_k h_l h_m.$$

Here

$$A = (3Q'/Q-1)/2,\ B = (1-Q'/Q)/2;$$
$$A_1 = (1-2Q'/Q + Q''/Q)/8,$$
$$B_1 = (6Q'/Q-5Q''/Q-1)/8,$$
$$C_1 = (35Q''/Q-30Q'/Q + 3)/8,$$
$$Q = 4\pi \int_0^1 exp\ \sigma_* x^2 dx$$

where the functions $L_n(\xi)$ are determined on the basis of the recurrent relationships $(L_0 = 1) L_{n-1} - L_{n+1} = (2n + l)L_n / \xi$; and we obtain

$$\omega_k = [2A_1 + 5B_1 + C_1 - A(A+B)] L_2 h_i h_k \lambda_{ik}$$

$h_i h_k \lambda_{ik}$ is determined as a result of convolution of the ratios (6.9) and (6.10) with $h_l h_m$ which gives

$$h_l h_m \lambda_{\lambda m} = \frac{h_l h_m < n_l n_m > -(A + B - 2AL/\xi) + 2A\xi_m d/d\xi(L/\xi)}{2a_1 + 5b_1 + c_1 - AL_2(A + B - 2AL/\xi)}$$

Here

$$a_1 = A_1 + 2B_1 L/\xi + C_1 L_2/\xi^2, b_1 = B_1 L_2 + C_1 L_3/\xi, c_1 = C_1 L_4$$

Consequently, introducing the relaxation time τ_B, we obtain the following relaxation equation $(h_l h_m < n_l n_m > 0 = A + B - 2AL/\xi)$

$$\frac{\partial}{\partial t} h_l h_m < n_l n_m > = -\frac{1}{\tau_B}\left(h_l h_m < n_l n_m > - h_l h_m < n_l n_m >_0 + 2A\xi_m \frac{d}{d\xi}\left(\frac{L}{\xi}\right)\right)$$

where

$$\tau_B = \frac{\alpha(2a_1 + 5b_1 + c_1 - AL_2(A + B - 2AL/\xi))}{4kT(A + B - 2AL/\xi - (2a_1 + 5b_1 + c_1))}$$

From this, for $h_l h_m <n_l n_m>$ in the alternating magnetic field $\xi_m = \xi_{m0} \exp(-i\,\omega t)$ we have

$$h_l h_m <n_l n_m> = h_l h_m <n_l n_m>_0 - (1 - i\,\omega\tau_B)^{-1} 2A\xi_m d/d\xi(L/\xi),$$

and for the bulk magnetostriction we obtain the equation:

$$\omega_k = -\frac{(2A_1 + 5B_1 + C_1 - A(A + B))L_2 i\omega\tau_B A\xi_m d/d\xi(2L/\xi)}{(1 - i\omega\tau_B)(2a_1 + 5b_1 + c_1 - AL_2(A + B - 2AL/\xi))} \quad (6.11)$$

In the case of high frequency ($\omega\tau_B \gg 1$), the dependence of the amplitude of the ultrasound oscillations on the strength of the magnetising field, as shown by the relationship (6.11), is determined by the multiplier

$$\frac{L_2 d/d\xi(L/\xi)}{(2a_1 + 5b_1 + c_1 - AL_2(A + B - 2AL/\xi))} = F(\xi)$$

Dependence $F(\xi)$ for certain values of the parameters of magnetic anisotropy of the ferroparticles σ_* is shown in Fig. 6.14 (curve 1 – $\sigma_* = 1$, the curve 2 – $\sigma_* = 9$, curve 3 – $\sigma_* = 16$). The intermittent line in the figure shows the Langevin magnetisation curve $L(\xi)$.

In the range of small fields for relatively high σ_* ($\sigma_* > 10$), when $\tau_N > \omega^{-1}$, the curves have only the formal meaning because they were obtained assuming the equilibrium with respect to the Néel mechanism.

The data presented in Fig. 6.14 shows the qualitative agreement with the experimental results for the dependence of the amplitude of the ultrasound oscillations on the strength of the magnetising field. Thus, the assumption on the excitation of the ultrasound oscillations of the magnetic fluid by the bulk magnetostriction of the ferroparticles taking into account their thermal rotational motion does not contradict qualitatively the experimental data for the excitation of the oscillations by the homogeneous field.

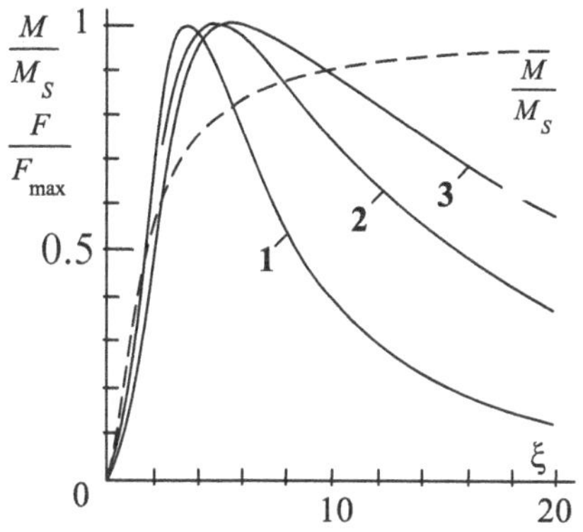

Fig. 6.14. Dependences $L(\xi)$ and $F(\xi)/F(\xi)_{max}$.

However, the problem with the given model is that the bulk magnetostriction for massive ferromagnetics is very small (several orders of magnitude smaller than linear). Unfortunately, there are no reliable data for single-domain nanoparticles. It is possible that additional information for this problem will appear with time. Therefore, it should not be assumed that the discussion of this mechanism has been completed.

6.7. Magnetocalorific effect as a possible mechanism of excitation of elastic oscillations

In the alternating homogeneous magnetic field, the temperature of the magnetic fluid oscillates around the equilibrium value as a result of the magnetocalorific effect [188]. The effect of thermal expansion also causes oscillations in the volume of the fluid. Therefore, it has been assumed that the magnetocalorific effect determines the excitation of the elastic oscillations in the experimental conditions. We determine the contribution of this process to the generation of elastic oscillations.

Let it be that the magnetic field, whose time dependence is given by the equation $H = H_0 + H_m \cdot \cos \omega t$, is directed along the plane-parallel layer of the magnetic fluid. The amplitude of the alternating component of the magnetic field is so small that the condition $H_m \ll H_0$ is fulfilled. In the isothermal application of the magnetic field, for which $H_n = 0$, there is no pressure gradient at the fluid–vacuum interface. However, in the case of the adiabatic process, a pressure 'jump' forms at the magnetic fluid–vacuum interface and, consequently, the fluid is deformed at the free surface of the interface. This will be described. The enthalpy differential $\tilde{H}$ can be described in the form [57]

$$d\tilde{H} = T\cdot dS + V\cdot dp - \mu_0 V(d)\,(\mathbf{M}\cdot d\mathbf{H}) \tag{6.12}$$

Consequently, we obtain

$$\frac{(V-V_0)}{V} = -\mu_0\left\{\frac{\partial}{\partial p}\left[V\int_0^H M\cdot dH\right]_{S,H}\right\}\Big/V$$

The amplitude of deformation of the fluid $\delta V_m/V$ in the application of the field H_m can be determined as follows:

$$\frac{\delta V_m}{V} = -\frac{\mu_0}{V}\frac{\partial}{\partial p}\left\{V\int_0^{H_0+H_m} M\cdot dH - \int_0^{H_0} M\cdot dH\right\}_s =$$

$$= -\frac{\mu_0}{V}\frac{\partial}{\partial p}\left\{V\int_0^{H_0+H_m} M\cdot dH + \int_{H_0}^{0} M\cdot dH\right\}_s =$$

$$= -\frac{\mu_0}{V}\frac{\partial}{\partial p}\left[V\int_{H_0}^{H_0+H_m} M\cdot dH\right]_S = \mu_0\beta_S\left[\int_{H_0}^{H_0+H_m} M\cdot dH - \frac{\partial}{\partial p}\int_{H_0}^{H_0+H_m} M\cdot dH\right]_S$$

Because of the inequality $H_m \ll H_0$ it may be assumed that $M = \text{const} = M_0$,and consequently

$$\frac{\delta V_m}{V} = \mu_0\left\{\beta_S M_0 - \left[\frac{\partial M_0}{\partial p}\right]_S\right\}H_m \tag{6.13}$$

We transfer from p to density ρ, using the linear equation of state

$$\frac{\delta V_m}{V} = \mu_0\beta_S\left\{M_0 - \rho\left[\frac{\partial M_0}{\partial p}\right]_S\right\}H_m \tag{6.14}$$

In the absence of relaxation, the increase of magnetisation at the selected orientation of the vector **H** can be described by

$$\delta M_0 = -(nM_n + \gamma_* M_T)\frac{\partial u}{\partial x}$$

Using the continuity equation in the form $\frac{\partial u}{\partial x} = -\frac{\delta\rho}{\rho}$, we obtain

$\delta M = (nM_n + \gamma_* M_T)\frac{\delta\rho}{\rho}$ and, consequently

$$\left(\frac{\partial M}{\partial \rho}\right)_S = (nM_n + \gamma_* M_T)/\rho \tag{6.15}$$

Substituting the expression (6.15) into equation (6.14) we obtain

$$\frac{\delta V_m}{V} = \mu_0 \beta_S \left\{ M_0 - nM_n - \gamma_* M_T \right\}_S H_m \quad (6.16)$$

If $M_0 \sim n$, which is fulfilled, for example, for the system of the ferroparticles whose magnetisation is described by the Langevin equation, then $nM_n = M_0$ and

$$\frac{\delta V_m}{V} = -\mu_0 \beta_S \gamma_* M_T H_m \quad (6.17)$$

We use the relationship for the magnetocalorific effect available in the magnetism theory [41]

$$dT = -\mu_0 T (M_T)_H \frac{dH}{\rho C_p} \quad (6.18)$$

therefore, the equation (6.17) can be presented in a different form

$$\frac{\delta V_m}{V} = -\mu_0 T M_T q H_m / \rho C_p = q \cdot dT$$

Oscillations of the volume of the fluid with a frequency of variation of the field take place in the alternating magnetic field is a result of the magnetocalorific effect on the thermal expansion typical of the liquid.

The magnetocalorific effect has in this case the function of one of the possible mechanisms of electromagnetic excitation of the elastic oscillations in the magnetic fluid. The dependence of the amplitude of the oscillations on H is determined by the multiplier M_T in equation (6.17). Assuming that magnetisation of the liquid takes place in accordance with the Langevin function, the obtained

$$M_T = (M_S / T)(\xi sh^{-2}\xi - \xi^{-1})$$

and, consequently

$$\frac{\delta V_m}{V} = \mu_0 q H_m M_s D(\xi) / \rho C_p \quad (6.19)$$

where $D(\xi) \equiv \xi^{-1} - \xi / sh^2 \xi$.

The broken line in Fig. 6.15 shows the curve of the dependence of the relative variation of the function $D(\xi)$, denoted as $\beta_\xi(H)$ and plotted assuming that $m_* = 2.5 \cdot 10^{-19}$ A·m^2 and $T = 290$ K. Function

$D(\xi)$ has a single maximum at the point $H = 25.4$ *kA/m*, and its numerical value is 0.348.

To evaluate $\delta V_m/V$, we use the values $D(\xi) = 0.35$, $M_S = 50$ kA/m, $H_m = 1.5$ kA/m, $q = 0.64\cdot10^{-3}\ K^{-1}$, $\rho = 1250$ kg/m^3, $C_p = 2100$ J/kg·K [35], substituting these values into equation (6.19) we obtain $\delta V_m/V = 0.8\cdot10^{-8}$.

However, the experimental value of static deformation $\Delta h/h$ is $0.8\cdot10^{-7}$, which is an order of magnitude higher than the estimate of the deformation as a result of the magnetocalorific effect. The solid line in Fig. 6.15 shows the experimental curve of the dependence of the relative amplitude β_a on the strength of the magnetic field H, plotted using the data for the specimen of the magnetic fluid produced using magnetite and MVP oil. The positions of the maxima of the theoretical and experimental curves are similar to each other but the falling parts of the graphs greatly differ from each other.

Evidently, the magnetocalorific effect in the magnetic fluid based on magnetite can not play the controlling role in the electromagnetic excitation of the elastic oscillations in the megahertz frequency range [188].

In addition to the effect of generation or absorption of heat in magnetic reversal of the fluid, determined by the alignment of the ferromagnetic dipoles in the field, another phenomenon is the generation or absorption of heat caused by the intrinsic magnetocalorific effect of the ferromagnetic phase. Its contribution to the change of the volume of the fluid will be estimated.

The amount of heat, generated in magnetisation of a single particle Q_{T1} is determined using the expression (6.18) which, after transformations, has the following form

$$Q_{T1} = -\mu_0 V_G T(\partial M_G / \partial T)_H \cdot H_m$$

where G is the index of the solid ferromagnetic.

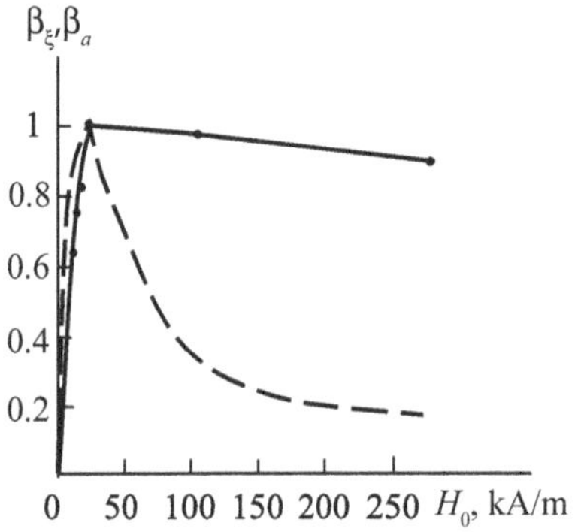

Fig. 6.15. Dependence of the relative variation of β_ξ (broken curve) and of the amplitude of elastic oscillations of the magnetic fluid β_a (solid curve) on the strength of the field H_0.

The volume of the solid phase of the magnetic fluid is $V_G = \varphi V$ and, therefore, the heat, generated in the volume of the dispersed system, can be determined from the expression

$$Q_{T1M} = -\mu_0 \varphi V T (\partial M_G / \partial T)_H \cdot H_m$$

Or, in calculation of the unit mass of the fluid

$$Q_{T1M} = -\mu_0 \varphi T (\partial M_G / \partial T)_H \cdot H_m / \rho \qquad (6.20)$$

The expression (6.20) can be used to determine the increase of the temperature of the fluid

$$\delta T_M = Q_{T11} / C_{pH} = -\mu_0 \varphi T (\partial M_G / \partial T)_H \cdot H_m / \rho C_{pH}$$

where C_{pH} is the specific heat capacity at p = const and H = const.

The relative increase of the volume is

$$\delta V_M / V = -\mu_0 \varphi q T (\partial M_G / \partial T)_H \cdot H_m / \rho C_{pH} = \rho_G C_{pG} \varphi q \cdot \delta T_G / \rho C_{pH}$$

The magnetocalorific effect in the ferromagnetics reaches the maximum value at the Curie point θ_k [41]. If, for example, the magnetic phase is represented by gadolinium (Gd) for which θ_k = 293 [42, 52], we can obtain a large magnetocalorific effect in the vicinity of room temperature. According to the data of K.P. Belov *et al* [42] at H_0 = 200 kA/m, $\Delta T/\Delta H = 0.25 \cdot 10^{-5}$ K·m/A, $\rho = 7.98 \cdot 10^3$ kg/m^3, C_{pG} = 320 J/(kg·K). Consequently, at H_m = 1.5 kA/m, ΔT_G = 3.75 10^{-3} K and $\delta V_m/V = 0.35 \cdot 10^{-6}$ which is almost two orders of magnitude greater than the result provided by the magnetocalorific effect of the alignment of the dipoles in the fluid based on magnetite and kerosene.

6.8. Other possible mechanisms

A characteristic feature determined by the experiments is the fact that the maximum of the field dependence of the relative amplitude of the excited oscillations for different magnetic fluids is obtained at 20–30 kA/m which is possibly associated with the formation of chain-like aggregates. E.E. Bibik [43] explains the magnetoacoustic effect in the megahertz frequency range by attributing the main role to the processes of generation of alternating mechanical stresses as a result of the 'swinging' of the chains around the constant component of the magnetic field under the effect of its alternating component. However, this concept has not been extended to obtaining specific dependences characterising the adequacy of the model.

The probable mechanisms of the magnetoacoustic effect include the dipole–dipole interaction of the ferroparticles, dispersed in the fluid–carrier. The dipole–dipole mechanism of generation of the ultrasound oscillations in the magnetic fluid was investigated by A.I. Lipkin [208] on the basis of the dynamic approach used previously in studying the effect of the acoustic paramagnetic resonance. The region of strong magnetic fields ($\xi \gg 1$) assuming $\mathbf{H}_0 \perp \mathbf{H}_m$ and $\mathbf{H}_0 \gg \mathbf{H}_m$ was discussed. It was shown that the driving force, determined by the dynamic mechanism of interaction of the ferroparticles, in the region close to the magnetic saturation, is inversely proportional to the strength of the magnetising field H_0, and the deformation of the layer of the fluid is $\sim 3 \cdot 10^{-10}$ m. However, the conclusion according to which the condition of excitation of the oscillations $\mathbf{H}_m \| \mathbf{H}_0$ is 'considerably less favourable' than $\mathbf{H}_m \perp \mathbf{H}_0$ is not confirmed by experiments. Nevertheless, to explain the model, it is convenient to consider the primary source.

The number of probable mechanisms, free from some of the previously mentioned shortcomings, may include the mechanism of densening of the medium in the vicinity of the ferroparticles during its rotational oscillations in the alternating magnetic field [260]. The rotational oscillations of the ferroparticles result in periodic changes of the orientation of the elongated rod-like molecules of the shielding shells (according to the available data, the number of these molecules is $\sim 10^3$) and, consequently, the density of their molecular packing periodically changes. In the vicinity of the particle the fluid undergoes periodic stretching and changes of the volume (Fig. 6.16), synchronously in all particles of the dispersed phase.

If the period of the oscillations of the magnetic field is relatively small (shorter than the relaxation time of restoration of the equilibrium orientation of the molecules of the stabiliser), the rotational oscillations of the ferroparticles lead to oscillations of the volume of the fluid as a whole.

In the magnetising magnetic field $\mathbf{H}_0$, the mean statistical (at the given temperature) magnetic moment $\langle \mathbf{m}_* \rangle$ forms the angle $\langle \theta \rangle$ with the direction of the field. In the application of the coaxial alternating

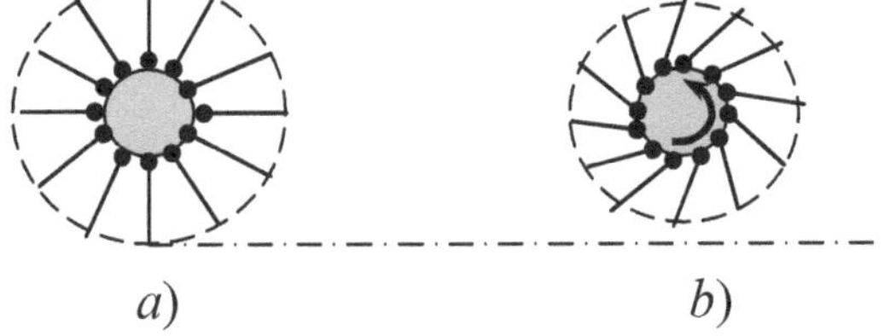

Fig. 6.16. Orientation of the molecules of the stabiliser: *a*) in the equilibrium state; *b*) in the non-equilibrium state.

field $H_m \ll H_0$ the angle $\langle\theta\rangle$ varies in the range from $\langle\theta\rangle_{min}$ to $\langle\theta\rangle_{max}$. The magnetic moment $\langle \mathbf{m}_* \rangle$, 'frozen' in the ferroparticles deviates in a single period of the oscillation of the field deviates by some angle on both sides from the equilibrium value. The rotation of the spherical particle in the viscous fluid–carrier causes oscillations of the volume of the fluid with double frequency.

Because of the non-linearity of the equation of the magnetic state, the deviation $\langle \mathbf{m}_* \rangle$ from the equilibrium direction will not be completely symmetric: it will be greater with the reduction of the strength of the magnetic field and smaller with the increase of the field which in turn leads to the appearance of the harmonics of the elastic oscillations with the frequency of the alternating field.

According to the results of investigations of the acoustic birefringence and absorption of the ultrasound waves in castor oil [63], which consists mostly of the rod-like molecules of the ricinoleic acid, the relaxation time of restoration of the equilibrium orientation is $\tau = 1.5 \cdot 10^{-7}$ s. The molecules of the surfactant in the shielding shell may be characterised, as a result of interaction with the surface of the solid particle, by the long duration of structural rearrangement in comparison with this time and the duration of Brownian rotational relaxation of the ferroparticles. Therefore, it may be expected that the proposed mechanism of excitation of the oscillations will be most efficient at frequencies of $\nu \geq 10^6$ Hz, i.e., in the megahertz frequency range.

Thus, the ponderomotive force is used as the elastic component and the driving force of the oscillatory system when the oscillations of the magnetic fluid elements are accompanied by its flow with constant volume (magnetic fluid–bridge, magnetic fluid–droplet) but if the excitation takes place in the form of oscillations of the bulk of the liquid, the mechanism of structural nature may be an alternative of the ponderomotive mechanism.

7

Magnetic fluid compacting as an oscillatory system

7.1. The magnetic fluid membrane

From the viewpoint of development of the magnetic fluid emitter, operating at the lower boundary of the sonic frequency range, it is interesting to consider the studies [27–30, 214–217] describing the oscillatory systems with controlled magnetic fluid inserts. In particular, in [28–30, 215–218] studies were carried out of the oscillatory system with the magnetic fluid inertia element, 'spring-loaded' by an isolated gas cavity and the elasticity of the ponderomotive type. Such a system can be regarded as the magnetic fluid membrane (MFM), with the property of receiving–emitting sound and electromagnetic pulses.

The MFM is a droplet of a magnetic colloid overlapping the cross-section of the tube with the inner diameter of ~1.5 cm as a result of the stabilising effect of the heterogeneous magnetic field. If the tube contains a bottom plate, the magnetic fluid bridge isolates the air cavity situated below it. In this case, the magnetic fluid operates as an incompressible medium, and the properties of the fluid such as the magnetic controllability of the free surface, fluidity, inertness, become very important [36].

Since the shape of the surface of the magnetic fluid droplet in the conditions of absence of gravitation and capillary forces is determined by the parameters of the magnetic field [58, 59], the forced rupture of the bridge (for example, as a result of the formation of a pressure gradient) is followed by the restoration of its continuity. Consequently, in contrast to the 'conventional' fluid films, the MFM has the capacity for self restoration.

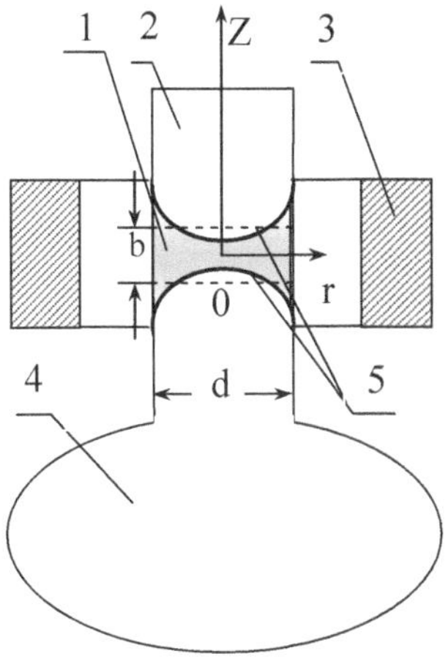

Fig. 7.1. Schematic of the oscillatory system.

The oscillatory system is shown schematically in Fig. 7.1. The droplet of the magnetic fluid 1 overlaps the cross-section of the glass tube 2 under the effect of the ponderomotive force of the heterogeneous magnetic field created by the circular magnet 3, magnetised along the axis. The tube with the diameter *d* is soldered to the glass vessel 4, filled with air (glass tubes of different lengths, sealed on one end, are also used). Both free surfaces of the magnetic fluid bridge 5 have the form of a concave meniscus as a result of the heterogeneity of the magnetic field in the radial direction.

On both sides of the bridge there are conical peaks, determined by the instability of the surface of the magnetic fluid in the transverse field. According to our results, 1–5 approximately identical peaks can form, with the height of the peaks being 1–2 mm (Fig. 7.2).

The experiments were carried out using the magnetic fluid prepared by the standard procedure using magnetite and kerosene. The physical parameters of the magnetic colloid are presented in Table 7.1.

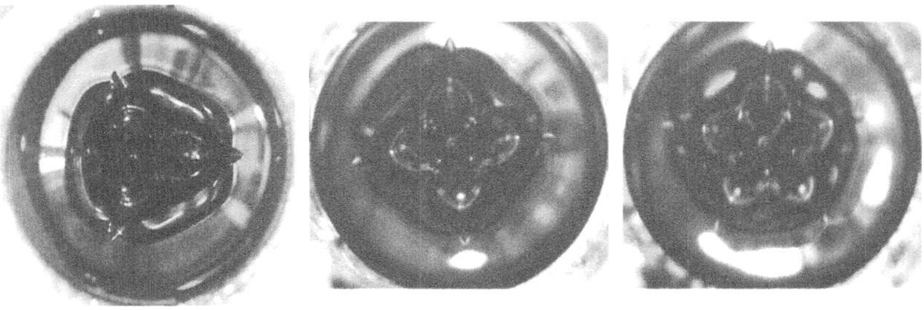

Fig. 7.2. Photographs of the surface of the magnetic fluid bridge.

Table 7.1

ρ, kg/m^3	φ, %	η_s, Pa·s	M_s, kA/m
1499	16.2	$8\cdot10^{-3}$	60±1

To rupture the magnetic fluid bridge, it is sufficient to change the volume of the gas-containing cavity by ~0.1% by moving the magnetic system along the tube or the piston inside the tube.

The pressure gradient, formed during restoration of the continuity of the bridge, displaces the oscillatory system from equilibrium.

The damping of the oscillations is indicated by the induction method using an induction coil placed coaxially inside the circular magnet. An electromagnetic pulse is supplied to the input of the oscilloscope operating in the waiting mode. The oscillograms are transferred to a computer for subsequent processing and analysis. The oscillogram appears on the screen of the monitor and is used to determine the frequency ν and the damping factor of the oscillations β. The error of measurement of ν and β by this method is respectively 5 and 10%, with the fiducial probability of 0.95.

The dots in Fig. 7.3 show the results of the measurements of the oscillation frequency in dependence of the volume of the air cavity V_0.

Table 7.2 shows the results of measurement of the damping factor of the oscillations of the system β.

The experimental results were analysed on the basis of the model of the oscillatory system with concentrated parameters.

It is assumed that both free surfaces of the fluid are flat and the distance between them is b (Fig. 7.1). The fluid is non-viscous,

Table 7.2

ν, Hz	28	29	35	37	40	49	63	190	310
β, s^{-1}	10	10	20	22	26	28	45	70	100

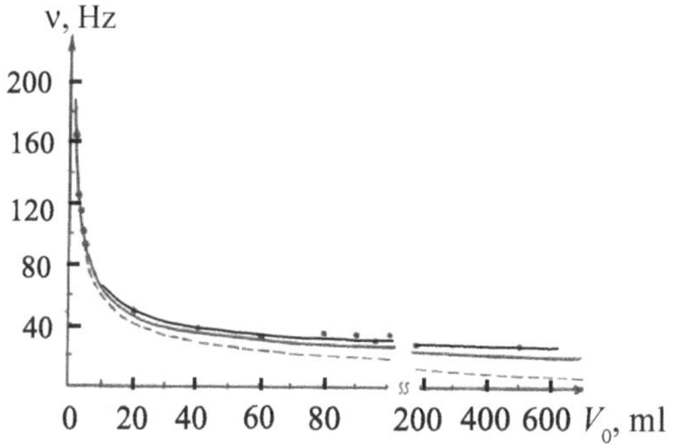

Fig. 7.3. Dependence of the frequency of oscillations on the volume of the air cavity.

incompressible and has no heat conductivity. The oscillations of the gas density are equilibrium.

The elasticity of the oscillatory system is formed by three mechanisms: thermal motion of the gas molecules in the isolated cavity (gas elasticity); interaction of the magnetised magnetic fluid with the heterogeneous magnetic field (ponderomotive mechanism); the mechanism associated with the presence of an interface in the two-phase system (surface tension elasticity).

Therefore, the elasticity coefficient of the system k is determined by the sum

$$k = k_g + k_p + k_\sigma \tag{7.1}$$

were k_g, k_p and k_σ are the coefficients of gas elasticity, ponderomotive elasticity and surface tension elasticity, respectively.

The expression for k_g for the adiabatic process has the form [71]:

$$k_g = \rho_g c^2 \frac{S^2}{V_0} \tag{7.2}$$

where ρ_g is the density of the gas (air in the present case); c is the speed of sound in air; S is the cross-sectional area of the tube; V_0 is the volume of the isolated gas cavity.

If the isolated chamber is part of the cylindrical tube with the height h_0 then

$$k_g = \frac{\gamma \pi d^2 p_0}{4h_0} \tag{7.3}$$

where p_0 is the gas pressure in the cavity in the absence of oscillations; d is the tube diameter; γ is the ratio of the heat capacities.

The elasticity coefficient of surface tension $k\sigma$ will be estimated using the simplified calculation model, shown in Fig. 7.4.

The cross-section of the tube with a radius R is overlapped by the plane-parallel layer of the magnetic fluid. It is assumed that when taking into account the viscosity of the fluid on the side surface of the magnetic fluid bridge, the adhesion condition is satisfied and, consequently, the free surfaces of the bridge periodically bend acquiring the form of the spherical segment.

At small oscillations, the height of the arrow h_a is small, i.e., $h_a \ll R$, and, therefore, the volume of the spherical segment can be expressed by the equation:

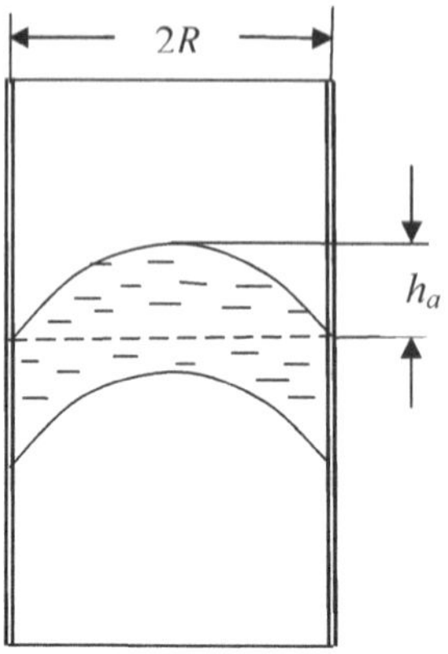

Fig. 7.4. The calculation model.

$$V_a = \frac{1}{6}\pi h_a(3R^2 + h_a^2) \approx \frac{1}{2}\pi h_a R^2$$

Taking into account the equality of the volumes

$$\pi R^2 \Delta Z = \frac{1}{2}\pi h_a R^2$$

where ΔZ is the displacement of the centre of gravity of the magnetic fluid bridge, we obtain:

$$h_a = 2\Delta Z$$

The work, determined by the increase of the area of the free surface of the magnetic fluid bridge (both lower and upper), can be determined by the equation:

$$\Delta A = \sigma \cdot \Delta S = \sigma \cdot 2\left[\pi\left(R^2 + h_a^2\right) - \pi R^2\right] = 2\sigma\pi h_a^2 = 8\sigma\pi(\Delta Z)^2$$

where σ is the surface tension coefficient of the magnetic fluid.

Taking into account the analogy with a spring pendulum, we obtain the equation for the rigidity coefficient, determined by surface tension:

$$k_\sigma = 16\pi\sigma \tag{7.4}$$

Thus, in the proposed model k_σ is independent of the inner diameter of the tube.

We compare the coefficients k_g and k_σ, forming the ratio k_σ/k_g:

$$\frac{k_\sigma}{k_g} = \frac{64\sigma h_0}{\gamma d^2 p_0}$$

Let it be that $\gamma = 1.4$, $d = 1.5$ cm, $p_0 = 10^5$ Pa, $\sigma = 28 \cdot 10^{-3}$ N/m. To evaluate k_g, we use intentionally a large volume of the gas cavity in order to obtain the minimum value of k_g for the experimental conditions: $h_0 = 1$ m. Then $k_\sigma / k_g \cong 0.05$.

Consequently, the contribution of coefficient k_σ to the elasticity of the investigated oscillatory system can be ignored.

The losses of energy for overcoming the forces of surface tension and the appropriate compression of the gas cavity in displacement of the centre of mass of the bridge by dZ will be described by the equations:

$$\Delta A_\sigma = 8\pi\sigma(\Delta Z)^2$$

$$\Delta A_g = p_0 \Delta V = p_0 \pi R^2 \Delta Z$$

The ratio of the energy losses at $R = 10^{-2}$ m, $\Delta Z = 10^{-3}$ m is:

$$\frac{\Delta A_\sigma}{\Delta A_g} = \frac{8\sigma\Delta Z}{p_0 R^2} \cong 2.2 \cdot 10^{-5}$$

Thus, the energy losses for the compression of the cavity are considerably greater than the energy losses for the increment of the free surface of the magnetic fluid droplet.

The increase of pressure in the gas cavity as a result of surface tension Δp_σ and compression of the gas Δp_g is obtained from the expressions:

$$\Delta p_\sigma = \frac{k_\sigma \Delta Z}{\pi R^2} = \frac{16\pi\sigma}{\pi R^2}\Delta Z = \frac{16\sigma}{R^2}\Delta Z, \; \Delta p_g = \frac{\gamma p_0}{h_0}\Delta Z$$

The ratio of the increase of the pressure in the gas cavity $\Delta p_\sigma / \Delta p_g$ has the form:

$$\frac{\Delta p_\sigma}{\Delta p_g} = \frac{16\sigma h_0}{R^2 \gamma p_0}$$

For the same assumptions $\Delta p_\sigma / \Delta p_g \approx 0.03$.

Consequently, in the investigated approximation the contribution of surface tension to the numerical value of pressure in the gas cavity can be ignored [295].

We now examine the mechanism of ponderomotive elasticity. The magnetic field of the circular permanent magnet used in the investigations has been investigated by us in the experiments and theoretically, and the results are presented later in the section 7.6.3. The centre of mass of the magnetic fluid droplet, having the form of a disc having the radius R and thickness b, carries out small oscillations along the Z axis around the equilibrium position at the point $z = 0$ (Fig. 7.1). The axial component of the force in the approximation of the 'weakly magnetic' medium is [29, 36]

$$\Delta f_z = 2\pi\mu_0 \int\limits_{-\frac{b}{2}+\Delta Z}^{\frac{b}{2}+\Delta Z} \int\limits_0^R \left[M_r \frac{\partial H_z}{\partial r} + M_z \frac{\partial H_z}{\partial z} \right] r \cdot dr \cdot dz \tag{7.5}$$

where M_r and M_z is the radial and axial component of the magnetisation of the liquid, respectively.

Taking into account in accordance with the topography of the magnetic field (Fig. 7.27 *a* and 7.27 *b*) that $M_z \gg M_r$, as a result of the symmetry of the magnetic field in relation to the plane $Z = 0$ we obtain for $\Delta \ll b$:

$$\Delta f_z = -2\mu_0 S \left(M_z \frac{\partial H_z}{\partial z} \right)_{z=-\frac{b}{2}} \Delta z$$

Consequently

$$k_p = 2\mu_0 S \left(M_z \frac{\partial H_z}{\partial z} \right)_{z=-\frac{b}{2}} \tag{7.6}$$

However, if the magnetic fluid is magnetised to saturation, then

$$k_p = 2\mu_0 S M_s \left(\frac{\partial H_z}{\partial z} \right)_{z=-\frac{b}{2}} \tag{7.7}$$

It is assumed that in the conditions of this problem as a result of the small value of the capillary constant [219] the surface tension forces can be ignored in comparison with the ponderomotive forces.

The frequency of oscillations taking equations (7.2) and (7.6) into account is determined from the following expression:

$$\nu_m = \frac{1}{2\pi} \sqrt{\frac{\rho_g c^2 S}{\rho b V_0} + \frac{2\mu_0 M_z}{\rho b} \cdot \frac{\partial H_z}{\partial z}} \tag{7.8}$$

If the magnetic fluid is not magnetised to saturation, then the additional perturbation of the magnetic pressure will form in the field at the boundary normal to the surface of the fluid. This perturbation is associated with the rupture of the normal component of the strength of the magnetic field, and the equations (7.6) and (7.8) have the following form

$$k_p = 2\mu_0 S M_z \left(\frac{\partial H_z}{\partial z} + \frac{\partial M_z}{\partial z} \right)_{z=-\frac{b}{2}};$$

$$\nu_m \frac{1}{2\pi} \sqrt{\frac{\rho_g c^2 S}{\rho b V_0} + \frac{2\mu_0 M_z}{\rho b} \cdot \left(\frac{\partial H_z}{\partial z} + \frac{\partial M_z}{\partial z} \right)} \tag{7.9}$$

In the absence of the magnetic field

$$\nu_m = \frac{c}{2\pi} \sqrt{\frac{\rho_g S}{\rho V_0 b}} \tag{7.10}$$

The dependence of the frequency ν_m, calculated from the equation (7.8), on the volume of the air cavity V_0 is shown graphically in Fig. 7.3 by the solid thick line. The following experimental data are used: $M = 45$ kA/m, $\partial H_z / \partial z = 4.6 \cdot 10^6$ A/m², $S = 2 \cdot 10^{-4}$ m², the volume of the magnetic fluid $V = 3$ cm³ and the available numerical values of ρ_g, c. The thin line shows the dependence of the frequency calculated for the same parameters taking into account the variation of the magnetic pressure (7.9) under the condition of the linear dependence of magnetisation on the strength of the magnetic field $M_z = \chi H_z$. The dashed line in Fig. 7.3 shows the curve of the dependence $\nu'(V_0)$, obtained using equation (7.10).

The elasticity of the investigated oscillatory system in the upper range of the given frequency range is determined by the elasticity of the gas cavity, and in the lower range by the elasticity of the ponderomotive type.

In magnetic fluid hermetizers (MFH) (magnetic fluid sealants (MFS)), used widely in engineering [220], the droplet of the magnetic fluid overlaps the gap between the shaft and the sleeve as a result of the retaining action of the magnetic field is concentrated in the region of the gap. Regarding the introduced model of the ponderomotive elasticity as the first approximation, we will use the model to evaluate the resonance frequency ν_p of the MFS. This will be carried out using the expression for the critical pressure Δp_{cr} of the 'single-tooth' sealant

$$\Delta p_{cr} = \mu_0 M_s (H_{\max} - H_{\min}) \tag{7.11}$$

where $H_{\max}$ and $H_{\min}$ is respectively the maximum and minimum strength of the magnetic field on the free surfaces of the magnetic fluid bridge.

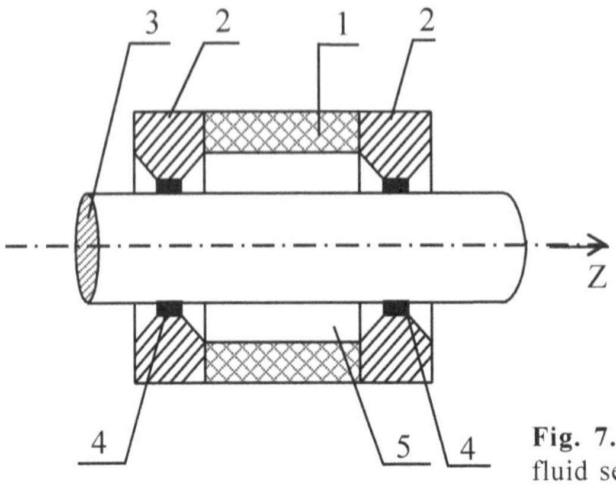

Fig. 7.5. Schematic of the magnetic fluid sealant.

Taking into account only the ponderomotive elasticity, calculated from equation (7.7), we obtain

$$\nu_r = \frac{1}{2\pi b}\sqrt{\frac{2\Delta p_{cr}}{\rho}} \tag{7.12}$$

If Δp_{cr} = 0.75·10^5 Pa, b = 2 mm, ρ = 1.5·10^3 kg/m^3, then $\nu_r \approx$ 800 Hz.

The sealants with the symmetric distribution of the sealing elements are used in most cases [220, 290]. The design of the simplest magnetic fluid sealant of this type is shown schematically in Fig. 7.5.

The pole terminals 2, encircling the shaft of the magnetic material 3, are connected with the circular magnet 1. The magnetic fluid 4 is introduced into the gaps between the pole terminals on the shaft. The resultant cavity 5 is filled with air. This cavity is used as the elastic binding element between the two identical magnetic fluid bridges.

Each magnetic fluid bridge is affected by the following force:

$$\begin{aligned} \rho S_r b\frac{d^2 Z_1}{dt^2} &= -k_g(Z_1 - Z_2) - k_p Z_1; \\ \rho S_r b\frac{d^2 Z_2}{dt^2} &= -k_g(Z_2 - Z_1) - k_p Z_2 \end{aligned} \tag{7.13}$$

where S_r is the area of the annular gap; Z_1 and Z_2 are the displacements of the left and right bridge respectively from the equilibrium position.

The system of equations (7.13) is the well-known system of two connected oscillators [222].

Such an oscillatory system has two normal frequencies:

$$\omega_1 = \sqrt{\frac{k_p}{\rho S_r b}} \text{ and } \omega_2 = \sqrt{\frac{k_p + 2k_g}{\rho S_r b}} \tag{7.14}$$

The inequality $2k_g/k_p \ll 1$ determines the condition of the weak bond. Taking into account the expressions (7.2) and (7.12) it is reduced to the following form

$$V_0 >> \frac{\rho_g c^2 S_r b}{2\Delta p_{cr}} \tag{7.15}$$

Assuming that $S_r = 5 \cdot 10^{-5}$ m^2, we find the restriction for the volume of the closed cavity: $V_0 \gtrsim 300$ mm^3.

When the inequality (7.15) is fulfilled and for the initial conditions $Z_1 = Z_2 = 0$ and $Z' = v_0$, the solutions of the system of equations (7.13) have the following form

$$Z_1 \approx \frac{v_0}{\omega_1} \cdot \cos\Omega t \cdot \sin\omega_1 t ,$$

$$Z_2 \approx \frac{v_0}{\omega_1} \cdot \sin\Omega t \cdot \cos\omega_1 t ,$$

where $\Omega \equiv k_g / (2\rho S_r b\omega_1)$.

In these conditions, the magnetic fluid bridges carry out oscillations with frequency ω_1 the amplitude of which changes in accordance with the harmonic law with the low frequency Ω and this is accompanied by the periodic exchange of energy between them. If the frequency of the external periodic force, determined by for example the eccentricity of the shaft [290], coincides with one of the normal frequencies (7.14), resonance occurs. The amplitude of the oscillations in the investigated dissipation-free approximation increases without bounds.

The presence in the magnetic fluid membranes (MFM) of a number of unique properties is the prerequisite for their application in practice. For example, in electroacoustics use is made of the effect of generation of the electromagnetic response – the damping low-frequency electromagnetic pulse, formed immediately after the rupture of the magnetic fluid bridge, displaced from the region of the maximum magnetic field. Some of the physical, physical–biological and pharmaceutical technologies use the processes of metered supply of the gas into a reactor. In this respect, it is interesting to consider the possibility of using the MFM as a valve capable of transmitting specific portions of gas with appropriate signalling in the form of acoustic and electromagnetic pulses [221, 254]. In some situations, it is preferred to use the MFM as the main element of the pump – the piston [255]. Therefore, it is necessary to describe in greater detail the special features of the formation and functioning of the membranes based on the magnetic fluid.

We consider the results of experimental examination of the effect of the parameters of the magnetic fluid and the conditions of excitation of the oscillations on the elastic (the coefficients of ponderomotive and gas elasticity, the frequency of oscillations, the critical pressure gradient), electrodynamic (the amplitude of the electromagnetic response, the dynamic range, sensitivity) and kinetic (the gas flow rate in the orifice of the bridge, the lifetime of the orifice, the mass of the transmitted portion of gas) properties of the MFM [32].

7.2. Elastic and electrodynamic properties of the magnetic fluid membrane

The expression for the coefficient of ponderomotive elasticity in the studies [28–30] was derived for the bridge – disc model in which the magnetic field is symmetric in relation to the plane of symmetry of the circular magnet, and the equilibrium position of the magnetic fluid bridge coincides with this plane. In a general case, the magnetic field can be non-symmetric, and the centre of mass of the magnetic fluid bridge is displaced from the region of the maximum field. This situation occurs, for example, in magnetic fluid seals at a specific configuration of the poles and the presence of a pressure gradient [220].

The approximation of the 'weakly magnetic' medium, in which the demagnetising field is neglected, is used in [214, 215] in deriving the coefficient of ponderomotive elasticity on the basis of the following simple scheme (Fig. 7.6). Inside the tube 1 with the cross-section S there is the magnetic fluid bridge 2, with the height b. As a result of the pressure gradient in the gas cavities 3, the centre of mass of the bridge is displaced to the point with the coordinate $z = a$. The small displacement of the centre of mass by δz results in the increase of the volume of the bridge at a point with the coordinate

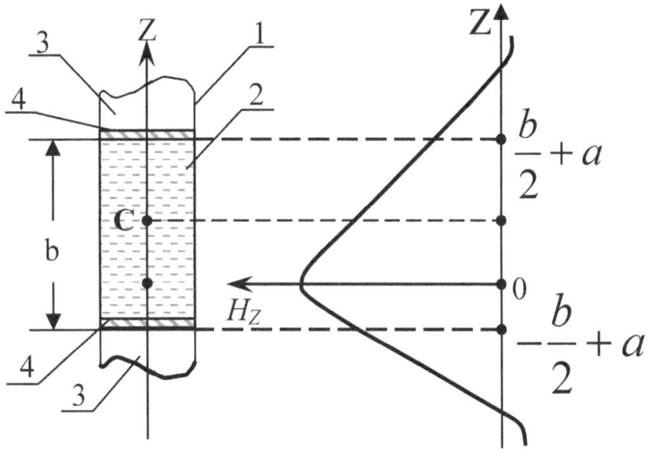

Fig. 7.6. The calculation diagram.

$z=\frac{b}{2}+a$ and $S\cdot\delta z$ and also in a decrease of its volume by the same value at the point $z=-\frac{b}{2}+a$. The appearance (disappearance) of the 'virtual' magnetic fluid disc 4 in the vicinity of the upper and lower open surfaces of the magnetic fluid droplet determines the perturbation of the ponderomotive force

$$\delta f_1=\mu_0 S\left(M_z\frac{\partial H_z}{\partial z}\right)_{z=b/2+a}\cdot\delta z-\mu_0 S\left(M_z\frac{\partial H_z}{\partial z}\right)_{z=-b/2+a}\cdot\delta z \qquad (7.16)$$

At the point $z=\frac{b}{2}+a$, the value $\frac{\partial H_z}{\partial z}<0$, $M_z>0$, and at the point $z=-\frac{b}{2}+a$, the value $\frac{\partial H_z}{\partial z}>0$, $H_z<0$. In addition to this, the normal component of the magnetic field on the upper and lower open surfaces shows a break leading to the formation of the magnetic pressure force. The perturbation of this force can be described by the equation:

$$\delta f_2=-\mu_0 S\left[\left(M_z\frac{\partial M_z}{\partial z}\right)_{z=-\frac{b}{2}+a}-\left(M_z\frac{\partial M_z}{\partial z}\right)_{z=\frac{b}{2}+a}\right]\cdot\delta z$$

or, assuming that $M_z=\chi H$, where χ is the local magnetic susceptibility

$$\delta f_2 = -\mu_0 S\left[\left(\chi M_z \frac{\partial H_z}{\partial z}\right)_{z=-\frac{b}{2}+a} - \left(\chi M_z \frac{\partial H_z}{\partial z}\right)_{z=\frac{b}{2}+a}\right]\cdot\delta z \quad (7.17)$$

Taking into account the equations (7.16) and (7.17), the equation for the driving force of the ponderomotive type is written in the form

$$\delta f_p = \delta f_1 + \delta f_2 =$$

$$= \mu_0 S\left\{\left[(1+\chi)M_z\cdot\frac{\partial H_z}{\partial z}\right]_{z=b/2+a} - \left[(1+\chi)M_z\cdot\frac{\partial H_z}{\partial z}\right]_{z=-b/2+a}\right\}\cdot\delta z$$

Consequently, for the coefficient of ponderomotive elasticity

$$k_p = \mu_0 S\left\{\left[(1+\chi)M_z\cdot\frac{\partial H_z}{\partial z}\right]_{z=-\frac{b}{2}+a} - \left[(1+\chi)M_z\cdot\frac{\partial H_z}{\partial z}\right]_{z=\frac{b}{2}+a}\right\} \quad (7.18)$$

The frequency of the oscillations of the system with only the coefficient k_p taken into account is calculated from the following equation

$$\nu_p = \frac{1}{2\pi}\sqrt{\frac{k_p}{\rho S b}} =$$

$$= \frac{1}{2\pi}\sqrt{\frac{\mu_0}{\rho b}\left\{\left[(1+\chi)M_z\cdot\frac{\partial H_z}{\partial z}\right]_{z=-\frac{b}{2}+a} - \left[(1+\chi)M_z\cdot\frac{\partial H_z}{\partial z}\right]_{z=\frac{b}{2}+a}\right\}} \quad (7.19)$$

Taking into account the coefficient of elasticity of the gas cavity k_g:

$$k = k_g + k_p = \rho_g c^2\cdot\frac{S^2}{V_0} +$$

$$+ \mu_0 S\left\{\left[(1+\chi)M_z\cdot\frac{\partial H_z}{\partial z}\right]_{z=-\frac{b}{2}+a} - \left[(1+\chi)M_z\cdot\frac{\partial H_z}{\partial z}\right]_{z=\frac{b}{2}+a}\right\} \quad (7.20)$$

where ρ_g and c is the density of gas and the speed of sound in gas, respectively; V_0 is the volume of the isolated gas cavity.

$$\nu_p = \frac{1}{2\pi}\sqrt{\frac{\rho_g c^2 S}{\rho b V_0} + \frac{\mu_0}{\rho b}\left\{\left[(1+\chi)M_z\frac{\partial H_z}{\partial z}\right]_{z=-\frac{b}{2}+a} - \left[(1+\chi)M_z\frac{\partial H_z}{\partial z}\right]_{z=\frac{b}{2}+a}\right\}} \quad (7.21)$$

Under the condition of symmetry of the magnetic field in relation to the plane $Z = 0$ it may be assumed that the absolute values are equal:

$$\left(M_z \cdot \frac{\partial H_z}{\partial z}\right)_{z=\frac{b}{2}-a} = \left(M_z \cdot \frac{\partial H_z}{\partial z}\right)_{z=\frac{b}{2}+a}$$

and

$$\left(\chi M_z \cdot \frac{\partial H_z}{\partial z}\right)_{z=-\frac{b}{2}+a} = -\left(\chi M_z \cdot \frac{\partial H_z}{\partial z}\right)_{z=\frac{b}{2}-a}$$

In particular, at $a = 0$ equation (7.21) assumes the form of (7.9).

To determine the width of the dynamic range, experiments were carried out with the MFM [216, 217]. In the experiments, the magnetic fluid bridge overlaps the cross-section of the tube which is the neck of a glass flask with the volume of 0.5 l. The diameter of the neck of the flask is 16.5 mm.

When the flask is lifted to the height Δz above the base and secured in this position by light compression, the bridge is displaced in relation to the equilibrium position by δz. In the linear approximation

$$\delta z = \frac{k_g}{k_g + k_p} \Delta z$$

where k_g and k_p are the coefficients of gas and ponderomotive elasticity.

In rapid return of the flask to the initial position, as a result of inertia the bridge remains displaced in relation to the equilibrium position by δz and this also predetermines the development of the oscillatory process. At the moment of passage of the bridge through the equilibrium position, the maximum value of the EMF ε_m is recorded. The rapid displacement of the flask is the result of the impact of a body with a mass of 125 g on a teflon plug, closing the neck of the flask non-hermetically. The mass of the flask with the plug is $m = 90$ g. The height h' from which the weight was dropped was varied in the range 9–20 mm.

Table 7.3

Specimen	ρ, kg/m³	η_s, Pa·s	M_s, kA/m	χ
MF-1	1294	$3.2 \cdot 10^{-3}$	52±1	6.2
MF-2	1499	$8.1 \cdot 10^{-3}$	60±1	7.5
MF-3	1424	—	43±1	5.0

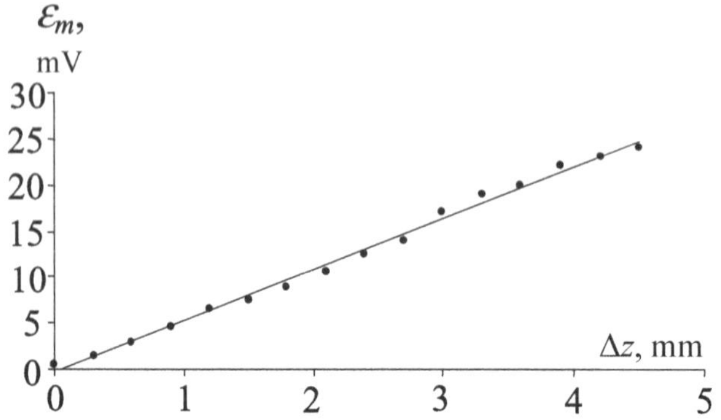

Fig. 7.7. Dependence ε_m (Δz) for MF-1.

Table 7.4

Colloid	h', mm	β, mV/mm	ε_{m0}, mV	Colloid	h', mm	β, mV/mm	ε_{m0}, mV
	9.0	4.6	0.5		10.8	2.5	0.7
MF-1	14.6	4.9	0.5	MF-2	20.3	2.6	0.5
	14.6	5.3	0.5				

Experiments were carried out with the magnetic fluids used in engineering, in the form of a colloidal solution of single-domain magnetite particles Fe_3O_4 in kerosene (MF-1 and MF-2) and in organic silicon (MF-3). The physical parameters of the magnetic colloids are presented in Table 7.3.

The density of the liquid was measured using a pycnometer and the viscosity by the capillary method. The saturation magnetisation of the colloid M_S was determined by the extrapolation of the dependence $M = f(H^{-1})$ to the range of strong magnetic fields. Magnetic susceptibility was determined from the slope of the tangent to the curve $M(H)$ in the initial section.

The fluid was supplied into the neck of the flask for the formation of the magnetic liquid bridge using a syringe, and the weight of the syringe with a portion of the magnetic fluid was weighed on an analytical balance prior to and after charging.

Figure 7.7 shows the dependence $\varepsilon_m(\Delta z)$, obtained for the MSM on the basis of the colloid MF-1. Linear approximation was carried out using MS Excel. For Δz = 3.5 mm for MF-2 and Δz = 4.5 for MF-1, the dependence ε_m (Δ_z) differed from linear.

The sensitivity (to displacement) of the device β is the tangent of the angle of inclination of the approximated curve, and the value of the amplitude of the first oscillation at $\Delta z = 0$ is the initial response ε_{m0}. The presence of the initial response is the consequence of excitation of the elastic oscillations of the walls of the flask at the moment of impact.

Table 7.4 shows the values of β and ε_{m0} obtained in experiments with different height of fall of the weight h'.

Parameter β is almost doubled if the concentrated colloid MF-2 is replaced by the colloid MF-1. It may be assumed that this result is determined by the negative role of the viscous friction forces which result in a reduction of the amplitude of the initial displacement of the bridge from the equilibrium position of the moment of impact. The small increase of β with the height of fall of the weight h', characteristic mainly of the bridge of MF-1, is evidently caused by the inert properties of the bridge.

The experiments carried out to determine the parameter k_p in [223] were carried out using the method of 'the attached cavity'. The principle of the method may be described as follows. Measurements are taken gradually of the frequency of oscillations with a tube open at one end ν_1 and a tube closed at both ends ν_2.

The equivalent mechanical model of the oscillatory system with the convicted cavities is shown in Fig. 7.8. With the tube open at one end, the magnetic fluid bridge with the mass m_f is 'spring-loaded' by the elasticity of the isolated gas cavity k_g and the elasticity of the ponderomotive type k_p. Thus,

$$\nu_1 = \frac{1}{2\pi}\sqrt{\frac{k_g + k_p}{m_f}}, \quad \nu_2 = \frac{1}{2\pi}\sqrt{\frac{k_g + k_p + k_{ad}}{m_f}}.$$

Solving the resultant system of the equations with respect to k_p, we obtain

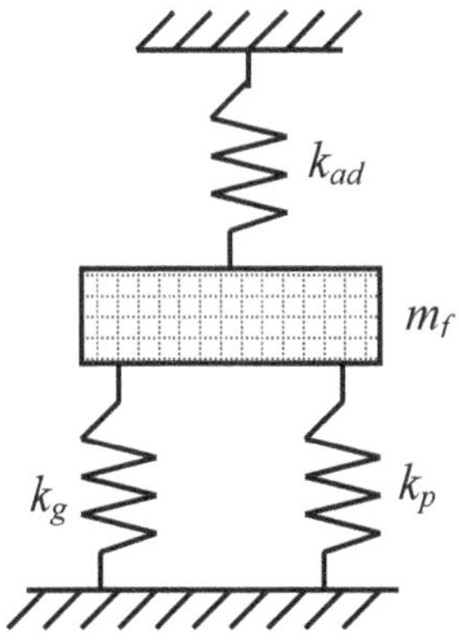

Fig. 7.8. The mechanical model of the oscillatory system with the attached cavity.

$$k_p = \frac{\pi^2 \rho_g c^2 d^4}{16 V_{ad}} \cdot \left[\frac{1}{n^2 - 1} - \frac{V_{ad}}{V_0} \right]$$

where V_{ad} is the volume of the attached cavity, $n \equiv \nu_2 / \nu_1$.

At a relatively high value of the ratio V_0 / V_{ad} the value of n equals several units. Approximately, may be assumed that you are

$$k_p \cong \frac{\pi^2 \rho_g c^2 d^4}{16 V_{ad} (n^2 - 1)} \tag{7.22}$$

If the attached cavity is part of the tube with a constant cross-section, equation (7.22) has the following form

$$k_p \cong \frac{\pi^2 \rho_g c^2 d^2}{4 h_{ad} (n^2 - 1)} \tag{7.23}$$

The error of measurement of k_p by the method of the attached cavity:

$$\frac{\Delta k_p}{k_p} = \frac{\Delta \rho_g}{\rho_g} + \frac{2 \cdot \Delta c}{c} + \frac{2 \cdot \Delta d}{d} + \frac{\Delta h_{ad}}{h_{ad}} + \frac{2n \cdot \Delta n}{n^2 - 1}$$

The largest contribution to the area is provided by the last two terms, and the sum of these terms is in the range 10–15%.

In the experiments, the magnetic fluid bridge is situated in the cylindrical neck of the glass flask. A hermetically sealed plug is used for the formation of the attached cavity. The results of the preliminary measurements of the field dependence of the magnetisation of magnetic colloids gave the data required for calculating $(k_p)_{th}$. For the bridge based on MF-1, if b = 1.68 cm, and M = 34.5 kA/m, $G = 4.6 \cdot 10^6$ A/m^2, $\chi = 0.25$, the calculated value $(k_p)_{th}$ = 100 N/m. The experimental value of $(k_p)_{ex}$, determined from the results of measurements at h_{ad} = 4.64 *cm*, ν_1 = 24 Hz, ν_2 = 86 Hz, was $(k_p)_{ex}$ = 98 N/m. For the bridge based on MF-2 at b = 2 cm, M = 42 kA/m, $G = 4.6 \cdot 10^6$ A/m^2, $\chi = 0.4$, h_{ad} = 6.3 cm, n = 1.9 the value $(k_p)_{th}$ = 136 N/m, and the experimental value $(k_p)_{ex}$ = 137 N/m.

Evidently, in this case, the good agreement between the calculated and measured values of k_p is partially due to the mutual compensation of the errors of determination of the individual parameters.

Equations (7.18) and (7.19) show that the displacement of the bridge because of the symmetry of the magnetic field of the circular magnet in relation to the plane $Z = 0$ (Fig. 7.1) should not lead to any significant changes of magnetic elasticity and the frequency of

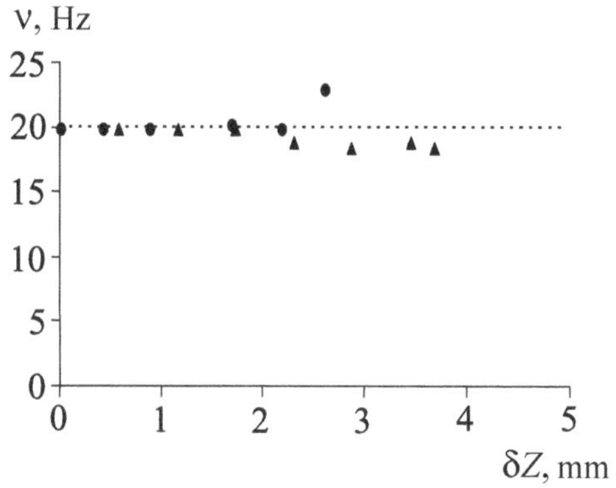

Fig. 7.9. Dependence of the frequency of oscillations ν on the displacement of the magnetic fluid bridge from the region of the maximum field δZ.

oscillations. However, as the free surface of the bridge comes closer to the plane of the maximum field $\partial H_z / \partial z \to 0$, the values of k_p and ν decrease. This position of the bridge of the magnetic fluid sealant is regarded as critical [220] because a further increase of the pressure gradient results in rupture of the bridge.

The experimental modification of these conclusions in the theory of the model was carried out on two colloids of the type of magnetite in kerosene [214]. The magnetic fluid overlapped the cross-section of the neck of the flask. The oscillations excited by a mechanical push in the vertical direction. The displacement of the equilibrium position of the bridge was achieved by lifting the circular magnet rigidly connected with the kinematic section of the cathetometer, to the height Δ*Z*, fixed with the accuracy of 0.01 mm. The experiment temperature was 21°C. The displacement of the bridge δ*z* was calculated taking into account the values of k_p and k_g.

Figure 7.9 shows the results of measurement of the frequency of oscillations ν at different displacements of the centre of mass of the bridge from the equilibrium position δ*Z*. For the first of the specimens of the magnetic fluid we used notations in the form of crosshatched triangles, for the second – in the form of crosshatched circles. Both the dependences have almost the same form of the straight line section, parallel to the abscissa (dotted line). The prediction of the model theory in this part of the experiment is confirmed. However, in both cases during several intervals of 0.5 mm displacements to fracture of the bridge the oscillations acquired distinctive non-linear properties: the oscillograms are initially saw-shaped and then the second harmonic appear, followed by doubling of the oscillation frequency. The conclusion of the model theory concerning the rapid reduction of the frequency of oscillations in the vicinity of the critical position of the magnetic fluid bridge could not be confirmed or rejected in these experiments.

7.3. Non-linear oscillations of a thin magnetic fluid bridge

Because of the large number of applications of magnetic fluid sealants, it is necessary to examine in considerable detail the oscillations of the volumes of the magnetic fluid, overlapping the air channel, in the field of the magnet, sustaining the fluid. The part of the section 7.2, concerned with the experimental study of the dependence of the oscillation frequency of the thin magnetic fluid bridge, displaced from the equilibrium position, shows that at relatively slowly changing oscillation frequencies with high initial displacements from the centre of symmetry of the field, the oscillations acquire the distinctive non-linear form.

Different non-linear modes of the oscillations of the magnetic fluid bridge were discussed in [248]. A thin magnetic fluid bridge, placed in the field of a circular magnet and displaced by δz in relation to the equilibrium position (Fig. 7.6), was investigated. Let it be that the bridge was displaced by Δz. The axial component of the magnetic force is

$$f_z = 2\pi\mu_0 \int_{z_1+\Delta z}^{z_2+\Delta z} M_z \frac{\partial H_z}{\partial z} r dr dz$$

$$f_z = \mu_0 S \int_{z_1}^{z_2} M_z \frac{\partial H_z}{\partial z} dz + \mu_0 S \int_{z_2}^{z_2+\Delta z} M_z \frac{\partial H_z}{\partial z} dz - \mu_0 S \int_{z_1}^{z_1+\Delta z} M_z \frac{\partial H_z}{\partial z} dz \tag{7.24}$$

Thus, the restoring force is equal to:

$$\Delta f_z = \mu_0 S \left(\int_{z_2}^{z_2+\Delta z} M_z \frac{\partial H_z}{\partial z} dz - \int_{z_1}^{z_1+\Delta z} M_z \frac{\partial H_z}{\partial z} dz \right) \tag{7.25}$$

For small displacements equation (7.25) is transformed to the form

$$\Delta f_z = \mu_0 S \left\{ M_z \frac{\partial H}{\partial z}\bigg|_{z_2} - M_z \frac{\partial H}{\partial z}\bigg|_{z_1} \right\} \Delta z \tag{7.26}$$

(indexes z are omitted). If the initial non-symmetry in the position of the magnetic fluid bridge is small, then $z_1 = -z_2 - 2\delta z$ and, consequently

$$M(z_1)\frac{\partial H}{\partial z}\bigg|_{z_1} = M(-z_2 - 2\delta z)\frac{\partial H}{\partial z}\bigg|_{-z_2-2\delta z} =$$

$$= \left(M(-z_2) - 2\left(M(-z_2) - 2\frac{\partial M}{\partial z}\bigg|_{z_2} \delta z \right)\delta z \right) \times \left(\frac{\partial H}{\partial z}\bigg|_{-z_2} - 2\frac{\partial^2 H}{\partial z^2}\bigg|_{z_2} \delta z \right) \tag{7.27}$$

Taking into account the symmetry $H_z(z)$ and $M_z(z)$ in relation to $z = 0$, from (7.27) we obtain:

$$M\frac{\partial H}{\partial z}\bigg|_{z1} = -M\frac{\partial H}{\partial z}\bigg|_{z_2} - 2\left[M\frac{\partial^2 H}{\partial z^2} + 2\frac{\partial M}{\partial z}\frac{\partial H}{\partial z} \right]_{z_2} \delta z \tag{7.28}$$

Substituting (7.28) into (7.26) we obtain the expression for the restoring force

$$\Delta f_z = 2\mu_0 S \left\{ M\frac{\partial H}{\partial z} + \left[M\frac{\partial^2 H}{\partial z^2} + 2\frac{\partial M}{\partial z}\frac{\partial H}{\partial z} \right] \delta z \right\}_{z_2} \Delta z \tag{7.29}$$

Here, the second term in the brackets takes into account the effect of the non-symmetry of the equilibrium position of the bridge.

If the magnetic fluid is not magnetised to saturation, it is also necessary to take into account the variation of magnetic pressure when the bridge is displaced from the equilibrium position

$$\Delta f_z^p = \mu_0 S \left\{ M\frac{\partial M}{\partial z}\bigg|_{z_2} - M\frac{\partial M}{\partial z}\bigg|_{z_1} \right\} \Delta z$$

Using the procedure described above, we obtain

$$\Delta f_z^p = 2\mu_0 S \left\{ M\frac{\partial M}{\partial z} + \left[M\frac{\partial^2 M}{\partial z^2} + 2\frac{\partial M}{\partial z}\frac{\partial M}{\partial z} \right] \delta z \right\}_{z_2} \Delta z \tag{7.30}$$

The complete expression for the force is obtained on the basis of (7.29) and (7.30)

$$\Delta f_m = 2\mu_0 S \left\{ \begin{array}{l} M\left(\dfrac{\partial H}{\partial z} + \dfrac{\partial M}{\partial z} \right) + \\ \left[M\left(\dfrac{\partial^2 H}{\partial z^2} + \dfrac{\partial^2 M}{\partial z^2} \right) + 2\dfrac{\partial M}{\partial z}\left(\dfrac{\partial H}{\partial z} + \dfrac{\partial M}{\partial z} \right) \right] \delta z \end{array} \right\}_{z_2} \Delta z \tag{7.31}$$

In the weakly magnetic approximation $M = \chi H$:

$$\Delta f_m = 2\mu_0 S(1+\chi)\left\{ H\frac{\partial H}{\partial z} + \chi\left[H\frac{\partial^2 H}{\partial z^2} + 2\left(\frac{\partial H}{\partial z}\right)^2 \right]\delta z \right\}_{z_2} \Delta z \quad (7.32)$$

If the magnetic fluid is magnetised to saturation

$$\Delta f_m = 2\mu_0 S\left\{ M\frac{\partial H}{\partial z} + M\frac{\partial^2 H}{\partial z^2}\delta z \right\}_{z_2} \Delta z \quad (7.33)$$

The coefficient of ponderomotive elasticity

$$k_p = 2\mu_0 S\left|\left\{ \begin{matrix} M\left(\frac{\partial H}{\partial z} + \frac{\partial M}{\partial z}\right) + \\ \left[M\left(\frac{\partial^2 H}{\partial z^2} + \frac{\partial^2 M}{\partial z^2}\right) + 2\frac{\partial M}{\partial z}\left(\frac{\partial H}{\partial z} + \frac{\partial M}{\partial z}\right) \right]\delta z \end{matrix} \right\}\right|_{z_2} \quad (7.34)$$

is expressed through the coefficient of ponderomotive elasticity of the symmetric case k_p^0,

$$k_p = k_p^{(0)}\left(1 + \left[\frac{\frac{\partial^2 H}{\partial z^2} + \frac{\partial^2 M}{\partial z^2}}{\frac{\partial H}{\partial z} + \frac{\partial M}{\partial z}} + \frac{2}{M}\frac{\partial M}{\partial z} \right]_{z_2} \delta z \right) \quad (7.35)$$

For a sufficiently wide column of the magnetic fluid ($\delta z \ll h$, where h is the height of the column), the oscillations remain linear, regardless of the variation of frequency. However, if the column is relatively thin, then it is necessary to take into account the non-symmetry of the field on the base of the column even in the absence of the initial non-symmetry at the displacements, since $\delta z = \Delta z$. In this case $\Delta f_z = k_p^{(0)}\Delta z + \kappa(\Delta z)^2$, where

$$\kappa = 2\mu_0 S\left[M_z\left(\frac{\partial^2 H_z}{\partial z^2} + \frac{\partial^2 M_z}{\partial z^2}\right) + 2\frac{\partial M_z}{\partial z}\left(\frac{\partial H_z}{\partial z^2} + \frac{\partial M_z}{\partial z}\right) \right]\Bigg|_{z_2}$$

The equation of the oscillations becomes non-linear

$$\rho V\frac{\partial^2 u}{\partial t^2} + \alpha\frac{\partial u}{\partial t} + \left(k_p^{(0)} + k_g\right)u + \kappa u^2 = 0 \quad (7.36)$$

Here k_g is the gas elasticity coefficient.

We now introduce the following notations: $2\beta = \alpha/\rho V$ – the damping factor, $\omega_0^2 = \left(k_p^{(0)} + k_g\right) / \rho V$ – the square of frequency, $b = \kappa/\rho V$ – the non-linearity coefficient. Consequently, equation (7.36) has the form

$$\frac{d^2u}{dt^2} + \omega_0^2 u + bu^2 + 2\beta\frac{\partial u}{\partial t} = 0 \qquad (7.37)$$

In the absence of damping (7.37) is simplified:

$$\frac{d^2u}{dt^2} + \omega_0^2 u + bu^2 = 0 \qquad (7.38)$$

where

$$b = -\frac{2\mu_0 S}{\rho V}\left[M\left(\frac{\partial^2 H}{\partial z^2} + \frac{\partial^2 M}{\partial z^2}\right) + 2\frac{\partial M}{\partial z}\left(\frac{\partial H}{\partial z} + \frac{\partial M}{\partial z}\right)\right]_{z_2} \qquad (7.39)$$

It is well known that the equation of type (7.38) permits the detailed analytical investigation of the oscillation conditions in dependence on the parameter b and constant C, using the equation

$$C - f(u) = 0 \qquad (7.40)$$

Consequently, the potential energy of the system is

$$f(u) = \frac{\omega_0^2}{2}u^2 + \frac{b}{3}u^3$$

We discuss these conditions, following [249]:

1. $b = 0$. Equation (7.38) is the equation of harmonic oscillations and its solution is well known.

2. $b < 0$. Movement is determined by the value of the constant C.

2.1. $C > \omega_0^6/6b$ or $C < 0$ – movement is infinite in the range from the value u equal to the unit root of equation (7.40) to $+\infty$.

2.2. $C \geq \omega_0^6/6b$ and $C \geq 0$ – the equation (7.40) has three roots $u_1 < 0 < u_2 < u_3$. The movement between u_1 and u_2 is finite, and at $u \geq u_3$ it is infinite. At $u_1 < u < u_2$ the solution is expressed by the cylindrical function:

$$u(t) = u_1 - (u_1 - u_2)\cdot sn^2\left[\sqrt{\frac{b}{6}(u_1 - u_3)t}, s\right]$$

and the oscillation period is

$$T = \frac{2K(s)}{\omega} = \sqrt{\frac{24}{bA}}s\, K(s)$$

where $K(s)$ is the total elliptical integrals the first kind with the notations

$$A = u_1 - u_1,\ s^2 = \frac{u_1 - u_2}{u_1 - u_3},\ \omega = \sqrt{\frac{b}{6}(u_1 - u_3)} = \sqrt{\frac{bA}{6s^2}}$$

Here A is the oscillation amplitude, s is the modulus of the elliptical function, determining the degree of distortion of the form of the oscillations $u(t)$ in comparison with the sinusoidal form; ω is frequency.

2.3. $C = \omega_0^6/6b$ – the equation (7.40) has three roots, two of which coincide: $u_1 = \omega_0^2/2b, u_2 = u_3 = -\omega_0^2/b$. This state is unstable: on falling to the point $u = u_2 = u_3$ the system completely loses kinetic energy and can both move in the opposite direction or move by infinite motion to $+\infty$.

3. At $b > 0$ the modes are determined by the same procedure.

3.1. $C > \omega_0^6/6b$ or $C < 0$ – equation (7.40) has a single root. The movement is infinite in the range from the value u, equal to the single root of the equation (7.40), to–∞.

3.2. $C < \omega_0^6/6b$ and $C \geq 0$ – equation (7.40) has the three roots $u_1 < u_2 < 0 < u_3$. The movement between u_2 and u_3 is finite, and at $u \leq u_1$ it is infinite. At $u_2 < u < u_3$, the solution has the form

$$u(t) = \frac{A}{3s^2}\left(1 + s^2 - \sqrt{1 - s^2 + s^4}\right) - A\, sn^2(\omega t, s),$$

where A and ω were determined previously, and

$$S^2 = \frac{u_2 - u_3}{u_2 - u_1}$$

3.3. At $C = \omega_0^6/6b$ the situation is identical to 2.2 – the given state is unstable.

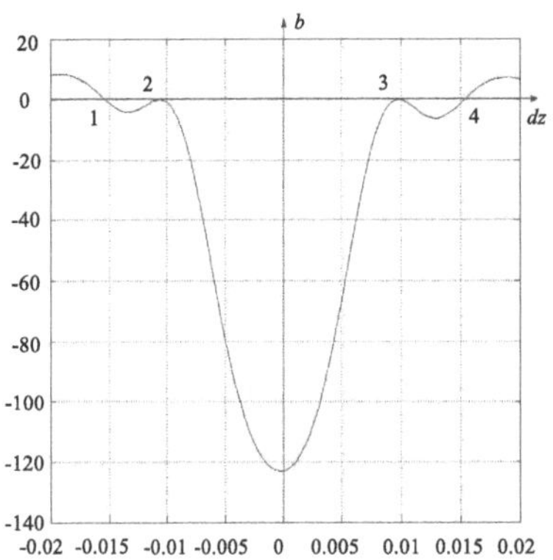

Fig. 7.10. Dependence $b(\delta z)$.

Figure 7.10 shows the graph of the dependence of the characteristic parameter b (δz) on the initial displacement. The magnetic field was calculated using the algorithm derived in [78]. The graph provides the information on the boundaries in which the magnetic fluid bridge moves in a specific mode. It may be seen that the points 1, 2, 3 4 correspond to the mode 1. In the sections 1–2, 2–3, 3–4, the mode of movement of the magnetic fluid bridge is determined by the value of the constant C and, depending on it, corresponds to one of the three variants of the conditions 3.

3.4. At $C = \omega_0^6 / 6b$ the situation is identical to that in 2.2 – the given state is unstable.

7.4. The kinetic properties of the magnetic fluid membrane

We consider the results of measurement of the critical pressure gradient p_k, ensuring the rupture of the magnetic fluid bridge, and the amplitude of a single oscillation of the electromagnetic response for three magnetic colloids [33]. In this case, the magnetic fluid membrane was produced using a glass tube with a flat bottom, with a length of 350 mm and the internal diameter of 13.5 mm. In the experiments, the bridge was formed by the method of 'self-capture' of the portion of the magnetic fluid by the circular magnet, introduced through the bottom of the tube containing the colloid, and lifted to a certain height h above the level of the liquid. The circular magnet was connected to the kinematic section of the cathetometer.

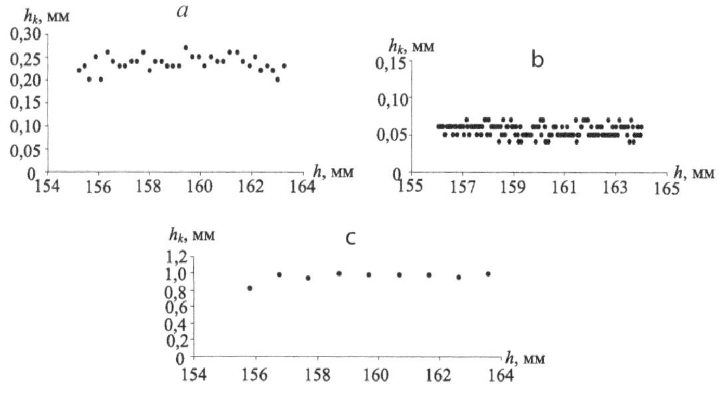

Fig. 7.11. Dependence $h_k(h)$ for the magnetic fluid membrane based on: a) MF-1; b) MF-2; c) MF-3.

Table 7.5

	h, mm	$\bar{h}_k$, mm	p_k, kPa	Δm, mg	ε_{m1}, mV	k_g, N/m	k_p, N/m	h_d, mm	v_g, m/s	τ, ms
MF-1	161	0.24	0.09	0.042	6.6	96	145	0.096	18.8	1.7
	177	0.25	0.088	0.045	6.0	88	145	0.094	17.2	1.9
	194	0.26	0.086	0.045	6.0	80	145	0.092	17.7	2.0
	204	0.26	0.084	0.046	5.9	76	145	0.089	17.5	2.0
MF-2	90	0.06	0.033	0.010	4.8	173	166	0.031	29.7	0.3
	164	0.06	0.023	0.010	3.5	94	166	0.022	25.2	0.3
	181	0.05	0.018	0.009	3.0	86	166	0.017	24.3	0.3
MF-3	150	0.97	0.330	0.17	7.5	103	107	0.476	28.7	4.6
	161	0.96	0.314	0.17	7.3	96	107	0.454	28.2	4.7
	170	1.03	0.327	0.18	8.2	81	107	0.473	27.8	5
	181	1.05	0.322	0.19	6.8	86	107	0.468	27.2	5.4

Figure 7.11 shows graphically the results of measurements of h_k (h_k is the distance between two consecutive breaks of the bridge) in relation to the height of the isolated (by the fluid) air column h for the colloids MF-1, MF-2, MF-3 (Table 7.5).

The scatter of the values of h_k is associated partly with the absence of special measures of vibro-, acoustic and thermal insulation of the magnetic fluid membrane. The individual results for h_k can not be used to study the tendency for the change of this parameter with the increase of the height of the air column. The form of the dependence can be detected only by averaging the experimental data over a large number (no less than 50 for MF-1, 150 for MF-2) in a narrow range of the displacement of the magnetic head from $h + \Delta h$ (Δh = 1 cm) for several greatly differing values of the height of the air column h. Thus, the mean value $\bar{h}_k$ is obtained. The appropriate data are presented in Table 7 5.

The breaks in the magnetic fluid membrane are not detected during the displacement of the magnetic head in the reverse direction within the region of specific width 2Γ, where Γ is the distance between the initial equilibrium position in the first displaced equilibrium determined using the cathetometer. For MF-1, MF-2 and MF-3 the resultant values were Γ_1 = 1.77, Γ_2 = 3.04 and Γ_3 = 4.53 (mm).

Figure 7.12 shows the model of the thermodynamic process in the coordinates $P(z)$ taking place in the gas cavity assuming the

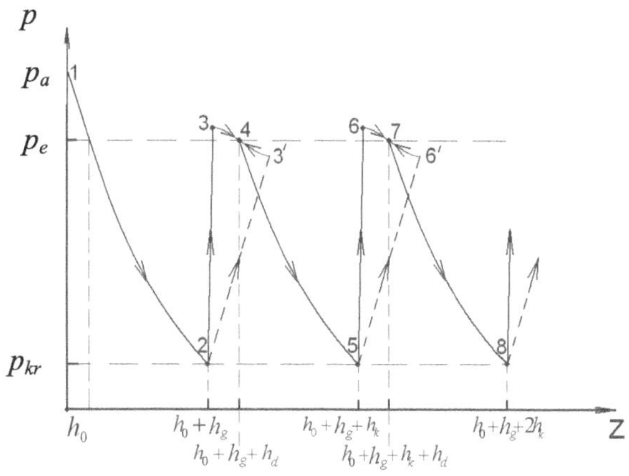

Fig. 7.12. Thermodynamic process in the gas cavity.

'slow' lifting of the magnetic head along the pipe with a constant cross-section.

In the initial positioning of the bridge, the height of the column h_0 and the pressure in the gas cavity p_a: $p_a = p_0 + \rho g b$, where p_0 is the external (atmospheric) pressure.

As an example, the numerical values of b = 1.3 cm, $p_0 = 10^5$ Pa, ρ = 1294 kg/m^3 and $\rho g b$ = 165 Pa will be considered.

In this situation, the bridge is situated in the region of the maximum magnetic field, and the displacement of the magnetic head both above and below prior to the first break of the bridge will be the same and equal to Γ, and

$$\Gamma = h_g + h_d \tag{7.41}$$

where h_g is the increase of the height of the air column; h_d is the displacement of the bridge in relation to the magnetic head.

The isothermal expansion of the gas cavity takes place in the sections 1–2, 4–5 and 7–8.

The critical pressure gradient, which determines the break of the bridge during its displacement from the initial equilibrium value, to the first break is:

$$p_{k1} \equiv p_a - p_{kr} \tag{7.42}$$

where p_{kr} is the pressure in the gas cavity at which the bridge breaks. At the isothermal expansion of the gas cavity

$$p_a h = p_{kr}(h + h_g)$$

from each

$$p_{kr} = p_a \frac{h}{h + h_g} \tag{7.43}$$

As a result of the third Newton law

$$h_g k_g = h_d k_p$$

Here k_g is the isothermal coefficient of elasticity of the gas cavity: $k_g = \frac{p_a S}{h}$; coefficients k_g and k_p have a constant value.

Taking into account (7.41), we obtain

$$h_g = \Gamma \frac{k_p}{k_p + k_g} \tag{7.44}$$

After substituting (7.44) to (7.43

$$p_{kr} = p_a \frac{h}{h + \tilde{\gamma}\Gamma} \tag{7.45}$$

where $\tilde{\gamma} \equiv \frac{k_p}{k_p + k_g}$.

Substituting (7.45) into (7.42), we obtain

$$p_{k_1} = \frac{\tilde{\gamma} \rho_a \Gamma}{h + \tilde{\gamma}\Gamma} \tag{7.46}$$

The calculated values of k_g and k_p are presented in Table 7.5.

The values of p_k for the investigated specimens of MF-1, MF-2 and MF-3 equal respectively $5.7 \cdot 10^2$ Pa, $10.4 \cdot 10^2$ Pa and $12.4 \cdot 10^2$ Pa.

The experiments carried out to determine p_{k_1} by the hydrostatic method resulted in approximately ~30% lower values of this parameter [233, 234]. The difference between the experimental results is explained by the effect of the dependence of the elasticity of the ponderomotive type on special features of the viscous flow of the magnetic fluid in the process of displacement of the bridge at different rates of application of pressure, and also by the non-linear rates. The comparison of the resultant values makes it possible to evaluate the static values of the parameters $\tilde{\gamma}$ and k_p.

In the conditions 2, 5 and 8 the continuity of the bridge is disrupted as a result of the formation of a circular orifice in its central part. Under the effect of the pressure gradient, the air is directed into the orifice resulting in a jump-like increase of pressure. In this stage of the process the bridge is displaced in the direction of the equilibrium position, i.e., in the direction ∇H which, on the one hand, results in a small increase of the volume of the gas cavity and, the other hand, in the formation of suitable conditions for 'shutting' of the cavity.

The change in the state of the gas in the gas cavity during the existence of the orifice ('lifetime' of the orifice τ) can develop by one of the two variants of transition to the new state with the equilibrium pressure p_e, each of which ensures the excitation of the natural oscillations of the magnetic fluid membrane.

In the first variant (transitions 2–3; 5–6, etc) the displacement of the bridge is small as a result of its inertia and the existence of the 'rigid connection' of the restoration of continuity with the topography of the magnetic field, and the increase of pressure in the cavity is caused by pumping the air through the orifice which takes place in accordance with the adiabatic law [33].

In the second variant (transitions 2–3', 5–6', etc., shown by the dotted line in Fig. 7.12) the bridge during time τ travels through the equilibrium position and at the moment of closure of the orifice (i.e., 3', 6') its movement is interrupted and then moves in the reverse direction to the equilibrium position. This variant is hypothetically possible if there is no 'rigid' connection of the continuity of the bridge with the topography of the magnetic field, and the process of deceleration of the displacement and closure of the bridge is determined mainly by the gas dynamic effect of the increase of the force of resistance of the gas flow to movement with increase of its speed.

After completion of the damping oscillations and the establishment of the thermodynamic equilibrium, the gas in the cavity is in the state indicated by the points 4 and 7 in the graph. With further lifting of the magnetic head, the break of the bridge takes place at a smaller increase of pressure in the gas cavity.

We derive the equation for the calculation of the increase of pressure in the gas cavity at subsequent breaks of the bridge p_k:

$$p_k = p_e - p_{kr} \tag{7.47}$$

At isothermal expansion of the gas

$$p_e = p_a \frac{h}{h + h_g + h_d - h_k}, \tag{7.48}$$

where h_d is the displacement of the bridge in relation to the magnetic head.

In this stage of expansion of the gas cavity

$$h_k = h'_g + h_d,$$

where h'_g is the increment of the height of the gas cavity.

Substituting equation (7.45) for p_{kr} into (7.47) and the expression (7.48) for p_e, and also taking into account the condition of the quality of the forces $h'_g k_g = h_d k_d$, we obtain

$$p_k = p_a h \frac{h_k + \tilde{\gamma}\Gamma - h_g - h_d}{(h + h_g + h_d - h_k)(h + \tilde{\gamma}\Gamma)}, \tag{7.49}$$

Taking into account (7.44) and the relationship

$$\frac{h_d}{h_k} = \frac{k_g}{k_g + k_p}, \tag{7.50}$$

the equation (7.49) can be written in the new form

$$p_k = p_a \frac{\tilde{\gamma} h h_k}{(h + \tilde{\gamma}\Gamma - \tilde{\gamma} h_k)(h + \tilde{\gamma}\Gamma)} \tag{7.51}$$

Taking into account these notations, the expression for p_e (7.48) can be presented in the following form:

$$p_e = p_a \frac{h}{h + \tilde{\gamma}(\Gamma - h_k)} \tag{7.52}$$

All quantities, included in the equations (7.51) and (7.52), are determined directly or indirectly using the experimental data.

In the experimental conditions $h \gg \tilde{\gamma}\Gamma$, therefore in the given case the equation (7.51) can be simplified:

$$p_k \cong p_a \tilde{\gamma} \frac{h_k}{h} \tag{7.53}$$

When comparing the expressions (7.46) and (7.53), we obtain $p_{k1} \gg p_k$, i.e. $\Gamma \gg h_k$.

The results of calculation of the parameters h_d, p_k, carried out using the value $\overline{h_k}$ and equations (7.50) and (7.53), are presented in Table 7.5.

We can note the small decrease of p_k with the increase of the height of the gas cavity h for the MF-1 and MF-3. A larger decrease of p_k, obtained for the specimen MF-2, is evidently associated with the reduction of the mass of the bridge as a result of the consumption of the bridge on the internal surface of the tube, as recorded by visual studies.

The magnetic fluid membrane, produced using the colloid MF-1, characterised by the lower concentration of the magnetic phase and the correspondingly smaller values of the parameters M_S and χ, in comparison with the other colloids with the identical carrier, MF-2, is characterised nevertheless by the considerably higher increment of the pressure p_k (although the parameter p_{k1} is characterised by the reversed relationship) which at first sight is unexpected. The physical nature of the result may be determined by the more rigid connection of the continuity of the membrane on the basis of the magnetic fluid with a high value of χ with the topography of the magnetic field. The restoration of its continuity takes place at a small displacement in the direction ∇H.

The mass of the gas portion Δm, transmitted by the magnetic fluid membrane, during a single break of the bridge within the framework of the anticipated thermodynamic process (Fig. 7.12) is obtained from the equation of state of the ideal gas written for two adjacent states:

$$p_e V = \frac{m}{\mu} RT \text{ and } p_e\left(V + \frac{\pi d^2}{4} - \overline{h_k}\right) = \frac{m + \Delta m}{\mu} RT,$$

where μ is the molar mass of the gas; R is the universal gas constant; T is absolute temperature.

The given system of equations gives

$$\Delta m = \frac{\mu p_e \pi d^2 \overline{h_k}}{4RT} \tag{7.54}$$

at $h >> \tilde{\gamma}(\Gamma - h_k)$, the equation (7.52) gives $p_e = p_a$.

Accepting for air μ = 30 Г/mole, p_e = 10^5 Pa, $d = 1.36 \cdot 10^{-2}$ m, and T = 298 K, we obtain $\Delta m = 1.76 \cdot 10^{-4}\ \overline{h_k}$.

Table 7.5 gives the values of Δm for the investigated magnetic fluid membranes. The lowest value Δm_{min} = 0.009 mg was obtained for the magnetic fluid membrane on the basis of MF-2, the highest value Δm_{max} = 0.17 mg is typical of the magnetic fluid membrane on the basis of MF-3. Regulating the amount of the colloid, introduced into the bridge, it is possible to slightly expand the range of the values of Δm. For example, the large decrease of Δm takes place as a result of the reducing the amount of MF-2 in the bridge to the minimum value. In this case in the absence of special measures for thermostatic control and vibration and acoustic insulation, the process of rupture–restoration of the magnetic fluid membrane may become uncontrollable.

The speed of the airflow through the orifice will be estimated on the basis of the relationship linking the pressure gradient in the orifice Δp_g and the speed in the area of the maximum compression of the flow v_g [224]:

$$\Delta p_g = \frac{1}{2}\rho_g v_g^2 \xi\left[\frac{\sigma}{S}\right],$$

where σ is the area of the orifice; S is the cross-sectional area of the tube; ξ is the hydraulic resistance coefficient which depends on the area of the orifice σ and the Reynolds number.

Using the results obtained in [224], we accept for the case $\sigma \ll S$, $\xi = \xi_0 = 2.9$, which leads to

$$v_g = \sqrt{\frac{2\Delta p_g}{\rho_g \xi_0}} \tag{7.55}$$

The pressure gradient

$$\Delta p_g = p_a \frac{p_e + p_{kr}}{2} = \frac{1}{2}\left(p_a - p_e + p_a - p_{kr}\right)$$

Taking into account the previously derived equations (7.45) and (7.52), we carry out elementary transformations:

$$\Delta p_g = \frac{p_a}{2}\left(1 - \frac{h}{h + \tilde{\gamma}\left(\Gamma - h_k\right)} + 1 - \frac{h}{h + \tilde{\gamma}\Gamma}\right)$$

Since $\Gamma \gg h_k$, then at $h \gg \tilde{\gamma}\Gamma$ we can write:

$$\Delta p_g \cong \frac{\tilde{\gamma}\Gamma}{h} p_a \tag{7.56}$$

Substituting (7.56) to (7.55), we obtain

$$v_g \approx \sqrt{\frac{2\tilde{\gamma} p_a \Gamma}{\rho_g \xi_0 h}} \tag{7.57}$$

The 'lifetime' of the orifice τ can be estimated on the basis of the following equation:

$$\tau = \frac{\Delta m}{p_g \sigma v_g} \tag{7.58}$$

The direct measurement of σ was not carried out in these experiments. In approximate calculations it may be assumed that the diameter of the orifice is in the range 0.1÷0.3 cm. Estimating τ 'at the top', calculations will be carried out using the smallest of the given values.

Table 7.5 shows the kinetic parameters v_g and τ for the investigated magnetic fluid membranes. The lifetime of the orifice proved to be shorter than the oscillation period of the bridge (10 ÷ 15 ms) so that the first variant of transition of the magnetic fluid membrane to the equilibrium state (Fig. 7.12) can be regarded as more probable. An argument favouring this conclusion is the fact that the increase of the potential energy of the oscillatory system at the moment of break of the bridge, calculated from the equation

$$\Delta E_p = 0.5\left(k_p h_d^2 + k_g h_g'^2\right)$$

equals only $2 \cdot 10^{-6}$ J (in the experiments with MF-1), whereas the kinetic energy, calculated from the mean speed of displacement of the bridge (according to the second variant) would be:

$$E_k \geq \frac{\pi \rho d^2 b h_d^2}{4\tau^2} \approx 2.6 \cdot 10^{-5} \text{ J}$$

7.5. Comparison of two methods of measuring the critical pressure drop

The method of investigation of the strength properties of the magnetic fluid membrane, described in this section, is based on using the results of measurement of the critical pressure drop obtained by two different methods on the same sample: acoustic–thermodynamic and hydrostatic [234, 241, 247].

The acoustic–thermodynamic method is applied using the process of isothermal constriction of the gas cavity.

Figure 7.13 shows the thermodynamic process in the coordinates $p(z)$ assuming the 'slow descent' of the magnetic head along the tube with a constant cross-section (the axis z coincides with the axis of the tube).

In the initial position, the height of the air column h_0 and the corresponding pressure p_0:

$$p_0 = p_a + \rho g b$$

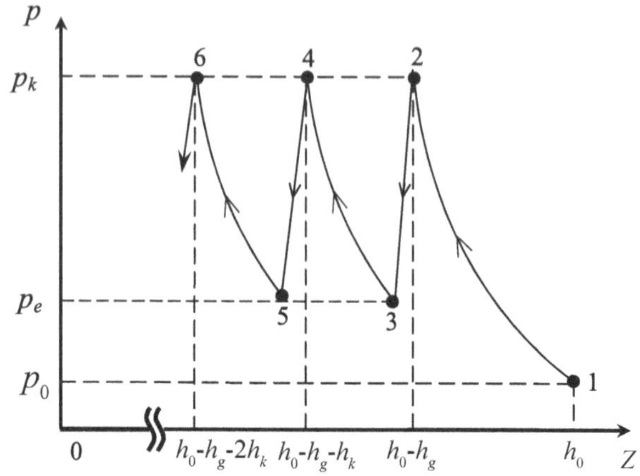

Fig. 7.13. Thermodynamic process in the gas cavity.

where p_a is the external (atmospheric) pressure; ρ is the density of the magnetic membrane; b is the thickness of the disc overlapping the cross-section of the tube and having the volume equal to the volume of the droplet of the magnetic fluid.

In this situation, the bridge is situated in the region of the maximum magnetic field, and the displacement of the magnetic head both upwards and downwards to the first break of the bridge will be practically the same and equal to Γ, and

$$\Gamma = h_g + H_d \tag{7.59}$$

where h_g is the increase of the height of the air; H_d is the displacement of the bridge in relation to the magnetic head.

In the sections 1–2, 3–4 and 5–6 isothermal compression of the isolated gas cavity takes place to pressure p_{kr} which is higher than the atmospheric pressure.

The critical pressure gradient, determining the rupture of the bridge during its displacement from the initial equilibrium position to the first break, is:

$$p_{k1td} \equiv p_{kr} - p_0 \tag{7.60}$$

where p_{kr} is the pressure in the gas cavity at which the bridge breaks.

On the basis of the law of isothermal compression of the gas cavity:

$$p_0 h_0 = p_{kr}(h_0 - h_g)$$

from which

$$p_{kr} = p_0 \frac{h_0}{h_0 - h_g} \tag{7.61}$$

Because of the third Newton law, assuming the elasticity coefficients being constant

$$h_g k_g = H_d k_p$$

where k_g and k_p are the coefficients of gas and ponderomotive elasticity, respectively.

Taking into account (7.59), we obtain

$$h_g k_g = k_p(\Gamma - h_g)$$

or

$$h_g(k_g + k_p) = k_p \Gamma$$

Thus,

$$h_g = \Gamma \frac{k_p}{k_p + k_g} \tag{7.62}$$

after substituting (7.62) into (7.61) we obtain:

$$p_{kr} = p_0 \frac{h_0}{h_0 - \alpha\Gamma} \tag{7.63}$$

where

$$\alpha \equiv \frac{k_p}{k_p + k_g} \tag{7.64}$$

Substituting (7.63) into (7.60) we obtain

$$p_{k1td} = \frac{\alpha p_0 \Gamma}{h_0 - \alpha\Gamma} \tag{7.65}$$

In the case of expansion of the gas cavity, the – sign in the numerator is replaced by the + sign and, therefore, taking into account both directions of movement of the magnetic head, we obtain

$$p_{k1td} = \frac{\alpha p_0 \Gamma}{h_0 \pm \alpha\Gamma}$$

According to the experimental conditions $h_0 \gg \alpha\, \Gamma$, and, therefore

$$p_{k1td} \approx \frac{\alpha p_0 \Gamma}{h_0} \tag{7.66}$$

We obtain the equation for calculating the increase of pressure in the gas cavity at consecutive raptures of the bridge:

$$p_k = p_{kr} - p_e \tag{7.67}$$

In condition 4 the bridge is displaced in relation to the magnetic head by H_d, in the condition 3 by $H_d - h_d$, where h_d is the displacement of the bridge at displacement of the magnetic head by h_k, consequently, the coordinate of the condition 3 $z = h_0 - h_g - h_d$ and, therefore, for the section of the process 3–4, we have

$$p_{kr}(h_0 - h_g - h_k) = p_e(h_0 - h_g - h_d)$$

from which

$$p_e = p_{kr}\frac{h_0 - h_g - h_k}{h_0 - h_g - h_d} \tag{7.68}$$

Substituting equation (7.67) for p_k to (7.68) gives:

$$p_k = p_{kr}\left(1 - \frac{h_0 - h_g - h_k}{h_0 - h_g - h_d}\right) = p_{kr}\frac{h_k - h_d}{h_0 - h_g - h_d} \approx$$

$$\approx p_{kr}\frac{h_k - h_d}{h_0} = p_0\frac{h_0}{h_0 - \alpha\Gamma}\cdot\frac{h_k - h_d}{h_0} = p_0\frac{h_k - h_d}{h_0 - \alpha\Gamma}$$

or

$$p_k = p_0\frac{h_k - h_d}{h_0 - \alpha\Gamma} \cong \frac{p_0 h_k}{h_0}\left(1 - \frac{h_d}{h_k}\right)$$

since $h_0 \gg \alpha\Gamma$.

According to the assumption of the elastic linear deformation

$$\frac{h_d}{h_k} = \frac{k_g}{k_d + k_g}$$

and therefore

$$p_k \cong p_0\alpha\frac{h_k}{h_0} \tag{7.69}$$

The equation for calculating the coefficient of gas elasticity for the isothermal process has the form:

$$k_g = \frac{\rho_g c^2 S^2}{V_0\gamma} \tag{7.70}$$

where p_g is gas density; S is the cross-sectional area of the tube; c is the speed of sound in the gas; $\gamma = 1.4$ is the Poisson coefficient; V_0 is the volume of the isolated gas cavity.

To determine k_p measurements were taken of the frequency of the free oscillations of the magnetic fluid membrane in the equipment shown schematically in Fig. 7.14. The flask 1 is secured in the vertical position so that its neck, having the form of a tube, is coaxial with the circular magnet 2. The magnetic colloid is poured into the region of the maximum magnetic field in the strictly determined volume using a measuring tube, forming a continuous magnetic fluid bridge 3. The piston 4 is lowered to the level restricted by the position of the open orifice 5. When the orifice 5 is close, the piston is pulled out. The formation of the pressure difference results in the displacement of the magnetic fluid bridge from the equilibrium position, its displacement under the effect of the ponderomotive factor in the reverse direction and subsequent oscillations. Using the inductance coil 6, inserted in the magnet, the variable EMF, corresponding to the oscillatory process, is recorded. The signal is transferred to the input of the oscilloscope 7, is recorded by digital camera 8 and travels to the personal computer 9 for processing. The values of k_p are calculated using the equation

$$k_p = 4\pi^2 \nu^2 m_f - k_g \tag{7.72}$$

where m_f is the mass of the magnetic colloid forming a continuous bridge; ν is frequency of oscillations.

As a result of the large volume V_0 the correction for the gas elasticity k_g, calculated from equation (7.70), is smaller than 5%.

The flow diagram of the experimental equipment, used for the hydrostatic measurement method, is presented in Fig. 7.15. The

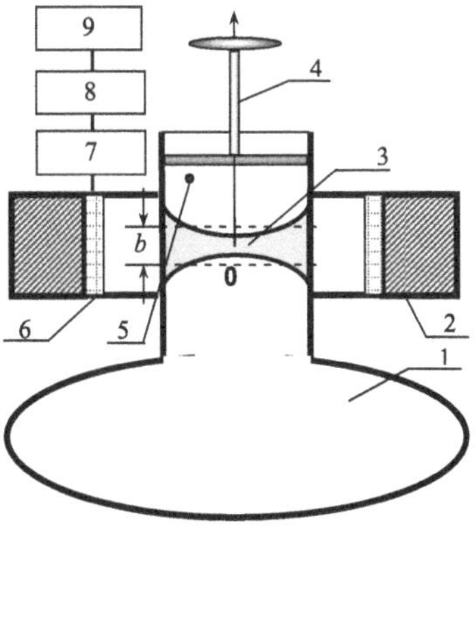

Fig. 7.14. The block diagram of equipment for determining the elasticity coefficient.

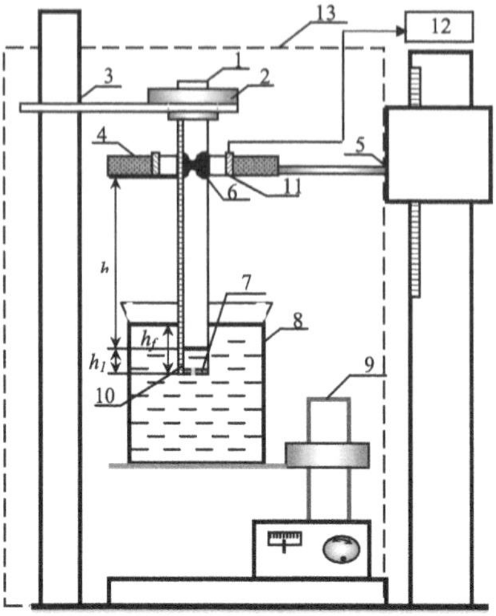

Fig. 7.15. Equipment for hydrostatic measurements.

magnetic fluid membrane is produced using a glass tube 1 open on both sides. Using the holder 2, rigidly secured to the support 3, the tube is positioned in the vertical position inside the circular magnet 4 connected with the kinematic section of the cathetometer 5. The magnetic colloid 6 overlaps the cross-section of the tube. The fluoroplastic plug 7, with an orifice with a diameter of 1 mm, is inserted tightly into the base of the tube. The lower end of the tube is lowered into the container with distilled water 8. Thus, an isolated gas cavity with the height of h_0 forms inside the tube. At the start of the experiment the height of the water column inside the tube h_{10} is equal to the height of the external water column h_{f0}. At slow lifting of the container with water using the lifting device 9 the height of the external water column h_f and, consequently, the internal water column h_1, counted from the lower end of the tube on the scale 10, increases. When the critical pressure difference p_{k1gs} is reached, the magnetic fluid membrane ruptures.

Thus,

$$p_{k1gs} = \rho_w g(h_f - h_1) \tag{7.73}$$

where ρ_w is the density of distilled water in the container.

The rupture of the magnetic fluid membrane is recorded using the inductance coil 11, coaxially installed in the ring-shaped magnet. The recorded signal is transferred to the input of the oscilloscope 12 operating in the external synchronisation mode. The experimental equipment is placed in the thermostat 13. The plug with the orifice 7 is designed for restricting the rate of filling of the air cavity with water so that auto-oscillations cannot form – series of 6–8 ruptures, following the first break of the magnetic fluid membrane. The relative error of determination of the critical pressure difference p_{k1gs} is 2–3%.

Taking into account the proportionality of p_{k1} and α, we can write

$$\frac{p_{k1td}}{p_{k1gs}} = \frac{\alpha}{\alpha_{gs}}$$

Taking into account (7.64) we obtain:

$$\frac{p_{k1td}}{p_{k1gs}} = \frac{\dfrac{k_p}{k_p + k_g}}{\dfrac{k_{ps}}{k_{ps} + k_g}}$$

Consequently, the following equation is obtained for the evaluation of the static coefficient of ponderomotive elasticity k_{ps}:

$$k_{ps} = \frac{p_{k1gs} k_p k_g}{p_{k1td} \cdot (k_p + k_g) - p_{k1gs} k_p}$$

Using the static coefficient of ponderomotive elasticity, we can evaluate the displacement of the bridge (in the bridge–disc model) or the increase of the volume of the gas cavity as a result of the displacement and distortion of the bridge at the moment of fracture using the following equations:

$$\Delta z = \frac{S p_{k1gs}}{k_{ps}} \text{ and } \Delta V = \frac{S^2 p_{k1gs}}{k_{ps}}$$

The studies [241, 247] presented the results of experimental determination of the critical pressure difference for a magnetic fluid membrane obtained on the basis of the acoustic–thermodynamic and hydrostatic methods. This was carried out using magnetic fluids employed in engineering consisting of the colloidal solution of single-domain particles of magnetite Fe_3O_4 in kerosene with different density (Table 7.6).

The results of determination of the value of Γ, obtained using the MF-2 sample in the magnetic fluid membrane, for different heights h_0, are presented in Fig. 7.16.

The increase of Γ with the increase of h_0 corresponds to the law of isothermal compression of the gas. The experimental data for and the values of p_{k1td} calculated on the basis of these data for three

Table 7.6

Specimen	ρ, kg/m^2	k_g, N/m	k_p, N/m	k_{ps}, N/m	Γ, mm	p_{k1td}, Pa	p_{k1gs}, Pa	$\bar{h}_k$ mm	p_k, Pa	p_{ks}, Pa	Δz, mm	ΔV, 10^{-6} m^3
MF-1	1587	92	72	52	2.44	680	560	0.11	30	25	1.5	0.21
MF-2	1600	92	74	42	2.84	810	570	0.17	50	33	1.5	0.21
MF-3	1946	92	107	61	2.29	790	580	0.06	20	15	1.3	0.18

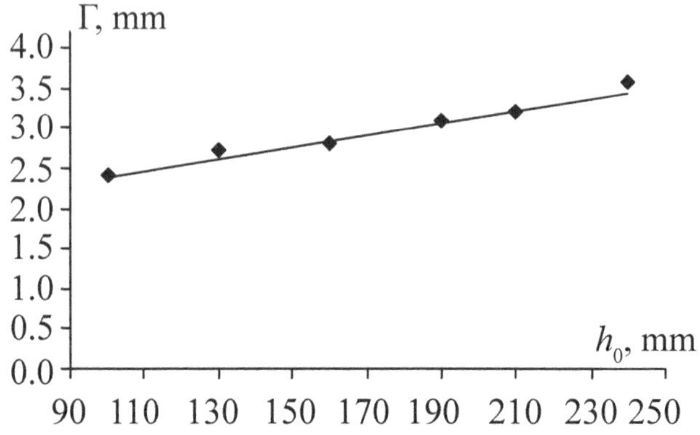

Fig. 7.16 Linear approximation of the dependence $\Gamma(h_0)$ for the magnetic fluid membrane based on MF-2.

samples of the magnetic fluid at h_0 = 160 and 16 mm are presented in Table 7.6.

By averaging the experimental data h_k (MF-1 – 50 points; MF-2 no less than 30 points; MF-3 more than 50 points) in the range of displacement of the magnetic head from h_0 to $h_0 + \Delta h$ (h_0 = 160 mm, $\Delta h \approx 1$ cm) for each specimen of the magnetic fluid we obtained $\bar{h}_k$ and the numerical values of this parameter are presented in Table 7.6. The results of calculation of the parameter p_k, carried out using the value $\bar{h}_k$ from equation (7.67), are presented in Table 7.6.

The results of experimental determination of the critical pressure gradient p_{k1td} and p_{k1gs} for the magnetic fluid membrane based on MF-2 for different values of the initial height of the isolated gas cavity h_0 are presented in the graphical form in Fig. 7.17.

In the error range of measurements the given parameters are constant which indicates that the resultant data are reliable.

Figure 7.18 shows the linear approximation of the dependence $p_k(h_0)$ for the magnetic fluid membrane based on MF-2. The reduction of p_k with the height h_0 corresponds to the previously obtained equation (7.67).

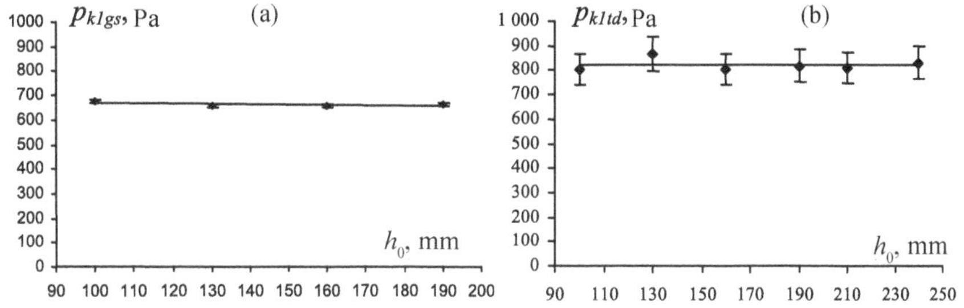

Fig. 7.17. Strength parameters for the magnetic fluid membrane based on MF-2: *a*) dependence p_{k1gs} (h_0); *b*) dependence p_{k1td} (h_0).

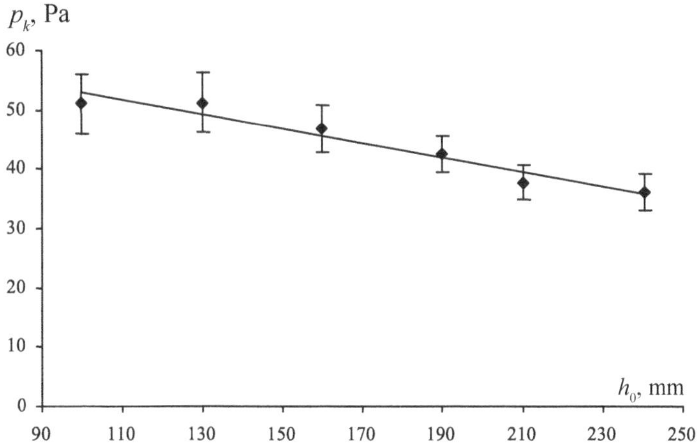

Fig. 7.18. Linear approximation of the dependence $p_k(h_0)$ for the magnetic fluid membrane based on MF-2.

According to the data in Table 7.6, the values of p_{k1td} are approximately 30% higher than the values of p_{k1gs} for the investigated magnetic fluid membranes. This is outside the range of the total error of measurement (~7%). This result may be explained by the fact that the coefficient α in equation (7.64) in the conditions of quasi-static increase of pressure from p_0 to p_{k1} does not remain constant and its average value is lower in comparison with the value calculated on the basis of the considered model theory.

The values of α were calculated using the coefficient of ponderomotive elasticity k_p. It may be assumed that the coefficient k_p, obtained from equation (7.72), is higher than the coefficient of ponderomotive elasticity characteristic of the quasi-static regime of increase of pressure k_{ps} because at 'rapidly changing' oscillatory movement the viscous friction forces block the mechanisms of flow of the liquid and the changes in the area and the form of the open surface of the bridge during its displacement of the magnetic field. Consequently, the fact that k_p is higher than k_{ps} is associated with the relaxation nature of the elasticity of the ponderomotive type.

Table 7.6 presents the results of calculating k_{ps} for the investigated magnetic fluid membranes. The data show that:

$$k_p / k_{ps} > 1$$

In the conditions of quasi-static experiments, the ponderomotive elasticity is evidently not linear ('Hookean') and, therefore, it may be expected that the value of k_{ps} depends on the displacement of the bridge from the equilibrium position.

The static coefficient of ponderomotive elasticity was used to determine the displacement of the bridge (in the bridge–disc model) and the increase of the volume of the gas cavity as a result of the displacement and distortion of the bridge at the moment of fracture. The results of calculations of Δz and ΔV are presented in Table 7.6. The table also gives the values of the parameter p_k, calculated from equation (7.67), where k_p was replaced by the value k_{ps} which improved the reliability of the estimate.

Several conclusions can be made as a result of summarising these results:

1. The results showing that the coefficient of ponderomotive elasticity k_p, measured in the dynamic regime, is higher than the coefficient of elasticity k_{ps}, obtained in the quasi-static experiment, and this is due to the relaxation nature of the elasticity of the ponderomotive type. It may be expected that in the conditions of the quasi-static experiment as a result of the non-linearity of ponderomotive elasticity the value k_{ps} will depend *on* the magnitude of the displacement of the bridge.

2. The value of the static elasticity coefficient k_{ps} can be used for evaluating the displacement of the bridge at the moment of its fracture.

3. The coefficient of the ponderomotive elasticity of the magnetic fluid membrane, the magnetic fluid sealing agent or valve depends strongly on the amplitude–rate regime of application of the pressure difference and this may be regarded as the reason for the difference in the estimates of the values of the critical pressure, obtained by different methods.

7.6. Investigation of the kinetic–strength properties of the magnetic fluid membrane by the optical method

At the present time, the magnetic fluids are used mostly in magnetic fluid hermetizers (MFH) and also magnetic fluid sealants (MFS). The strength and kinetic properties of these devices are of considerable importance but the investigation of these properties is a complicated experimental task. It is therefore interesting to investigate the magnetic fluid membrane which may be used as a model of the MFH and MFS [220, 290]. In addition to this, the magnetic fluid membranes are used as independent devices [221, 259, 261].

This section contains the results of investigation of the most important kinetic–strength parameters of the magnetic fluid membranes – the diameter D and the lifetime of the orifice τ in the magnetic fluid bridge, determined by direct measurements. This possibility was utilised using the optical methods [255, 295].

7.6.1. Measurement method

The experimental determination of the diameter of the orifice in the magnetic fluid bridge was carried out in equipment shown in Fig. 7.19.

Under the glass tube 1 there is the laser module 2 and the light from the module is transmitted by a continuous beam. The thin lens 3 is placed above the tube. When the bridge 4 ruptures, the light passes through the resultant orifice, falls on to the lens 3 and subsequently on the screen 5 situated at a certain distance from the lens. Consequently, the enlarged image of the orifice is displayed on the screen.

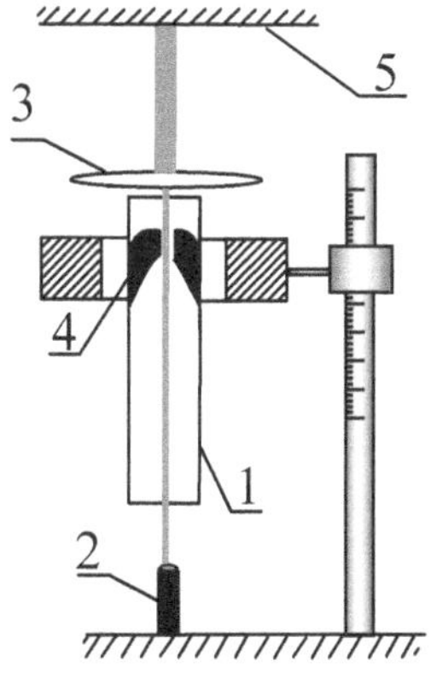

Fig. 7.19. Diagram of experimental equipment for determining the diameter of the orifice.

Graph paper is placed on the screen for the measurement of the diameter of the image of the orifice. The image of the orifice on the paper is recorded in a digital camera for subsequent processing in a computer. The dimensions of the orifice are calculated using the well-known relationship of geometrical optics

$$\beta = \frac{h}{H} = \frac{f}{d} \tag{7.74}$$

where β the linear magnification of the lens; h is the linear size of the image of the object; H is the linear dimension of the object; f is the distance from the lens to the image; d is the distance from the object to the lens.

In the experiments, the screen is placed at a distance of f = 77.5 cm from the lens. Therefore, to produce the orifice in the magnetic fluid bridge, the bridge must be displaced, and the value d changes in the range from 2.42 cm to 2.52 cm. Correspondingly, the linear image of the orifice changes in the range from 30.8 mm to 32.0 mm.

The diameter of the orifice is calculated from the equation

$$D = D_r / \beta, \tag{7.75}$$

where D_r is the diameter of the image of the orifice on the screen.

The mean value of β, equal to 31.4, is substituted into equation (7.75).

However, determination of the diameter of the image of the orifice, obtained from the screen, requires additional clarification. The point is that the image of the orifice is not strictly circular because the laser beam, transmitted through the orifice in the magnetic fluid bridge, does not have a circular cross-section and is slightly elongated. In addition to this, the intensity of light in the cross-section of the light beam is distributed non-uniformly (sections of the light beam with low intensity can not be displayed on the screen because of the absorption of light on the bottom of the tube and on the lens).

A relative method is used to increase the accuracy of calculating the diameter of the image of the orifice: in computer processing of the experimental results, a circle is constructed in the MS Word software which describes most efficiently the image of the orifice on the screen. This is followed by producing the 'reference' section – the section whose ends coincide with the ends of the centimetre section of the graph paper on the photograph. Consequently, the diameter of the image of the orifice (in cm) is determined from the equation

$$D_r = \ell \cdot D_c / \ell_c,$$

where ℓ is the length of the reference section in the scale 1:1 (1 cm); D_c is the diameter of the computer circle in points (determined with the accuracy to 0.01 points); ℓ_c is the length of the reference section, expressed in the same measurement units as D_c.

The error of determination of D_r is calculated from the equation

$$\frac{\Delta D_r}{D_r} = \sqrt{\left(\frac{\Delta l}{l}\right)^2 + \left(\frac{\Delta D_c}{D_c}\right)^2 + \left(\frac{\Delta l_c}{l_c}\right)^2}$$

and equals ~11%. The error of the method of determining the diameter of the orifice in the bridge, calculated from the equation:

$$\frac{\Delta D}{D} = \sqrt{\left(\frac{\Delta D_r}{D_r}\right)^2 + \left(\frac{\Delta \beta}{\beta}\right)^2}$$

in the experimental conditions does not exceed 13%.

To explain the method, Fig. 7.20 shows a photograph of the enlarged image of the orifice (in the face of the maximum size) in the magnetic fluid membrane, produced on the basis of the specimen of the magnetic fluid MF-3, with the elements of computer processing.

The experimental determination of the duration of existence of the orifice in the magnetic fluid bridge is carried out in equipment whose flow diagram is shown in Fig. 7.21.

The laser module 2, connected to the pulsed generator 7, displaced below the glass tube 1. The light pulses from the generator propagate in the vertical direction on the magnetic fluid bridge 3.

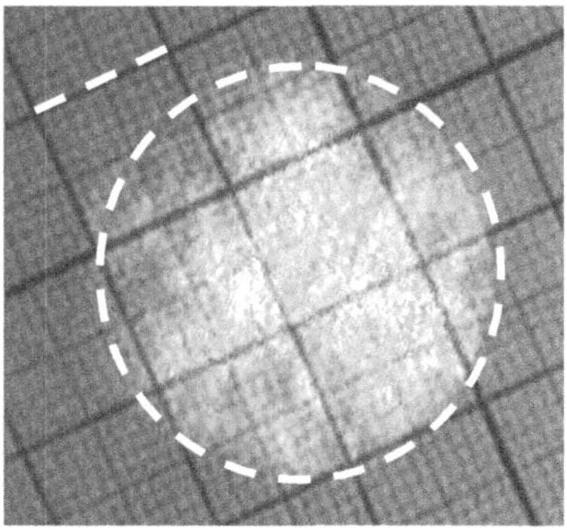

Fig. 7.20. Photograph of the orifice in the magnetic fluid membrane.

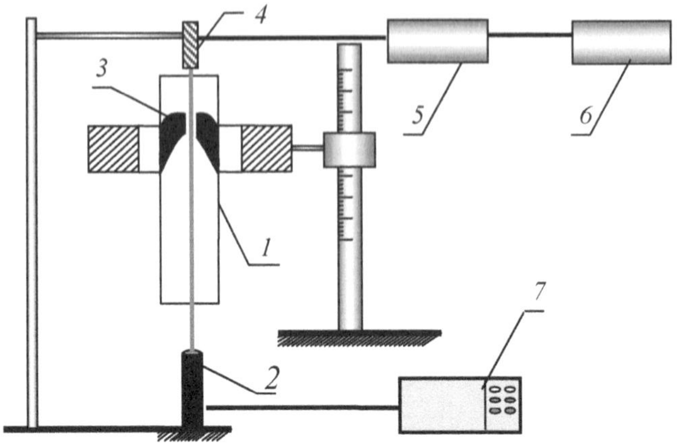

Fig. 7.21. Diagram of experimental equipment for determination of τ.

The photodiode 4 is installed above the tube and the signal from the photodiode travels to the analogue–digital converter 5 and computer 6. When the bridge fractures, a sequence of light pulses, falling on the photodiode, passes through the resultant orifice. A group of electrical pulses forms on the screen of the monitor, and the envelope curve of the pulses has a specific shape.

To determine the duration of existence of the orifice, it is necessary to calculate the number of light pulses transmitted in a single act of fracture. This number is then multiplied by the duration of a single pulse. These experiments were carried out using the laser module with a power of 3 mW and a photodiode FD-651. The error of measurement of τ is determined by the repetition frequency of light pulses. This frequency is limited by the duration of a single pulse reproduces by the device. In the experiments, the regime of generation of the pulses with the repetition frequency ν = 1250 Hz was used.

To demonstrate the method, Figs. 7.22 *a* and *b* show the dependence of the amplitude of the voltage on the photodiode on time obtained in the experiments carried out to determine the lifetime of the orifice in the magnetic fluid bridge using different regimes of generation of light pulses.

Table 7.7 shows the data on the fluid–carrier and the main physical parameters of the specimens of the magnetic fluid used in the formation of the magnetic fluid membrane.

The notations of the physical parameters, shown in Table 7.7; are: ρ – density, φ – concentration of the solid phase, M_s – saturation

Table 7.7

Specimen	Fluid-carrier	ρ, kg/m²	φ, %	M_s, kA/m	χ	η_s, Pa·s
MF-1	kerosene	1440	14.5	60	5.4	$5.4\cdot10^{-3}$
MF-2	polyethylensiloxane (PES-4)	1500	—	56	4.2	0,3
MF-3	kerosene	1192	8.9	37	3.1	$2.3\cdot10^{-3}$
MF-4	kerosene	1022	5	22	1.4	$1.1\cdot10^{-3}$

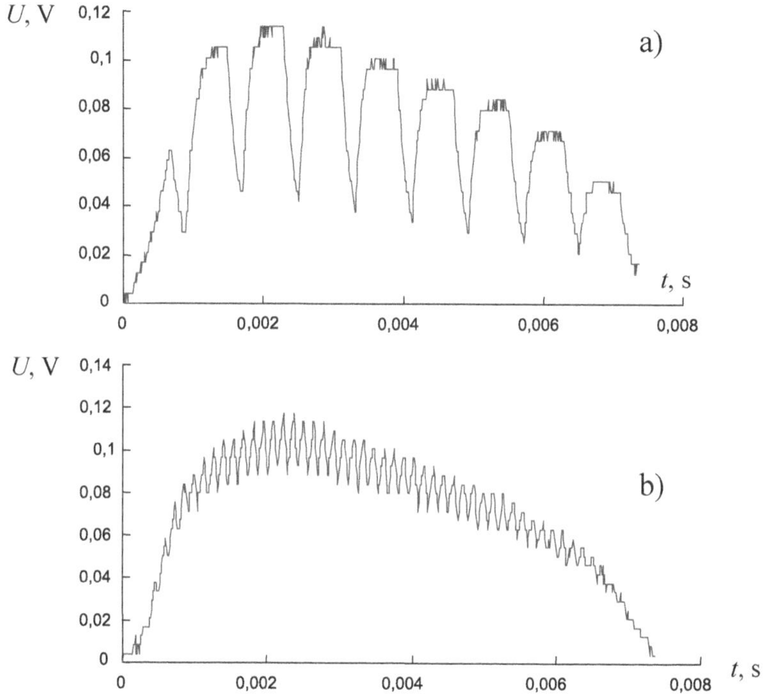

Fig. 7.22. Dependence of *U* on τ, *a*) pulse repetition frequency 1.2 k kHz; *b*) pulse repetition frequency 7.2 kHz.

magnetisation, χ – the initial magnetic susceptibility, η_s – static shear viscosity of the magnetic fluid.

These fluids consist of a colloidal solution of single-domain particles of magnetite Fe_3O_4 in hydrocarbon (kerosene) and organic silicon (polyethylsiloxane) media, stabilised with oleic acid.

The density of the magnetic fluid is determined by a pycnometer. The samples are weighed on the Ohaus RV-214 cl electronic balance with the accuracy of 0.0001 g. However, as a result of the formation of a meniscus on the surface of the pycnometer and the absence of magnetic transparency in the magnetic fluid, there is a certain indeterminacy in the determination of the level of the fluid in the neck of the pycnometer. The error in the determination of the level of the magnetic fluid in the pycnometer is ~1.5–2 mm. Thus, at the internal diameter of the neck of the pycnometer d = 5.4 mm, the error of measurement of ρ is ~0.5%.

The volume concentration of the solid phase φ is determined from the equation for mixing (3.6) used in colloidal chemistry.

The kinematic viscosity ν in the non-magnetised specimens of the magnetic fluid is measured using the VPZh-2 capillary viscometer, with the capillary diameter of 1.31 mm. In measurements, because of the optical non-transparency of the magnetic fluid, there are difficulties in the determination of the position of the level of the fluid in relation to the reference marks of the viscometer and, consequently, the scatter in time of the discharge of the fluid from the capillary according to the data from 10 experiments is ~10%, which may be used as an estimate of the error of measurement of viscosity. The values of the kinematic viscosity ν and density ρ are used to determine the static shear viscosity η: $\eta = \nu\rho$.

The initial magnetic susceptibility χ and saturation magnetisation M_s were obtained by studying the magnetising curve of the magnetic fluid. The magnetising curve was plotted using the ballistic method, described in section 1.3. Figure 1.3 shows the diagram of this method.

7.6.2. *Measurement results and analysis*

The dependence of the duration of existence of the orifice in the magnetic fluid membrane τ on the volume of the isolated gas cavity V_0 in the volume range 0.007–0.036 l is determined using a glass tube with the inner diameter of 13.5 mm. A number of experiments were carried out using flasks with the volume of 0.135, 0.3 and 0.5 l in which the necks were represented by the tubes of the same diameter. For the investigated specimens and each volume of the gas cavity V_0 the value of τ was measured at least 20–25 times. The mean statistical error of determination of τ was not greater than 10%. It should be mentioned that the diameter of the orifice in the magnetic fluid membrane decreases with the increase of the volume of the colloid introduced into the tube and, in addition to this, at the moment of fracture droplets splash out from the bridge and settle on the walls and bottom of the tube. This results in a reduction of the width of the

Table 7.8

Specimen	$V_{f\,min}$, ml
MF-1	2.0
MF-2	2.0
MF-3	1.8
MF-4	1.5

Table 7.9

Specimen	D, mm	
	Expansion of gas cavity	Compression of gas cavity
MF-1	0.76	0.55
MF-2	0.75	0.53
MF-3	1.05	1.00
MF-4	1.13	0.97

light beam passing in the process of repeated measurements through the orifice and this in turn increases the scatter of the experimental values of τ and D. For this reason, the duration of existence of the orifice is determined at the minimum volume of the magnetic fluid $V_{f\,min}$, sufficient for the formation of the magnetic fluid bridge. Table 7.8 gives the values of $V_{f\,min}$ for the formation of the magnetic fluid membrane in the selected sample of the magnetic fluid.

Table 7.9 shows the experimental results of determination of the maximum diameter of the orifice, obtained for the magnetic fluid membrane based on the investigated specimens in the regimes of expansion and compression of the isolated gas cavity. The random error of determination of the diameter of the orifice was not greater than 6%.

The experimental results of determination of the maximum orifice in the magnetic fluid membrane, presented in Table 7.9, can be interpreted as follows. For less concentrated samples of the magnetic fluid, the maximum diameter of the orifice D is greater than for the more concentrated samples. This is explained by the more efficient interaction of the concentrated specimens with the magnetic field.

The results showing that the size of the orifice at expansion of the gas cavity is slightly greater than its size in the process of compression are explained by the effect of the gravitational force which displaces the bridge at expansion of the cavity into the region with the lower strength of the magnetic field.

The theoretical evaluation of the duration of existence of the orifice in the magnetic fluid membrane is carried out using the equation:

$$\tau = \frac{\Delta m}{\rho_g \sigma v_g}$$

in which $\sigma = \pi D^2$ is the area of the orifice in the magnetic fluid membrane, determined from the experimentally obtained value of the orifice parameter D (Table 7.9), v_g is the rate of discharge of the gas.

The mass of the gas portion can be determined on the basis of the gas laws [241] using equation (7.54). The flow rate of air, passing through the orifice, is calculated from equation (7.57).

Table 7.10 shows the values of the dynamic Earth is of the ponderomotive elasticity required for the calculations and determined on the basis of the results of measurements of the frequency of oscillations of the magnetic fluid membrane at a sufficiently large volume of the gas cavity.

Table 7.11 shows the values of the coefficient of gas elasticity k_g and α , calculated for different volumes of gas cavity

In [236] the authors obtain the equation for the theoretical calculations of the duration of complete discharge of the gas through the orifice in the case of a quasi-stationary flow, permitting the application of the Bernoulli equation

$$\tau = \frac{V_0}{\sigma c}\sqrt{\frac{2\Delta p}{\gamma p_0}} \tag{7.76}$$

where c is the speed of sound in air; γ is the Poisson coefficient; Δp is the variation of pressure in the gas cavity, equal to p_h in the investigated case; p_0 is the initial pressure in the gas cavity, equal to p_a in the investigated case.

Figure 7.23 shows graphically the dependence of the duration of existence of the orifice of the magnetic fluid membrane τ on the

Table 7.10

Specimen	k_p, N/m
MF-1	49
MF-2	67
MF-3	59
MF-4	28

Table 7.11

V_0, l	k_g, N/m	α			
		ML-1	ML-2	ML-3	ML-4
0.523	4	0.92	0.94	0.94	0.88
0.300	7	0.88	0.91	0.89	0.80
0.135	15	0.76	0.81	0.79	0.65
0.036	58	0.46	0.54	0.50	0.33
0.029	73	0.40	0.48	0.45	0.28
0.021	97	0.34	0.41	0.38	0.22
0.014	145	0.25	0.32	0.29	0.16
0.007	291	0.14	0.19	0.17	0.09

Table 7.12

Specimen	τ, ms		τ_t, ms		τ_∞, ms	
MF-1	15.0	7.8	11.1	16.9	68.5	104.6
MF-2	37.8	31.2	16.9	13.3	183.2	275.1
MF-3	16.7	8.0	8.1	7.5	551	50.8
MF-4	18.7	13.2	5.7	10.8	69.6	89.6

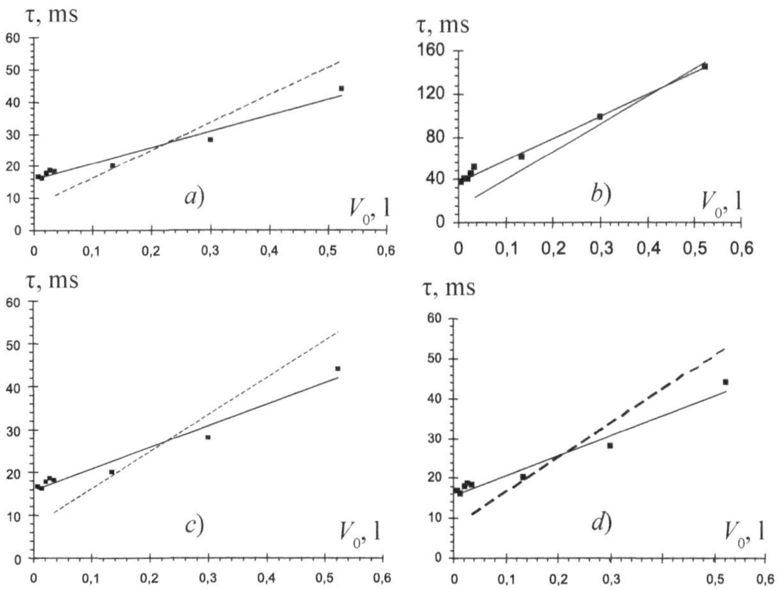

Fig. 7.23. Dependence of τ on V_0.

volume of the isolated gas cavity V_0. The points show the results of experiments, the solid lines – the linear approximation of the experimental results, the dotted line – the theoretical curves, obtained using (7.76). There is a completely satisfactory agreement between the results of the theory and experimental data.

Table 7.12 shows the values of the duration of existence of the orifice in the magnetic fluid membrane: τ – results of measurements, t_t – the value obtained from equation (7.58), using the values calculated from the equations (7.54) and (7.55), τ_∞ – the value calculated from equation (7.76).

The longest lifetime of the orifice τ was recorded for the magnetic fluid membrane based on MF-2, characterised by the highest viscosity. Evidently, the viscosity of the fluid reduces the flow rate both in the processes of formation of the orifice and during its closure. The calculated value of τ_t is approximately half the value of τ, which reflects the approximate nature of the calculation model.

The duration of complete discharge of the gas from the orifice τ_∞, calculated from equation (7.76), is approximately an order of magnitude greater than the value of the 'lifetime' of the orifice presented in Table 7.12.

This result is in agreement with the previously obtained data according to which the pressure in the cavity p_e, determined after closure of the orifice, is higher than the external pressure p_0 and indicates the partial discharge of the excess gas from the cavity. It may be assumed that this mechanism also operates in the case of the rupture and restoration of the bridge in the 'single-tooth' magnetic fluid hermetizers and magnetic fluid sealants [220]

7.6.3. The cavitation model

In the previous sections, we discussed the methods and results of measurement of the kinetic–strength parameters of the magnetic fluid membrane – the duration of existence of the orifice in the magnetic fluid bridge and its dimensions. However, the physical nature of the factors, which determine the processes of fracture and restoration of the orifice and predetermine the numerical value of these parameters, has not been clarified.

The exact analytical solution of this problem, based on the application of the magnetohydrodynamic equations taking into account the forces of surface tension, the specific geometry of the magnetic field on the free surface of the magnetic fluid bridge, is an extremely complicated task [290]. It is therefore interesting to consider the results of theoretical and experimental problems examined in considerable detail and being physically identical with the task described here. It is believed that it is highly promising to use the model of rupture–restoration of the magnetic fluid bridge based on the results of studies of acoustic cavitation [296–300, 313].

When discussing the reasons for the rupture and closure of the orifice in the magnetic fluid, it is necessary to consider the special features of the topography of the magnetic field of the magnet used in the experiments. Therefore, the magnetic field of the circular magnet, magnetised along the axis, has been studied in detail in experiments and theoretical studies (results were also used in solving a number of other problems [32, 241]).

The dashed line in Fig. 7.24 approximates the results of measurement of the strength of the magnetic field along the axis using a Hall-type teslameter. The abscissa gives the distance from the centre.

The theoretical analysis of the magnetic field was carried out using the model in which the circular magnet was magnetised with the magnetisation *M*, constant in the volume and directed along its axis. The components of induction of the magnetic field are determined by the equation [57]: **B** = –grad ψ, where the scalar potential has the following form:

$$\psi = -\frac{M}{2\pi}\left(\int_{R_1}^{R_2} k(k_1)\frac{k_{1q}}{\sqrt{qr}}dq\right) - \frac{M}{2\pi}\left(\int_{R_1}^{R_2} k(k_2)\frac{k_{2q}}{\sqrt{qr}}dq\right)$$

where $k_1 = 2\sqrt{qr \,/\, ((q+r)^2 + (z-\ell)^2)}$; $k_2 = 2\sqrt{qr \,/\, ((q+r)^2 + (z+\ell)^2)}$;

R_1, R_2 is the inner and outer radius of the magnet; ℓ is its half thickness; $K(k)$ is the elliptical integral of the first kind.

The magnetisation was determined from the value of the induction of the magnetic field, measured in the centre of the magnet. The solid line (Fig. 7.24) shows the mean value of the axial component *H* in the cross-section of the tube obtained using this model.

The difference between the average data and the measurement results does not exceed 8.5%. Figure 7.25 shows the distribution of the lines of force of the magnetic field in the plane passing through the axis of the tube. The arrows indicate the relative value in the direction of the vector of induction of the magnetic field.

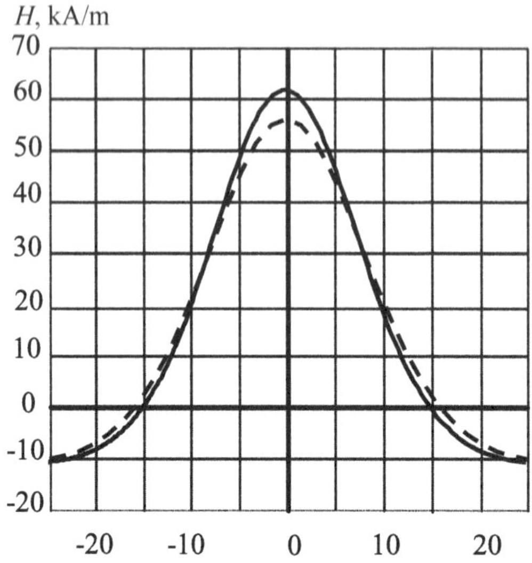

Fig. 7.24. Graph of the dependence of the strength of the magnetic field on the coordinate of the point on the axis of the magnet.

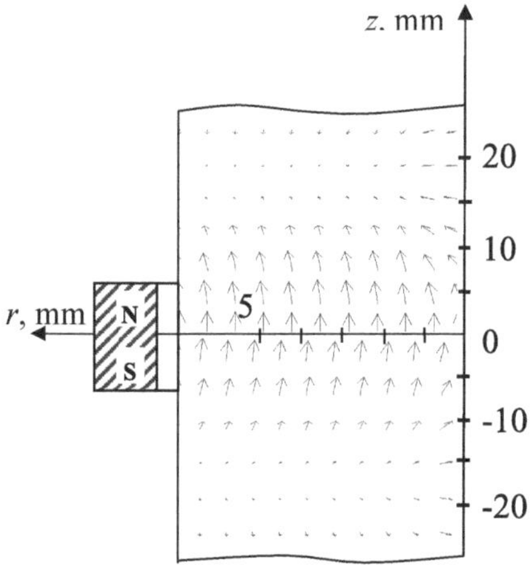

Fig. 7.25. Distribution of the lines of force of the magnetic field.

Figures 7.26 *a* and *b* shows the isolines of respectively the axial H_z and radial H_r components of the magnetic field. The vertical line restricts the section of the field inside the tube.

Special attention is given to the part of the active zone of the magnetic field, restricted by the cylindrical surface with a radius R, with the centres of the base having the coordinates $z = 0$ and $z = b$. Taking into account the diameter of the tube and the volume of the magnetic fluid droplet, introduced into the tube, we assume that $R = 7$ mm and $b = 7$ mm. The area of the investigated section of

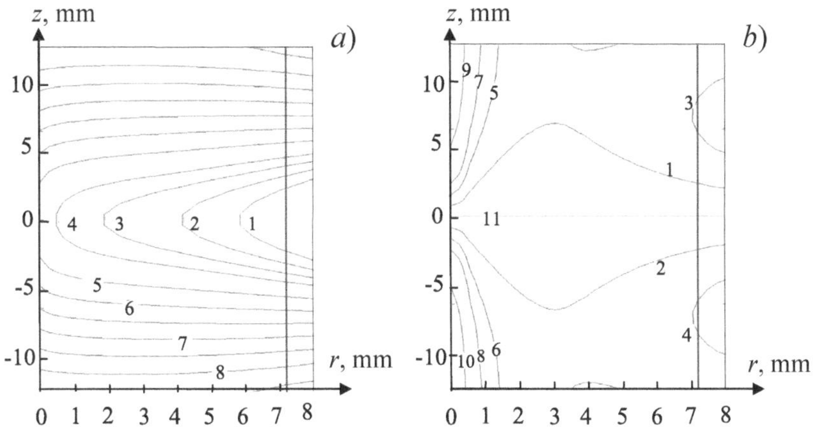

Fig. 7.26. Isolines of the magnetic field: a) the isolines of the axial projection of the induction of the magnetic field: 1) 90, 2) 86, 3) 81, 4) 77, 5) 68, 6) 60, 7) 42, 8) 25 mT; b) the isolines of the radial projection of the induction of the magnetic field: 1) –3, 2) 3, 3) –7) 4) 7, 5) –7, 6) 7, 7) –10, 8) 10, 9) –23, 10) 23, 11) 0 mT.

the magnetic field is divided by the coordinates of the lines $z = 0$; 1; 2; ...7 (mm) and $r = 0$; 1; 2; ...7 (mm) into square cells ($\Delta z = 1$ mm and $\Delta r = 1$ mm). The results of investigation of the topography of the magnetic field are presented in the form of matrices (tables) with each table having 7 lines and 7 columns. The elements of the matrix a_{ij} have the specific value of one of the parameters of the magnetic field which include: the modulus of the vector of the strength of the magnetic field $\mathbf{H}_{ij}$, the axial $(H_z)_{ij}$ and radial $(H_r)_{ij}$ components of the vector $\mathbf{H}$, the axial $(\Delta H_z / \Delta z)_{ij}$ and radial $(\Delta H_z/r)_{ij}$ components of the gradient of the strength of the field. The numerical values a_{ij} represent the values of the parameters of the field, averaged in the scale of the cell.

The magnetic field of the other part of the active zone, restricted by the cylindrical surface with the centres of the bases at the points z = 0 and $z = -b$ is a mirror image of the first part as a result of the cylindrical symmetry of the field.

Analysis of the resultant data set leads to a number of characteristic special features of the magnetic field of the investigated circular magnet.

Above all, it is important to note the presence of the cylindrical symmetry of the field in relation to the axis Z and the heterogeneity of the axial component H_z in both the axial and radial directions. In the active zone (in the area in which the magnetic fluid bridge is found) the axial component H_z is considerably greater than the radial component H_r. The modulus **H** slightly increases in the radial direction (~10%) then decreases in the direction of increase of z. The absolute value of the axial component $\Delta H_z/\Delta z$ increases in the direction of the axis Z on average from 0 to ~$5\cdot 10^6$ A/m^2, and in the initial section (0–4 mm) the increase is more rapid and then with increasing z the rate of increase decreases. The radial component $\Delta Hz/\Delta r$ decreases with increasing r. For example, in the plane $Z = 0$ this component in the section $0 \div 3$ mm decreases quite rapidly from 8.7 to 1.25 and then increases to ~2.5 ($\times 10^{-6}$ A/m^2).

The increase of H_z in the radial direction results in the overflow of the fluid to the region in the vicinity of the wall and in the outflow of the fluid from the axis of symmetry to the walls of the tube. This predetermines the fracture of the fluid film in its central part.

The increase of H_z in the direction to the plane of symmetry of the circular magnet on the other hand results in the outflow of the excess fluid from the wall of the tube and, consequently, in closure of the orifice in the bridge. The effect of the latter mechanism slightly weakens as a result of the 'jump' of the normal component of magnetisation M_z at the air–magnetic fluid interface.

The components of the pressure acting on the base and the side surface of the magnetic fluid cylinder with a radius of 7 mm and a height of 7 mm will be estimated. These components include the hydrostatic pressure, the pressure of ponderomotive force, the magnetic jump of pressure on the free surface, and the pressure below the distorted surface of the fluid.

Calculations were carried out using the data for the sample MF-4, presented in Table 7.7. The calculated results will now be presented:

1. Hydrostatic pressure $p = \rho g h = 71.5$ Pa;

2. Approximate calculation of the pressure of ponderomotive force acting on the lower base of the cylinder, is carried out using the following simple procedure. In the initial stage, we calculate the pressure increment Δp_i in the i-th layer with a thickness $\Delta z = 10^{-3}$ m using the equation

$$\Delta p_i = \nabla p_i \cdot \Delta z = \mu_0 \cdot M \cdot (\nabla H)_i . \Delta z .$$

The value of the gradient of the strength of the magnetic field $(\nabla H)_i$ is obtained by averaging with respect to all elements of each line: $(\nabla H)_i = \sum_{j=1}^{j=7} (\Delta H_z / \Delta z)_{ij} \Big/ 7$. Magnetisation M for the range of the strength of the magnetic field in the active zone in accordance with the magnetisation curve is assumed to be equal to 20 kA/m. The pressure on the base determined by the summation of the values of Δp_i with respect to all eight layers which in the present case gives $p = 4.9 \cdot 10^2$ Pa.

3. The pressure of the ponderomotive forces on the side surface of the cylinder is estimated using the same procedure. The pressure increment Δp_j in the j-th column with a thickness $\Delta r = 10^{-3}$ m is carried out by averaging, over the column, the values of the radial gradient of the strength of the magnetic field $(\nabla H)_i = \sum_{j=1}^{j=7} (\Delta H_z / \Delta z)_{ij} / 7$.

The required pressure, determined by summation of the values of Δp_j over all columns is $p = 2.35 \cdot 10^2$ Pa.

4. The values of the magnetic 'jump 'of pressure will be evaluated using the model proposed in [53] according to which its value, calculated from the equation $p = 1/2 \cdot \mu_0 M^2$ is ~$2.5 \cdot 10^2$ Pa. It should be mentioned that the magnetic jump of pressure and the pressure of ponderomotive forces on the lower base of the cylinder have the opposite signs.

5. The pressure below the distorted free surface of the column of the magnetic fluid is calculated using the Laplace equation:

$$p_\sigma = 2\sigma / R,$$

where R is the radius of the curvature of the surface; σ is the surface tension coefficient of the magnetic fluid.

Assuming that $R = 7$ mm, $\sigma = 20 \cdot 10^{-3}$ N/m, we obtain $p_\sigma = 8$ Pa.

Thus, the pressure of the ponderomotive forces on the base and the walls of the cylinder is approximately equal and at the same time each value is considerably higher than the gravitational and capillary components of pressure.

The data obtained for the distribution of the parameters of the field in the active cell can be used for carrying out evaluation calculations of the geometry of the free surface of the magnetic fluid bridge and for making a number of predictions regarding the change of this geometry with the decrease (increase) of the height of the layer of the magnetic fluid, restricted at the bottom by the surface $Z = 0$. This will be carried out using the algorithm for calculating the geometry of the free surface of the magnetic fluid in the magnetic field of a straight conductor with the current described in [53].

The interface between the fluid and the gas is capable of deforming under the effect of the volume forces, applied to the fluid. The equilibrium form of the surface is determined by the balance of the normal stresses and capillary forces and is determined using the equation:

$$\rho_{gz} - \Phi + \sigma(1/R_1 + 1/R_2) - 1/2\mu_0(\boldsymbol{Mn})^2 = C$$

where $z = \xi\ (x, y)$ is the equation of the interface, $\Phi = \mu_0 \int_{H_0}^{H} MdH$ is the potential of the bulk ponderomotive magnetic force, σ is the surface tension coefficient, R_1, R_2 are the radii of curvature, $\boldsymbol{n}$ is the normal to the surface.

Assuming the absence of the surface tension and of the magnetic 'jump' of pressure, the equation of the free surface has the form [53]:

$$z = \Phi / (\rho g) + \text{const.}$$

Taking into account the fact that the magnetisation is the increasing function of the strength of the magnetic field, the value of the potential of the magnetostatic force Φ is always greater in areas in which the strength of the field is higher. Therefore, the free surface of the magnetic fluid in the field of the gravitational force has always the height of lift greater in the areas where the strength of the field is greater.

Analysis of the matrix with the elements $(H_z)_{ij}$ shows that the variation of this element in the lines (i = const) in the direction of radius r can be approximated by the linear dependence:

$$(H_z)_i = (H_{z0})_i + c_i \cdot r,$$

where c_i is the tangent of the slope of the straight section.

As shown by direct calculation, coefficient c_i slightly increases when i decreases from 7 to 1. The numerical value of the axial combine of the field $(H_{z0})_i$ also changes (Fig. 7.25). In the framework of the approximate evaluation model, they will be replaced by the mean values of $\overline{c_1}$ and $\overline{(H_{z0})_i}$ by averaging each of them over all lines from the maximum value $i = m$ to $i = 1$:

$$\overline{c}_m = \sum_{i=1}^{i=7} c_i \Big/ m,\ \overline{(H_{z0})}_m = \sum_{i=1}^{i=7} (H_{z0}) \Big/ m.$$

In this case, we can write

$$\overline{(H_z)}_m = \overline{(H_{z0})}_m + \bar{c}_m \cdot r.$$

For $m = 7$ using the results, we have $\overline{(H_{z0})}_m = 49$ kA/m, $\bar{c}_m =$ 1.4 MA/m^2.

In the investigated zone of the magnetic field, the magnetisation of the 'test' sample of the magnetic fluid is 19–21 kA/m, i.e., slightly differs from the saturation magnetisation $M_s = 22$ kA/m. Therefore, it can be assumed that the specimen is in the saturated magnetic field. According to [53] in the fluid magnetised to saturation $\Phi = \mu_0 M_s H$ + const, and the value const is assumed to be equal to 0.

Taking these results into account, we ignore the effect of the gravitational and capillary forces on the droplet of the magnetic fluid. Also, the presence of the magnetic 'jump' of pressure on the free surface of the column of the magnetic fluid will be disregarded. In this situation, the axial component of the ponderomotive force will play the function identical to the function of the gravitational force. However, the characteristic feature of the ponderomotive force is that its potential does not remain constant and depends on the height of the layer of the magnetic fluid, i.e., on the value of the maximum number of the line $i = m$. Therefore, assuming the absence of surface tension, the forces of gravity and the magnetic jump of pressure the equation of the free surface acquires the following form [53]:

$$z_m \overline{\Phi}_i \Big/ (\mu_0 \cdot M \cdot \overline{G}_{tm}),$$

where $\overline{G}_{tm}$ is the average value of the gradient of the axial component of the field with respect to each line:

$$\overline{G}_i = \sum_{j=1}^{j=7} (\Delta H_z / \Delta z)_{ij} \Big/ 7$$

and with respect to lines from $i = 1$ to $i = m$, respectively, the height of the column of the magnetic fluid in integers from 1 to 7 mm

$$\overline{G}_{tm} = \sum_{i=0}^{i=m} \overline{G}_i / m\cdot,$$

$\overline{\Phi}_m$ is the potential of the ponderomotive force, presented in the framework of the approximate model as $\overline{\Phi}_m = \mu_0 \cdot M \cdot (\overline{(H_{z0})}_m + \bar{c}_m \cdot r)$. After simple transformations we obtain:

$$z_m(r) = (\overline{(H_{z0})}_m + \bar{c}_m r) \Big/ \overline{G}_{tm}.$$

Thus, the dependence $z_m(r)$ has the form of a straight line and its slope depends on the value m, i.e., on the given height of the column of the magnetic fluid: $\mathrm{tg}\,\alpha_m = \bar{c}_m / \bar{G}_{tm}$. Secondly, in the investigated approximation the free surface of the droplet of the magnetic fluid is conical. It should be noted that the tangent of the slope of the straight-line is not affected by the value of the parameter $(\overline{H_{z0}})_m$. The value of $\bar{c}_m$ increases with the reduction of the height of the layer of the magnetic fluid, and the value $\bar{G}_{tm}$ decreases and, consequently, the increments of these values provide a positive contribution to the increase of the tangent of the slope of the straight-line.

The parameters of the magnetic field, presented in the matrix form, can be used for relevant calculations giving the values of the slope of the generating line of the cone α. For example, at $m = 7$ and $m = 3$ the values of α are equal to respectively 15 and 30 deg. Modelling estimates predict the deepening of the 'funnel' on the surface of the magnetic fluid droplet (more extensive spreading of the fluid on the walls of the tube) as the free surface of the droplet approaches the plane $Z = 0$.

Investigations of the acoustic and hydrodynamic cavitation started in the study by Rayleigh [296] where attention was given to the processes of closure of a spherical (not filled with gas) cavity in the ideal fluid. A.D. Pernik published one of the first monographs [297] with the systematic explanation of the problem of cavitation flows. The currently available data for acoustic cavitation were presented in a book by M.G. Sirotyuk [298]. Part of the material of the book is concerned with description of the factors affecting the cavitational strength of water (surface tension, viscosity, temperature, preliminary degassing, presence of impurities and surfactants). I.G. Mikhailov and V.M. Polunin [299] published the resuls of experimental studies of the effect of wettability of a solid surface on the separation of cavitation bubbles from it.

The diagram of the cavitation model of the process of fracture–restoration of the magnetic fluid bridge is shown in Fig. 7.27. The black crosshatching shows the magnetic fluid bridge with the circular orifice with a radius R in the centre. The upper and lower open surfaces of the bridge have the form of a conical surface with the axis Z. The magnetic field with the induction **B** is formed by a circular magnet, magnetised in the axial direction and positioned coaxially in relation to the axis of the glass tube in the plane of

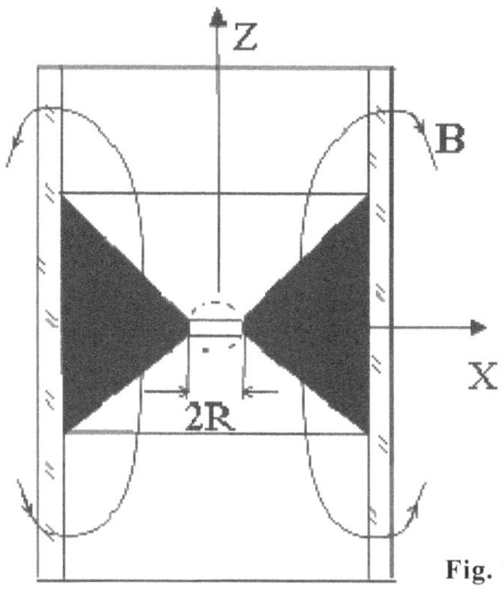

Fig. 7.27. Cavitation model.

symmetry of the magnetic fluid bridge (not shown in the figure). The simulated cavitation cavity is indicated by the dot and dash line. Thus, two spherical segments (top and bottom) are excluded in transition to the magnetic fluid bridge from the 'standard' spherical symmetric scheme of closure of the cavitation cavity.

A specific feature of the proposed model is the assumption that the function of the hydrostatic pressure in this case is played by the pressure of ponderomotive origin from the side of the heterogeneous magnetic field of the circular magnet. This assumption is made on the basis that the orifice in the magnetic fluid bridge in contrast to the cavitation cavity is not a closed surface. It is also assumed that the effect of the ponderomotive forces on both surfaces of the magnetic fluid bridge results in the distribution of pressure in the fluid and the equivalent distribution of pressure in the spherically symmetric flow [298]. In the examined model, as in the formulation of the problem by Rayleigh [296], the effect of the surface tension and gravitational forces is neglected.

We use the results of investigation of the topography of the magnetic field of the circular magnet for evaluating the ponderomotive effects of the magnetic field on the magnetic fluid bridge.

Equation (7.5) in the approximation $M_z \gg M_r$ has the form

$$\Delta f_z = -2\mu_0 S\left(M_z \frac{\partial H_z}{\partial z}\right)_{z=-\frac{b}{2}} \delta z$$

where δz is the displacement of the bridge in relation to the magnet after fracture and discharge of the portion of air ($\delta z \ll b$), i.e. $\delta z = h_d$.

Correspondingly, the estimate of the excess pressure on the magnetic fluid bridge p_0 can be obtained from the expression:

$$p_0 = -2\mu_0 \left(M_z \frac{\partial H_z}{\partial z} \right) h_d \tag{7.77}$$

The equation most frequently quoted in the acoustic cavitation literature is the equation of motion of the surface of the solitary gas-filled spherical cavity in the 'infinite' liquid with density ρ_0 [300]:

$$R\frac{d^2R}{dt^2} + \frac{3}{2}\left(\frac{dR}{dt}\right)^2 + \frac{1}{\rho_0}[p_\infty - p(R)] = 0 \tag{7.78}$$

where R is the actual radius of the cavity; t is time; p_∞ is the pressure in the fluid at infinity; $p(R)$ is the pressure on the surface of the cavity.

Using the initial conditions, according to which if $t = 0$ then $R = R_0$ and the speed of expansion of the cavity $U = 0$, integration of (7.78) at $p_\infty = p_0$ (p_0 is hydrostatic pressure) leads to the expression [297]:

$$U^2 = \frac{2}{3}\frac{p_0}{\rho}\left(1 - \frac{R_0^3}{R^3}\right) \tag{7.79}$$

where ρ_0 is the density of the fluid.

The maximum value of the rate of propagation of the cavity U_{max} is obtained at $R \to \infty$:

$$U_{max} = \sqrt{\frac{2}{3}\frac{p_0}{\rho_0}} \tag{7.80}$$

Equation (7.80) contains an important result according to which the rate of expansion of the sphere in the initial period of movement rapidly increases approaching its asymptotic value.

Figure 7.28 shows in graphical form the dependence of the relative speed $\beta = U/U_{max}$ on the radius, expressed in units of R_0. If, for example, the initial radius (the radius of the cavitation nucleus) is $R_0 = 10$ mm, and the maximum radius $R_m = 1$ mm, then, as indicated by Fig. 7.28, the mean value of the rate of expansion of the spherical cavity $\bar{U} \approx U_{max}$.

Rayleigh studied the case of a hollow cavity [$p(R) = 0$] and constant pressure at infinity, when $p_\infty = p_0$. Under these conditions from Eq. (7.78) we can obtain the rate of closure of the cavity U':

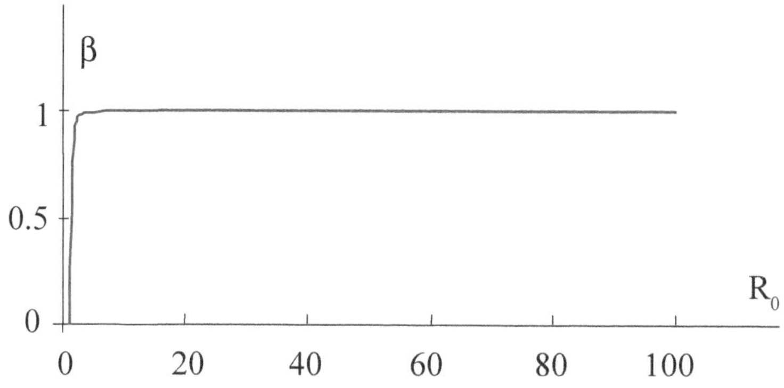

Fig. 7.28. The speed of the boundary of the expanding bubble at constant tensile stress.

$$U^{1/2} = \sqrt{\frac{2}{3}\frac{p_0}{\rho_0}\left(\frac{R_m^3}{R^3} - 1\right)} \tag{7.81}$$

where R_m is the initial (maximum) radius of the cavity; p_0 is hydrostatic pressure.

Integration of (7.81) using the Γ-functions leads to the well-known Rayleigh equation for the closure time of a hollow cavity in the hydrostatic pressure field [296]:

$$\tau_m = 0.915 \cdot R_m \cdot (\rho_0 / p_0)^{\frac{1}{2}} \tag{7.82}$$

Taking into account the previously obtained results (Table 7.3, sample MF-1, Table 7.5, Table 7.9, Fig. 7.22*b*, Fig. 7.25) it is assumed that M_z = 45 kA/m, h_d = 0.1 mm, $\partial H_z/\partial z = 4.6 \cdot 10^6$ A/m². After substituting these values into (7.77) we obtain $p_0 \approx 50$ Pa.

The evaluation of the closure time of the orifice using equation (7.82) when substituting the resultant value of p_0, and also $R_m = 10^{-3}$ m, $\rho_0 = 10^3$ kg/m³ gives $\tau_m \approx 4.5$ ms which is in good agreement with the experimental data (according to the data in Fig. 7.22 τ_m = 5 ms).

The rate of propagation of the cavity U_{max}, calculated from equation (7.80), is 0.18 m/s. The average rate of opening of the cavity, obtained from the experimental data, is calculated from the equation

$$\overline{U} = R_{max} / \tau$$

where τ is the duration of the process from its start to reaching the maximum of the curve of the dependence of U on τ (Fig. 7.22*b*). Consequently:

$$\overline{U} = 0.24 \text{ m/s}.$$

Thus, the approximate equality $\overline{U} \approx U_{max}$ is fulfilled. These examples show that the results of experimental examination of the process of expansion–closure of the orifice in the magnetic fluid bridge and the calculations based on the cavitation model are in satisfactory agreement with each other.

The rupture strength of the fluid is determined by the presence in the fluid of cavitation nuclei and by their distribution in the volume [298]. Therefore, it is interesting to study the process of displacement of microscopic gas bubbles under the effect of ponderomotive forces.

The air bubble is affected by the force [53]

$$F_b = -4\pi\mu_0 M\nabla H \cdot R_b^3 / 3$$

from the side of the heterogeneous magnetic field in the magnetic fluid. Here R_b is the radius of the cavitation nucleus.

In the plane $Z = 0$, the bubble is situated in the inhomogeneous magnetic field with the radial component $\Delta H_z / \Delta r$, and consequently, the bubble travels in the viscous liquid to the centre of symmetry of the magnetic field (into the region with the lower strength of the field) with the average speed:

$$\overline{v} = 2\mu_0 M\overline{\nabla H} R_b^2 / 9\eta$$

The bubble travels from the wall to the centre during the time

$$\overline{\tau} = \frac{d}{2\overline{v}} = \frac{9d \cdot \eta}{4\mu_0 M\overline{\nabla H} R_b^2}$$

The radius of the neck of the bubble r_b at separation of the bubble from the wall is obtained from the equation:

$$r_b = 2\mu_0 M\nabla H R_b^3 / 3\sigma$$

where σ is the surface tension coefficient of the magnetic fluid.

Assuming that $\overline{\nabla H} = 3.91 \cdot 10^6$ A/m², $M = 40 \cdot 10^3$ A/m, $\eta = 4.7 \cdot 10^{-3}$ Pa·s, $\sigma = 28 \cdot 10^{-3}$ N/m, $R_b = 10$ μm, we obtain: $\overline{v} = 9.3 \cdot 10^{-4}$ m/s, $\overline{\tau} = 7$ s, $r_b = 4.7 \cdot 10^{-3}$ μm.

Therefore, the central part of the magnetic fluid membrane, representing the thinnest part of the magnetic fluid bridge as a result of the overflow of the fluid into the near-wall region (the region with the higher strength of the magnetic field), is characterised by the minimum mechanical strength as a result of the buildup of cavitation nuclei.

It may be assumed that the data from this area, in particular, the information on the physical–chemical nature of the cavitation strength of the fluids, can be used for magnetic fluid hermetizers and magnetic fluid sealants [220].

8

Acoustomagnetic spectroscopy

8.1. Dispersion of the speed of sound in the fluid – cylindrical shell system. Description of the problem

The distinctive features that characterize the wave processes in the fluid in a pipe are the dispersion of the speed of sound due to the influence of the pipe walls and also the initiation at certain frequencies of higher oscillation modes having phase velocities that exceed ten times the speed of sound in the free liquid. Identification of the various modes of vibrations occurring in the liquid–shell system is the essence of acoustomagnetic spectroscopy which is described in this chapter.

The speed of sound in a fluid filling the tube may substantially differ from the speed of sound in the same but unlimited liquid. One of the first problems of the propagation of elastic waves in the liquid–shell system was studied by Korteweg [271].

In contrast to a pipe with gas, to solve the problem of propagation of sound in the pipe it can be assumed that the walls of the pipe are absolutely rigid, that is, the normal component of particle velocity on the walls is vanishingly small [71], and in a pipe filled with a fluid, the walls also oscillate. This leads to a marked decrease in the speed of sound. Changing the speed of sound as a result of compliance of the walls depends on the geometrical parameters of the pipe and mechanical characteristics of the wall material and also on the frequency of the excited oscillations.

In [272], a formula was derived to calculate the speed of sound in the fluid filling the pipe at a known speed of sound in the free liquid. If the frequency of propagating oscillations in the fluid–tube is very low, i.e. the sound pressure varies slightly along the axis of the pipe, the pipe stiffness in bending in the longitudinal direction can be neglected. As a result, we obtain a formula for the speed of sound waves in the fluid–pipe system at low frequencies, known as the Korteweg equation. Let us examine this question in more detail.

In the case of low frequencies the pipe will expand and contract in time with oscillations in the fluid as a flexible sleeve. So we may consider the separately taken annular element of the pipe with width dx, echoing the vibrations of the entire pipe as a whole.

We express the compressibility of the liquid β from the formula for speed sound in an infinite medium c_0:

$$\beta = 1/\rho_f c_0^2$$

where ρ_f is the density of the fluid.

The compliance of the pipe walls leads to increased compressibility of the fluid, or in other words, to additional compressibility β'. Under the influence of sound pressure P the volume of the fluid disk inside the annular tube element is changed to the following value:

$$dV = P\beta\pi R_2^2 dx$$

where R_2 is the inner radius of the tube.

Furthermore, due to compliance of the walls there is an additional change in volume. Force f, with which the internal pressure acts on the half of the pipe ring is:

$$f = 2PR_2\, dx$$

The are of the contact surface of the half rings S is:

$$S = 2hdx$$

where $h = R_1 - R_2$ is the wall thickness; R_1 is the outer radius. The tangential stress σ in the material of the walls is equal to:

$$\sigma = \frac{f}{S} = \frac{PR_2}{h}$$

The radius of the tube under the effect of this stress changes by

$$dR_2 = \frac{PR_2}{hE'} R$$

where R is the average radius; E' is the modulus of elasticity of the wall material, as presented

$$E' = E(1 - \nu'),$$

and wherein E and ν' are the Young's modulus and the Poisson ratio of the wall material, respectively.

E' is defined as follows:

$$E' = \rho_t c_p^2,$$

where ρ_t – the density of the pipe wall, c_p – the speed of propagation of longitudinal waves in the pipe wall material.

The volume change associated with the compliance of the walls:

$$dV' = 2\pi R_2 dR_2 dx = 2\pi R_2 \frac{PR_2 R}{hE'} dx.$$

The total change in volume

$$dV + dV' = \pi R_2^2 P \beta \left(1 + \frac{2}{\beta E'} \frac{R}{h}\right) dx.$$

Due to the finite compliance of the material of the walls of the tube the fluid filling it acquires additional compressibility β', so that the total compressibility will be equal to:

$$\beta + \beta' = \frac{dV + dV'}{VP} = \beta\left(1 + \frac{2}{\beta E'} \frac{R}{h}\right).$$

The result is a formula for the speed of the acoustic wave in the fluid–pipe system at low frequencies, known as the Korteweg formula:

$$c^2 = \frac{1}{\rho_f(\beta + \beta')} = \frac{1}{\rho_f \beta(1 + \frac{2}{\beta E'} \frac{R}{h}} = \frac{c_0^2}{1 + \frac{2}{\beta E'} \frac{R}{h}}. \tag{8.1}$$

Neglecting the mass of the pipe element and taking into account only its elasticity, we the following formula relating c and c_0:

$$c = \frac{c_0}{\sqrt{1 + \frac{2}{\beta E'} \cdot \frac{(R_1 / R_2)^2 + 1}{(R_1 / R_2)^2 - 1}}},$$

where R_1 is the outer radius of the pipe.

In [272] the Korteweg equation is also derived for the case of higher frequencies at which the pipe walls begin to move faster

$$c = \frac{c_0}{\sqrt{1 + \frac{2}{\beta E'} \cdot \frac{(R_1 / R_2)^2 + 1}{(R_1 / R_2)^2 - 1}\left[1 - \left(\nu/_{\nu_0}\right)^2\right]^{-1}}}, \tag{8.2}$$

where ν is the frequency of sound vibrations introduced into the system; ν_0 is the radial resonance frequency of the ring element tube:

$$\nu_0 = \sqrt{\frac{E'}{\rho_t R R_2 4\pi^2}} \approx \frac{c_p}{2\pi R}.$$

This classic formula is widely used in the calculations for pipes. However, as shown by the studies carried out, it can be used in a limited frequency range and for a very limited range of the parameters of shells. The restrictions are due to the neglect in the derivation of the equation of the flexural rigidity of the pipe or, in other words, the elastic coupling between the individual annular elements.

In [273, 274] it is shown that there are many resonance frequencies in different shells instead of one, as follows from Korteweg' theory.

Boyle and Field [84] believe that the stiffness of the wall takes an intermediate value between the limiting cases of the absolutely soft and absolutely rigid walls. They obtained the following dispersion relation for the speed of sound in the liquid filling the pipe:

$$\left(\frac{\rho_t}{\rho_f} - \frac{E'}{R\omega^2\rho_f}\right)\frac{h}{R} = \begin{cases} \dfrac{J_0(x)}{xJ_1(x)}, x \geq 0 \\ -\dfrac{I_0(x)}{xI_1(x)}, x \leq 0 \end{cases}. \tag{8.3}$$

Here ρ_t and ρ_f are density of the pipe wall and the fluid respectively; ω is the angular frequency of oscillations; $J(x)$ and $I(x)$ are the Bessel functions of zero and first order of the real and imaginary argument; $x = kR$, where k is the wave number determined from the boundary conditions.

The propagation of multiple modes of vibrations at the same time in a steel shell filled with water was studied by Fay [275]. When the pipe contains two or more simultaneous modes of vibrations, the sound pressure along the axis of the shell changes in a complicated manner. The author examines the full oscillations of a shell, representing a superposition of longitudinal and transverse waves. However, the conclusions of this work contradict the known experimental data and also our results [79, 92].

Guelke and Bunn [85–87] developed a theory that describes the dispersion phenomena in tubes with a gas and a fluid, based on the method of electroacoustic analogies. Their formula for low-frequency coincides with Korteweg's formula for low frequencies.

Very often there are cases where an incompressible fluid fills a compliant cylindrical shell and these cases were studied in the works of Moodie and Haddow [276, 277]. The assumption of the incompressibility of the fluid can be accepted if its compressibility is much smaller than the compliance of the pipe walls. For example, for water in a rubber tube or the blood in the arteries this condition is well satisfied.

The propagation of sound in the fluid that fills the shell with very thick (in the limit infinitely thick) walls, was considered in Safaai-Jazi *et al* [278], in which the equations are solved for the vector potential in the shell and in the fluid with the transverse waves in the shell taken intp account.

Jacobi [88] studied experimentally and theoretically the propagation of sound waves in a fluid filling cylindrical shells with different boundary conditions without taking losses into account. In particular, the following were considered: a fluid cylinder with absolutely rigid walls; a fluid cylinder with absolutely soft walls and a liquid cylinder immersed in an unlimited amount of the fluid; a fluid cylinder with liquid walls; a fluid cylinder bounded by a thin elastic membrane. The following dispersion relations were obtained for the speed of sound:

$$1-\frac{1}{\omega*^2}+\frac{h^2\omega*^2}{12R^2}\left(\frac{1}{c^{*2}}-\frac{1}{\gamma^2}\right)^2=\begin{cases}\dfrac{\alpha^2 J_0(x)}{xJ_1(x)}, x\geq 0\\[2ex] -\dfrac{\alpha^2 I_0(x)}{xI_1(x)}, x\leq 0\end{cases}. \tag{8.4}$$

Here

$$x=\omega*\left(1/\gamma^2-1/c^{*2}\right)^{1/2}, \omega*=\frac{2\pi\nu R}{c_p}, \gamma^2=c_0^2/c_p^2, \alpha^2=\frac{\rho_f R}{\rho_t h}, c^*=c/c_p$$

The distribution of the pipe with the fluid of modes of the zero, first and second order was studied in detail by V.N. Merkulov, V.Yu.Prikhod'ko and V.V. Tyutekin [279]. In this paper, the various harmonics of each mode are called normal waves. A distinctive feature of the work is that it addresses not only axially symmetric modes of oscillations, but also the non-axisymmetric modes. The wave field in the fluid–tube system is described by the following dispersion relation:

$$xJ_n'(x)+g_nJ_n(x)=0 \tag{8.5}$$

where n is the order of the Bessel function,

$$g_n = \frac{\rho_f \omega^{*2} R}{\rho_t \det\|C_{jk}\| h}(C_{11}C_{22} - C_{12}^2),\, C_{11} = \omega^{*2} - \alpha^2 - \frac{1-v'}{2}n^2,$$

$$C_{12} = C_{21} = -\frac{1-v'}{2}n\alpha,\, C_{13} = C_{31} = iv'\alpha,\, C_{23} = i\left(n + \frac{h^2 v'}{8R^2(1-v')}(n-n^3)\right),$$

$$C_{32} = in,\, C_{22} = \omega^{*2} - n^2 - \frac{1-v'}{2}\alpha^2,$$

$$C_{33} = -\omega^{*2} + 1 + \frac{h^2}{24R^2(1-v')}\left[2(1-v')(\alpha^2+n^2)^2 + 2 + v' - n^2(4-v')\right]$$

where $\alpha = kR$.

The propagation of axially symmetric oscillations in cylindrical shells containing a liquid was studied in Lin and Morgan [118]. The approach to solving the problem is based on 'stitching' cylindrical shell oscillation equations with the equations of fluid motion. As a result, the following dispersion relations are obtained:

$$1 - \frac{1}{\omega^{*2}} + \frac{v'^2}{\omega^{*2}(1-c^{*2})} - \frac{\xi\omega^{*2}(1-N_*c^{*2})}{c^{*2}(c^{*2}+\xi\eta\omega^{*2}(1-N_*c^{*2}))} = \begin{cases} \dfrac{\alpha^2 J_0(x)}{xJ_1(x)}, x \ge 0 \\ -\dfrac{\alpha^2 I_0(x)}{xI_1(x)}, x \le 0 \end{cases} \quad (8.6)$$

$$x = \omega^*(1/\gamma_p^2 - 1/c^{*2})^{1/2};\, \omega^* = \frac{2\pi v R}{c_p}, c^* = c/c_p, \xi = \frac{h^2}{12R^2};$$

$$\eta = \frac{2}{(1-v')K}; \gamma_p^2 = c_0^2/c_p^2; \alpha^2 = \frac{\rho_f R}{\rho_t h}.$$

The value of N_* characterizes the contribution of the rotational inertia of the shell element. If $N_* = 1$, the rotational inertia is taken into account, if $N_* = 0$, then it is neglected. The value of η corresponding to the bending force. The parameter K is a constant for small adjustments of formulas and is assumed to be 0.99.

Equations (8.4) and (8.6) differ in left-hand parts. They have different third terms and the fourth term in (8.6), associated with the rotational inertia and bending force, is not found in (8.4). This is due to the fact that Jacobi used simpler equations of oscillations of a shell than Lin and Morgan.

Equation (8.5) for $n = 0$ is similar to equations (8.2)–(8.4) (8.6). From (8.6) it follows that there are two axial modes that exist in the entire frequency range. One of them, propagating at speeds lower than the speed of sound in a free liquid, is characterized by the fact that the oscillations are mainly localized in the fluid. Our method is mainly designed to indicate this mode. The other mode is localized mainly in the membrane and our method does not record it.

The listed theoretical constructions are used form comparison with experimental data obtained by the method based on the acousto-magnetic effect (AME).

There is a large body of theoretical work in this area, based on different assumptions and lead to different models describing the dispersion of the speed of sound in such systems. Already from the above list of works spanning more than a hundred years we see the obvious interest in this area of physical acoustics from scholars around the world.

However, these experimental studies of the speed of sound in the fluid filling a pipe in the available literature are sparse and unsystematic. Therefore, it is not possible to select the most appropriate of the existing theories which leads to the task of conducting this kind of research on a new methodical basis.

8.2. The experimental technique based on the AME

Experimental studies of the speed of propagation of sound in a fluid–pipe system, made by traditional methods, are few and unsystematic. From the methodological standpoint they have significant drawbacks, making it difficult to choose an adequate theoretical model.

R.W. Boyle and G.S. Field [84] determined the phase velocity in the pipe, measuring the distance between the rows of cavitation bubbles formed in the acoustic wave nodes. The cavitation method, based on the appearance of air bubbles in the sound wave nodes with measuring of the distance between them, is applied using sufficiently powerful sound waves, followed by heating of the studied object and the non-linear distortion of the sound field. In addition, the error of this method increases due to the fact that in the cavitation zone the fluid compressibility increased significantly and the speed of sound decreases accordingly. The authors of [84] did not evaluate this error. The generality of the method is also limited by the fact that it is not possible to use opaque shells and the experimental equipment used to create high-intensity ultrasound is complicated.

Guelke and Bunn [85–87] measured the speed of sound through microphones in glass, perspex and rubber tubes, filled with water. The microphone was moved along the pipe, alternately passing through the nodes and antinodes of the standing wave. The use of the microphone introduces diffraction distortions in the wave field but the impact of the field on the results of the experiment was not determined. Furthermore, the use of the microphone in the tube is rather difficult and in capillary tubes it is virtually impossible. Jacobi [88] published data for a copper pipe filled with water. The method for determining the speed of sound was not described but it is likely tht the method is based on using the microphone.

The authors of the experimental works do not explain how they made measurements in cases where two acoustic modes propagated simultaneously in the pipe. However, this issue is important, since the propagation of the two modes at the same time results in a complex interference pattern that is difficult to decipher.

A common feature, typical of previously used methods to measure the speed of sound in tubes with a fluid, id a large measurement error due to the different the factors discussed above. This is partly due to the lack of experimental data on the measurement of the speed of sound in the fluid–pipe system, cited in the literature. Therefore there is a clear need for the use of a new method, free of these shortcomings, in measurements in such systems.

V.M. Polunin and I.E. Dmitriev [79, 92] published the experimental results for the dispersion of the speed of sound in the magnetic fluid–cylindrical shell system on the basis of identification and comparison of these findings with the existing model theories.

Qualitative improvement of the experimental technique is achieved using as a receiver of sound waves a magnetic head instead of a hydrophone. The magnetic head is placed outside the pipe and is not acoustically connected to the fluid [89]. This makes it possible to move the receiver along the pipe with the fluid over long distances without disturbing this fluid. There is no need to take into account the diffraction perturbation introduced by the microphone.

The magnetic head consists of a coil and a DC magnetic field source which can be either a permanent magnet or an electromagnet.

Figure 8.1 shows the experimental setup. The source of elastic waves is the emitter 3, containing a piezoelectric plate with a resonance freqency of 1 MHz. Measurements are taken at frequencies not higher than 200 kHz, i.e. away from the resonance frequency

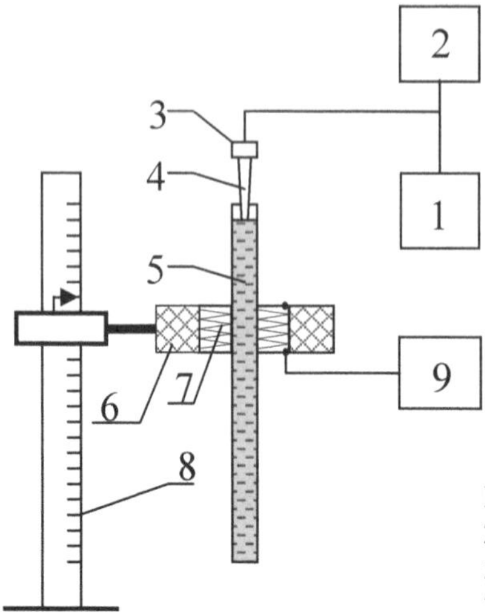

Fig. 8.1. Experimental setup: 1 – oscillator; 2 – frequency meter; 3 – emitter; 4 – waveguide 5 – tube with magnetic fluid, 6 – annular magnet, 7 – inductor, 8 – cathetometer, 9 – oscilloscope.

so that its frequency characteristic can therefore be considered linear (plateau shaped). The variable EMF source is oscillator 1. The elastic waves pass through the waveguide 4 to the MF 5 filling the pipe. The magnetic head includes annular permanent magnet 6, magnetized axially, and coil 7 disposed within the magnet and rigidly connected therewith. Between the coil and the tube there is an air gap. The annular magnet at the location of the coil magnetizes the fluid preferentially along the tube axis (detailed description of the magnetic field of the annular magnet is provided in section 7.6.3). The variable EMF, induced in the coil during propagation of the sound wave in the fluid, is sent to the oscilloscope 9. The magnetic head is placed on the kinematic unit of the cathetometer 8 and is free to move along the tube.

With the convenient design of fastenings the tube with the magnetic fluid can be easily replaced. If the sound wave propagate in the fluid filling the tube, the acoustomagnetic effect excites variable EMF in the coils. The EMF induced in the magnetic head arrives at the receiver input. The oscilloscope can be replaced by a voltmeter, but the oscilloscope shows the time base signal and can be used to control its sinusoidal form. This allows one to see the excitation of parasitic frequency, distorting the shape of the main signal and eliminate the causes of this frequency (e.g., the sound emitter touches the tube or the pipe loose, etc.).

When pouring the fluid into the pipe which already has the annular magnet with a measuring coil, the fluid either strongly bubbles (in the case of low concentration MF and a weak annular magnet), or is delayed by the magnet without leaking below its level. In this variant variation of the installation the annular magnet is composed of two half-rings which are mounted on the magnetic head already after pouring a fluid into the tube after a time required by the air bubbles to exit from the fluid. The strength of the annular magnet field at its centre is about ~115 kA/m so that the fluid is in a state close to the magnetic saturation.

The magnetic head is rigidly connected to the carriage of the cathetometer that allows to move it to a distance of 1 m with an accuracy of 0.01 mm. Therefore, the lower limit of the possible test frequency range is ≈ 4 kHz.

In the experiments conducted to investigate the low-frequency (piston) oscillation mode, the sound oscillations were introduced in the fluid not from the top, as shown in Fig. 8.1, but at the bottom using a transducer is pressed against the lower surface of the tube bottom.

The schematic of the interferometer is shown in Fig. 8.2.

The glass tube 1 is filled to a certain level with MF-2. At the bottom the tube is closed by a thin partition 5 to which a piezoceramic plate 6, being the source of sound oscillations, is pressed using an acoustic contact fluid. The source of a constant magnetic field is annular magnet 3, magnetized along the axis. The magnet is mounted coaxially with the tube on a special carriage. The disturbances of the magnetic flux, caused by the sound wave, are recorded by the induction method using the measuring coil 4 constituting an air gap with the outer surface of the tube and placed

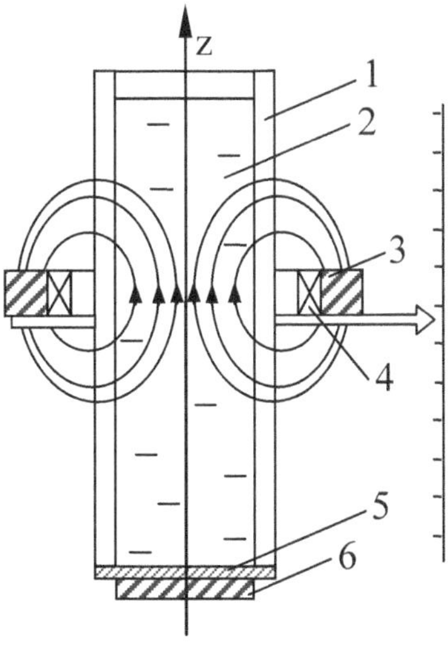

Fig. 8.2. Magnetic fluid interferometer.

in the cradle inside the annular magnet. Movement of the magnetic head is provided and secured by a cathetometer, conditionally depicted as a linear scale and the arrow–pointer. The fundamental resonant frequency of the piezoplate is 2.5 MHz. The frequency of variable EMF, supplied to the piezoplate, is 10–100 kHz, so we can assume that the excitation of the vibrations by the piezoplate takes place on a plateau of the resonance curve.

In experiments with separate frequencies the oscillograms were distorted; one possible reason could be the presence of natural oscillations of the membrane transition element.

Reflected from the free surface of the magnetic fluid, the sound wave is superimposed on a direct wave, resulting in a system of standing waves formed at certain frequencies in the liquid column. In the antinodes of the pressure there is the largest amplitude modulation of the magnetic flux penetrating through the measuring coil, and in the nodes – the minimum modulation due mainly to the presence of a travelling wave. So if the carriage is moved along the column of the MF there is a periodically repeating pattern of highs and lows of the EMF induced in the measuring coil. As in the conventional ultrasonic interferometer, the measurement of the speed of sound in the fluid is reduced to finding the number of standing waves N_s, fitting into the length ΔL. The calculation is performed using the formula:

$$c = 2\Delta L v / N_s \cdot$$

8.3. Some of special features of the study of oscillation modes

It is known that the amplitude of the excited vibration modes in the tube (normal waves) depends on the mode of excitation. In the piston excitation of sound, mainly a flat wave propagates in the system. At point introduction of sound into the fluid any oscillation mode can propagate in the fluid–pipe system, i.e. the point source excites the entire spectrum of oscillations, while the transducer pressed against the bottom, excites mainly the piston mode. This phenomenon causes a transition to introducing the sound into the tube from the top.

An important characteristic is also the size of the receiver coil. If the linear dimension of the coil along the axis of the tube is equal to the length of the wave, the EMF in it will not be excited as a result of compensation of each other by oscillations with opposite phases.

Therefore, the linear dimension of the coil must not exceed at least half the wavelength of the sound.

The inductor with the linear size of 5 mm was used initially. A simple estimate shows that in this case, the speed of sound in the MF is 1000 m/s and the upper limit of the test frequency range is 100 kHz. Since the higher modes of oscillations which should be studied are excited at higher frequencies, a coil with a linear size of 1 mm was made. Regardless of reducing the number of turns, due to the use of the high-sensitivity oscilloscope, the amplitude of the excited EMF is quite sufficient for measurements.

In the experiment, when the frequency increases beyond a certain value higher modes oscillations are excited in the pipe with the MF. When each successive mode is excited, its speed is very high, and with a further increase in frequency rapidly at first and then more slowly falls. In the case where the amplitudes of the oscillation modes propagating in the tube are close, it is impossible to determine the speed of the modes. The spatial distribution of the amplitude of the oscillations differs from sinusoidal characteristic of the piston mode, and represents a complicated interference pattern.

Due to the complexity of analysis, in [79, 92] the investigated frequency range is divided into monochromatic bands, in which the amplitude of one of the modes of oscillation prevails over the others: the piston mode band, the band of the first radial mode, etc. In the frequency range in which the amplitudes of the modes are close, analysis has not been carried out.

A limitation of this method is the fact that the pipe containing the fluid must be made of a non-magnetic non-conductive material, which precludes the study of metal shells but almost all other materials can be studied.

Dependence $e_m(Z)$ is characterized by the periodicity corresponding to the distribution of sound pressure in the standing wave. However there is a danger that a permanent magnetic field of the annular magnet (~15 kA/m) leads to a distortion of the data due to the dependence of the speed of sound in the MF on the strength of the magnetic field. This concern is not substantiated; many of our measurements and measurements of other researchers show that the speed of sound in MF of the first and second type, i.e. in stable fluids, does not depend in the 1–2% measurement error range on the strength of the magnetic field, up to about ~ 500 kA/m.

The speed of sound is calculated as follows. At some distance from the bottom of the tube, not less than $3\lambda/2$, we fix the minimum amplitude of oscillations of the induced EMF and measured its coordinate x_d. The coordinate is measured on a scale of a V-630 cathetometer with an accuracy of 0.01 mm. 30 counts of the coordinate x_d are made. After counting the sample coordinates of the lower minimum x_d the slide of the cathetometer is moved up to a certain upper minimum value of the amplitude of the EMF x_u, with the number of half-waves N between x_d and x_u recorded. When using the oscilloscope mode in the external synchronization mode the phase shift is visible on the screen during the passage of the magnetic head through the nodes of the standing wave.

The coordinate of the upper minimum was also determined 30 times. In the case of determining the coordinates of the minimum of the standing sound waves there was some blurring of the minimum associated with the fact that the sinusoid varies slightly near the minimum, and the inaccuracy of observation on the oscilloscope is the most significant factor contributing to the measurement error. Therefore, this error substantially exceeds the error of the cathetometer.

To generate a hypothesis about the nature of the distribution of the error of measurement the experimental data were used to construct histograms. The form of histograms was used to proposed a hypothesis of the uniform error distribution with borders, equal to the maximum value of the deviation of the coordinates from the calculated average. The Pearson criterion was used verify compliance of the chosen distribution with the reality.

To construct the histogram, the following value was calculated

$$\chi^2 = \sum_{i=1}^{k} \frac{(n_i - np_i)^2}{np_i},$$

where n_i is the number of observations in the division interval; n is the sample size (in our case $n = 30$); p_i is the theoretically calculated probability of observations in a given interval.

The value of χ^2 was compared with the Pearson criterion χ^2_{1-p}, taken from tables. For a confidence level of 0.99 (level of significance 0.01) and the number of degrees of freedom $f = k$–3, where k is the number of intervals into which the histogram was divided, the hypothesis should be rejected when $\chi^2 \geq \chi^2_{1-p}$.The calculations show the validity of the hypothesis of the uniform distribution of the error in determining the coordinates of the minima

The most probable value of the coordinate is the average from the sample, and the error is half the length of the interval of uniform distribution. From the measured values of the coordinates x_u and x_d we calculate the wavelength of sound by the formula:

$$\lambda = 2(x_u - x_d) / N$$

and its error

$$\delta\lambda = \frac{1}{n}\sqrt{\sum \delta x_u^2 + \sum \delta x_d^2}.$$

Then we calculated the speed of sound c in a magnetic fluid filling the tube:

$$c = \lambda\nu.$$

The frequency of sound waves ν, introduced into the system, was measured with the relative error of 0.005%. Despite the small value of $\delta\nu$, this value is included in the calculation formulas, not to limit the generality of consideration. Error velocity measurement determined by the formula calculation error of indirect measurements:

$$\varepsilon_c = \frac{\delta\lambda}{\lambda} + \frac{\delta\nu}{\nu}.$$

According to the estimates, the error in determining the speed of sound in the MF in the tube does not exceed 1%.

The formulas (8.3)–(8.6), used for the construction of dispersion curves, are transcendental equations containing Bessel functions of various orders. A fairly rapid solution of these equations is possible only by numerical methods with the use of computers. Therefore, appropriate programs, enabling calculation of the dispersion curves by setting the parameters of the pipes and of the fluid filling, them were developed.

8.4. On the influence of inhomogeneity of the magnetic field

There may be also the question of the legality of the use of the source of the inhomogeneous magnetic field for measuring the speed of sound in the MF since there are theoretical studies which allow their authors to suggest the presence of a field dependence of the speed of sound waves solely on the basis the magnetic field inhomogeneity.

In a theoretical paper [252] the results were used to find frequency-dependent expressions for the speed of propagation and the absorption coefficient of sound reflecting the force effect on ferroparticles by inhomogeneous magnetic fields. The resulting rather cumbersome mathematical expressions of these parameters are considered as general expressions for the acoustic waves in magnetic fluids taking into account the restructuring of the fluid in a wide range of density, concentration, pressure, temperature, the frequency of oscillations and the external inhomogeneous magnetic field. The MF model used does not consider the rotation of the ferroparticle around its axis, the presence of aggregates of the ferroparticles, and Néel relaxation time is assumed to be zero. The numerical calculation was carried out of the frequency dependence of the absorption coefficient and the speed of sound in the MF, prepared on the basis of kerosene and Fe_3O_4 magnetic particles for values of the gradient of the strength of the field of 10 A/m^2 and 10^3 A/m^2 in the frequency range 10^5–10^{13} Hz. According to the data presented in graphical form the speed of sound monotonically increases with frequency (the absolute value of the increment of the velocity of sound is not shown). For a uniform magnetic field, as regards this dependence the theory predicts a zero effect.

In contrast, the study [311] shows the dependences of the anisotropy of the speed of ultrasound in the magnetic fluid based on water with a mass concentration of magnetite particles $\varphi_m = 0.4$, magnetization saturation $M_S = 30$ G and a density of $\rho = 1.38$ g/cm^3 in a uniform magnetic field. Measurements of the anisotropy of the speed of ultrasound were performed for four values of the magnetic field $B_1 = 50$ mT, $B_2 = 150$ mT, $B_3 = 250$ mT, $B_4 = 350$ mT at a fixed temperature of 25°C. The MF sample was placed in a measurement cell, in which the distance between the piezoelectric transducers was 32 mm, ultrasonic frequency was 1 MHz. The experimental results are represented by graphic dependences of the relative changes in the speed of ultrasonic on the angle α between the wave vector and the direction of the magnetizing field. The angle α was varied in range 0°–180°. It should be noted that the experimental curves are not symmetrical with respect to $\alpha = 90°$, which, according to the authors, due to flaws in the experiment setup. In the experimental study of the speed of ultrasound in unstable magnetic fluids a very important parameter is the holding time at a certain value of the magnetizing field. Since the formation of chain aggregates is diffusive in nature, this time depending on the concentration of magnetic material particles, may range from tens to hundreds of hours. The structure of the MF formed as a result of these processes is highly variable, leading to the partial destruction of chain aggregates with the change of the angle α. Therefore, a certain time is needed to establish the dynamic equilibrium in the structure of the MF. Experimentally, the holding time was only 1 min after the change of angle α. As a result, after placing the MF in the magnetizing field, the values of the speed of ultrasound in the MF in the absence of an external magnetic field become different and depend on the size of the field.

In the monograph [260] in the discussion of the physical nature of the field dependence of the speed of sound the magnetic fluids are considered as dispersion media, which in addition to the mechanisms characteristic of the continuum (induced magnetic field inhomogeneity field in the sound wave [12]), there are also mechanisms associated with features of the structure of magnetic colloids ('slippage' particles relative to the liquid matrix [126, 128]), with the latter leading to a significant change in the speed of sound in the magnetic field. It is shown that the magnetic field inhomogeneity may affect the speed of sound in the magnetic fluid with a stable structure at the level of modern measurement error (~1 m/s) only in strongly inhomogeneous magnetic fields.

Putting the question open, we have attempted to determine this dependence by experiments [301]. For this purpose, experimental studies of the speed of the sound wave in the MF of the degree of inhomogeneity of the magnetic field of a permanent magnet were carried out. The diagram of the experimental setup is given in Fig. 2.10.

The magnetic fluid fills the glass tube, the lower part of which is located between the poles of a permanent magnet. The magnetic fluid column height was 355 mm. The parameters of the glass tube: glass HC–3, Young's modulus $E = 7.26 \cdot 10^{10}$ Pa, Poisson ratio $\nu = 0.21$, density $\rho_t = 2400$ kg/m^3, the velocity of longitudinal waves $c_p = 5500$ m/s; $R_1 = 8$ mm, $R_2 = 6.9$ mm, $h = 1.1$ mm.

Ultrasound was introduced into the column of the magnetic fluid using the acoustic cell schematically shown in Figure 9.4. When moving the coil along the surface of the tube the oscilloscope screen showed periodically alternating maxima and minima of the amplitude of the received signal corresponding to the nodes and antinodes of the standing wave. The relative error in determining the speed of sound in this case did not exceed 1%.

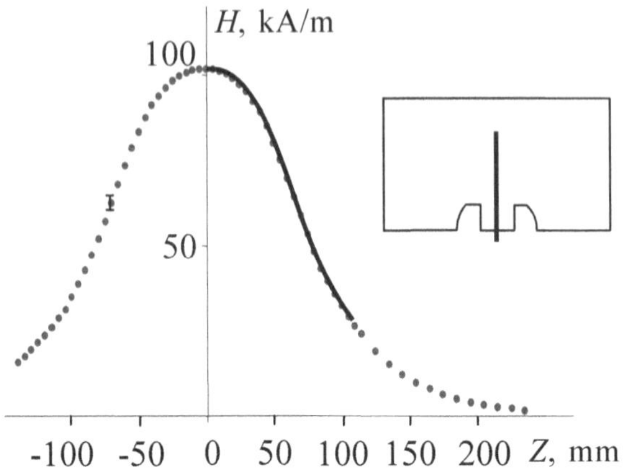

Fig. 8.3. Comparison of the calculated and experimental values of the projection the magnetic field strength.

The system consists of a permanent magnet 12 (Fig. 2.10) the magnetic field of which in the pole gap and its surroundings has been sufficiently studied in detail. Measurements were taken of the component of the magnetic field strength normal to the vertical axis $0Z$ and positioned symmetrically to the middle of the pole tips. The experimental data and the lines along which the graph was plotted on the background of the overall computational domain, are shown in Fig. 8.3.

Theoretical analysis of the magnetic field was carried out on the basis of a model in which the magnet tips are magnetized with magnetization M constant in the volume and directed along its axis.

The components of the magnetic field are determined by the formula [57, 301]: $\mathbf{B} = -\operatorname{grad}\psi$, where the scalar potential satisfies the elliptic differential equation $\nabla^2\psi = 0$ with zero boundary conditions of the second kind $(\operatorname{grad}\psi)\cdot\mathbf{n} = 0$ on the surface of the magnet, given by the normal vector $\mathbf{n}$. Moreover, the condition $\psi = 0$ at infinity was applied.

Calculations were carried out in an axially symmetric geometry in the cylindrical coordinates with the axis coinciding with the axes of both magnetic tips. The dimensions and shape of the tips, and the gap between, used for calculations, are the same as the measured actual values. The calculation domain is selected so the measured experimental values magnetic field at its edges (and its scalar potential) can be taken equal to zero within the accuracy of measurement.

To determine the magnetization, as well as clarify the forms of the curved back surface of the magnetic tip (the introduction of an additional effective curvature allows taking into account the influence of the magnetic material in which the tip is secured) comparison was made with the experimentally measured value of the projection B_z of the induction of the magnetic field. The corresponding graph with the comparison is shown in Figure 8.3; the points on it are the results of direct measurements.

The calculations are performed using a program of numerical solutions of differential equations in partial derivatives FlaxPDE v. 5.0.22, based on the finite element method with the adaptive triangulation mesh.

Figure 8.4 shows the distribution of lines of force of the inhomogeneous magnetic field in a plane passing through the axis of the tube. The bold line schematically shows the permanent magnet pole tips. The scale of the vertical and horizontal is the same. The arrows indicate the relative magnitude and direction of the vector of the induction of the magnetic field. Because the tube is positioned symmetrically with respect to the pole tips (along axis Z), the lines of force intersect the side surface of the tube preferably along the normal to the axis of the tube. The main parameter of this problem is the gradient of the magnetic field used $\Delta H/\Delta Z$ calculated on the basis of the experimental data with the step $\Delta Z = 5$ mm.

Figure 8.5 shows the dependence of the gradient of the magnetic field on the coordinate along the axis $0Z$.

Two MF samples based on kerosene and magnetite and stabilized with oleic acid were studied. The physical parameters of the samples are presented in Table 8.1.

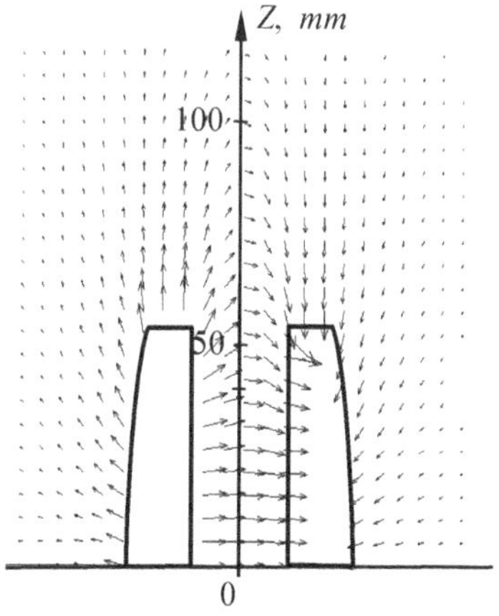

Fig. 8.4. The lines of the magnetic field.

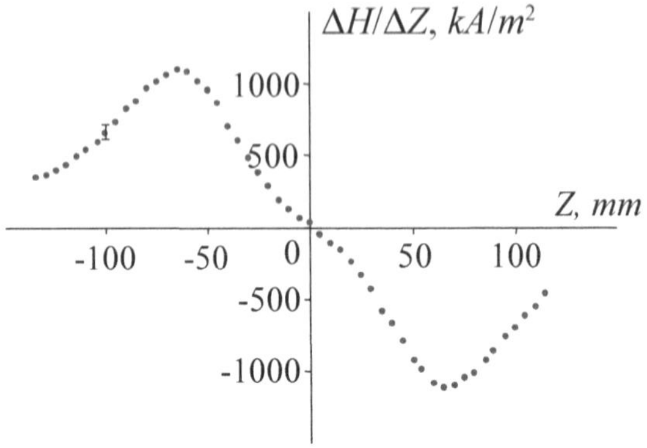

Fig.8.5. The dependence of the gradient of the strength of the magnetic field on the coordinate.

Table 8.1

Specimen	MF-1	MF-2
ρ, kg/m^3	1360	1028
φ, %	13.0	5.4
M_s, kA/m	57	25
c, m/s	930	1030

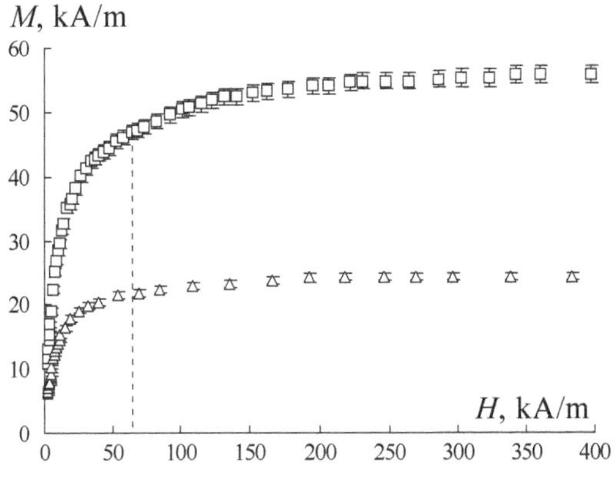

Fig. 8.6. The *M(H)* dependence: □ – MF1, Δ – MF2.

Figure 8.6 shows the curves of magnetization for MF-1 and MF-2 obtained by the ballistic method in the installation shown in Fig. 1.3. In this case the length of the ampoule with the MF greatly exceeds its diameter so that the demagnetizing field can be neglected. The vertical dotted line shows the section of the magnetic field with the highest field strength gradient.

Table 8.2

Number of half waves (distances between maxima)	21
Wavelength	24.58 mm
Speed of sound	1030 m/s
Absolute error of measurement of the coordinates of the upper and lower maxima	0.4 mm
Absolute error of determination of wavelength	0.08 mm
Absolute error of determination of speed of sound	3 m/s
Relative error of determination of speed of sound	0.5 %

The results of measurement of the speed of sound in the MF-2 sample at a frequency of 41.9 kHz at 32°C are presented in Table 8.2.

The shaded triangles in Fig. 8.7 show the results of measuring the speed of sound in MF-2 and the shaded squares – in MF-1 for different values of the gradient of the magnetic field strength.

Measurements were carried out in the direction of decreasing strength of the magnetic field. The wavelength was calculated by measuring the distance between two adjacent nodes, followed by doubling of the data.

The speed of sound measurements averaged over the entire length of the magnetic fluid column ($N = 21$) are presented in Table 8.1 and shown in Fig. 8.7 by straight lines, and the top line was obtained for MF-2, the bottom line – for MF-1.

The Korteweg formula (8.1) is used to calculate c_0 – speed of sound in an unbounded magnetic fluid. Compressibility of the liquid β is expressed by the relation $c_0 = \sqrt{\dfrac{1}{\rho_f \beta}}$, and after simple transformations we obtain

$$c_0 = c\sqrt{\frac{E' h}{E' h - 2R\rho_f c^2}}$$

When substituting the values of the parameters in this formula, the following values of the speed of sound in an infinite medium c_0 are obtained: for MF-1 – 1093 m/s, for MF-2 – 1194 m/s, which is similar to the earlier results for this type of MF [260].Thus, the data obtained, the sound speed within the measurement error is independent of the gradient magnetic field. This fact does not

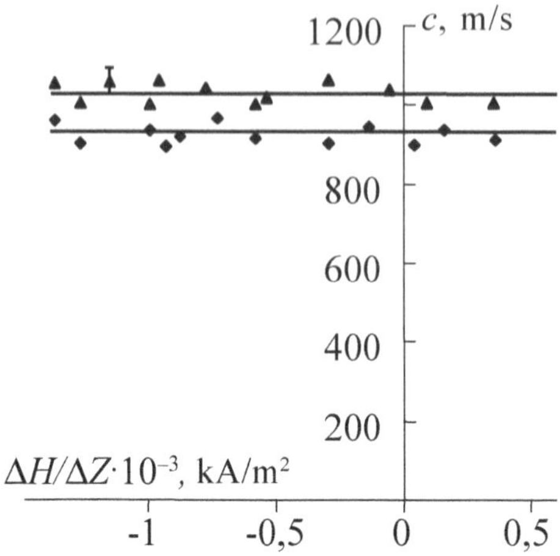

Fig. 8.7. Dependence of the speed of sound on the gradient of the strength of the magnetic field.

completely exclude the possibility of the existence of the dependence of the speed of sound on the gradient of the magnetic field strength predicted by the theory [252], but not allow us to agree with the statement of the crucial role of this parameter.

However, this result is consistent with estimates of the possible impact of the inhomogeneity of the magnetic field on the speed of propagation of sound waves in the MF [260] (see also section 3.9). The largest contribution to the increase in the speed of sound in the nanodispersed MF due to the inhomogeneity of the magnetic field can be obtained in a high-gradient magnetic field as a result of the pressure drop Δp formed in the fluid.

The static equilibrium conditions in this case takes the form:

$$\nabla p = \mu_0 M \nabla H + \rho g$$

Neglecting the hydrostatic pressure and considering that the magnetization of the samples at the points of the maximum gradient of the magnetic field according to the data (Figure 8.6) is $M_{\text{MF-1}}$ = 47 kA/m and $M_{\text{MF-2}}$ = 21 kA/m, we find for two MFs $\nabla p_{\text{MF-1}} = 0.66 \cdot 10^5$ Pa and $\nabla p_{\text{MF-2}} = 0.29 \cdot 10^5$ Pa. According to [112], the baric coefficient of the speed of sound in the MF of this type depending on the concentration of the dispersed phase is in the range (0.34–0.38)$\cdot 10^{-5}$ m/s·Pa, i.e. an average of 0.36 m/s·Pa. Therefore, under these experimental conditions in the direction ∇H the average speed along the sample increases by Δc increments, equal respectively to ~0.2 m/s and about 0.1 m/s, which is within the measurement error range.

The dominant contribution to the increment of the speed of sound in the magnetization of the MF (up to ~10 m/s) is provided by the effect of the relative motion of the phases of the dispersion system ('slippage' of the particles) taken into account by the dynamic theory [126, 128].

The presented Δc values presented should be considered as an upper estimated, obtained on the basis of the model theory.

8.5. Experimental results and analysis

At the initial stage of the study measurements were taken of the speed of the reciprocating (piston) mode by the interferometer schematically shown in Figure 8.2.

These experiments were carried out using glass tubes with different diameters and wall thicknesses filled with MFs of several concentrations. Studies were carried out in the range 10–60 kHz. At higher frequencies there were complex wave phenomena, which are the result of the superposition in the pipe with the fluid of higher oscillation modes. At higher frequencies the amplitude of the EMF excited in the coil also decreased because the wavelength was comparable to the vertical size of the coil. In these experiments, the sound was introduced into the pipe through the bottom – the membrane using the apparatus shown in Fig. 9.4, which provided the almost piston-like excitation of acoustic waves in the system.

In the pipes with a sufficiently large wall thickness at low frequency the relationship between the speed of sound c in the fluid filling the pipe and the speed of sound in an unbounded liquid medium c_0 is sufficiently well described by Korteweg' formula (8.1).

This expression was used for calculating the speed of sound in the unlimited magnetic fluid medium from the measured velocity of propagation of elastic oscillations in the magnetic fluid filling sufficiently a thick-walled tube. The criterion for determining whether the wall thickness of the tube is sufficiently large is the absence of the dispersion of the speed of sound in the fluid filling the tube.

The magnetic fluid used in the studies is a magnetic colloid in which the dispersed phase is magnetite, Fe_3O_4, the dispersion medium – kerosene and stabilizer – oleic acid.

The dependence of the speed of sound on the frequency in MF-1 with a density of 2070 kg/m^3 was studied. The fluid was poured into a tube with parameters $2R_1 = 13.7$ mm, $2R_2 = 9.7$ mm, $h/R = 0.41$. Furthermore, the density of the MFs used in the experiments was

1890 kg/m^3, 1730 kg/m^3, 1620 kg/m^3, 1450 kg/m^3, 1260 kg/m^3. The speed of sound in an unlimited MF with a density of 1260 kg/m^3, measured by the pulsed at a frequency of 2 MHz at 29°C is (1065±5) m/s, temperature coefficient of the speed in the MF was 3 m/(s·K).

The error of density measurements was determined by the error of measurement of the MF mass poured into a pycnometer, followed by weighing on the analytical scales with the measurement error of volume, given by the pycnometer. The relative error in determining the density of the MF in this case was $2 \cdot 10^{-3}$.

The dispersion of the speed of sound in the MF–thin-walled compliant cylindrical shell was studied in two thin-walled glass tubes, the first of which had the outer diameter of 11.0 mm, inner diameter 9.7 mm, h/R = 0.13, while for the second tube the values of these parameters were 13.3 mm, 12.4 mm and 0.07 respectively. The experiment was carried out at a temperature ranging from 20°C to 21.5°C. The experimental results were compared with theoretical calculations made using the Korteweg formula. c_0 used in the calculations was the value obtained in experiments with the thick walled pipe.

The higher modes of oscillations in the MF, filling elastic cylindrical shells with different parameters, were studied using a coil with a larger diameter but smaller length. The inner diameter of the coil was 24 mm, height – 1 mm. This made it possible to measure the wider tubes and extend the frequency range of the experiment on 200 kHz. To increase the amplitudes of the higher modes and also increase their contribution to the resulting sound oscillations, in further experiments the sound introduced into the tube through the top using the transducer shown in Fig. 8.1. The speed of sound was measured in glass tubes with various parameters filled with the MF-2 with a density of 1279 ± 3 kg/m^3.

In order to cover the widest possible range of elastic moduli and geometric parameters of the studied systems, experiments carried out to determine the frequency dependence of the speed of sound for the reciprocating (piston) mode and higher oscillation modes were conducted on several types of pipes made of different materials: glass, ceramics, plastics, rubber, acrylic plastic, teflon and rubber. The parameters of some of these materials are given below.

1. Pipes made of HC-3 glass. The parameters of the glass (according to the manufacturer) are: Young modulus $E = 7.26 \cdot 10^{10}$ Pa, Poisson ratio $\nu_p = 0.21$, density $\rho_t = 2400$ kg/m^3, the speed of longitudinal waves $c_p = 5500$ m/s.

2. Pipes made of ebonite, with the parameters (according to reference data): ρ_t = 1200 kg/m^3, c_p = 2400 m/s.

3. Pipes made of plastics. Since the certificate data for plastics were not available, measurements were taken of the speed of sound in the plastics (in UZIS equipment) and their density was also determined. Due to the significant damping of sound in a plastic, the speed of sound was determined with a large error: c_p = 2000 ± 400 m/s. Density measurements yielded results: ρ_t = 1100 ± 100 kg/m^3. The data given in handbooks for polystyrene: ρ = 1050 kg/m^3, c_p = 2350 m/s, $E = 2.5 \cdot 10^9$ Pa.

4. Fluoroplastic tubes having parameters (according to the handbook data): ρ_t = 2200 kg/m^3, $E = 0.45 \cdot 10^9$ Pa, c_p = 1340 m/s.

The experimental data for these tubes are shown in Figs. 8.8–8.11. For all the pipes there is a range of frequencies in which the speed of sound in the piston mode (zero mode) with growth rate slightly decreases or remains unchanged with increasing frequency. In these figures this fall is almost imperceptible due to the selected scale. Theoretical models [84, 88, 118, 271] satisfactorily describe the dispersion of the reciprocating (piston) mode in the range 10–60 kHz.

According to the classical Korteweg formula, in the tube with the fluid there should be a resonant frequency and when approaching this

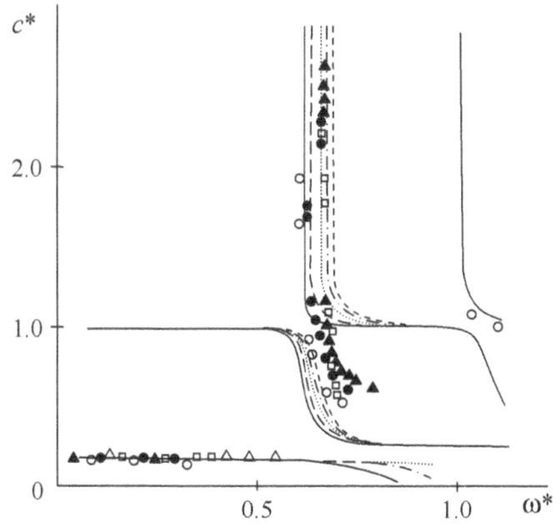

Fig. 8.8. Glass pipes.

—— ○ – R_i = 6.2 mm, h = 0.5 mm;
– – ● – R_i = 5.1 mm, h = 0.7 mm;
······· ▲ – R_i = 5.9 mm, h = 1.9 mm;
–·– △ – R_i = 4.6 mm, h = 2.0 mm;
---- □ – R_i = 5.5 mm, h = 2.5 mm.

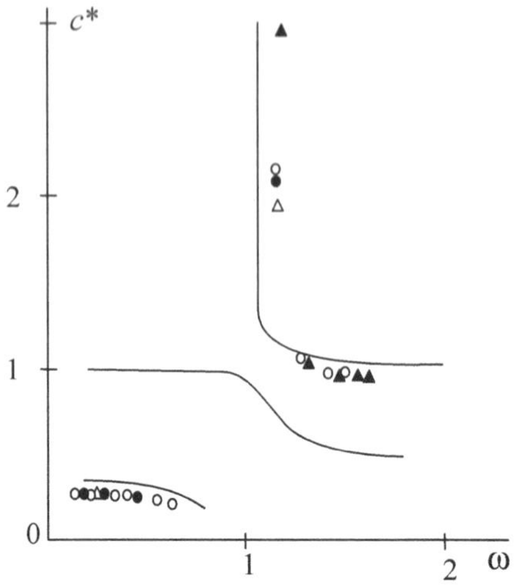

Fig. 8.9. Ebonite pipe.

○ – R_i = 5.0 mm, h = 5.1 mm;
● – R_i = 4.0 mm, h = 4.3 mm;
▲ – R_i = 6.0 mm, h = 4.0 mm;
△ – R_i = 4.0 mm, h = 2.0 mm.

frequency from the left the speed of sound tends to zero, and on the right from it increases indefinitely. Thus, both above and below the resonance frequency in the pipe there is, according to Korteweg's theory, only one piston mode of oscillations, for which the speed of sound experiences dispersion due to the influence of the pipe walls. Experiments have shown that this is not the case. There was no frequency region in which the speed of sound in the fluid filling the pipe would tend to zero.

In addition to the piston mode for all tubes in the frequency range 80–100 kHz and above there were high values of the phase velocity of the sound waves, which in this model are interpreted as the higher modes of oscillations. The results indicate the existence of an abnormally strong dependence of the phase velocity on frequency for higher modes of oscillations.

Analysis of the data on the dispersion of higher-order modes shows that the best agreement is found with the experimental curves based on the model proposed by Lin and Morgan. In order to avoid unnecessary complexity of the information provided, Figs. 8.8–8.11 show the dispersion curves calculated using only the formulas (8.6).

Plots of the dependence of the dimensionless speed c^* on dimensionless frequency ω^* for five glass tubes with different inner radii R_i and wall thickness h are displayed in Fig. 8.8. The curves coincide well with the experimental points on the vertical axis. The displacement of the theoretical graphs horizontally is due to the difference of geometric parameters of shells.

The fact that the experimental plots for higher oscillation modes fall below the theoretical ones can be explained by the shortcomings of calculations by Lin and Morgan, in which there is no criterion of 'the thin wall thickness' of the pipe. In this connection we can assume that the model [118], as well as other theoretical constructions, insufficiently takes into account the stiffness of the shell.

The theory [118] is also in good agreement with the experimental data obtained on tubes filled with MFs with different magnetic parameters. Effect of fluid properties from this point view on the correctness of the theoretical calculations were found.

It should be noted that the formulas (8.3) and (8.6) give better agreement with the experiment in the glass tubes than other formulas.

The measurements of the dependence of the speed of sound on frequency for the hard-rubber (ebonite) tubes are displayed in the form converted for dimensionless frequency in Fig. 8.9. The theoretical curves calculated for the model proposed by Lin and Morgan, are practically the same for the different tubes, as their parameters are similar.

The observed slight deviation of the theory of the experiment on the horizontal and vertical can be explained by the discrepancy of the real elastic moduli of ebonite in comparison with the reference data.

The theoretical curves converted to dimensionless frequency and obtained from the model proposed by Lin and Morgan, as well as the experimental results for the plastic pipes filled with the MF, are shown in Fig. 8.10. The curves are in good agreement on the vertical. The horizontal displacement of the theoretical curves relative to each other is due to the difference of the geometrical parameters of the pipes.

Higher oscillation modes in plastic tubes are described by the formulas (8.6) more accurately than in glass tubes, as is evident from a comparison of Figs. 8.8 and 8.10. There are no discrepancies caused by insufficient rigidity of the tube walls. On the other hand, the piston mode of the oscillations is described by the model (8.3) in the plastic tubes far less accurately than in the glass tubes. This may be due to the fact that the substitution of the parameters corresponding to the plastic walls in these equation gives an understated inadequate model stiffness of the walls of the tubes at

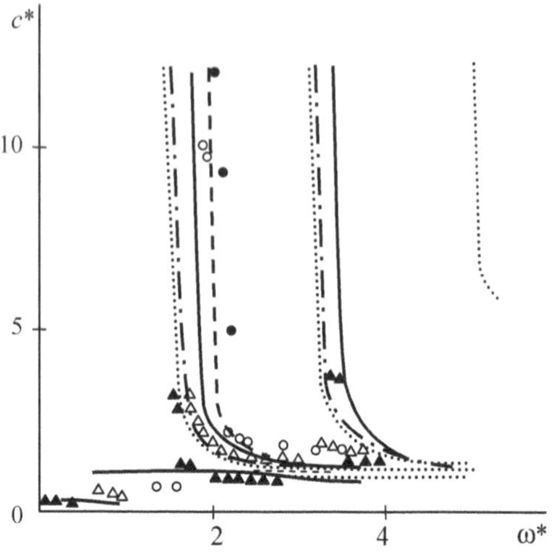

Fig. 8.10. Plastic tubes.

— ○ – R_i = 6.8 mm, h = 1.2 mm;
– –● – R_i = 4.1 mm, h = 1.7 mm;
·····▲ – R_i = 9.1 mm, h = 1,7 mm;
–·–Δ – R_i = 9.2 mm, h = 2.0 mm.

the short-wave fluctuations (piston mode). Apparently, at short scales, corresponding to the piston mode, the plastic pipe behaves somewhat stiffer than at the large scale (higher oscillation modes). Nevertheless, the amplitude of oscillations in the plastic (more compliant) tubes is much smaller than in the glass tubes (less compliant).

In the fluoroplastic tube for which the data are shown in Fig. 8.11, three oscillation modes were recorded: piston and two higher modes. The zero mode of the oscillations is represented by fairly sketchy data, which is likely to be due to the yielding of the tube wall. All theoretical constructions give too low values of the speed of the piston mode of the oscillations compared with the experimental values.

The first higher mode of oscillations is satisfactorily described by the models proposed by Lin and Morgan [118], Jacobi [88] and Boyle and Field [84].

There are deviations for the second experimentally identified higher oscillation mode. A simple variation of the elastic parameters of the shell (the actual parameters may differ from the reference data) can not be used to achieved the simultaneous satisfactory description of both higher modes of oscillation. This may be indicative of the inadequacy of these model theories.

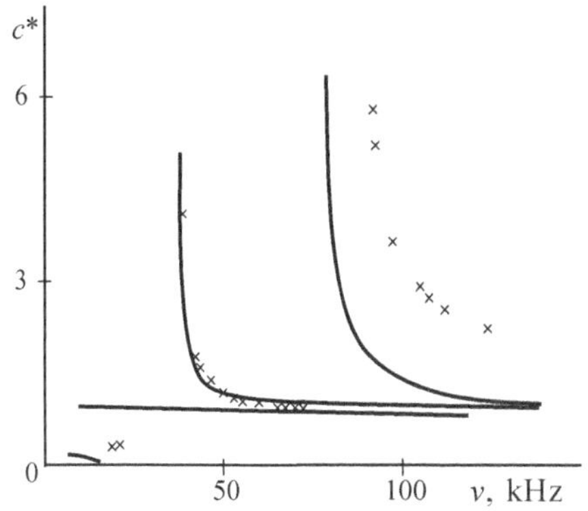

Fig. 8.11 The data rate measurements PTFE tube in sound parameters with R_i = 8.7 mm, h = 1.5 mm.

Data on the absence of the dispersion of the speed of sound in the MF–tube system in the frequency range 3.7–94.6 kHz, obtained in [90], correspond to the results presented in Fig. 8.8 for tubes with the parameters R_i = 4.6 mm, h = 2.0 mm, for which no higher modes were observed in the frequency range to 100 kHz.

According to the results of this work, it can be considered as confirmed that the Korteweg formulas (8.1)–(8.2) can be used for simple approximate calculations of the frequency dependence of the speed of flat wave in a tube with a fluid. However, the higher modes of vibration of the theory can not be described.

The model proposed by V.N. Merkulov, V.Yu. Prikhod'ko and V.V. Tyutekin, represented by the expression (8.5) describes the higher vibration modes of not only zero but also higher orders.

Better agreement with experiment is provided by the models based on 'stitching' shell oscillation equations with the equations of motion of the fluid (8.3), (8.4), (8.5) and (8.6). Of these, the most appropriate model is the one proposed by Lin and Morgan (8.6), which takes into account the rotational inertia of the shell element. However, the existing theories apparently do not sufficiently take into account the rigidity of the shells.

For a more informed conclusion about the adequacy of the theoretical models it is advisable in our views to extend the application of the acoustomagnetic indication method, in particular, due to the transverse magnetization of the fluid column and the appropriate placement of the miniature coil outside the tube [301] as well as through the wider variation of geometric, elastic and thermal parameters of cylindrical shells. It is advisable to supplement the experimental data by the results of spectral analysis of the dependence of the induced EMF on the coordinates of the magnetic head in the case of simultaneous excitation of different modes of oscillations.

However, the experimental data obtained for the dispersion of the speed of sound in the fluid–shell system using shells with a wide range of geometrical and elastic parameters show the effectiveness of using the acoustomagnetic method for the study of such systems. The proposed experimental technique can be applied for measurements in various systems, including those in which it is difficult to obtain theoretical results.

9

Acoustic granulometry

9.1. Prologue

A new area of physical acoustics – molecular acoustics [63] that was formed in the 60s of the previous century, is based on the methods for examination of the previously not studied properties of matter and special features of its molecular structure (the nature of intermolecular forces, the kinetics of molecular processes, the theory of non-equilibrium processes, etc).

In the monograph by I.S. Kol'tsova [280], concerned with the propagation of ultrasound waves in the suspensions and emulsions, it is noted that the acoustic methods of investigating dispersed systems are highly promising. Studying the acoustic fields, one can obtain information on both the set of the properties of heterogeneous media and on the individual characteristics of the media.

The acoustics of the nanodispersed media is a comparatively new direction of investigations which can be regarded as the extension of the methods of molecular acoustics to the studies of the nanoscale structure of matter.

At the same time, the acoustic measurements of the physical parameters of the dispersed nanosystems open new approaches to the examination of matter and represent an addition to the available experimental methods: electron diffraction, field methods, scanning probe microscopy, *x*-ray, Mössbauer and optical spectroscopy [237, 238, 253, 270, 310]. If we compare the acoustic granulometry (AGM) of the nanodispersed magnetic fluids with magnetic granulometry in direct and alternating magnetic fields (MGM), with the electron and atomic force microscopy, we note the following facts.

The MGM in the direct field imposes very stringent requirements on the mechanism of positioning and rotation of the measuring cell in the magnetic field, on the method of pouring the fluid – sample in the cell and its sealing (it is necessary to prevent the possibility of trapping air bubbles, etc.). The currently available methods of investigating the magnetic and geometrical parameters of the dispersed ferroparticles in the alternating magnetic fields (in 'simple' and 'combined', in particular in 'crossed') are based on measuring the magnetic susceptibility of the colloid and its frequency dependence. The applicability of these methods is restricted to the dispersed systems with low viscosity. The rotation of the ferroparticles in the viscous medium under the effect of the alternating field becomes impossible with increasing duration of Brownian motion.

The electron microscopy methods can be used for investigating samples in the hardened condition (the replica method) in which the kinetic properties of the ferroparticles in the processes of their aggregation, taking place in the magnetic fluid during magnetisation, are blocked. The limitation of the atomic force microscopy methods is associated with the same circumstance and with the fact that the method does not make it possible to produce high contrast images of nanoparticles and nanoaggregates because of the 'blurring' of the electrical field surrounding them.

The method of acoustic granulometry is based on the analysis of the field dependence of the EMF amplitude, induced in the circuit by the acoustomagnetic effect [263, 289]. In contrast to the situation with the unlimited beam of flat sound waves at $\mathbf{H}||\mathbf{k}$, the increments of magnetisation and the demagnetising field do not compensate each other, i.e. $\delta M \neq -\delta H_p$, and $\delta B = \mu_0(1-N_d)\,\delta M \neq 0$. The EMF, induced in the circuit, is proportional to the amplitude of oscillations of the magnetisation of the fluid determined mainly by the oscillations of the concentration of the dispersed phase particles. By analogy with equation (2.25) in the propagation of the longitudinal sound wave in the tube with the magnetic fluid placed in the transverse the magnetic field, the EMF is induced in the conducting frame on the outer surface of the tube and its amplitude can be described by the equation:

$$e_0 = -\mu_0(1-N_d)N_k L\cdot h\cdot\cos\varphi\cdot M\cdot\ddot{u}_0\left[M'_\beta+(\omega\tau)^2\right]\cdot c^{-1}\cdot\left[1+(\omega\tau)^2\right]^{-1}, \tag{9.1}$$

where L is the length of the frame; φ is the angle between the normal in the centre of the frame and $\mathbf{H}$; h is the height of the frame; N_d is the dynamic demagnetising factor.

As shown later, the analysis of the relative variation of the amplitude of the EMF, induced in the circuit as a result of the AME (further the amplitude of the AME) with the strength of the magnetic field, proportional to the relative variation of the magnetisation of the magnetic fluid in the sound wave, produces information on the geometrical and magnetic parameters of the dispersed particles. In the process of measurements it is not necessary to ensure any displacement, calibration or setting of the functional elements of experimental equipment, i.e., to carry out actions which represent an additional source of errors.

The principal difference of the acoustic methods of investigating the magnetisation mechanism is the superposition on the investigated object of alternating pressure whilst fulfilling the condition of adiabatic nature of the process. This advantage of the method enables the method to be used for probing the baric and thermal dependence of the magnetic parameters of the colloid and of the nanoparticles dispersed in it.

In some cases, advantages are offered also by the fact that the proposed method of measuring the physical parameters of the magnetic particles that the require preliminary calibration, i.e., it is an absolute method.

9.2. Mechanism of perturbation of magnetisation in the magnetic field transverse to the soundwave

If the direct magnetic field is normal to the sound beam with the restricted site surface, the perturbation of the magnetisation is accompanied by the perturbation of the demagnetising field $\delta H = -N_d\, \delta M$ which in turn influences the intensity of magnetisation. Therefore, the expression for the equilibrium magnetisation is written by analogy with equation (2.9) in the form:

$$M_e = M_0 - (nM_n + \gamma_* M_T)\frac{\partial u}{\partial x} - N_d M_H \cdot \delta M, \qquad (9.2)$$

where N_d is the dynamic demagnetising factor at perturbation of magnetisation of the magnetic fluid by the sound wave in the magnetic field transverse to the fluid column; δM is the increment of magnetisation of the magnetic fluid under the effect of adiabatic deformation of the medium in the sound wave and the demagnetising field: $\gamma_* = qTc^2C_p^{-1}$, $M_T \equiv (\partial M / \partial T)_{H,n,m*}$, $M_H \equiv (\partial M / \partial H)_0$.

The course of the process of establishment of the equilibrium magnetic state in time is described by the equation of relaxation of magnetisation of the fluid by a flat sinusoidal sound wave described by the equation (1.26).

Carrying out mathematical operations, as in deriving equation (2.10), we obtain the equation for the magnetisation increment δM:

$$\delta M = -\frac{(nM_n + \gamma * M_T)/(1 + N_d M_H) + i\omega\tau M_0}{1 + i\omega\tau} \cdot \frac{\partial u}{\partial x}, \tag{9.3}$$

where $\tau \equiv \tau_1 (1 + N_d M_H)^{-\S}$.

Separating the real part of the equation (9.3), we obtain the equation for the amplitude of perturbation of magnetisation at the points of the central circular section of the cylindrical column of the magnetic fluid, magnetised in the transverse direction

$$\Delta M = M_0[M_\beta + (\omega\tau)^2] \cdot \dot{u}_0 \cdot \left(c[1 + (\omega\tau)^2]\right)^{-1} \tag{9.4}$$

where $\dot{u}_0$ is the amplitude of the oscillatory speed, $M_\beta \equiv (nM_n + \gamma_* M_T)\, [M_0\, (1 + N_d\, M_H)]^{-1}$, N_d is the parameter characterised by the specific dependence on the shape parameter P which will be described later.

The slipping of the ferroparticles or aggregates in the fluid–carrier with the relative speed β_V is taken into account by substituting M_β by:

$$M'_\beta \equiv (\beta_V nM_n + \gamma_* M_T)[M_0(1 + N_d M_H)]^{-1}.$$

(Index 0 at the notation of the magnetisation of the magnetic fluid in the non-perturbed state will be omitted in subsequent considerations).

If the magnetisation M is proportional to the concentration of the magnetic phase n, then:

$$M'_\beta \equiv \frac{M \cdot \beta_V + \gamma_* M_T}{M(1 + N_d M_H)}$$

The process of magnetisation of the magnetic fluid is determined mainly by two mechanisms of the orientation of the magnetic moments of the ferroparticles along the magnetic field: Brownian motion of the particles in the liquid matrix and the Néel mechanism of thermal fluctuations of the magnetic domain inside the particle itself. Of the two mechanisms of relaxation of magnetisation, the important mechanism is the one which is characterised by the shorter duration of rotational diffusion. For the magnetite particles with the diameter $d < 10$ nm, dispersed in kerosene, the shorter duration of rotational fluctuations is typical of the Néel mechanism in which the duration of thermal fluctuations $\beta_n < 10^{-9}$ s. Experiments carried out in the frequency range of sound oscillations of 10–50 kHz and, consequently, in equation (9.4) it may be assumed that $(\omega\tau)^2 \to 0$. The amplitude of perturbation of magnetisation can be described by the equation:

$$\Delta M = \frac{M \cdot M'_\beta \cdot \dot{u}_0}{c}.$$

In the fields close to magnetic saturation of the magnetic fluid, $M \to M_s$, $M_T \to 0$, $M_H \to 0$, $M'_\beta \to \beta_V$, and therefore, the maximum increment of magnetisation is:

$$\Delta M_{\max} = \frac{M_s \cdot \beta_V \cdot \dot{u}_0}{c}.$$

Ignoring the slipping of the nanoparticles of the disperse phase in relation to the fluid matrix (complete 'dragging' of the particles by the fluid, $\beta_V = 1$), we obtain:

$$\Delta M_{\max} = \frac{M_s \cdot \dot{u}_0}{c}.$$

The relative increment of the amplitude of the magnetisation $\beta_\xi = \Delta M / \Delta M_{\max}$ is expressed by the relationship

$$\beta_\xi = \frac{M + \gamma_* M_T}{M_s[1 + N_d M_H]}. \qquad (9.5)$$

Since the amplitude of AME e_0 is proportional to ΔM, then denoting its maximum value as $e_{0\max}$, we obtain the relationship:

$$\frac{e_0}{e_{0\max}} = \beta_\xi.$$

The amplitude of AME in relative units can be described in the theory of superparamagnetism as the function of the parameter ξ:

$$\beta_\xi = \frac{L(\xi) + \dfrac{\gamma_*}{M_s} \cdot M_T}{1 + N_d \cdot M_H} = \frac{L(\xi) - k' \cdot D(\xi)}{1 + k'' \cdot \xi^{-1} \cdot D(\xi)}, \qquad (9.6)$$

where $L(\xi) = \mathrm{cth}(\xi) - 1/\xi$ is the Langevin function; $D(\xi) = 1/\xi - \xi \cdot \mathrm{sh}^{-2}(\xi)$; $k' = qc^2 C_p^{-1}$; $\xi = \dfrac{\mu_0 m_* H}{k_0 T}$; $k'' = \dfrac{N_d \mu_0 n m_*^2}{k_0 T} = \dfrac{N_d \mu_0 M_s m_*}{k_0 T}$.

Expression (9.6) can also be written in the form of the explicit function of the parameter ξ:

$$\beta_{\xi} = \frac{cth\xi - \xi^{-1} - k'\xi(\xi^{-2} - sh^{-2}\xi)}{1 + k''(\xi^{-2} - sh^{-2}\xi)}. \qquad (9.7)$$

Strictly speaking, the Langevin model of magnetisation can be used in the system of non-interacting magnetic nanoparticles whereas in real magnetic fluids such an interaction does not always take place (for example, as a result of the dipole–dipole interaction of the ferroparticles), and this imposes certain restrictions on the applicability of this model. However, in a number of cases, we will be interested in the initial and final (the region of magnetic saturation) sections of the magnetisation curve where these processes are either insignificant or almost completed.

We obtain the approximation of the function $\beta\xi$ (ξ) in weak and strong magnetic field, i.e., at $\xi \to 0$ and $\xi \to \infty$:

at $\xi \to 0$:

$$\beta_{\xi} = \frac{(1-k')\cdot\xi}{3+k''};$$

at $\xi \to \infty$:

$$\beta_{\xi} = ((1-\xi^{-1}) - k'\cdot\xi^{-1})/(1+k''\cdot\xi^{-2}) \approx 1-(1+k')\cdot\xi^{-1}.$$

These equations will be expressed by the strength of the field H:

at $H \to 0$:

$$\beta_H = \mu_0 \frac{(1-k')}{3+k''}\frac{m_*}{k_0 T}\cdot H, \qquad (9.8)$$

at $H \to \infty$:

$$\beta_H = 1 - \frac{(1+k')k_0 T}{\mu_0 m_*}\cdot\frac{1}{H}. \qquad (9.9).$$

In particular, equation (9.9) shows that the results of measurement of the relative amplitude of the AME in strong magnetic fields can be used to determine the magnetic moment of the particle without taking any additional measurements of the magnetisation curve, the initial magnetic permeability and the concentration of the magnetic phase.

9.3. Calculation of the dynamic demagnetising factor

We estimate the value of the dynamic demagnetising factor, using the

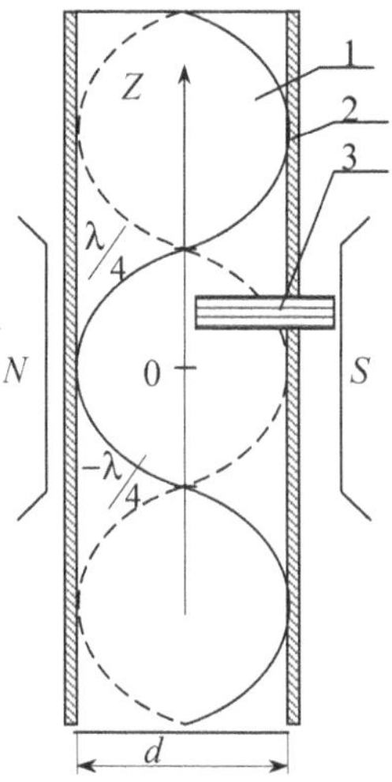

Fig. 9.1. Diagram of the problem for evaluation of N_d.

scheme of the problem shown in Fig. 9.1. The magnetic fluid 1 fills the glass cylinder 2, and the oscillations are indicated by the narrow (in comparison with the wavelength) inductance coil 3. A system of standing waves is excited. The outer magnetic field $\mathbf{H}_e$ is normal to the axis of the cylinder (axis Z). Let it be that the geometry of pole terminals is such that the magnetic field within the limits of a single standing wave is homogeneous and outside this range rapidly attenuates. This distribution of the magnetic lines of force can be obtained if we use plane-parallel pole terminals with bevelled edges. In this case, to calculate the demagnetising field, it is sufficient to investigate the region of the magnetic liquid: situated between the pole terminals in which a single standing wave forms. The antinode of the standing wave is situated in the centre of the magnetic field, and the nodes at the boundaries of the field.

The perturbation of magnetisation of the magnetic fluid in the phase of compression of the medium is given by the equation:

$$\Delta M = \Delta M_m \cdot \cos kz.$$

In the non-perturbed liquid, filling the 'long' cylindrical vessel, the strength of the homogeneous magnetic field H_i, orthogonal to the axis of the cylinder, is determined from the equation (see section 1.3):

$$H_i = H_e - 0.5M.$$

Here M is the magnetisation of the magnetic fluid in the field H_i. The increment of magnetisation in the layer of the magnetic fluid with a thickness dz is:

$$dM = -k\Delta M_m \cdot \sin kz \cdot dz. \tag{9.10}$$

We introduce a new variable $P \equiv \frac{z-(-z)}{d} = \frac{2z}{d}$ – the form parameter.

$P \geq 0$ for z varying from 0 to $\frac{\lambda}{2d}$, and, consequently, equation (9.10) has the form:

$$dM = -\frac{kd}{2}\Delta M_m \cdot \sin\frac{kPd}{2} \cdot dP.$$

The increment of magnetisation in the cylinder with the coordinates of the base $\pm z$ results in the increment of the demagnetising field image:

$$dH = -N_c dM = -\frac{kd}{2}N_c \Delta M_m \cdot \sin\frac{kPd}{2} \cdot dP,$$

where N_c is the demagnetising factor for the magnetic in the form of a cylinder magnetised in the magnetic field transverse to the axis.

The demagnetising factor is the function of the form parameter, i.e., $N_c = f(P)$. The limiting values of N_c at $P \to 0$ and $P \to \infty$ are equal to respectively 0 and 0.5. This dependence is approximated by the equation

$$N_c = 0.5(1 - e^{-aP}), \tag{9.11}$$

and the coefficient a is selected on the basis of optimum matching of the dependence with the available values of N_c for the bodies of the ellipsoidal shape.

The solid line in Fig. 9.2 shows the graph of the dependence $N_c(P)$ at a = 1.2, and the crosshatched squares show the tabulated values of N_c, taken from [244].

The demagnetising field, determined by the perturbation of magnetisation in the central circular cross-section of the column of the magnetic fluid in the range of a single standing wave, limited by the coordinates $z = \pm\frac{\lambda}{4}$, is determined from the expression

$$\Delta H_p = \frac{kd}{2}\Delta M_m \int_0^{\frac{\lambda}{2d}} N_c \sin\frac{kpd}{2} dp.$$

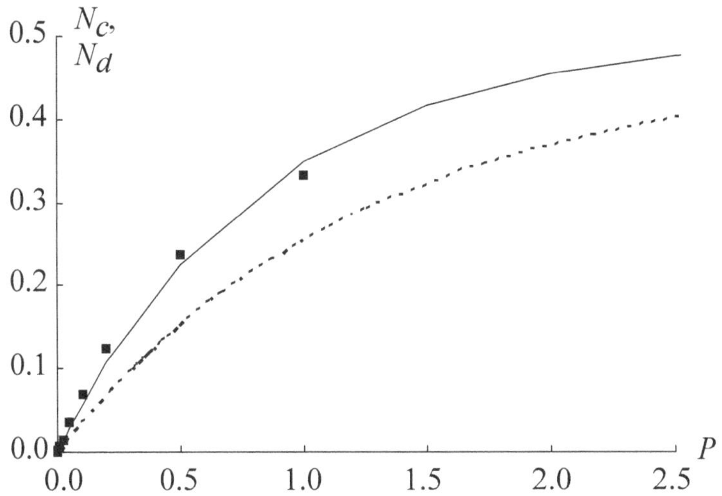

Fig. 9.2. Dependence of the demagnetising factors on the form parameter.

From this equation, we obtain the following equation for the dynamic demagnetising factor represented as $N_d = \dfrac{\Delta H_p}{\Delta M_m}$:

$$N_d = \frac{kd}{2} \cdot \int_0^{\frac{\lambda}{2d}} N_c \sin\frac{kPd}{2} dP. \qquad (9.12)$$

Substituting the value of N_c from equation (9.11) into (9.12) we obtain:

$$N_d = \frac{kd}{4} \int_0^{\frac{\lambda}{2d}} (1 - e^{-aP}) \cdot \sin\frac{kPd}{2} dP = \frac{\pi d}{2\lambda} \int_0^{\frac{\lambda}{2d}} (1 - e^{-aP}) \cdot \sin\frac{\pi P d}{\lambda} dP.$$

After carrying out integration, we have:

$$N_d = 0.5\left[1 + \pi d\left(a \cdot e^{\frac{a\lambda}{2d}} - \pi d / \lambda\right)\left(\lambda \cdot \left\{a^2 + (\pi d / \lambda)^2\right\}\right)^{-1}\right]. \qquad (9.13)$$

The graph of the dependence N_d (P) is shown in Fig. 9.2 by the dotted curve. The values of N_d and P for different frequencies of oscillation for a tube with the inner diameter d = 23 mm and the magnetic fluid, with the speed of sound in the fluid equal to c = 950 m/s, are presented in Table 9.1.

Table 9.1

ν, kHz	5	10	15	20	25	30	35	40	45	50	55	60
P	4.35	2.17	1.45	1.09	0.87	0.72	0.62	0.54	0.48	0.43	0.40	0.36
N_d	0.46	0.38	0.32	0.27	0.23	0.21	0.18	0.17	0.15	0.14	0.13	0.12

If the level of magnetisation of the magnetic fluid in the non-perturbed state is assumed to be equal to 0, the increment of magnetisation in the phase of tensile loading of the sound wave will be negative, i.e., $dM < 0$. The sign of the expression for dM (9.10) changes to opposite. In this case, the 'demagnetising' field becomes 'magnetising'. However, the increments of the magnetisation in the demagnetising field (dM and dH) have the opposite signs and, therefore, as in the phase of compression of the soundwave, the relative amplitude of the EMF, induced in the inductance coil, is calculated from the equation (9.6).

9.4. Magnetic granulometry

Superparamagnetism is interesting not only as a unique magnetic phenomenon but also as a specific method of determining the dimensions, concentration and distribution of the magnetic nanoparticles. This method is referred to as magnetic granulometry (MGM).

The MGM is the simplest representation of the method of granulometric analysis of the composition of the dispersed medium which, without using the function of the size distribution of the particles, is restricted to determining the 'maximum' and 'minimum' dimensions of the particles (respectively – the values of the magnetic moment $m_{*\max}$ and $m_{*\min}$) using the data from linear approximation of the magnetization curve in the initial section and in the magnetic saturation range.

The investigated magnetic fluids consist of the colloid solution of magnetite in the hydrocarbon medium – kerosene, stabilised with oleic acid. The investigated liquids are characterised by high homogeneity and stability over a long period. In this study, experiments were carried out with samples of magnetic fluids No. 1–No. 4. The samples No. 2–No. 4 were prepared by diluting the colloid No. 1.

The density of the magnetic fluid sample is was measured using the method based on a glass pycnometer with a volume of 10 ml.

The saturation magnetisation M_s and initial magnetic susceptibility were determined on the basis of the results of measurement of the magnetisation curve by the ballistic method described in section 1.3.

The maximum and minimum magnetic moments of the particles m_{*max} and m_{*min} were obtained using the magnetic granulometry method. In accordance with the theory of superparamagnetism

$$m_{*\max} = \frac{3k_0T\chi}{\mu_0 M_S}, \tag{9.14}$$

$$m_{*\min} = \frac{k_0T}{\mu_0 M_S}.(M / H^{-1})^{-1}, \tag{9.15}$$

where M/H^{-1} is the tangent of the slope of the straight section of the curve $M(H^{-1})$ at $H \to \infty$.

The particle size is determined from the expression:

$$d = \sqrt[3]{6m_* / \pi M_{S0}} = 0.016\sqrt[3]{m_*}, \tag{9.16}$$

where M_{S0} is the saturation magnetisation of the dispersed phase (M_{s0} = 477.7 kA/m for magnetite).

9.5. Acoustic granulometry of magnetic nanoparticles

9.5.1. Calculation relationships

The sound and ultrasound waves propagate quite efficiently in the magnetic fluid so that we can examine the structure of real magnetic colloids on the basis of acoustic spectra of the methods [260, 263, 289, 303, 305, 309].

The initial aim of acoustic granulometry is the determination of the 'maximum' and 'minimum' dimensions of the magnetic nanoparticles using the results of linear approximation of the relative amplitude of AME in the initial section of the curve and in the range of magnetic saturation without taking into account the size distribution of the particles in the real magnetic fluid.

Equation (9.8) leads to the expression for the calculation of the maximum magnetic moment:

$$m_{*\max} = \frac{(3+k'')k_0T}{\mu_0(1-k')} \cdot \beta_H / H, \tag{9.17}$$

where β_H/H is the tangent of the slope of the initial section of the dependence $\beta_H\,(H)$.

Using the dependence $\beta_H(H^{-1})$ and also (9.8), we determine the minimum magnetic moment of the particles

$$m_{*\min} = \frac{(1+k')k_0T}{\mu_0} \cdot (\beta_H / H^{-1})^{-1}, \qquad (9.18)$$

where (β_H / H^{-1}) is the tangent of the slope of the straight section of the curve $\beta_H(H^{-1})$ at $H \to \infty$.

The expression for k'' includes: the parameter m_* determined on the basis of the results of measurements of the AME amplitude, and the parameter n – the concentration of the particles, which is presented in the form

$$n = \frac{M_{SO} \cdot \varphi_M}{m_*}$$

where φ_M – is the concentration of the magnetic phase.

After substitution of k'', expression (9.17) has the following form

$$m_{*\max} = \frac{(3k_0T + N_d\mu_0\varphi_M M_{SO} m_*)}{\mu_0(1-k')} \cdot \beta_H / H.$$

After simple transformations we obtain

$$m_{*\max} = 3k_0T \cdot \beta_H / H \cdot [\mu_0(1-k' - N_d\varphi_M M_{SO}\beta_H / H)]^{-1}. \qquad (9.19)$$

When φ_M is replaced by the volume concentration of the dispersed system – φ, we obtain $m_{*\max}$, value since $\varphi > \varphi_M$ [35]. Since the last term in the square brackets of the equation (9.19) for the low concentration magnetic fluids has only an approximate value, for calculations it can be assumed that $\varphi = a\varphi_M$, where $a = 1.25$ for the kerosene magnetic fluids.

In the equations (9.18) and (9.19) H is the value of the strength of the field inside a specimen (the index i is ignored), and in these experiments we measure directly the external field H_e.

The strength of the internal field is calculated by the standard procedure using the experimentally determined magnetization curve.

The magnetic moments of the particles $m_{*\max}$ and $m_{*\min}$ are determined from the field dependence of the AME. The AME amplitude is measured at the maximum of the standing wave in the region of the most homogeneous magnetic field.

The error of calculation of the magnetic moment consists of a large number of partial errors. They include: the error of determination of the volume concentration of the solid phase, which depends on the error of density measurements, and the error of determination of the amplitude of oscillations which in turn depends on the strength of the noise interference of the amplifier; the level of RF interference; the errors determined by the instability of temperature and voltage of the mains. In the development of experimental equipment measures were taken to minimize these errors. The values of m_{*max} and m_{*min} are determined from the dependence plotted in relative units and this reduces the value of the error from 11% to 6%, respectively.

9.5.2. Description of experimental equipment

The flow diagram of experimental equipment is shown in Fig. 9.3 [303]. The magnetic colloid fills the glass tube 1 in the vertical position. The sound oscillations are generated by the piezoelectric sheet 2 which receives from the generator 3 the alternating electrical voltage of the given frequency. Frequency is controlled with a frequency meter 4, the voltage with the voltmeter 5. The semi-circular inductance coil 6, placed in the immediate vicinity of the outer surface of the tube, is rigidly connected with the kinematic section of the cathetometer 7. The alternating EMF from the inductance coil travels to the input of the selective amplifier 8 and from the output of the amplifier it is transferred to the oscilloscope 9 and

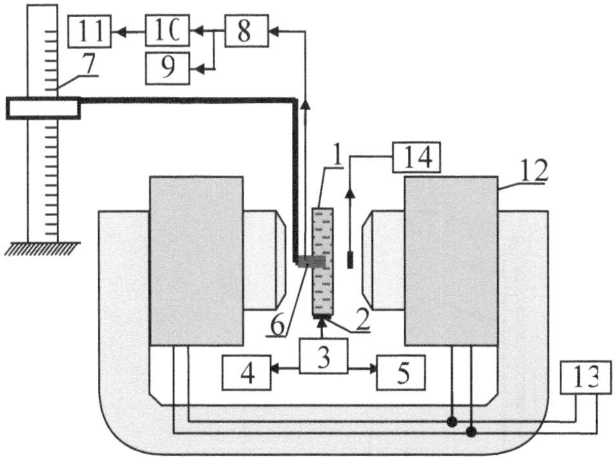

Fig. 9.3. Flow diagram of experimental equipment.

the analog–digital converter 10, connected to the notebook 11. The source of the magnetic field is the laboratory electromagnet 12 (FI-1 grade), connected to the power source 13. The magnetic induction is determined using the tesla metre 14 fitted with a Hall sensor. The software was produced in the NI LabView medium in which the resultant signal is decomposed into a spectrum for controlling the level of interference, calculating the measured parameters, and constructing the relevant dependences.

To prevent distortion of the open surface of the magnetic fluid under the effect of the normal component of the magnetic field, it is necessary to carry out the forced stabilisation of the fluid surface. For this purpose, the upper end of the tube after filling with the investigated samples of the magnetic fluid to the top is hermetically closed with a thin film or sheet.

The acoustic cell, used in the experimental equipment, transfers the alternating EMF to the piezoelectric sheet and provides its mechanical protection. In addition to this, the design of the acoustic cell allows: to fix the lower end of the tube; using a rubber seal to seal the filled cavity; prior to pouring in the next magnetic fluid sample to carry out dismantling, cleaning and assembling of the device [301].

Figure 9.4 shows the schematic of the acoustic cell. The magnetic fluid 1 fills the glass tube 2 with the inner diameter of 13.8 mm. The biasing ring 4, the cover 5, and the body 6 are produced from a non-magnetic material – duralumin, fastening screws 3 – from brass. The generator of sound oscillations 7 produces the alternating electrical voltage of the required frequency which travels, through the spring 10, to the piezoelectric sheet 9. The filled cavity is sealed using the rubber ring 8, partially immersed in the ring-shaped groove.

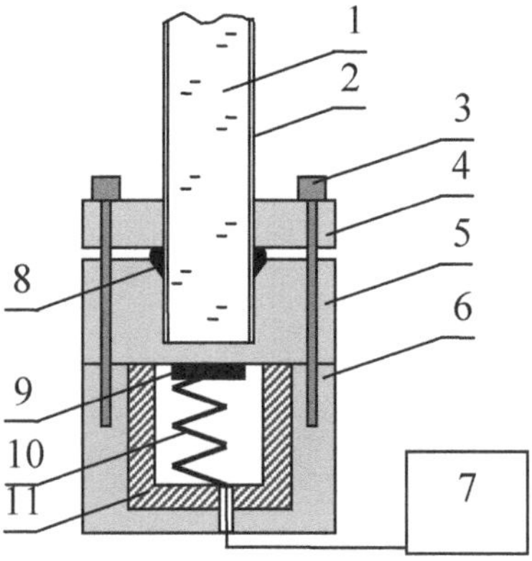

Fig. 9.4. Diagram of the acoustic cell.

The lower end of the spring rests on the bottom of the fluoroplastic vessel 11.

The piezosheet is compressed to the bottom of the cover from the lower end and, therefore, the elastic waves in this structure are introduced through the thin plane-parallel bottom of the cover into the magnetic fluid which results, with the restriction (2.29) satisfied, in the excitation of zero (piston) oscillation mode is in the fluid–cylindrical shell system. For more efficient passage of the sound wave into the fluid, the gap between the bottom and the piezosheet is filled with a thin layer of a contact lubricant.

The acoustic cell with the magnetic fluid is installed in the space between the poles of the electromagnet fulfilling the condition of perpendicularity of the lines of the strength of the homogeneous field and the axis of the tube with the magnetic fluid.

The acoustic cell is compact so that it is possible to reduce the size of the gap between the pole terminals to reach the region of magnetic saturation of the magnetic fluid.

9.5.3. Measurement results

The investigations were carried out on a specimen of MF-1, consisting of a magnetic colloid, in which the dispersed phase was magnetite Fe_3O_4, the disperse medium – kerosene, and the stabiliser – oleic acid.

Table 9.2 shows the physical parameters of the investigated sample (density – ρ, concentration of the solid phase – φ, initial magnetic susceptibility – χ, saturation magnetisation – M_s, the speed of sound in the magnetic fluid–glass tube system – c) obtained at a temperature of 31°C.

Table 9.2.

Sample	ρ, kg/m³	φ, %	χ	M_S, kA/m	c, m/s	$m_{*max}\cdot10^{19}$, A·m²	$m_{*min}\cdot10^{19}$, A·m²	d_{max}, nm	d_{min}, nm
							MGM		
MF-1	1315	12	3.4	45.8	930	7.52	2.74	14.6	10.4
							AGM		
						9.55	1.61	15.6	8.6

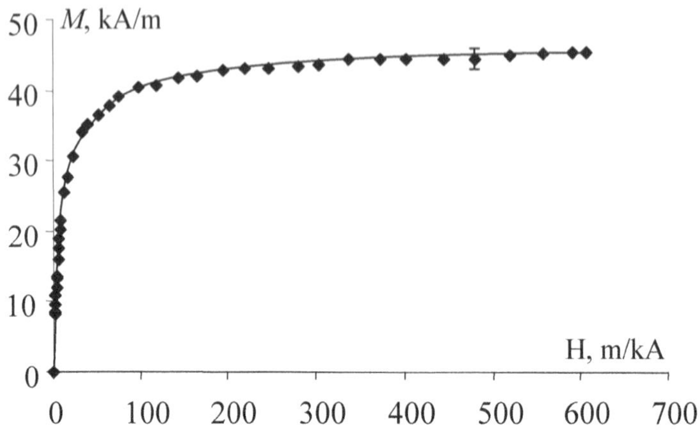

Fig. 9.5. Magnetisation curve of MF-1.

The density of the fluid is measured by the standard procedure using a pycnometer, and the concentration of the solid phase is calculated from the mixing equation (3.6.).

To determine the magnetic parameters χ and M_S, the magnetizing curve $M(H)$ is recorded using the ballistic method. The magnetic fluid fills the cylindrical ampoule; the length of the ampoule is considerably greater than is diameter so that the demagnetizing field can be ignored (section 1.3). The resultant magnetization curve of the magnetic fluid is shown in Fig. 9.5.

Initial magnetic susceptibility χ is determined from the slope of the initial (straight) section of the curve $M(H)$. Parameter M_S is

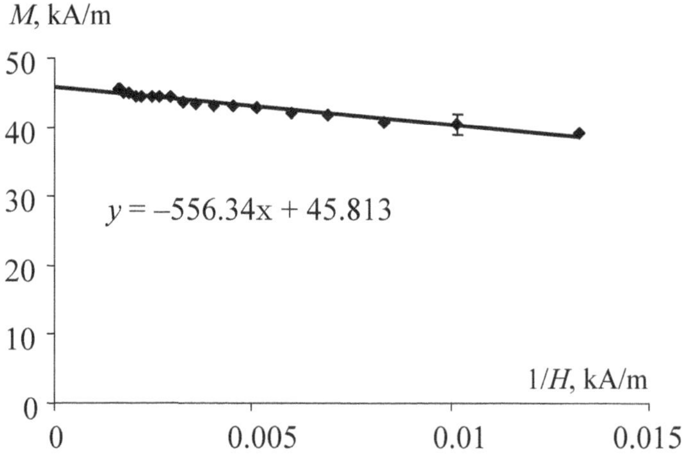

Fig. 9.6. Approximation of the dependence $M(H^{-1})$.

determined by linear approximation of the dependence $M(H^{-1})$ in the vicinity of $H^1 \approx 0$ and accept in the resultant straight-line to intersection with the ordinate (Fig. 9.6).

The speed of sound in the magnetic fluid, filling the glass tube, was measured using a heterogeneous magnetic field and the procedure described in section 8.4.

The dependences of the relative amplitude of the EMF, induced in the inductance coil, on the strength of the magnetic field β_H (H) normal to the tube with the magnetic fluid, were obtained at frequencies of 18, 24, 27, 33, 41, 47, 55 and 65 kHz. The width of the induction coil was 3 mm, the length of the standing wave in the magnetic fluid, filling the tube, at a frequency of 65 kHz was 7.2 mm. Experiments were carried out taking into account the optimum conditions of noise protection at a controlled temperature of $T = 31 \pm 0.2$°C and a constant voltage at the piezoelement of $U = 40 \pm 0.5$ V.

The dependence of the amplitude of the induced EMF e_0 on the strength of the magnetic field is measured at constant values of the temperature of the medium, the frequency of sound oscillations, the voltage on the piezoelement and the constant position of the inductance coil in relation to the antinode of the standing wave. This minimises the error of measurement of β_H.

The error of measurement of β_H includes: the error of measurement of the absolute value of the EMF induction using the ATsP device 1.22%, the error of measurement of the strength of the magnetic

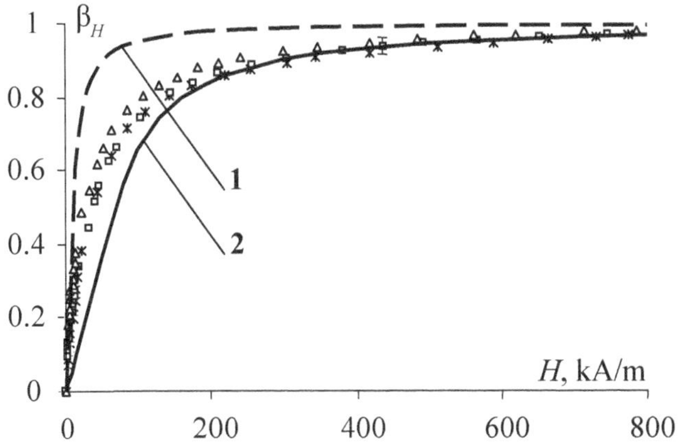

Fig. 9.7. Experimental data: * – 24 kHz, □ – 41 kHz, Δ – 65 kHz. Theoretical dependences $\beta_H(H)$: 1) for the particles d_{max} = 16.0 nm, 2) for the particles d_{min} = 8.5 nm.

field with a tesla metre, which according to the certificate data was 2%, and the error of determination of tg θ and ctg ψ, calculated by the least square method for the linear regression function 0.3%. The total error was ~3%.

The experimental results of measurement of the relative amplitude of the induction EMF for the frequencies of 24, 41 and 65 kHz are presented in Fig. 9.7. The values of the demagnetising factor N_d [263], used in the calculations for these frequencies equalled 0.31, 0.22 and 0.16, respectively. When plotting the graph of the dependence β_H (H), the abscissa was corrected for the presence of the demagnetising field using the equation (section 1.3):

$$H_i = \frac{H_e}{1+0.5\chi}.$$

The limiting value of the relative amplitude of the induced EMF (AME amplitude) is determined by extrapolating the experimental dependence β_H (H^{-1}) to the range $H \to \infty$.

The investigations of the AME includes the three relevant regions of the magnetic field: the initial section of the magnetisation curve, the region of 'moderate' magnetic fields (in this region $\partial\chi$ / ∂H (0), and the vicinity of magnetic saturation of the magnetic fluid.

In the initial section of the magnetisation curve and in the section, characterised by the large change of the curvature of the curve, the magnetic moments of the largest particles and aggregates are oriented with respect to the field. In addition to this, in the presence of the external magnetic field as a result of the dipole-dipole interaction of the ferroparticles, aggregation processes take place which can also be of the hysteresis nature. In the range of the 'moderate' magnetic fields different physical mechanisms of magnetisation of the magnetic fluid operate simultaneously and this is reflected in the field dependence of the amplitude of the induced EMF. The physically justified interpretation of the results can provide a relatively complete information on the participation of these mechanisms in the formation of this dependence and carry out the structural specification of the magnetisation process.

In the vicinity of the magnetic saturation of the magnetic fluid the processes of aggregation and alignment of the chain-like aggregates in the fields are completed, and further magnetisation takes place by the orientation of the magnetic nanoparticles with the minimum magnetic moment in the field. The degree of alignment of the magnetic moments of the 'fine 'nanoparticles in the field (and, correspondingly, also the magnetisation of the colloid) depend almost completely on the ratio of the potential energy of the nanoparticles the magnetic field to the energy of its thermal motion. The interpretation of the results of measurements in the vicinity of the magnetic saturation of the colloid is greatly simplified and, in addition to this, enables estimation of the minimum size of the domain of the dispersed ferromagnetic. Therefore, in the experimental section of the study special attention is given to plotting the magnetisation curve and the characteristic of the acoustic magnetic effect in the maximum wide range of the strength of the magnetic field.

Returning to Fig. 9.7, it should be mentioned that in the region of the 'moderate' magnetic fields there is a very weak (almost in the range of the measurement error) tendency for the increase of the amplitude of the EMF. In the initial section of magnetisation and in the vicinity of magnetic saturation there is no amplitude–frequency dependence of the *EMF*. Therefore, the assumption on the quasi-stationary nature of the AME in the MF-1 sample in the investigated frequency range can be regarded as acceptable.

9.5.4. Determination of the fiducial size of the magnetic nanoparticles

The magnetic moments of the 'large' m_{*max} and 'fine' m_{*min} particles, determined from the equations of the magnetic granulometry method (9.14), (9.15) and (9.16) and also from the equations of acoustic granulometry (9.18) and (9.19), were subsequently used for comparison with each other and for constructing the size distribution of the magnetic nanoparticles. Consequently, the resultant values

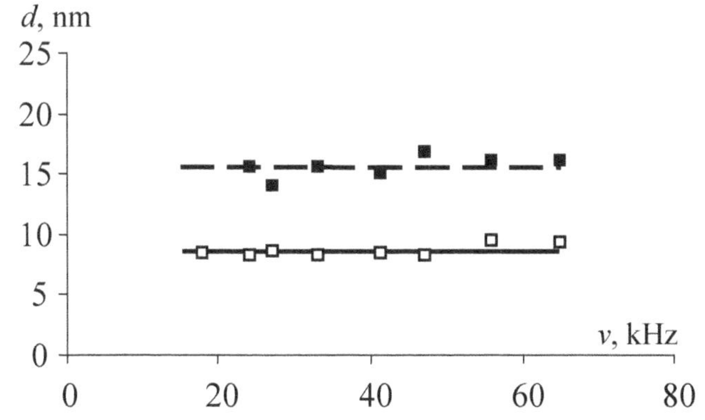

Fig. 9.8. Fiducial size of the MF-1 particles.

of m_{*max}, m_{*min} and the corresponding dimensions d_{max}, d_{min} will be referred to as fiducial values.

The crosshatched squares in Fig. 9.8 shows the values of d_{max}, obtained by the acoustic granulometry methods, and the open symbols show the values of d_{min}. The solid and dashed lines show the values of $\langle d_{min}\rangle$ and $\langle d_{max}\rangle$, averaged with respect to all frequencies investigated.

The values of the demagnetising factor, used in the calculations: at ν = 24.2 kHz, 41.4 kHz and 65.1 kHz respectively, are: $N_d = 0.31$, $N_d = 0.22$, $N_d = 0.16$. The mean deviation of the values of d_{max} and d_{min} from the results of averaging $\langle d_{max}\rangle$ = 16 nm and $\langle d_{min}\rangle$ = 9 nm is 0.7 nm.

The fiducial values m_{*max}, m_{*min}, d_{max} and d_{min}, determined on the basis of the MGM and AGM with averaging of the results with respect to all investigated frequencies, are also presented in Table 9.2.

The curves, constructed ffrom the experimental points, in Fig. 9.7 present the results of theoretical calculations, based on determination of the values of $\langle d_{max}\rangle$ (the dotted curve) and $\langle d_{min}\rangle$ (solid curve) from the results of the AGM data. Thus, these curves characterise the monodispersed magnetic fluids. Calculations were carried out using equation (9.7) and the following values of the quantities: N_d = 0.22, M_s = 45800 A/m, M_{S0} = 480000 A/m, T = 303 K, C_p = 1967 J/kgK, q = 0.00085 1/K, c = 930 m/s, ρ = 1315 kg/m^3.

The 'natural' special feature of the dotted and solid curves is the fact that the former are in good agreement with the experiments carried out in small magnetic fields, and the latter – in the fields close to saturating fields. The dotted curve, describing the 'large' magnetic nanoparticles, reaches the values close to unity in comparatively small fields (~200 kA/m). The solid curve on the same scale is characterised by a relatively large slope even in the fields of ~800 kA/m. Consequently, it may be concluded and in the fields close to magnetic saturation, magnetisation of the magnetic fluid takes place mainly as a result of 'fine' particles. At the same, attention should be given to the fact that in the fields, close to magnetic saturation, both curves come closer to each other.

The large difference in the theoretical and experimental values are β_H in the region of 'moderate' magnetic fields is determined in all likelihood by the fact that the model used in this case does not take into account the size distribution of the particles of the system. As shown later, the development of the theoretical model in this direction results in a relatively good agreement between the results of the theory and experimental data [303].

The values of the physical parameters of the magnetic nanoparticles β_H and d, obtained on the basis of the AGM, are in good agreement [303].

The values of the physical parameters of the magnetic nanoparticles m_* and d, obtained on the basis of AGM, are in satisfactory agreement with the data determined by other methods, for example, on the basis of magnetorelaxometry (MRX) according to which for the magnetite nanoparticles $d_{eff} = 9.2 \pm 2.5$ nm [312]. At the same time, attention should be given to the difference in the numerical values of the parameters m_* and d, presented in Table 9.2.

Values of $m_{*\max}$ and $d_{\max}$, obtained on the basis of AGM, are ~41% and ~7% higher than the respective values, obtained by the MGM method. Possibly, this difference is associated with the insufficient accuracy of the calculations of the dynamic demagnetising factor $N_{d'}$ based on the approximation (9.13) used in deriving the equation (9.11). In addition to this, in the denominator of the ratio for $m_{*\max}$ (9.19) there is the parameter $k' = qc^2C_p^{-1}$ as one of the terms, and its value for each specimen should be obtained separately, especially by the method of direct measurements of the quantities included in this term. The numerical values of the parameters q and C_p, used in this study, were obtained using the mixing equations which take into account the volume concentration of the components of the colloidal solution.

On the other hand, the values $m_{*\min}$ and $\langle m_{*\min}\rangle$ are slightly displaced in relation to each other in the reverse direction. The ratio $m_{*min}/\langle m_{*\min}\rangle = 1.7$. In this case, the effect of the dynamic demagnetising field on the measurement results is minimal because the measurements are carried out in the saturated magnetic field. It may be assumed that the difference of the numerical values of $m_{*\min}$ and $\langle m_{*\min}\rangle$ is associated with different possibilities of detecting the minimum dimensions of the magnetic nanoparticles using the methods of acoustic granulometry and magnetic granulometry analysis.

9.5.5. Detailing the mechanisms of perturbation of the magnetisation of the magnetic fluid in the sound wave

The concentration model of the AME [260] takes into account, in addition to the oscillations of the concentration of the magnetic nanoparticles of the magnetic field, also the oscillations of the temperature of the magnetic fluid in the adiabatic sound wave. In addition to this, under the condition of the limited side surface of the sound beam and depending on the mutual rotation of the wave vectors on the vector of the strength of the magnetic field, the 'dynamic' demagnetising field, specific for the given case, forms [263]. The subject of this section is the analysis of the processes of interaction of physical fields in the acoustic magnetic effect on the magnetic fluid, in particular, in the verification of the physical nature of the model theory.

The perturbation of the magnetisation of the disperse system under the effect of thermal oscillations is realised by the Brownian and Néel mechanisms of the orientation of the magnetic moments of the particles along the magnetic field. The latter is determined by the thermal fluctuations of the moment inside the hardest particle [34]. At a high viscosity of the fluid–carrier the Brownian rotational movement of the ferroparticles will be blocked, and the process of orientation of the magnetic moment of the ferroparticles takes place by the Néel 'solid-state' mechanism.

Below (section 9.7) we discuss the 'baric' solid-state mechanism of AME, associated with the effect of the alternating sound pressure on the crystal lattice of the magnetic nanoparticles. A distinguishing feature of the solid-state mechanism, investigated in this case, is its conditioning by the thermal oscillations in the soundwave.

Equation (2.8) reflects the effect of both the 'purely' concentration and of the temperature factor. Equation (2.8) can be written in the new form:

$$\Delta M_m = M_s k u_m(\operatorname{cth}\xi - \xi^{-1}) - M_S \frac{qc^2}{C_p} k u_m(\xi^{-1} - \xi \operatorname{sh}^{-2}\xi).$$

In the phase of compression of the sound wave the magnetisation of the magnetic fluid increases as a result of the increase of the concentration of magnetic particles (the first term) and decreases as a result of increase of temperature (the second term). The contribution of thermal oscillations to ΔM_m increases with the increase of $q \equiv -\rho^{-1}\dfrac{\partial \rho}{\partial T}$ – the temperature coefficient of expansion and, therefore, its value depends strongly on the selection of the fluid–carrier.

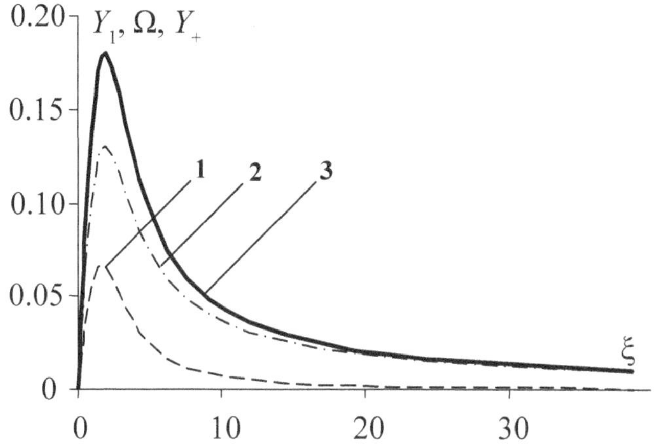

Fig. 9.9. Theoretical dependences: 1) Y_1; 2) Ω; 3) Y_+.

In the tensile loading phase the magnetisation of the magnetic fluid decreases as a result of a reduction of the concentration of the ferroparticles and increases as a result of decreasing temperature. However, in the framework of linear acoustics, the processes of compression and tensile loading take place symmetrically and, consequently, the field dependences of the induced EMF with the induction coil is situated in the antinode of the adjacent standing waves are the same.

The mechanisms, determining the amplitude of the perturbation of magnetisation, include the mechanism of induction of the dynamic demagnetising field by the sound wave in the magnetised liquid. Expression (2.8) was obtained assuming that the dynamic demagnetising factor N_d is equal to 0. Formally, the effect of this mechanism is reflected by the presence in the denominator of the equations (9.5) and (9.6) of the term with the coefficient N_d.

In Fig. 9.9, the ordinate 0Y gives the numerical values of the functions:

$$\Omega(\xi) \equiv k'D(\xi), \tag{9.20}$$

where $D(\xi) = \xi^{-1} - \xi \, \mathrm{sh}^{-2}\xi$;

$$Y_1(\xi) \equiv L(\xi) - \frac{L(\xi)}{1 + k''D(\xi)/\xi}; \tag{9.21}$$

$$Y_+(\xi) \equiv L(\xi) - \beta_\xi(\xi) = \frac{L(\xi) \cdot k''D(\xi)/\xi + \Omega(\xi)}{1 + k''D(\xi)/\xi}. \tag{9.22}$$

The function $\Omega(\xi)$ reflects the contribution, to the concentration mechanism of AME, of the process of perturbation of magnetisation of the magnetic fluid by thermal oscillations (in the absence of the dynamic demagnetising field). The reduction in the rate of growth of the relative amplitude of AME as a result of the dynamic demagnetising field is described by the function $Y_1(\xi)$ (in the absence of thermal oscillations). The contribution of these factors to the AME mechanism under the simultaneous effect of these mechanisms, is represented by the function $Y_+(\xi)$.

The numerical values of $\Omega(\xi)$, $Y_1(\xi)$ and $Y_+(\xi)$ are determined from equations (9.20), (9.21), (9.22), assuming the monodispersed magnetic fluid consisting only of the particles of the 'minimum' size whose magnetic moment is $m_{*\min} = 1.61 \cdot 10^{-19}$ A·m². The coefficient at $D(\xi)$, denoted by k', assuming that c = 930 m/s, C_p = 1967 J/kg K, q = 0.000765 1/K, is 0.34. The dynamic demagnetizing factor N_d was assumed to be equal to 0.31, k'' = 0.68.

Both mechanisms provide approximately the same contribution to the reduction of the rate of growth of the relative amplitude of AME. However, conditions can be created in which both factors, or one of them, are greatly weakened. For example, when using water as the fluid–carrier, with the thermal expansion of water in the vicinity of 4°C practically equal to 0, i.e., $q \approx 0$, the factor of thermal oscillations in the sound wave can be ignored. In greatly diluted samples of the magnetic fluid, the coefficient $k'' \approx 0$ so that the dynamic demagnetizing field can be neglected.

It should also be mentioned that the reduction of the rate of increase of the relative amplitude of AME as a result of the effect of the dynamic demagnetizing field, described by the function $Y_1(\xi)$, becomes smaller with the increase of the strength of the magnetic field and becomes insignificant in comparison with the reduction of the rate of increase caused by the thermal oscillations, i.e., the function $\Omega(\xi)$. Analysis of the equations (9.20) and (9.21) shows that in the 'strong' fields the function $\Omega(\xi)$ decreases in proportion to ξ^{-1}, and the function $Y_1(\xi)$ – in proportion to ξ^{-2}.

We transform the function $Y_+(\xi) \rightarrow Y_+(H)$, using the previously introduced notation $\xi = \dfrac{\mu_0 m_* H}{k_0 T}$

$$Y_+(H) = \frac{L(\mu_0 m_* H/(k_0 T)) \cdot N_d M_S H^{-1} + k'}{D^{-1}(\mu_0 m_* H/(k_0 T)) + N_d M_s H^{-1}}. \tag{9.23}$$

In the vicinity of magnetic saturation ($H \to \infty$) the functional Y_+ (H) has the form:

$$Y_+(H) = \frac{qc^2 k_0 T}{C_p \mu_0 m_*} \cdot \frac{1}{H} + \frac{N_d M_S k_0 T}{\mu_0 m_*} \cdot \frac{1}{H^2}. \qquad (9.24)$$

The first term of the equation (9.24) is $\Omega(H)$ – the contribution of thermal oscillations, the second term Y_1 (H) – the contribution of the dynamic demagnetising field to the reduction of the rate of increase of the relative amplitude of the induced EMF in the vicinity of magnetic saturation. Consequently, in the case of the monoparticle dispersed system in 'large' fields, the dependence $Y_+(H)$ is hyperbolic, and the dependence Y_+ (H^{-1}) has the form of a section of the straight line. The formation of the dependence in the experiments indicates that the particles with the narrow distribution of the magnetic and linear parameters take part in the process of magnetization in the investigated range of the strength of the magnetic fields.

Figure 9.10 shows in relative units the magnetizing curves $\beta_M(H)$ (crosshatched rhombs) and the field dependence of the amplitude of the EMF induced as a result of the AME $\beta(H)$ (triangles). The thick solid line indicates the 'difference' curve $Y_+^{(e)}(H)$ reflecting the total contribution to the perturbation of the magnetization of the magnetic fluid by the thermal oscillations and the dynamic

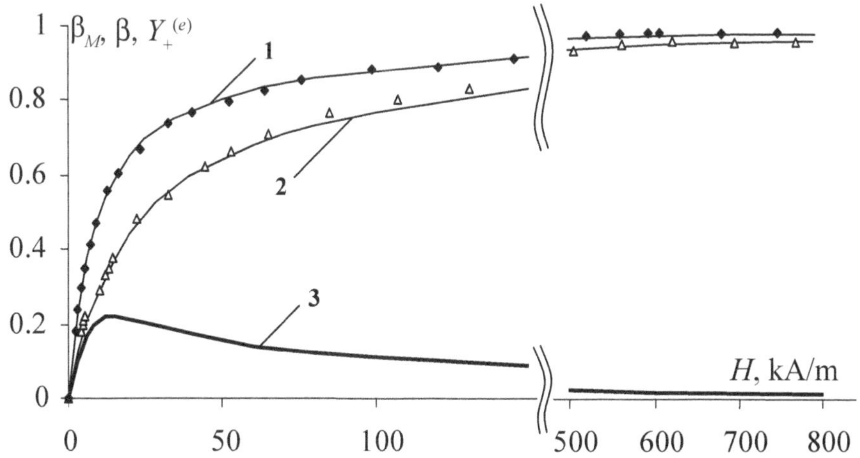

Fig. 9.10. Construction of the 'difference' curve from the experiments: 1) the relative value of magnetization $\beta_M(H)$, 2) the experimental values of the relative amplitude of AME $\beta(H)$ at ν = 65.1 kHz, 3) 'difference' curve $Y_+^{(e)} = \beta_M(H) - \beta(H)$.

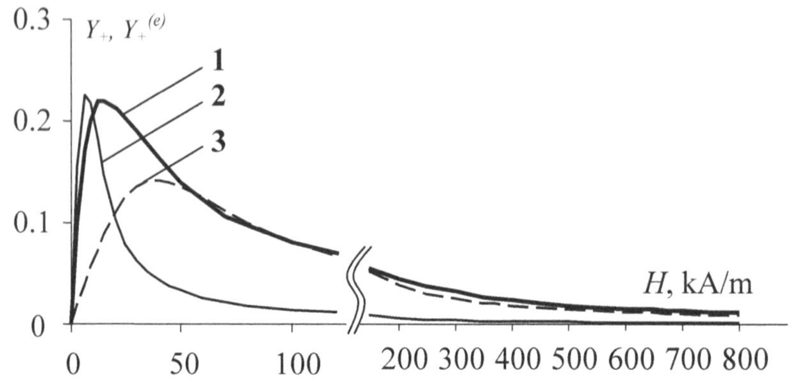

Fig. 9.11. Comparison of the depedences for the magnetic fluids: 1) – 'difference' curve, 2) $Y_+(H)$ at d_{max} = 16 nm; 3) $Y_+(H)$ at d_{min}= 9 nm.

demagnetizing field. The 'difference' curve is plotted by deducting the values of $\beta_M(H)$ from the values of $\beta(H)$ in the magnetic fields with the same strength, i.e. $\beta_M(H) - \beta(H)$. The experimental values of the dependences $\beta_M(H)$ and $\beta(H)$ are approximated by the smooth curves in MS Excel.

The 'difference' curve has a maximum and its value and position are determined by the total contribution to the perturbation of the magnetization of the magnetic fluid from the thermal oscillations and the dynamic demagnetizing field in the sound wave.

Figure 9.11 shows the 'difference' curve $Y_+^{(e)}$ (H) –1 and two calculated dependences $Y_+(H)$ at $\langle d_{max}\rangle > 16$ nm – 2 and at $\langle d_{min}\rangle$ = 8.5 nm – 3. The dependences $Y_+(H)$ were calculated using equation (9.23) and the previously mentioned values of the quantities included in this equation.

It is important to mention a number of special features of the results: satisfactory agreement of the function $Y_+(H)$ in the initial (rising) section of the 'difference' curve for the case $d = d_{max} = 16$ nm and in its falling section for the case $d = d_{min} = 8.5$ nm; the maxima of these dependences are far from each other with respect to the strength of the magnetic field; the right-hand branch of the curve 2 rapidly approaches the zero value with increasing strength of the magnetic field, and the right-hand part of the curve 3 decreases relatively slowly.

Thus, the 'difference' curve reflects the mechanism of competition of two opposite effects on the process of magnetisation of the colloid – the orientation effect of the magnetic field and the disorientation effect of thermal motion. If the start of the magnetisation process in which mostly 'large' magnetic nanoparticles take part is characterised by the dominance of the role of the thermal factor, then with a further increase of the strength of the magnetic field the degree of orientation of the magnetic moments of the 'fine' nanoparticles in the field increases. In the fields, close to magnetic saturation, magnetisation is controlled by the factor of the effect of the magnetic field.

It may be concluded that the formation of the 'difference' curve is an additional confirmation of the justification of the concentration model of the physical mechanism of AME.

9.5.6. Evaluation of the physical parameters of the finest magnetic particles

In the vicinity of the magnetic saturation of the magnetic fluid the 'difference' curve reflects the effect of thermal oscillations in the adiabatic soundwave primarily on the finest particles, because large particles are blocked by the magnetic field. Therefore, the right falling part of the curve characterises the process of straightening of the finest nanoparticles in the field.

This special feature of the 'difference' curve may be used for evaluating the size of the magnetic nanoparticles and the results obtained in this case are closer to the physically possible minimum result than the previously mentioned value of $d_{\min}$. In this case, we are concerned with the evaluation of the minimum linear size of the domain in the dispersed system of the particles determined by quantum constraint.

The theoretical estimate of the critical (minimum) size of the single-domain particle d_{cr} is based on the application of the indeterminacy relationship [253]: $\Delta p \cdot d_{cr} \approx \hbar$. The indeterminacy of the energy of the exchange interaction of the electron as a result of the quantum constraint is written in the form $\Delta\varepsilon \approx \hbar^2/(2md_{cr}^2)$. Equating this energy to the energy of the volume interaction which is mostly responsible for the formation of magnetic ordering, i.e., $\Delta\varepsilon \approx k_0 T_k$ where T_k is the Curie temperature of the 'massive' material, we obtain the equation for evaluating the critical size:

$$d_{cr} \approx 2 \cdot 10^{-8} T_k^{-1/2}.$$

For magnetite with the Curie temperature of 586 K the transition to the paramagnetic state takes place when the domain size reaches ~1 nm.

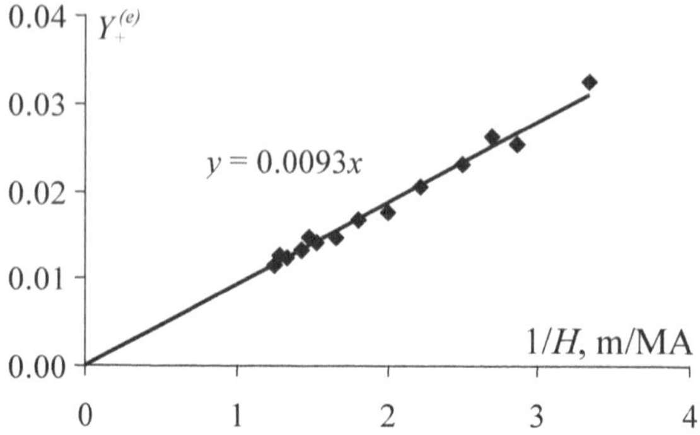

Fig. 9.12. Dependence $Y_+^{(e)}(H^{-1})$.

The proposed method is based on combining the right section of the function $Y_+(H)$ with the right section of the 'difference' curve by selecting the numerical values of parameter m_*. A simpler and reliable method is the determination of the tangent of the slope α of the section of the straight line $Y_+^{(e)}(H^{-1})$ in the vicinity of magnetic saturation, because, as follows from equation (9.24), $\mathrm{tg}\,\alpha = \frac{qc^2k_0T}{C_p\mu_0m_*}$ (The index (e) stresses the fact of using in this case the experimentally determined 'difference' curve). Finally, the essential condition for the reliability of the proposed method is the fulfilment of the requirements on the accuracy of measurements of both dependences $\beta(H)$ and $\beta_M(H)$.

Figure 9.12 shows the straight section of the dependence $Y_+^{(e)}(H^{-1})$ obtained for $H \geq 300$ kA/m. We obtain tg α = 9300 A/m which, assuming the previously accepted values of T, C_p, q and c, is fulfilled at $m_* = 1.2 \cdot 10^{-19}$ A·m². This leaves to the estimate of the minimum size of the single-domain particles of the disperse phase of the investigated specimen of the magnetic fluid as d = 7.8 nm. The quantitative difference between the resultant estimate d = 7.8 nm and the reference value $d_{\min}$ = 10.4 nm, provided by the MGM method, is outside the range of the maximum error of the proposed method, ≤5%.

The presence of the dipole–dipole interaction, in particular between the 'large' magnetic nanoparticles, weakens their contribution to the magnetization process in the vicinity of magnetic saturation. The nanoaggregates are blocked more extensively by the magnetic fields than the individual particles and the effect of thermal oscillations on them is less pronounced. Therefore, for the

Table 9.3

Sample	ρ, kg/m^3	φ, %	χ	M_s, kA/m	c, m/s	$m_{*max}\cdot 10^{19}$, A·m^2	$m_{*min}\cdot 10^{19}$, A·m^2	d_{max}, nm	d_{min}, nm
							MGM		
MFc	1852	24	6.8	89.6	864	7.6	2.64	14.5	10.2
							AGM		
						11.6	1.54	16.7	8.5

concentrated magnetic fluid in which the dipole–dipole interaction takes place between the magnetic particles because they are closely spaced, the formation of the right-hand branch of the dependence $Y_+ (H)$ takes place mostly as a result of the finest fraction of the magnetic particles. Consequently, the application of the proposed acoustomagnetic method for the concentrated magnetic fluid makes it possible to reduce the difference between the resultant values of m_{*min} and d_{min} and the lower physical boundary.

To verify these considerations, detailed investigations were carried out using a concentrated sample of the magnetic fluid (denoted by MFc). The MFc sample was produced by long-term drying of a part of the initially prepared colloid (the other part of the colloid was used as the MF-1 sample) with its surface open taking essential measures for protection against the penetration of dust and other contaminants.

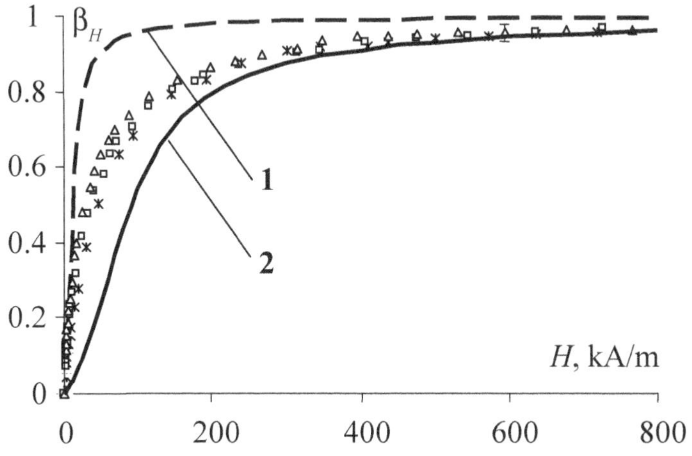

Fig. 9.13. Dependence $\beta_H (H)$ for the MFc, obtained at frequencies of 20, 33 and 52 kHz, and the theoretical dependences: 1) for the particle d_{max} = 16.7 nm, 2) for the particle d_{min} = 8.5 nm.

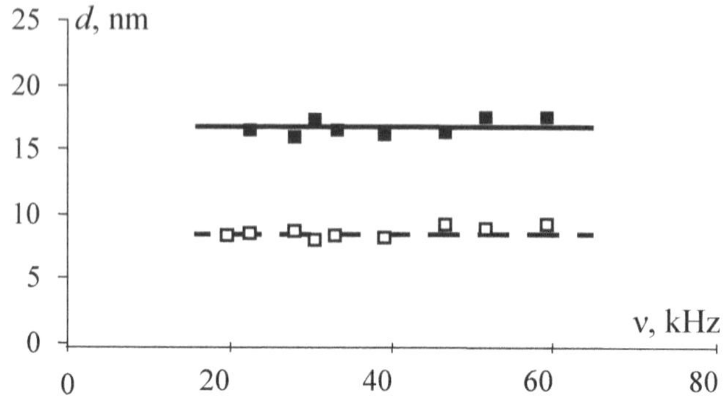

Fig. 9.14. Size of the MFc particles.

The physical parameters of the MFe sample are presented in Table 9.3.

Figure 9.13 shows the experimental results of measurement of the relative amplitude of induction EMF for the frequencies of 20, 33 and 52 kHz in the MFc, denoted correspondingly by the asterisk, the square and the triangle. The values of the demagnetizing factor N_d [263], using the calculations, equal 0.31, 0.22 and 0.16 in the order of frequencies. As in Fig. 9.7, which shows the results for the MF-1, the scatter of the experimental data fits the range of the experimental error $\pm 3\%$.

The crosshatched squares in Fig. 9.14 show the values of $d_{max,}$ determined by the acoustic granulometry methods, and the open symbols the values d_{min}. The solid and dashed lines show the values of $\langle d_{max} \rangle$ and $\langle d_{min} \rangle$, averaged with respect to all investigated frequencies. The average deviation of the resultant values of d_{max} and d_{min} from the averaged results $\langle d_{max} \rangle = 16.7$ nm and $\langle d_{min} \rangle = 8.5$ nm in the MFc sample is 0.6 nm and 0.3 nm, respectively.

The very small quantitative and uniform deviation of d_{max} and d_{min} from the averaged results, illustrated by Figs. 9.8 and 9.14, is explained by the quasi-stationary nature of the magnetization process in the initial section and in the section adjacent to the magnetic saturation of the magnetic fluid.

Returning to Fig. 9.13, it should be mentioned that the curves connecting the experimental points represent the results of calculations using the equation for the monodispersed magnetic fluid (9.7) for the values $\langle d_{max} \rangle$ (the dotted line) and $\langle d_{min} \rangle$ (solid curve). The following values included in the equation were used:

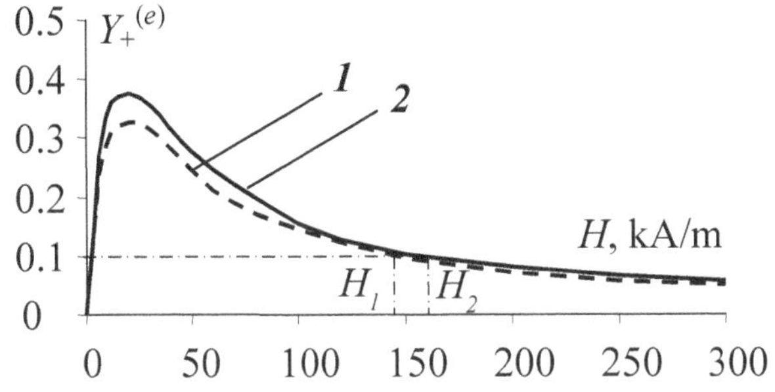

Fig.9.15. 'Difference' curves: 1) for MF-1, 2) for MFc.

$N_d = 0.22$, $M_S = 89600$ A/m, $M_{S0} = 480000$ A/m, $T = 305$ K, $C_p = 1787$ J/kg·K, $q = 0.00068$ 1/K, $c = 864$ m/s, $\rho = 1852$ kg/m^3.

As in the case of MF-1, the curve relating to the 'large' magnetic nanoparticles reaches the values close to unity in the relatively small fields (~200 kA/m). The curve obtained for the 'fine' particles on the same scale is characterized by a relatively large slope even in the fields ~800 kA/m. Here, the conclusion according to which the magnetization of the magnetic fluid in the fields close to magnetic saturation takes place mainly as a result of the 'fine' particles is again confirmed here.

The large difference between the theoretical and experimental values of β_H in the range of the moderate magnetic fields may be removed by taking into account the size distribution of the particles of the system [303].

The MFc sample is also characterized by a difference in the numerical values of the parameters m_* and d, obtained by the MGM and AGM methods (Table 9.3). The reasons for this difference were discussed previously.

The 'difference' curve was constructed using the magnetization curve $\beta_M(H)$ in relative units and the field dependence of the amplitude of the induced EMF $\beta(H)$, obtained for the MFc sample. The graph of the dependences $\beta_M(H)$, $\beta(H)$ and $Y_+^{(e)}(H)$ for the MFc sample is identical with the curves shown in Fig. 9.10.

Figure 9.15 shows the 'difference' curves for the MFc (solid line) and MF-1 (dotted line) samples. The left parts of the dependences $Y_+^{(e)}(H)$ are almost identical. The two right (falling) branches differ. The curve for the MF-1 sample falls more rapidly. The same figure shows the dot-and-dash straight line reflecting the 0.1 level of the total weakening of perturbation of the magnetisation of the magnetic

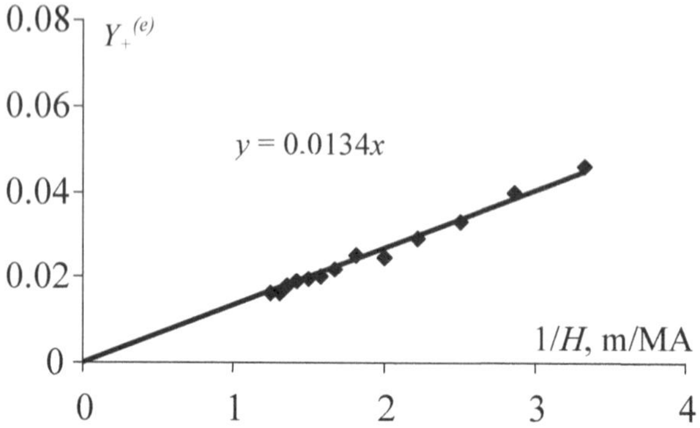

Fig. 9.16. Dependence $Y_+^{(e)}(H^{-1})$ for MFc.

fluid by thermal oscillations and the dynamic demagnetising field. The values H_1 and H_2, corresponding to the intersections of the dot-and-dash straight line with the curves $Y_+^{(e)}$ (H) for the MF-1 and MFc, are equal to 143 kA/m and 160 kA/m.

The results showing that H_2 is higher than H_1 confirm the assumption according to which in the concentrated magnetic fluid the formation of the right-hand part of the dependence $Y_+(H)$ takes place mostly as a result of the finest fraction of the magnetic particles. The straightening of the particles of the fine fraction 'in the fields' takes place in the fields with higher strength.

It should be mentioned that the agreement in the left parts of the dependences $Y_+^{(e)}(H)$ shown in Fig. 9.15 in the quantitative aspect is the consequence of two circumstances – identical nature of magnetization of both fluids (the dependences β_M (H) coincide in the measurement error range), and the large difference of the dependences $\beta(H)$ in the magnetic field $H \geq 40$ kA/m which is greater than the strength at the extremum $H \approx 20$ kA/m.

Figure 9.16 shows the straight section of the dependence $Y_+^{(e)}$ (H^{-1}) obtained for $H \geq 300$ kA/m. Here tg α = 13400 A/m which, taking into account the previously accepted values of T, C_p, q and c, is fulfilled at $m_* = 0.7 \cdot 10^{-19}$ A·m². This gives the estimate of the minimum size of the single-domain particles of the dispersed phase of the investigated magnetic fluid sample, obtained by the proposed procedure: d = 6.5 nm. In this case, comparison with the fiducial size gives $d \approx 0.6\ d_{\min}$.

It may be assumed that the blocking of the 'large' magnetic nanoparticles in the magnetic field will be intensified if viscous liquid media are used as the fluid–carrier or the temperature of the colloid is reduced. In both cases, the contribution of the 'fine 'particles to the magnetization of the magnetic fluid in the fields, close to magnetic saturation, will increase. The process of interaction and aggregation of the ferroparticles in the region of the 'moderate' magnetic fields has the same effect on the dependences $\beta_M(H)$ and $\beta(H)$ and it is therefore natural to assume that its effect on the shape of the right-hand part of the 'difference' curve will be minimal.

The advantage of this method of calculation of tg α value in comparison with the MGM method is that in the latter the small increments of magnetization as a result of the alignment of the fine particles in the strong fields are observed on the background of the considerably higher degree of magnetization of these particles resulting from the large and medium size. Consequently, in the saturation fields, the method of detection of the 'fine' particles based on the MGM is insufficiently effective.

A special feature of the proposed method of evaluating the physical parameters of the finest magnetic nanoparticles of the dispersed system is the efficient utilisation of the results of magnetic granulometry and acoustic granulometry measurements, and the fact that these methods can be used for magnetic colloids with different ferromagnetic phases.

It should be mentioned that in many studies of magnetic granulometric analysis (in particular, in [286, 306]) no attention is given to the possibilities of the large difference in the magnetic properties of the nanosized and massive ferromagnetics the parameters of which are usually used in the calculations. In the monograph published by A.I. Gusev [270] the studies concerned with this problem are reviewed. For example, the reduction of the saturation magnetisation with the reduction of the size of the nanoparticles of Fe, Ni and Co and ferromagnetic alloys is discussed. This circumstance, if it were related to the nanoparticles of magnetite Fe_3O_4, it would lead to the overestimation of the presented numerical values.

9.6. The size distribution of magnetic nanoparticles

The values of the physical parameters of the magnetic nanoparticles m_{*max}, d_{max}, m_{*min} and d_{min}, presented in Tables 9.2 and 9.3, are not only of the evaluation but also conditional nature because the magnetic fluid contains particles of different sizes and also aggregates of this particles and this is not taken into account in the proposed model. Therefore, the next aim of acoustic granulometry is the determination of the distribution of the magnetic nanoparticles, dispersed in the real magnetic fluids, as regards the linear dimensions and magnetic moments.

9.6.1. Distribution on the basis of atomic force microscopy data

Atomic force microscopy (AFM) is based on the strong dependence of the force of interaction of the molecules on the distance between them (van der Waals interaction). Molecules of two solids interact – the molecules of the surface of the investigated solid in the molecule of the probe (the tip of a needle), the so-called cantilever. The force interaction in the process of precision scanning with such a needle along the surface is recorded so that the device of this type is referred to as the scanning force microscope [292, 293].

The dimensions of the particles of the disperse phase of the magnetic fluid are measured in a scanning probe microscope Aist NT Smart Spm (Aist NT company) in the laboratory of the Regional Nanotechnology Centre, Southwest State University, Kursk.

The possibilities of atomic force microscopy in studying magnetic colloids have been investigated insufficiently. Therefore, the studies carried out in the preparation of samples for measurements will be described.

Substrates for the samples were prepared using the following procedure. Coating glass was cleaned to remove large dust particles with a brush, washed in an alcohol solution, and dried in the natural conditions for several hours. Subsequently, the coating glass was placed in the fluid–carrier (kerosene) for 10 min and then dried for several hours.

Several methods of depositing the magnetic fluid on the substrate have been tested:

1. The substrate is placed on a horizontal surface and then a droplet of the magnetic fluid is deposited on it and spreads on the glass surface. The liquid dries up in the natural conditions. The balance of the surface tension forces results in slight spreading of the droplet on the surface of the glass resulting in the formation of a thick hardened layer of the magnetic fluid.

2. After depositing on the substrate, the magnetic fluid droplet is spread with another degreased coating glass and then dried for several hours. Scanning in the AFM shows that the resultant surface of the solidified magnetic fluid has irregularities of several hundreds

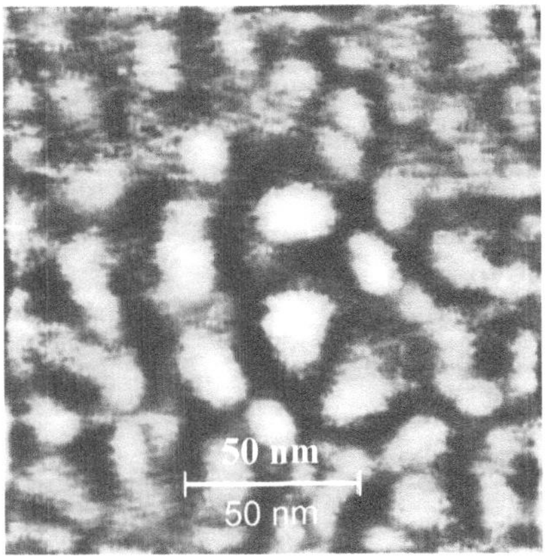

Fig. 9.17. Scan of MF-2 in the atomic force microscope.

of nanometres and this is unacceptable for analysis of the disperse composition of the magnetic fluid.

3. To reduce the thickness of the hardened layer, the coating glass, held for 10 min in kerosene, is placed on the horizontal surface. Subsequently, a droplet of the magnetic fluid of the minimum possible size is deposited on the substrate. Since the glass surface, wetted with kerosene is characterised by higher wettability for the investigated colloid, spreading of the droplet on the surface of the glass takes place prior to the formation of the minimum possible thickness of the layer of the magnetic fluid. The resultant system is dried in the natural conditions for 20 h. The results of investigation of the surface of the MF-2 sample, produced by this method, are presented in Fig. 9.17.

The dimensions of the particles of the disperse phase of the magnetic fluid are analysed using the software for AFM scanning Gwyddion 2.15. The particles are separated using the watershed algorithm, installed in the given program. This is followed by the calculation of the distribution of the radii of the discs, equivalent to the area of projection of the grain. The size of the particles is corrected, taking into account the diameter of the tip of the needle of the cantilever, using the equation [292]:

$$r_c = 2\sqrt{r \cdot R},$$

where r is the radius of the nanoparticle; r_c is the radius of the AFM image; R is the radius of the needle of the cantilever, determined from the parameters of the image of the largest particle: $R = (w^2 + h^2)/2h$; h is the height of the image of the nanoparticles, w is the width of the image.

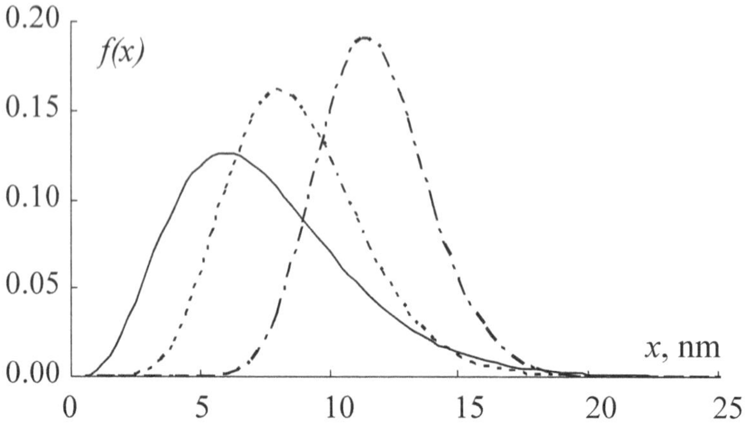

Fig. 9.18. Size distribution of the particles of MF-2: ------ MGM,----- AGM, -.-.-.- AFM.

The size distribution of the particles, obtained by the AFM method, is shown in Fig. 9.18.

9.6.2. Distribution of the results of acoustic granulometry and magnetic granulometry

The size distribution of the magnetic nanoparticles on the basis of acoustic granulometry was obtained for the first time by P.A. Ryapolov [281, 282]. Some of the investigations in this direction were reviewed in [263] and continued in [283, 285, 303, 304] – in studies carried out by the author of the present book and colleagues, together with P.A. Ryapolov.

The disperse composition of the magnetic particles of the magnetic fluid, obtained on the basis of the results of acoustic magnetic measurements, is described by the two-parameter Γ-distribution [286]:

$$f(x) = \frac{x^{\alpha}\exp(-x/x_0)}{x_0^{\alpha+1}\Gamma(\alpha+1)}, \tag{9.25}$$

where $\Gamma(\alpha + 1)$ is the gamma function; x is the nanoparticle diameter; x_0, α are the distribution parameters, determine from the equations:

$$\frac{(\alpha+5)^3-\alpha-5}{(\alpha+2)^3-\alpha-2} = \frac{\langle m_*^2\rangle}{\langle m_*\rangle^2}, x_0^3 = \frac{6\langle m_*\rangle}{\pi M_{s0}(\alpha+1)(\alpha+2)(\alpha+3)}, \tag{9.26}$$

where $\langle m_*^2 \rangle = 3kT\chi_L / \mu_0 n$ is the mean square of the magnetic moment; $\langle m_*^2 \rangle = M_s / n$ is the mean magnetic moment; M_{s0} is the saturation magnetisation of the dispersed phase (for magnetite M_{S0} = 480 kA/m); $M_L(H)$ = $M_s L(H)$; χ_L – is the 'Langevin' magnetisation and initial magnetic susceptibility [286].

The equation (9.7) is transformed to the form:

$$\beta_\xi = \frac{\int_0^\infty L(\xi) f(x)dx - k' \int_0^\infty \xi(\xi^{-2} - \mathrm{sh}^{-2}\xi) f(x)dx}{1 + N_d \mu_0 n / k_0 T \int_0^\infty m_*^2(x)(\xi^{-2} - \mathrm{sh}^{-2}\xi) f(x)dx}. \tag{9.27}$$

At $H \to 0$ we have

$$\langle m_*^2 \rangle = 3k_0 T M_S \cdot \mathrm{tg}\theta / \left[\mu_0 n(1 - k' - N_d M_S \cdot \mathrm{tg}\theta)\right], \tag{9.28}$$

where θ is the slope of the initial section of the dependence β_H (H); n is the concentration of the particles.

At $H \to \infty$ we obtain

$$\langle m_* \rangle = (1 + k')k_0 T / \left[\mu_0 \cdot \mathrm{tg}\,\Omega\right] \tag{9.29}$$

where Ω is the slope of the section of the curve $\beta_H(H^{-1})$ at $H \to \infty$.

The system of equations (9.25)–(9.29) can be used to determine the size distribution of the particles of the magnetic fluid without taking into account the interparticle interactions on the basis of acoustic magnetic experiments. The interparticle interactions in the concentrated magnetic fluids were taken into account using the modified model of the effective field MMF2 [287, 288]:

$$M(H) = n \int_0^\infty m_*(x) L(\xi_e) f(x)dx, \tag{9.30}$$

$$\xi_e = \frac{\mu_0 m_*(x)}{kT}\left(H + \frac{M_L(H)}{3} + \frac{1}{144} M_L(H) \frac{dM_L}{dH}\right),$$
$$\chi = \chi_L\left[\chi_L + \chi_L / 3 + \chi_L^2 / 144\right], \tag{9.31}$$

where χ is the initial magnetic susceptibility determined by experiments.

Substituting the value of the parameter ξe, determined from the equation (9.30), into the dependence (9.7), we obtain the equation for the relative amplitude of the AME β_ξ with the interparticle interaction in the polydispersed magnetic fluid taken into account:

$$\beta_\xi = \frac{\int_0^\infty L(\xi_e) f(x) dx - k' \int_0^\infty \xi_e (\xi_e^{-2} - sh^{-2}\xi_e) f(x) dx}{1 + \frac{N_d \mu_0}{k_0 T} n \int_0^\infty m_*^2(x)(\xi_e^{-2} - sh^{-2}\xi_e) f(x) dx}. \quad (9.32)$$

At $H \to 0$ we have

$$\chi = M_s \cdot tg\theta / \left[1 - k' - N_d M_s \cdot tg\theta\right]. \quad (9.33)$$

At $H \to \infty$ we obtain

$$\beta_H = \left(1 - \frac{(1+k')nk_0T}{\mu_0 M_s H}\right)\left(1 + \frac{(1+k')nk_0T}{3\mu_0 H^2}\right). \quad (9.34)$$

The physical parameters of the particles of the nanodispersed phase for the investigated MF-2 sample, calculated on the basis of the field dependence of the AME (frequency 15 kHz, temperature 25°C) are presented in Table 9.4. The results obtained by the MGM method are also presented there.

The system of the equations (9.25) (9.26), (9.30), (9.33) and (9.34) can be used to determine the size distribution of the nanoparticles of the disperse phase of the magnetic fluid taking the interparticle interactions into account.

The magnetic fluid of the 'magnetite in kerosene' type with oleic acid as the surfactant was investigated. The MF-2 sample was synthesised at the Department of Physics of the Southwest State University in Kursk, by the chemical condensation method.

The density of the magnetic fluid was measured with a 10 ml glass pycnometer. The volume concentration of the solid phase φ was calculated by the mixing equation. The saturation magnetisation M_s and the initial magnetic susceptibility χ were determined using the results of measurements of the magnetisation curve by the ballistic method.

Table 9.4

Fluid-carrier	ρ, kg/m^3	φ, %	χ	M_s, kA/m	Method	$\langle m_* \cdot 10^{19} \rangle$ A·m^2	$\langle m_*^2 \rangle \cdot 10^{38}$ (A·m^2)^2	$\langle m_* \rangle$ nm	σ
Kerosene	1360	13.0	4.2	58	MGM	2.2	8.7	8.8	0.29
					AME	1.9	12.6	7.6	0.46

The physical parameters of the nanodispersed phase of the MF-2 sample were evaluated by acoustic magnetic measurements in the following sequence:

- Examination of the section of the dependence $\beta_H(H^{-1})$ in the vicinity of magnetic saturation of the colloid – the concentration n is calculated using the theoretical dependence (9.34);
- The slope of the straight section of the dependence β_H (H), using the dependence (9.33), is used to determine the value χ;
- The values $\langle m_*^2\rangle$ and $\langle m_*\rangle$ are determined and, using equations (9.26), are used to calculate by the numerical methods the distribution parameters x_0, α.

The mean diameter of the particle is determined from the equation $\langle x\rangle = x_0\ (\alpha + 1)$, and the dispersion of the size distribution of the particles is determined from the expression $\sigma = (\alpha + 1)^{-1/2.}$

The dispersion composition of the specimens of the magnetic fluid is also determined by the method of semi-contact atomic force microscopy (AFM).

The size distribution curves of the particles, obtained using the AFM and MGM data and the investigations based on AGM (using the model MMF2), are presented in Fig. 9.18.

It may be concluded that the results can be compared. The value of the mean diameter of the particles, obtained using the AFM, is $\langle x\rangle_{AFM}$ = 11.6 nm, which is slightly higher than the value of the diameter of the magnetic nucleus $\langle x\rangle$, determined by AME studies (Table 9.4). Evidently, this difference is determined by the presence

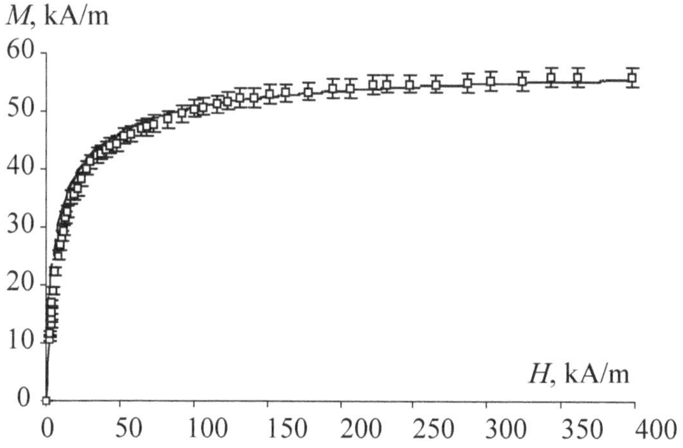

Fig. 9.19. The magnetisation curve of MF-2. The points – experiments, the curve – AME.

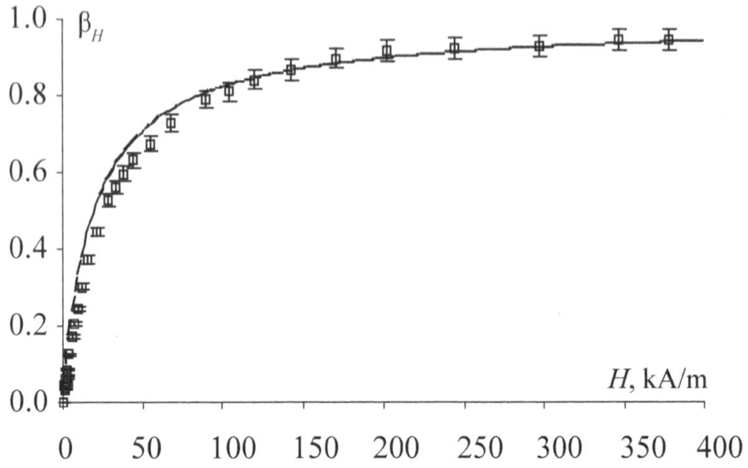

Fig. 9.20. Theoretical dependence of the amplitude of AME on the strength of the magnetic field for MF-2. The points – experiments.

of the stabilising shell and a thin non-magnetic layer on the surface of the magnetic particle.

Examples of the calculation of the magnetisation curves using the results of acoustic magnetic experiments and the theoretical dependence of the relative amplitude of AME on the strength of the magnetic field, obtained in the NI Labview software for the investigated sample of MF-2 (using equations (9.30), (9.34)) are presented in Fig. 9.19 and 9.20. These graphs also show the results of experimental measurement of the $M(H)$ and $\beta_H(H)$ curves. In this case, we can see the satisfactory agreement between the calculation results obtained by the modelling theory and the experimental data.

Comparison of the results of calculations using the modelling theory and experimental data (the values m_*, d, 'difference' curve) shows that in the colloid with the magnetic dispersed phase the other possible mechanism of AME, considered together with the concentration mechanism – solid state (baric), plays a secondary role.

It should also be mentioned that the concentration model of AME is also in agreement in other aspects with the experimental results (the field dependence of AME in the longitudinal and transverse magnetic fields, dependence on the amplitude of sound oscillations, on the phase of the wave). Therefore, evidently, taking into account the 'solid-state' mechanism will only be useful for making corrections.

9.7. Motivation of studies of the 'solid-state' mechanism of magnetisation of the magnetic fluid

The unique special feature of the acoustic magnetic methods of investigating the mechanisms of magnetisation of the dispersed system, based on the application of AME, in comparison with other well-known methods, including the wave methods, is the possibility of probing the baric dependence of these mechanisms.

The magnetic nanoparticles, dispersed in the magnetic fluid, in propagation of the elastic (sonic, ultrasound, hypersonic) waves in the fluid are subjected to the effect of the uniform pressure with changing sign. The sound pressure varies in accordance with the harmonic law:

$$\delta p = \delta p_0 \cdot \cos(\omega t - kx).$$

The pressure drop in the sound wave at the distance equal to the diameter of the magnetic nanoparticles D:

$$\Delta p = \frac{\partial(\delta p)}{\partial x} \cdot dx = \delta p_0 \cdot kD.$$

The ratio $\Delta p/\delta p_0 = kD = 2\pi \nu D/c$ at $D = 10^{-8}$ m, $\nu = 20 \cdot 10^3$ Hz, $c = 10^3$ m/s, equals $\sim 10^{-6}$. Consequently, the sound wave exerts mostly the effect of uniform tension–compression on the particle and to a considerably lesser extent the effect of in the longitudinal direction.

The concentration mechanism of AME applied to the diluted magnetic fluid explains the nature of its field dependence, the dependence on the amplitude of sound oscillations, on the phase of the wave, the displacement of the amplitude maximum of the AME from the central section of the magnetic fluid cylinder to its base with the increase of the frequency of oscillations, and the presence of the 'difference' curve. However, the question of the contribution of the 'solid-state' (baric) mechanism in the AME, determined by the effect of the alternating sound pressure directly on the crystal lattice of the magnetic nanoparticles, has not been answered. The role of this mechanism should become more important with the increase of the concentration of the particles with the minimum ('critical') size.

The ferromagnetic particles have a domain structure which is advantageous from the viewpoint of energy because of the closure of the magnetic fluxes inside the massive ferromagnetic. The magnetic properties of the nanostructures greatly differ from those of the massive material [238, 253, 270]. The more significant contribution is provided by the dimensional defects, the effect of the surface, interparticle interactions the interaction of the particles with the matrix, intraparticle or interparticle (cluster) organization. With the reduction of the size of the ferromagnetic the closure of the magnetic fluxes in it becomes less advantageous from the viewpoint of energy. After reaching some critical size, the ferromagnetic solids become single-domain and a further reduction of the size reduces the coercive force to 0 and leads to transition to the superparamagnetic state. However, the domain, as a quantum object, has also the lower critical size d_{cr}, at $d < d_{cr}$ the particles will be in the magnetically disordered state.

The number of magnetic moments in the particle should be sufficient to ensure that the total energy of the volume interaction exceeds the energy of thermal oscillations $k_0 T$. The intensity of the exchange interactions is influenced by different factors of the chemical or geometrical origin and, therefore, the exact calculation of d_{cr} is a complex task.

One of the special features of the physical state of the nanoparticles is the effect on heat of a high surface tension pressure $p_\sigma = 2\sigma/R$. Thus, at $\sigma = 1$ J/m^2, $R = 2$ nm, $p_\sigma = 10^9$ Pa.

I.P. Suzdalev, using the thermodynamic model of magnetic phase transitions, published the following relationship [253]:

$$T_{k0} = T_0(1 - \beta_* \lambda_* p_\sigma),$$

where T_{k0} is the Curie temperature of the nanoparticles; T_0 is the Curie temperature of the substance with the crystal lattice not subjected to compression; β_* is bulk compressibility; λ_* is the magnetostriction constant of the substance. Thus, as a result of the pressure p_σ T_{k0} decreases in comparison with T_0.

The increase of external pressure of the magnetic nanoparticles from p_0 to $p_0 + \Delta p$ results in a change of the temperature of the magnetic transition by the value $\Delta T_{k0} = -T_k \beta_* \lambda_* \Delta p$.

Therefore

$$\frac{\Delta T_{k0}}{T_{k0}} = \Delta p \left(\frac{2\sigma}{R} + p_0 - \frac{1}{\beta_* \lambda_*} \right)^{-1}.$$

The ratio $\Delta T_{k0}/T_{k0}$ is proportional to Δp. The proportionality coefficient strongly depends on the difference $(2\sigma/R - 1/(\beta_* \lambda_*))^{-1}$.

For the nanoparticles with the specific combination of the size, compressibility and magnetostriction we may expect a large increase of this coefficient and, consequently, a large increase of the sensitivity to excess pressure. Consequently, a relatively low pressure is capable of transferring the particle with the 'almost critical' size from the magnetically ordered state to the non-magnetic (paramagnetic) state.

The Villardi effect is well-known in the physics of magnetism [41]. This effect is represented by the rapid change of the shape of the magnetisation curve under the effect of one-sided tension or compression on the massive ferromagnetic. The existence of the magnetoelastic waves, determined by the relationship between the magnetic end elastic parameters of the magnetically ordered medium, is also well-known [197, 294].

The nanosized crystals are characterised by the higher anisotropy of the physical (in particular, elastic and magnetic) properties and their dependence on size. Therefore, we assume that the application of the alternating uniform pressure may result in the oscillation of the magnetic moment of the nanoparticles which in turn leads to the question of the existence of the previously not considered AME mechanism. Here, we do not specify the physical nature of the dependence of this type, taking into account the fact that the orientation of the magnetic moment $\langle m_* \rangle$, averaged out with respect to Langevin, is determined in the final analysis by the minimisation of the total energy of the ferroparticles in the magnetic field. This energy includes in particular the magnetostriction energy associated with the change of the equilibrium distances between the nodes of the lattice and the linear dimensions of the domain [41].

Taking these considerations into account, we discuss the approaches to the evaluation of the contribution of the 'solid-state' mechanism of magnetisation of the magnetic fluid to the perturbation of magnetisation by the sound wave.

The oscillations of the uniform pressure, having a periodic effect on the crystal lattice, cause the perturbation of the vector of spontaneous magnetisation of the nanoparticles δm_*:

$$\delta m_* = \left(\frac{\partial m_*}{\partial p} \right)_T \cdot \delta p$$

Consequently, the expression for M_e (9.2) contains another term:

$$\left(\frac{\partial M}{\partial m_*}\right)_{H,T,n} \cdot \left(\frac{\partial m_*}{\partial p}\right)_T \cdot \delta p$$

which can also be presented in the following form:

$$\frac{1}{n\beta_s} \cdot \left(\frac{\partial M}{\delta m_*}\right)_{H,T,n} \cdot \left(\frac{\partial m_*}{\partial p}\right)_T \cdot \delta n = \frac{\rho c^2}{n} \cdot \left(\frac{\partial M}{\delta m_*}\right)_{H,T,n} \cdot \left(\frac{\partial m_*}{\partial p}\right)_T \cdot \delta n$$

Introducing the notations $\left(\frac{\partial M}{\delta m_*}\right)_{H,T,n} \equiv M_{m*} \left(\frac{\partial m_*}{\partial p}\right)_T \equiv m_{*p}$, equation (9.2) can be written in the new form

$$M_e = M_0 + \left(M_n + \frac{1}{n\beta_s} \cdot M_{m*} \cdot m_{*p}\right) \cdot \delta n + M_T \cdot \delta T + M_H \cdot \delta H \quad (9.35)$$

If we denote: $M_n^* \equiv M_n + \frac{1}{n\beta_s} \cdot M_{m*} \cdot m_{*p}$ (9.30)

the equation (9.35) can be written in the form of equation (9.2):

$$M_e = M_0 + M_n^* \cdot \delta n + M_T \cdot \delta T + M_H \cdot \delta H \quad (9.36)$$

Consequently, equation (9.3) can be presented in the form:

$$\delta M = -\frac{(nM_n^* + \gamma_* M_T)/(1 + N_d M_H) + i\omega\tau M_0}{1 + i\omega\tau} \cdot \frac{\partial u}{\partial x} \quad (9.37)$$

In the approximation of the equilibrium magnetisation of the life and concentration magnetic fluids, using the Langevin equation $L(\xi) = \text{cth}(\xi) - 1/\xi$, we obtain

$$M_{m*} = n\left(\text{cth}\,\xi - \frac{\xi}{\text{sh}^2\xi}\right)$$

Introducing the notations $K_{m*}(\xi) \equiv \text{cth}\,\xi - \xi \cdot \text{sh}^{-2}\xi$, we can write:

$$M_{m*} = n \cdot K_{m*}$$

By analogy with equation (9.4), the expression for the amplitude of oscillations of magnetisation has the form:

$$\Delta M = M_0\left[M_\beta + (\omega\tau)^2\right] \cdot \dot{u}_0 \cdot \left(c\left[1 + (\omega\tau)^2\right]\right)^{-1}$$

in which

$$M_\beta = (nM_n^* + \gamma_* M_T)\left[M_0(1 + N_d M_H)\right]^{-1}$$

In the low frequency range $(\omega\tau)^2 \to 0$:

$$\Delta M = \left[n \cdot \left(M_n + \frac{c^2\rho}{n} \cdot M_{m*} \cdot m_{*P} \right) + \gamma_* M_T \right] (1 + N_d M_H)^{-1} u_m k \quad (9.38)$$

where $k = \omega/c$, u_m is the displacement amplitude.

If the magnetisation of the magnetic fluid M is proportional to the concentration of the magnetic phase n, then

$$\Delta M = \left[M + c^2\rho m_{*p} M_{m*} + \gamma_* M_T \right] (1 + N_d M_H)^{-1} u_m k \quad (9.39)$$

The equation (9.39) has the form:

$$\Delta M = \left[M_s (L(\xi) + wK_{m*}(\xi)) + \gamma_* M_T \right] (1 + N_d M_H)^{-1} u_m k \quad (9.40)$$

where $w \equiv c^2\rho k_p, K_p \equiv \frac{1}{m_*}\left(\frac{\partial m_*}{\partial p}\right)_T$ is the baric coefficient of the magnetic moment of the nanoparticle.

In the fields close to magnetic saturation of the magnetic fluid, $L(\xi) \to 1$, $K_{m*}(\xi) \to 1$, $M_T \to 0$, $M_H \to 0$, and, consequently, the maximum increment of magnetisation is

$$\Delta M_{\max} = \frac{M_s(1 + w) \cdot \dot{u}_0}{c} \quad (9.41)$$

The relative increment of the magnetisation amplitude $\beta_\xi = \Delta M / \Delta M_{\max}$ is expressed by analogy with equation (9.6) by the relationship:

$$\beta_\xi = \frac{L(\xi) + wK_{m*} + \frac{\gamma_*}{M_s} \cdot M_T}{(1 + N_d M_H)(1 + w)} = \frac{L(\xi) + wK_{m*} - k'D(\xi)}{(1 + k'' \cdot \xi^{-1} D(\xi))(1 + w)} \quad (9.42)$$

Thus, the dependence of the amplitude of AME on ξ, expressed in the relative units, has the form of (9.42) or in the extended presentation:

$$\beta_\xi = \frac{L(\xi) + c^2\rho k_p \cdot K_{m*}(\xi) - \frac{qc^2}{C_p} \cdot D(\xi)}{(1 + \frac{N_d M_s \mu_0 m_* \cdot D(\xi)}{k_0 T \xi})(1 + c^2 p k_p)} \quad (9.43)$$

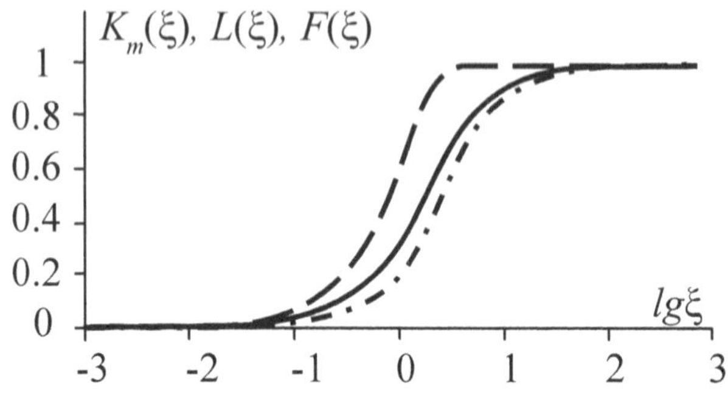

Fig. 9.21. The dependences $K_m(\xi)$, $L(\xi)$, $F(\xi)$. ----- $K_m(\xi)$; —— $L(\xi)$; -.-.-. $F(\xi)$.

To ensure that the contribution of the solid-state mechanism to the acoustomagnetic effect is comparable with that of the concentration mechanism, we should have $w \equiv c^2\rho k_p \approx 1$, i.e., at $c = 1000$ m/s, $\rho = 1000$ kg/m^3 we obtain: $k_p = 10^{-9}$ Pa^{-1}, i.e., the sensitivity of this method is quite high. The multiplier, enclosed in the square brackets of the equation (2.8), is denoted by $F(\xi)$, i.e. $F(\xi) = L(\xi) - 0.45D(\xi)$.

Figure 9.21 shows on the semi-logarithmic scale the functions $K_{m*}(\xi)$, $L(\xi)$ and $F(\xi)$, which determine the dependence $M_e(\xi)$.

It may be seen that the limiting value of the dependence $K_{m_*}(\xi)$ is equal to 1, and that this dependence is saturated earlier than the dependences $L(\xi)$ and $F(\xi)$.

It is assumed that the contribution of the solid-state mechanism is dominant, i.e., $w >> 1$, in this case $\beta_\xi \approx K_{m_*}(\xi)$. Consequently, the curve of the dependence $\beta_\xi(\xi)$ is steeper than the curve $L(\xi)$ and, correspondingly, the experimental dependence $\beta_H(H)$ is steeper than the magnetisation curve $M(H)$. Consequently, if the contribution of the 'solid-state' (baric) mechanism to the AME is dominant in comparison with the contribution of the concentration mechanism, the increase of the AME amplitude with the increase of the strength of the magnetic field will be faster than in the Langevin magnetisation curve.

However, the experimental data presented, in particular, in Fig. 9.10, indicate the reverse trend.

At $\xi \to 0$: $K_{m_*}(\xi) \approx (2/3)\xi$, consequently $\beta_\xi \approx (2/3)\xi$, and, therefore, $\xi = \dfrac{\mu_0 m_* H}{k_0 T} \approx 3/2\beta_\xi$, i.e. m_* is more than halved in comparison with the previous result.

Thus, under the assumption on the dominance of the 'solid-state' effect, the values of the parameter m_* are almost halved and this is also contradicts the experimental data.

Let it be that $w \le 1$, consequently (9.42) on approaching $\xi \to 0$:

$$\beta_\xi = \frac{1/3\xi + 2/3w\xi - 1/3k'\xi}{(1+w)(1+k''/3)}$$

in the final form

$$\beta_\xi = \frac{1+2w-k'}{(1+w)(3+k'')}\xi \tag{9.44}$$

Taking into account the controlling role of the mechanisms of the concentration model, it is natural to assume that $|w| << 1$. In this case, equation (9.41) can be written in the following form:

$$\xi \approx \frac{3+k''}{1-k'+w(1+k')}\beta_\xi \tag{9.45}$$

Equation (9.45) shows that by taking into account the solid-state mechanism there are some changes in the calculated values of m_{*max} and d_{max} with the direction of the change depending on the sign of w.

When the error of measurement of the dependence $\beta_\xi(\xi)$ is sufficiently low, it is possible to determine the sign and estimate the postulated parameter k_p. The acoustic method is highly efficient in this respect.

The role of the 'solid-state' mechanism should become more important with the increase of the concentration of the magnetic phase. In this case, it is necessary to answer the question of the linear transformation by the input oscillatory circuit (chapter 2, paragraph 2.7). The experimental data show that in the initial section of the magnetisation χ = const, and in the strong magnetic fields $\chi \approx 0$. The constancy of the inductance of the input oscillatory circuit in these ranges of magnetisation of the magnetic fluid indicates the absence of distortions of the results of the field dependence of the EMF by the receiving device.

The investigations of the highly diluted magnetic fluids have the advantage as regards the significance of the demagnetising fields, but it is necessary to use high-sensitivity reception equipment and carry out efficient screening of different types of interference.

In the application of the alternating uniform pressure on the nanoaggregates and nanoclusters, present in the structurized colloids, oscillations of the total magnetic moment of these formations may take place in the small and moderate magnetic fields as a result of the oscillations of the interparticle spacing and, consequently, the perturbation of the 'internal field'. To describe the process of magnetisation, the parameters of the 'purely solid-state' mechanism m_*, d and k_p, described previously should be adequately generalised to the case of structurized colloids.

9.8. Calculated value of the magnetic moment of the nanoparticles and the dynamic demagnetising factor from the data for highly concentrated magnetic fluids

To confirm the tendency of the reduction of the sizes of the 'fine' magnetic nanoparticles (as a result of the dipole–dipole interaction of the 'large' particles), mentioned in paragraph 9.5.6, it is convenient to expand the range of concentrations of the dispersed phase by moving it closer to the upper boundary. Long-term evaporation of the colloidal solution of the MFc produced a highly concentrated specimens of MFc with the volume concentration of the solid phase of 25.6%, the density 1934 kg/m^3, initial magnetic susceptibility 9.73 and saturation magnetisation 90.7 kA/m. The speed of propagation of sound in the MFc, filling the glass tube, was 811 m/s. The conditions and methods of measuring these parameters are the same as in paragraph 9.5.6.

As mentioned previously (sections 2.7 and 9.7), when obtaining the experimental data for the specimen with a high content of the magnetic component it is important to investigate the problem of the possible distortion of the results for the field dependence of the AME as a result of the resonance properties of the measuring inductance coil, which forms an oscillatory circuit with a specific resonance frequency with the capacitance of the input device. In this case, we are concerned with the 'effective' magnetic susceptibility of the measuring coil which is situated in the immediate vicinity of the surface of the tube with the MF. This topic will be discussed in greater detail.

The experimental equipment, designed for investigating the AME, was described in paragraph 9.5.1. The measuring unit of equipment is shown schematically in Fig. 9.22. The MFc 1 fills the glass tube 2, with the piezosheet 3 compressed to the bottom of the tube and used to excite sound waves. The lines of force of the magnetic field, indicated by the crosses in the circles, travel in the direction

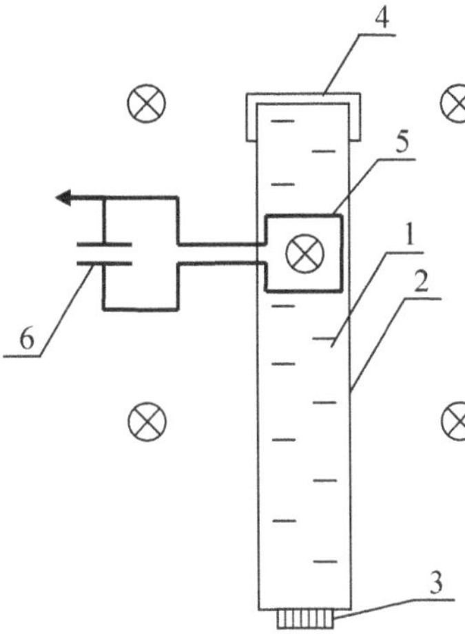

Fig. 9.22. The measuring unit of equipment.

normal to the axis of the tube. The upper surface of the magnetic fluid is 'forcefully' stabilised with the film 4. The alternating EMF is received by the measuring inductance coil 5 with 1100 turns, produced from a copper wire with a diameter of 0.07 mm. The inductance coil is circular and borders tightly with the outer wall of the tube. The input capacitance of the receiver–amplifier 6 and the inductance coil 5 form the oscillatory circuit, characterised by the resonance frequency ν. In this case

$$\nu = 1/(2\pi\sqrt{LC})$$

where L is the inductance, C is the capacitance of the circuit.

The inductance of the measuring coil is influenced by the presence of a liquid magnetic, placed in the tube. The 'effective' susceptibility χ_{eff}, brought in by the magnetic, is difficult to calculate. However, we can use the proportional dependence $L \sim (1 + \chi_{eff})$ so that the following equation can be written

$$\chi_{eff} = (\nu_g / \nu_f)^2 - 1 \qquad (9.46)$$

where ν_f is the resonance frequency at measurements in the tube filled with the magnetic fluid; ν_g is the resonance frequency of the circuit in the absence of the magnetic fluid in the chill.

Thus, using the equation (9.46), the value of χ_{eff} can be determined by experiments.

Analysis of the resultant amplitude–frequency characteristic of the oscillatory circuit shows: when the specimen of MFc fills the tube and is situated in the external magnetic field with the strength H = 17.2 kA/m, the resonance frequency of the circuit is equal to v_f = 101.16 kHz, and in the absence of the MF it is equal to v_g = 105.8 kHz. Substituting the resultant values of v_f and v_g into (9.46), we determine χ_{eff} = 0.084, which is considerably smaller than 1 and the initial susceptibility of the MFc. In addition to this, the range of the frequencies, used for investigating the AME (20–60 kHz), is situated far away from the resonance and is located on the almost flat section of the left branch of the amplitude–frequency characteristic.

In this case, the amplitude–frequency characteristic of the input oscillatory circuit is constant throughout the entire range of the strength of the magnetic field and does not change its curvature in the investigated frequency range.

When describing the process of magnetization of the nanodispersed magnetic fluids in section 9.5.6, it was proposed to use the 'difference' curve $Y_+^{(e)}(H)$. The 'difference' curve is plotted by deducting the values of the relative amplitude of AME $\beta(H)$ and the values of the relative magnetisation of the sample $\beta_M(H)$ in the magnetic fields of the same strength. The experimental values of the dependences $\beta_M(H)$ and $\beta(H)$ are approximated by the smooth curves in the MS Excel software.

A shortcoming of this method is the large error obtained as a result of deducting the experimental curves from each other. In addition to this, this method is characterised by a high labour content because it requires 'manual' processing of each dependence.

The methods proposed by A.M. Storozhenko [319] is more suitable for evaluating the parameters of the finest nanoparticles. The method is based on determining the angular coefficient directly from the equations of linear approximation of the final sections of the AME and magnetisation curves in the range $H \to \infty$, i.e., without

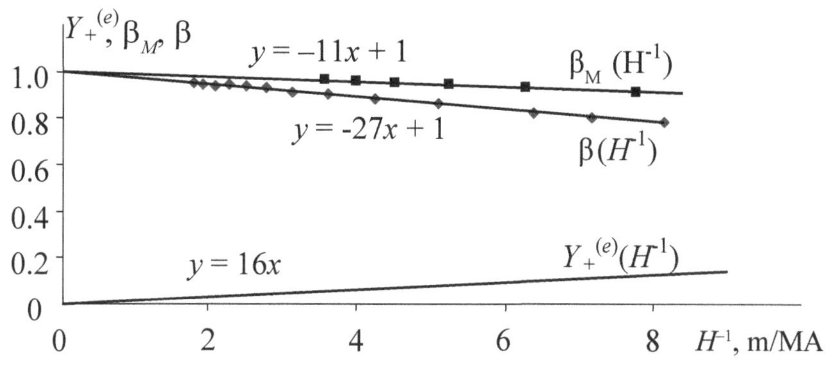

Fig. 9.23. Determination of the value tg α.

constructing the 'difference' curve (Fig. 9.23). In this case, tg α is determined by the simple deduction of the angular coefficients of the dependences $\beta(H)$ and $\beta_M(H)$. Subsequent calculations were carried out in accordance with [316]. The error of the method does not exceed 3%. The numerical value of $d_{\min}$ for the MFc equals 6.0 nm, which is ~10% lower than the calculated results obtained for MFc. Thus, the previously made conclusion on the existence of the tendency of the reduction of the values of the dimensions of the 'fine' magnetic nanoparticles with increase of the concentration of the investigated specimens of the MF is confirmed.

The curve of the dependence $Y_+^{(e)}(H)$, constructed on the scale of Fig. 9.15, for the MFc sample and also the curves for the samples MF-1 and MFc, has a maximum. The coordinate of the point of intersection of the dot-and-dash straight-line with the curve $Y_+^{(e)}(H)$ for MFc is $H_3 = 200$ kA/m. The 'difference' curve in the MFc decreases in a smoother manner and, consequently, alignment of the magnetic moments of the nanodomains 'in the field' is detected in the fields with higher strength as a result of the particles of the finest fraction: $H_1 < H_2 < H_3$.

The presence of the dipole–dipole interaction, in particular between the large magnetic nanoparticles, weakens their contribution to the magnetisation process in the vicinity of magnetic saturation. The nanoaggregates are blocked more efficiently by the magnetic field than the individual particles and are subjected to a lesser effect of the thermal oscillations. The conclusion according to which the conditions for intensification of the dipole–dipole interaction between the nanoparticles improve in the more concentrated specimens as a result of the proximity of the magnetic nanoparticles is confirmed.

We consider the question of the experimental and theoretical determination of the dynamic demagnetising factor N_d [321].

In the case of the magnetic fluid filling a cylindrical tube, with the magnetisation of the fluid modulated by the sound wave, the exact analytical solution with derivation of the equation for calculating N_d (in contrast to the homogeneous magnetised ellipsoidal magnetics [244]) is not possible. Therefore, new approaches to this problem are required which would be based on the application of modelling theories and the experimental measurement results.

For this purpose, we analyse the initial sections of the magnetisation curve $\beta_M(H)$ and the field dependence of the EMF amplitude, induced as a result of AME $\beta(H)$ in the highly

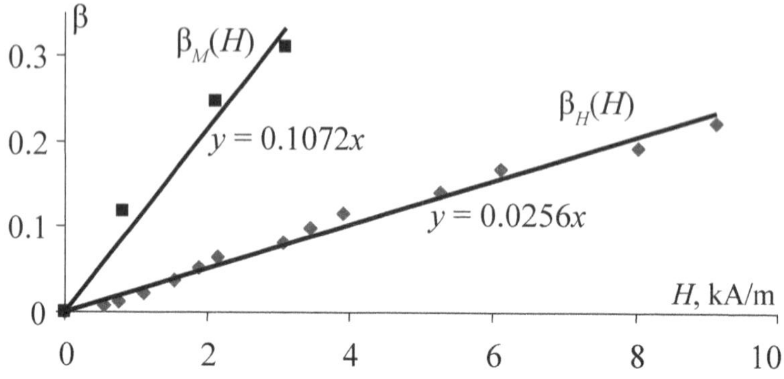

Fig. 9.24. Initial sections of the dependences $\beta_M(H)$ and $\beta_H(H)$.

concentrated sample of MFc, shown in Fig. 9.24. We use the limits of these dependences in the range $H \to 0$.

In weak fields, the Langevin equation has the form $L(\xi) = \xi/3$, so that we can write

$$\beta_M = \frac{\xi}{3} = \frac{\mu_0 m_*}{3k_0 T} \cdot H$$

Equation (9.8) was obtained for β_H in weak fields. We compare the partial β_M/β_H in the approximation $H \to 0$:

$$\frac{\beta_M}{\beta_H} = \frac{1 + k''/3}{1 - k'} \tag{9.47}$$

The ratio β_M / β_H does not depend on H but in the experimental determination of its value it is necessary to introduce a correction for the demagnetising field taking into account the fact that the demagnetising factor in the determination of $\beta_M(H)$ and $\beta_H(H)$ may differ (section 1.3).

The left part of equation (9.47) is written in the form of the ratio of the tangent of the slope of the initial section of the curve $\beta_M(H)$ and the curve $\beta_H(H)$:

$$\frac{\beta_M(H)}{\beta_H(H)} = \frac{\operatorname{tg} \beta_M}{\operatorname{tg} \beta_H}$$

Taking into account the values $q = 0.66 \cdot 10^{-3}$ K^{-1}, $c = 811$ m/s, $C_p = 1760$ J/kg·K, $N_d = 0.247$, $T = 303$ K, we obtain $k' = 0.248$, $k'' = 9.48$ and, correspondingly, the calculations give: $\beta_M \approx 5.5\ \beta_H$. According to the data in Fig. 9.4, the ratio of the experimentally determined values is $\beta_M \approx 4.3\ \beta_H$. The difference between the calculated and experimental data is determined mainly by the difference in the numerical values of $m_{*\max}$ according to the *MGM* and AGM data which, in turn, is associated with the inaccurate evaluation of N_d, based on the modelling theory. Nevertheless, the experiments and calculations show that the initial section of the magnetisation curve is steeper than the initial section of the field dependence of the EMF of AME.

The equation (9.47) can be presented in the extended form:

$$\frac{tg\beta_M}{tg\beta_H} = \frac{\mu_0 H m_{*\max 1}}{3k_0 T} \cdot \frac{\left(3 + \frac{N_d \mu_0 M_s m_{*\max 2}}{k_0 T} k_0 T\right)}{\mu_0 (1 - qc^2 C_p^{-1}) m_{*\max 2} H},$$

where $m_{*\max 1}$ and $m_{*\max 2}$ are the magnetic moments, calculated from the data obtained by the MGM and AGM methods.

Assuming that $m_{*\max 2} = m_{*\max 1}$, we determine

$$\frac{tg\,\beta_M}{tg\,\beta_H} = \frac{3k_0 T + N_d \mu_0 M_s m_{*\max 1}}{3k_0 T(1 - qc^2 C_p^{-1})}$$

Consequently, after elementary transformations we obtain the expression for calculating the dynamic demagnetising factor N_d:

$$N_d = \frac{3k_0 T \left[\frac{\text{tg}\,\beta_M}{\text{tg}\,\beta_H} (1 - qc^2 C_p^{-1}) - 1 \right]}{\mu_0 m_{*\max 1} M_s} \qquad (9.48)$$

To calculate N_d using equation (9.48), it is necessary to carry out detailed measurements of the field dependence of AME and the magnetisation curve of the investigated MF sample. Using the data considered here for frequencies ν = 21.9 kg (λ = 0.037 m, P = 1.34) and for frequency ν = 58.9 kHz (λ = 0.014 m, P = 0.5) we obtain respectively N_d = 0.29 and N_d = 0.16. Here, it was assumed that $m_{*\max 1} = 10.7 \cdot 10^{-19}$ A/m^2. Comparison of the resultant values of N_d with the data in Table 9.1 shows that the modelling theory, described in section 9.3, gives slightly lower values of N_d.

The best agreement with the experimental data is obtained if the coefficient α is selected on the basis of the optimum matching of the dependence (9.11) with the available values of N_C for a circular cylinder, presented in Table 1.1. In this case, $\alpha = 0.72$. It is important to mention that the modelling theory does not give only similar quantitative results but also accurately predicts the tendency for the reduction of the numerical values of the parameter N_d with the increase of the frequency of sound oscillations. At the same time, the scheme of the calculation of the demagnetising field, shown in Fig. 9.1, does not take into account the presence of adjacent half-waves so that the investigated model cannot be regarded as physical lyadequate in the range $\lambda/2d << 1$.

Equation (9.48) includes, as the measured value, the ratio $tg\ \beta_M/tg\beta_H$; the reliability of determination of this ratio increases with increasing concentration of the magnetic phase of the magnetic fluid. For this reason, precision measurements of the parameter N_d should be carried out on highly concentrated samples of the magnetic fluid.

To conclude the the chapter dealing with acoustic granulometry, we make several comments of principal nature. At first sight, it may appear that the magnetic granulometric and acoustic granulometric methods of evaluation of the physical parameters of the magnetic nanoparticles of the disperse phase of the magnetic fluid not differ from each other. All the more so that the magnetisation curve of the magnetic fluid and the curve of the dependence of EMF in AME on the strength of the magnetic field, presented in relative units, are identical: the linear section in the region of weak fields and asymptotic approach to 1 in the vicinity of magnetic saturation. However, the construction of the 'difference' curve, which also carry information on the dimensions of the magnetic nanoparticles, show that they greatly differ. In turn, the specific feature of the 'difference' curve is the physical nature of its appearance – in the presence of thermal oscillations in the sound wave and perturbation of the magnetisation of the magnetic fluid by these waves, whereas the role of perturbation of the demagnetising field is of the relative nature depending on the experiment conditions.

The geometrical parameters of the particles of the disperse phase of the magnetic fluid are often determined by analysis of the magnetisation curve [253]. The magnetisation curve is plotted by a large number of methods, for example, the ballistic method [315], using the vibrational magnetometer [318], by integration of the field dependence of magnetic susceptibility [286]. The magnetic moments and the dimensions of the nanoparticles of the magnetic fluid are calculated by the same relationships (9.14)–(9.16).

The acoustic granulometry methods, described in section 9.5, use the data on the physical parameters of the magnetic nanoparticles from the field dependence of the amplitude of EMF induced in the circuit as a result of the acoustic magnetic effect in the MF. This is carried out using the equations (9.17), (9.18) and (9.24), derived by analysis of the physical fields in the magnetised magnetic fluid during propagation of the adiabatic soundwave in it [319, 320].

9.9. Mechanism of thermal relaxation of magnetization of magnetic fluid

The sound wave is an adiabatic process. The temperature oscillations δT can be observed in the medium according to expression (2.2). The above estimation of temperature oscillations is made with the assumption that the strain amplitude in the sound wave is 10^{-4}, the intensity is $J = 10^4$ W/m^2, correspondingly. Taking an MF based on kerosene and magnetite as an example, the following values can be accepted: $C_p = 2\cdot10^3$ J/kg·K, $c = 1120$ m/s, $q = 0.53\cdot10^{-3}$ K^{-1}, $T = 300$ K, then $\delta T \approx 10^{-2}$ K. In the experiments with AME in order to avoid the non-linear effects, the cavitation processes and fluid heating, introduced into the MF sound oscillations, differed by 1–2 orders of magnitude with respect to distortion from those values used in the calculations. It can be assumed that the thermal oscillations in the case under study $\delta T \leq 10^{-3}$ K, i.e. they have ultralow values.

The studies on the effect of thermal relaxation of magnetization of magnetic fluid are rather promising, because, first, ultra-small thermal oscillations of the medium have no significant effect on the magnetic state of the object under study, and, secondly, this method makes it possible to evaluate the rotational mobility of nanoparticles determined by the rheology of the nearest molecular environment of a particle [321].

The relative increment in the magnetization amplitude $\Delta M/\Delta M_{\max}$ can be obtained using expressions (9.2) and (9.4):

$$\frac{\Delta M}{\Delta M_{\max}} = \frac{M_0}{M_S} \cdot \frac{\dfrac{nM_n / M_0 + \gamma_* M_T / M_0}{\left(1 + N_d M_H\right)} + (\omega\tau)^2}{1 + (\omega\tau)^2} \tag{9.49}$$

According to [322], in the initial part of the magnetization curve within the frequency range up to 10^5 Hz there are no noticeable effects of interaction of the particles in the magnetic fluid with the concentration of ≤19%. In such a case $nM_n=M_0=\chi H$ and we can rewrite the equation for the acoustomagnetic effect curve (9.49) as:

$$\Delta M / \Delta M_{\max} = (M_0 / M_S)\left[(1+\gamma_* M_T / M_0)\left[(1+N_d M_H)\right]^{-1} + (\omega\tau)^2\right] / \\ / \left[1+(\omega\tau)^2\right] \qquad (9.50)$$

The acoustomagnetic effect curve is presented in relative values. The $\mathrm{tg}\theta_A$ is the value of the slope of the initial part of this curve. The expression $\mathrm{tg}\theta_A$ is obtained from (9.50) by taking the derivative with respect to H:

$$tg\theta_A = (\chi / M_S)\left[(1+\gamma_* M_T / M_0)\left[(1+N_d M_H)\right]^{-1} + (\omega\tau)^2\right] / \left[1+(\omega\tau)^2\right] \qquad (9.51)$$

Using the relations:

$$\gamma_* M_T = k'T\frac{\partial M}{\partial T} = k'TH\frac{\partial \chi}{\partial T},\ k'TH\frac{\partial \chi}{\partial T} / M_0 = k'T\frac{1}{\chi}\cdot\frac{\partial \chi}{\partial T}$$

it is possible to rewrite (9.51) as:

$$\mathrm{tg}\theta_A = (\chi / M_S)\left[\left(1-k'\frac{T}{\chi}\cdot\frac{\partial \chi}{\partial T}\right)(1+N_d\chi)^{-1} + (\omega\tau)^2\right] / \left[1+(\omega\tau)^2\right] \qquad (9.52)$$

To characterize the initial part of the magnetization curve, we introduce the parameter $\mathrm{tg}\theta_M \equiv \chi/M_S$. The expression for ($\mathrm{tg}\,\theta_M/\mathrm{tg}\,\theta_A$) can be written as:

$$\frac{\mathrm{tg}\theta_M}{\mathrm{tg}\theta_A} = \frac{1+(\omega\tau)^2}{\left(1+k'\frac{T}{\chi}\cdot\frac{\partial \chi}{\partial T}\right)(1+N_d\chi)^{-1} + (\omega\tau)^2} \qquad (9.53)$$

The expression (9.53) can be used to compare experimental results in the frequency range covering the area of the thermal relaxation of the magnetization. Besides, an independent experiment should be carried out to determine not only χ, but also $\partial\chi/\partial T$. With reference to the low-concentration MF in a mono-dispersed approximation, it is possible to use a well-known generalization of Langevin theory of paramagnetism for superparamagnetics, see (1.3).

On substituting (1.3) into (9.53), we obtain:

$$\frac{\text{tg}\,\theta_M}{\text{tg}\,\theta_A} = \frac{1+(\omega\tau)^2}{\frac{1-k'}{1+k''/3}+(\omega\tau)^2} \tag{9.54}$$

The obtained expression (9.54) contains the dependence of $tg\theta_M/tg\theta_A$ on the frequency of sound oscillations ω and the thermal relaxation time τ of particle magnetization.

From the formula (9.54), if $(\omega\tau)^2=0$, and $(\omega\tau)^2=1$,we obtain, correspondingly:

$$\frac{\text{tg}\,\theta_M}{\text{tg}\,\theta_A} = \frac{1+k''/3}{1-k'} \tag{9.55}$$

$$\frac{\text{tg}\,\theta_M}{\text{tg}\,\theta_A} = \frac{1+k''/3}{1+k''/6-k'/2} \tag{9.56}$$

The conclusions of the model theory of thermal relaxation in the form of the formulas (9.54)–(9.56) will be compared below with the experimental results.

The samples of magnetic fluids based on the high-dispersive magnetite Fe_3O_4 stabilized with oleic acid in different carrier liquids were studied in the experiment:

- Sample MF-1, the carrier is kerosene;
- Sample MF-2, the carrier is polyethylsiloxane-2;
- Sample MF-3, the carrier is mineral hydrocarbon oil;
- Sample MF-4, the carrier is polyethylsiloxane-4;
- Sample MF-5, the carrier is synthetic hydrocarbon oil.

The physical parameters of the MF samples are listed in Table 9.5. The following notations are used: density ρ, volume concentration of the solid phase φ, volume concentration of the magnetic phase φ_M, initial magnetic susceptibility χ, saturation magnetization M_S, sound speed c in a 'MF–glass tube' system, thermal expansion coefficient q, specific heat capacitance at constant pressure C_p, plastic viscosity η, nanoparticle diameter d.

For the MF test samples, the time of the Brownian rotational diffusion of colloidal particles was determined from $\tau_B=3V\eta_{S0}/k_0T$, where V is the particle volume, η_{S0} is the static shear viscosity of

Table 9.5

	ρ, kg/m³	φ, %	φ_M, %	χ	M_S, kA/m	c, m/s	q, 1/K	C_p, J/kgK	η (Pa·s)	$m_*\cdot10^{19}$, A·m²	d, nm
MF-1	1252	10.6	8.5	2.8	35.1	937	0.91	1489	0.012	7.9	15
MF-2	1385	10.3	8.3	2.5	33.9	930	0.70	1375	0.125	7.1	14
MF-3	1282	10.3	8.2	2.6	33.6	1005	0.82	1283	0.368	7.7	15
MF-4	1405	10.2	8.2	2.6	34.1	954	0.69	1326	0.630	7.4	14
MF-5	1290	10.4	8.3	2.4	33.4	1024	0.82	1284	1.110	6.9	14

the carrier liquid. For MF-samples: for MF-1 τ_B= 2 μs, MF-2 τ_B= 6 μs, MF-3 τ_B= 21 μs, MF-4 τ_B= 26 μs, MF-5 τ_B= 46 μs.

The AME frequency-field dependence was studied using special research facility, which is described in 9.5.2. For all the studied

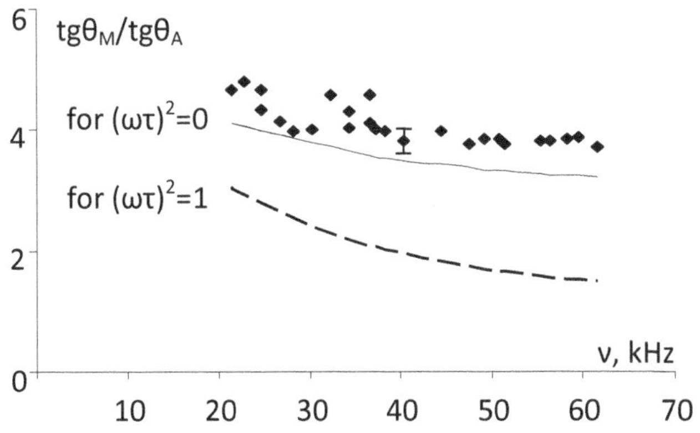

Fig. 9.25. Theoretical and experimental values of the $tg\theta_M/tg\theta_A$ relation for MF-1.

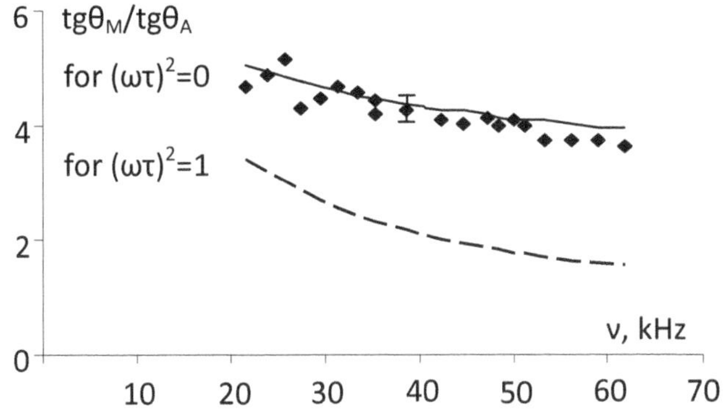

Fig. 9.26. Theoretical and experimental values of the $tg\theta_M/tg\theta_A$ relation for MF-5 sample.

MF samples the dependences of the AME relative amplitude on the magnetic field value within a frequency range of 20-60 kHz were obtained.

Figures 9.25 and 9.26 show the $tg\theta_M/tg\theta_A$ dependences on the frequency plotted according to formulas (9.55) and (9.56) (curves) for MF-1 and MF-5 samples and the corresponding experimental data (shaded diamonds). The continuous curves are the theoretical values of the relation $tg\theta_M/tg\theta_A$ when $(\omega\tau)^2 = 0$, the dot curves belong to the relation $tg\theta_M/tg\theta_A$ when $(\omega\tau)^2 = 1$.

It should also be noted that the relation $tg\theta_M/tg\,\theta_A=1$ is even more contradictory when $\omega\tau \gg 1$. The correspondence between experimental data and theoretical results calculated according to the formula (9.55) evidences of the absence of the thermal relaxation of the MFs magnetization in the initial part of the AME field dependence. The perturbation processes of magnetization with thermal oscillations in the sound wave occur in the pre-relaxation area inasmuch as the theoretical values of the slope of the initial part of this curve calculated with the assumption that $(\omega\tau)^2 = 0$ are in satisfactory agreement with the experimental data.

The relaxation frequency values $\nu_R \equiv 1/(2\pi\tau_B)$ for the MF-1, MF-2, MF-5, MF-3, and MF-4 samples are 80 kHz, 27 kHz, 7.6 kHz, 6.1 kHz and 3.5 kHz, respectively. Therefore, if $\omega\tau_B \approx 1$ is considered as the criterion of the thermal relaxation of the magnetization, then for the MF test samples (except the MF-1 sample) in the frequency range 20–60 kHz this phenomenon must be confirmed experimentally.

Thus, the fact of the absence of magnetization thermal relaxation in relation to the values of the slope of the initial parts of the magnetization curves and the acoustomagnetic effect has been experimentally proved. Consequently, the surrounding environment of the nanoparticles, in this case, exhibited the properties of non-Newtonian fluid, which is characterized by special relations between the strain rate and stress.

The medium shearing strain caused by the rotational oscillations of nanoparticles can be described by the Maxwell model, which is well-known in the viscoelastic media theory [63]. The Maxwell mechanical model consists of a series-connected spring and piston (Figure 9.27). In this case, the medium strain in the neighborhood of a nanoparticle is composed of two parts. The first part is an instantaneous elastic strain, which is characterized by the shear modulus G. The second part is retarded viscous strain caused by medium flotation characterized by the shearing viscosity η. The

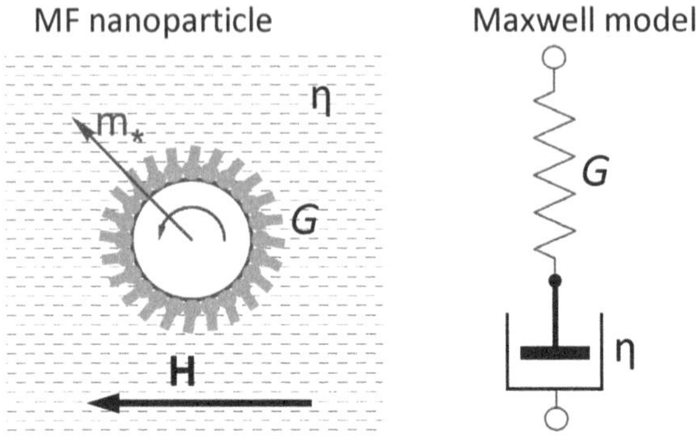

Fig. 9.27.

mono-molecular layer of the stabilizer serves as an elastic component in the model under study, and the surrounding dispersion medium serves as a damper. Besides, neighboring particles of the disperse phase can retard the establishment of the equilibrium orientation of the particle magnetic moment. This occurs due to steric interaction.

The isolation of the elastic properties of the stabilizer coating has the physicochemical nature: a molecule of oleic acid on its polar end O^-H^+ is attracted to the solid phase surface forming thereon a dense unimolecular film of $\delta \approx 2$ nm thickness. The relatively close-set molecular structure of the stabilizer causes a predominantly elastic shear strain of the coating, which is characterized by short relaxation time. The considerable viscosity of the dispersion medium blocks the flotation processes at low rotational oscillations of nanoparticles. The instantaneous shear strain of the stabilizer coating under ultra-low temperature oscillations provides significant decrease of the values of the magnetization thermal relaxation time τ compared to the Brownian rotational relaxation time τ_B.

Thus, the comparison of the conclusions of the model of thermal relaxation of magnetization and the experimental results makes it possible to obtain information about the features of the rheology for the nearest molecular environment of a particle – nanorheology [323].

10

Acoustometry of the shape of magnetic nanoaggregates and non-magnetic microaggregates

10.1. Column of the magnetic fluid in the tube as the inertial–viscous element of the oscillatory system

In the construction of many devices using magnetic fluids (MF) it is necessary to consider the impact of the magnetic field on a magnetic colloid droplet, i.e. ponderomotive forces and inertia forces is manifested in rapid displacement of the droplet from its equilibrium position. Such devices typically constitute the oscillatory system in which the MF performs the function of the viscous–inertial element. Interest in the study of such oscillatory systems is driven to some extent by the need to expand the measurement base of a structure-sensitive parameter of the magnetic colloid as 'rotational viscosity' ('magnetoviscous effect') which is the increment of the viscosity of MF in its magnetization.

Work in this area may be useful in terms of widening the possibilities of the controlled synthesis of magnetic colloids [302, 312]. The size of ferromagnetic particles dispersed in a magnetic fluid is determined by the technology of its production and is of the order of 10 nm. Experiments show that the application of the magnetic field may cause aggregation of the ferromagnetic particles The particles join together in aggregates – 'chains' oriented along the magnetic field lines of force. There is also a more complex phenomenon [242] at dilution of the concentrated magnetic fluid 'magnetite in kerosene' by solutions of oleic acid in kerosene of different concentration a process similar to 'phase transition' takes place – ferromagnetic particles condense into aggregates – 'drops' of up to ~200–1000 nm in size the magnetization of which is significantly greater than that of the magnetic fluid. The microdroplet aggregates, elongated along the magnetic field lines of force take a spherical shape when the field is switched off.

The physico–chemical properties of MF significantly depend on the number, size and shape of the aggregates, as well as their orientation in the magnetic field. This chapter examines the methodology of the acoustometry of the form of magnetic nanoaggregates and non-magnetic microaggregates in the reciprocating (oscillating) flow of the MF column in the tube.

We generalize the method of calculation of the coefficient of elasticity, described in chapter 7 (sections 7.1 and 7.2), for the case of the oscillatory systems with an inertial element in the form of a column of the magnetic fluid (design diagram is shown in Fig. 7.6).

The inertial element of the oscillatory system is the MF column in the tube that is held over a closed air cavity due to stabilization of the fluid–gas boundary by an inhomogeneous magnetic field. The upper exposed surface of the fluid is outside the magnetic field. The MF column can also be sustained by the film type overlapping or by hydrostatic balance in a U-shaped tube.

For the coefficient of ponderomotive elasticity k_p we use the results of the model with concentrated parameters. Under the conditions of this problem, expression (7.18) is converted to a form :

$$k_p = \mu_0 S \left[(1+\chi) M_z \cdot \frac{\partial H_z}{\partial Z} \right]_{z_f},$$

where z_f is the coordinate of the lower base of the MF cylinder.

Given the coefficient of elasticity of the gas density k_g (equation (7.2)):

$$k = \rho_g c^2 \frac{S^2}{V_0} + \mu_0 S \left[(1+\chi) M_z \frac{\partial H_z}{\partial Z} \right]_{z_f} \tag{10.1}$$

The frequency of free non-damping oscillations of the oscillatory system with the elasticity of this type is calculated according to the formula:

$$v_T = \frac{1}{2\pi\sqrt{h}} \sqrt{\frac{\rho_g c^2 S}{\rho V_0} + \frac{\mu_0 M_Z G_Z}{\rho}(1+\chi)} \tag{10.2}$$

where h is the height of the MF column.

If a closed gas cavity is part of the tube, the last expression reduces to:

$$v_T = \frac{1}{2\pi\sqrt{h}}\sqrt{\frac{\rho_g c^2}{\rho h_g} + \frac{\mu_0 M_Z G_Z}{\rho}(1+\chi)}, \qquad (10.3)$$

where h_g is the height of the air cavity.

Hydrostatic pressure remains constant and does not take 'direct' participation in the oscillatory process, but it has 'indirect' effect on the value of parameters k_p and k_g due to static displacement of the lower base of the MF-column. Furthermore, when the height of the MF-column reaches a certain height, air bubbles, overcoming the levitation forces, penetrate through the 'magnetic barrier'. resulting in impaired gas cavity insulation.

The findings of the theoretical model are consistent with the experimental data [226].

We consider the physical mechanisms of energy dissipation in the column of the magnetic fluid moving back and forth in a glass tube filled with the MF in the absence of the magnetic field.

In the shear flow the solid particle is affected by the moment of forces leads to its rotation which leads to an increase in the viscosity of the fluid. In the suspension the presence of spherical particles entails an increase in viscosity in accordance with the Einstein formula:

$$\eta = \eta_0\left(1 + \frac{5}{2}\varphi\right), \varphi = \frac{4\pi R^3}{3}n, \qquad (10.4)$$

where φ is a small ratio of the the total volume of all spheres to the volume of the suspension, n is the number of particles per unit volume.

In the presence of internal friction in the fluid the flow within the pipe at low speeds is laminar in the form of cylindrical layers moving at different speeds depending on the distance from the wall. The boundary layer of the wall is fixed and the axial layer moves at maximum speed. Additional losses form due to friction between the layers, moving at different velocities in the reciprocating fluid flow through the pipe.

In the reciprocating flow of the fluid through the tube of elastic energy dissipation due to the viscosity of the fluid is adequately interpreted on the basis of the concept of shear waves. introduced by Stokes. If the infinite plane placed in a fluid performs harmonic oscillations in a direction parallel to the plane, then near the flat surface there occurs a quasiwave process described by the function:

$$\dot{U} = \dot{U}_0 e^{\alpha(z-h)} \cdot \cos[\omega t + \alpha(z-h)]$$

where h is the distance from the surface, measured along the axis Z, perpendicular to the plane, $\dot{U}_0$ is the displacement velocity of the fluid layer at a distance h from the surface.

The propagation velocity c, the damping factor α and the length of the shear wave λ respectively are expressed by the formulas [63]:

$$c = \sqrt{\frac{2\omega\eta}{\rho}}, \ \alpha = \sqrt{\frac{\omega\rho}{2\eta}}, \ \lambda = 2\pi\sqrt{\frac{2\eta}{\rho\omega}}. \qquad (10.5)$$

The direction of oscillation of particles in the examined wave is perpendicular to the direction of propagation. At a distance of $\lambda/(2\pi)$ the amplitude is reduced by a factor e, i.e. 'the dept of penetration' of the viscous wave is $\sigma'' = \lambda/(2\pi)$.

We use the results of the theory of acoustic impedance applied to a sound wave propagating in a viscous fluid filling the tube [227].

If the length of the circle embracing the lateral surface of the fluid column is less than twice the length of the viscous wave. i.e. $\pi d<2\lambda'$, where $\lambda' = 2\sqrt{\pi\eta / v\rho}$ or $d < 4\sqrt{\eta / \pi v\rho}$, we have the following approximate expression for the impedance of the tube:

$$R' \approx 8\pi\eta b + i\frac{2}{3}\pi^2\rho b d^2 v. \qquad (10.6)$$

The expression $r' = 8\pi\eta b$ corresponds to the Poiseuille law for the drag coefficient in the laminar flow of a viscous fluid in a narrow tube. Poiseuille flow of the fluid is characterized by parabolic velocity distribution of the particles over the pipe section.

Here the rate of displacement of the column boundary $\dot{U}$ is the mean velocity of the particles in the tube cross section equal to half its maximum value. For narrow tubes active resistance in (10.6) is higher than reactive resistance and complete resistance is independent of frequency. In this model, the damping coefficient β' is calculated by the formula:

$$\beta' = \frac{16 \cdot \eta}{\rho d^2}. \qquad (10.7)$$

With the increase of d or v for a given η and ρ the approximate formula (10.6) becomes invalid.

The imaginary part of (10.6) is the inertial component $\left(\frac{4}{3}m2\pi v\right)$. Thus, this model predicts the presence of the added mass, equal to $m/3$, due to the shear viscosity.

For large values of d and ν, when $\pi d/2\lambda' > 10$, another approximate expression was derived for the complex impedance of the tube [227]:

$$R'' \approx db\sqrt{\pi^3\rho\eta v} + i\frac{\pi^2 d^2 \rho v}{2}\left(1 + \frac{2}{d}\sqrt{\frac{\eta}{\pi v \rho}}\right). \qquad (10.8)$$

The active resistance increases with increasing ν. The expression (10.8) for the drag coefficient was first obtained by Helmholtz

$$r'' \equiv db\sqrt{\pi^3\rho\eta\nu}, \qquad (10.9)$$

The second term in brackets in (10.8) is small in comparison with 1, so that the imaginary part can be represented as $m2\pi\nu$ where it follows that in the Helmholtz model has no added mass. The damping coefficient in this case is determined by the following expression:

$$\beta'' = \frac{2}{d}\sqrt{\frac{\pi\eta\nu}{\rho}}. \qquad (10.10)$$

The Helmholtz model predicts the increase in the damping coefficient with frequency.

Note that a simple derivation of the formula (10.10) can be given based on the expression for the energy dissipation per unit time per unit area of the oscillating plane provided in [219]

$$\Delta Q = -\frac{U_0^2}{2}\sqrt{\frac{\omega\rho\eta}{2}},$$

where U_0 is the amplitude of the oscillatory speed.

This formula is valid for high frequency ('short' viscous waves). The value of the energy dissipation for a single period in the area of the part the tube filled with the fluid:

$$\Delta Q_d = -\frac{U_0^2}{2}\sqrt{\frac{\omega\rho\eta}{2}} \cdot \frac{\pi \cdot d}{\nu}.$$

The logarithmic damping decrement δ [202]:

$$\delta = -0.5\Delta Q_d / \Delta Q_0,$$

where ΔQ_0 is the total mechanical energy of the oscillatory system. i.e. $\Delta Q_0 = \frac{mU_0^2}{2}$. Then $\beta'' = \delta \cdot \nu = \frac{2}{d} \cdot \sqrt{\frac{\pi\eta\nu}{\rho}}$.

Because of the small depth of penetration of the viscous wave σ'' the oscillatory motion of the fluid column is of the piston type. The fluid flow is concentrated in a thin boundary layer which increases the effective viscosity of the magnetic fluid with quasi-spherical aggregates commensurate with σ'' [148].

In support of the above we perform simple calculations. The lower limit of the frequency range under the experimental conditions is 20 Hz. Then the length of the viscous wave is

$$\lambda = 2\sqrt{\frac{\pi\eta}{\rho\nu}} = 2\sqrt{\frac{\pi \cdot 8.1 \cdot 10^{-3}}{1.5 \cdot 10^3 \cdot 20}} \cong 1.8 \cdot 10^{-3} = 1.8\ \text{mm},$$

and the 'penetration depth is $\sigma'' \cong 0.3$ mm. If $\nu = 80\ Hz$ then $\sigma'' \cong 0.15$ mm.

For a suspension of the particles in the form of ellipsoids of rotation the correction factor at φ in (10.4) takes the other numeric values. Thus, according to [219] the numerical values of coefficient L' in the formula

$$\eta = \eta_0 (1 + L'\varphi), \quad \varphi = \frac{4\pi ab^2}{3} n \qquad (10.11)$$

for several values of $a/b = S$ (a and $b = c$ – half axes of the ellipsoid) increases on both sides of the value $S = 1$ corresponding to spherical particles, as shown in Table 10.1:

Table 10.1

S	0.1	0.2	0.5	1.0	2	5	10
L'	8.04	4.71	2.85	2.5	2.91	5.81	13.6

If the MF column is held above an isolated air cavity, for example, due to the inhomogeneous magnetic field, the adiabatic compression and expansion of the gas in periodic displacement of the MF column from the equilibrium position result in heat exchange between the gas cavity on the one hand and the tube walls and the exposed surface of the MF on the other hand. Thanks to the low heat conductivity of the gaseous medium, heat transfer first occurs in a relatively narrow border area. and secondly is delayed with respect to oscillations of the MF column. This phase shift causes some additional damping of oscillations [71, 228].

Some contribution to the damping of oscillations come from the mechanism periodic displacement of the ferroparticles relative to the carrier fluid in an oscillating fluid column. Below we present evaluation of this contribution.

The damping factor is increased by the energy loss due to the emission of elastic waves by the oscillating system. The measurements of the damping coefficient of oscillations of the MF column in the tube are approximated by the dependence of the type $\beta_{ex} \sim \nu^n$, where $n \approx 0.6$, and in absolute value are several times higher than the data calculated by the formula (10.10) [233].

The magnetic field orients the magnetic moment of the particles and in the presence of a relationship between the moment of the particle and the particle makes complicates the free rotation of the particle. This leads to local fluid velocity gradients near the base particles and causes an increase in the effective viscosity of the MF ('magnetoviscous' effect) [140, 312]. For magnetic fluids with spherical particles the additional internal friction in strong fields is given by [34, 58]:

$$\Delta\eta_H = 1.5\varphi_g \eta_0 \cdot \sin^2\beta, \tag{10.12}$$

where $\Delta\eta_H$ is the variation of the coefficient of the magnetic fluid viscosity in a magnetic field; η_0 is the viscosity coefficient at $H = 0$; φ_g is the the hydrodynamic concentration of the MF; β is the angle between **H** and the angular velocity of the magnetic particle.

For a diluted magnetic fluid with a monoparticles dispersed phase in the flow in a flat capillary the viscosity increment due to the magnetoviscous effect is calculated using the formula:

$$\Delta\eta = \frac{3}{2}\varphi\eta\frac{\xi L^2}{(\xi - \mathrm{L})}, \tag{10.13}$$

where $L = \mathrm{cth}\xi - 1/\xi$ is the Langevin function; $\xi = \dfrac{\mu_0 m_* H}{k_0 T}$.

In a sufficiently strong magnetic field at the magnetic saturation $\xi >> 1$, $L \to 1$:

$$\Delta\eta = \frac{3}{2}\varphi\eta,$$

for the flow of the colloids in a circular capillary perpendicular to the axis of the capillary:

$$\Delta\eta = \frac{3}{4}\varphi\eta. \tag{10.14}$$

However, in the conditions of oscillations of the MF in a magnetic field there is also a vibration–rheological effect [266], which, as will shown below, reduces viscosity and hence improves the quality of the oscillatory system. We consider this question in more detail.

10.2. Vibration–rheological effect

In the works of V.V. Gogosov and co-workers [18, 126, 128] the expression were derived for the coefficient of absorption of ultrasonic waves and the algorithm of their application for calculation of parameters of the dispersed nanoparticles (diameter, the ratio of the half-axes of the ellipsoidal aggregates. the thickness of the stabilizing shell).

An alternative approach is to use a theoretical model to calculate the contribution of vibration–rheological effect to the effective viscosity of the colloid and the subsequent consideration of it in the analysis and processing of the results of measurement of the damping coefficient of the MF column in a magnetic field. For this purpose we use the expression for the Stokes ultrasound absorption coefficient (4.30).

This formula does not include the bulk viscosity η_v, so directly from it we obtain an expression for the static shear viscosity η_{so}:

$$\eta_{so} = 3\alpha_s\rho c^3 / 2\omega^2. \tag{10.15}$$

(In the following the indices *s* and *so* will be omitted).

The model proposed by Gogosov is attractive in the sense that it considers both the thermal Brownian motion of nanoparticles and the magnetization of the MF in the whole range of the strength of magnetic fields, including the saturation magnetization. Unlike the situation that arises during the propagation of a longitudinal plane wave in the dispersed system, in the case of the shear reciprocating flow of the fluid in the tube the process is isothermal on the macro- and microscale so the problem of internal heat exchange is fully removed.

In [126, 128] expressions were derived for the damping decrement of sound δ_Δ in two forms:

$$\delta_\Delta = 0.5 m_\Delta^2 \omega\tau h_{022} \Big/ (L_{011} \cdot L_{022} - L_{012}^2) \qquad (10.16)$$

and

$$\delta_\Delta = 0.5 m_\Delta^2 X \left(\sin^2\phi_0 + \frac{\cos^2\phi_0}{K} \right),$$

where $m_\Delta^2 = \varphi_0(1-\varphi_0)(\rho_{a0}-\rho_{f0})^2/(\rho_{a0}\cdot\rho_{f0})$; φ_0 is the volume concentration of the solid particles with the stabilizer shell; ρ_{a0} is the density of the aggregates containing surfactants; ρ_{f0} is the density of the carrier fluid;

$$X = \frac{\omega\tau}{K_\perp};\ \tau = \frac{2}{9}\frac{(R+\delta)^2}{\eta_f}\rho_{eff};\ \rho_{eff} = (1-\varphi_0)\frac{\rho_{a0}\cdot\rho_{f0}}{\rho_0};\ K = \frac{K_\parallel}{N_* K_\perp};$$

$K_\parallel$, $K_\perp$ are the correction factors presented in the form of a function of axes N_* in tabular form in [18]; R is the radius of the solid particles; δ is the thickness of the stabilizing shell; η_f is the viscosity of the carried fluid; $\rho_0 = (1-\varphi_0)\,\rho_{f0} + \varphi_0\rho_{a0}$ is the MF density φ_0 is the angle between the directions of the fluid flow velocity and the magnetic field strength vector; $N_* = l/(R + \delta)$ is the number of particles in the aggregate (its major half-axis).

The formulas for calculating the L_{011}, L_{022} and L_{012} were previously numbered respectively (4.25), (4.26) and (4.27). The working formula chosen :

$$\delta_\Delta = \frac{1}{2}\frac{\varphi_0(1-\varphi_0)(\rho_{a0}-\rho_{f0})^2\omega 2(R+\delta)^2(1-\varphi_0)\rho_{a0}\rho_{f0}}{\rho_{a0}\rho_{f0}K_\perp 9\eta_f\rho_0}\times$$

$$\times\left(\sin^2\phi_0 + \frac{\cos^2\phi_0}{K_\parallel}\cdot N_* K_\perp \right) =$$

$$= \frac{1}{9}\frac{\varphi_0(1-\varphi_0)^2(\rho_{a0}-\rho_{f0})^2\omega(R+\delta)^2}{K_\perp\eta_f\rho_0}\cdot\left(\sin^2\phi_0 + \frac{N_* K_\perp}{K_\parallel}\cdot\cos^2\phi_0 \right).$$

$$\delta_\Delta = \frac{1}{9}\frac{\varphi_0(1-\varphi_0)^2(\rho_{a0}-\rho_{f0})^2\omega(R+\delta)^2}{K_\perp\eta_f\rho_0}\cdot\left(\sin^2\phi_0 + \frac{N_* K_\perp}{K_\parallel}\cdot\cos^2\phi_0 \right).$$

(10.17)

We will assume that the stabilizing shell is absent, i.e. $\delta = 0$, $\rho_{a0} = 2$, $\rho_{f0} = \rho_1$, $\rho_0 = \rho$ is the density of the two-phase MF. In addition, we assume the condition $\varphi_0 << 1$, then the formula (10.17) is

$$\delta_\Delta = \frac{1}{9}\frac{\varphi_0(\rho_2-\rho_1)^2\omega R^2}{K_\perp \eta_f \rho}\cdot\left(\sin^2\phi_0 + \frac{N_* K_\perp}{K_{II}}\cdot\cos^2\phi_0\right). \quad (10.18)$$

The expression for the damping coefficient of sound $\alpha = \delta_\Delta\omega/c$ with a view of (10.18) becomes:

$$\alpha = \frac{1}{9}\frac{\varphi_0(\rho_2-\rho_1)^2\omega^2 R^2}{K_\perp \eta_f \rho c}\cdot\left(\sin^2\phi_0 + \frac{N_* K_\perp}{K_{II}}\cdot\cos^2\phi_0\right).$$

After substituting α into (10.5), we obtain the expression for additional viscosity η_r:

$$\eta_r = \frac{1}{6}\frac{\varphi_0(\rho_2-\rho_1)^2 c^2 R^2}{K_\perp \eta_f}\cdot\left(\sin^2\phi_0 + \frac{N_* K_\perp}{K_{II}}\cdot\cos^2\phi_0\right). \quad (10.19)$$

The experimental procedure is very simple at $\phi_0 = \frac{\pi}{2}$ and $\phi_0 = 0$.

We consider the case $\phi_0 = \frac{\pi}{2}$ and expression (10.19) gives:

$$\eta_{r\perp} = \frac{1}{6}\cdot\frac{(\rho_2-\rho_1)^2 R^2 c^2 \varphi_0}{K_\perp \eta_f} \quad (10.20)$$

To account for the field dependence we perform the migration:

$$K_\perp \to K_\perp^* = K_\perp\left(1 - \frac{\text{cth}\,\xi}{\xi} + \frac{1}{\xi^2}\right) + \frac{K_\parallel}{N_*}\left(\frac{\text{cth}\,\xi}{\xi} - \frac{1}{\xi^2}\right).$$

When $\xi \ll 1$ η_r is independent of ξ and the coefficient L_{022} is determined by

$$L_{022} = \frac{K_\parallel}{N_*}\sin^2\phi_0 + K_\perp\cos^2\phi_0 \text{ and } L_{011}\cdot L_{022} - L_{012}^2 = K_{II}K_\perp / N_*,$$

which after substitution into (10.16) and the subsequent transformations gives (10.19). When $\phi = 0$, the formula (10.19) is transformed to the expression

$$\eta_{r_{II}} = \frac{1}{6}\frac{(\rho_2 - \rho_1)^2 c^2 R^2 N_* \varphi_0}{K_{II}\eta_f}.$$

The ratio $\frac{\eta_{r\perp}}{\eta_{rII}} = \frac{K_{II}}{N_* K_\perp} = K$ depends only on the ratio $l/R = N_*$.
It was shown in [18] that

$$K\left(\frac{l}{R}\right) = \frac{N_* + \frac{2N_*^2 - 3}{\sqrt{(N_*^2 - 1)}} ln\left(N_* + \sqrt{(N_*^2 - 1)}\right)}{-2 \cdot N_* + \frac{2N_*^2 - 1}{\sqrt{(N_*^2 - 1)}} ln \frac{N_* + \sqrt{(N_*^2 - 1)}}{N_* + \sqrt{(N_*^2 - 1)}}}. \qquad (10.21)$$

Table 10.2, taken from [18], gives the values of the functions $K_\perp$, $K_{||}$ and K depending on the value N_*, which allows from the value K found by experiment to obtain the value 1/R.

The value $K_\perp$ monotonically decreases with increasing N^*. According to equation (10.18), $\eta_{r\perp}$ will monotonically increase.

We estimate $\eta_{r\perp}$ using (10.20). Let it be $\rho_2 = 5 \cdot 10^3$ kg/m³, $\rho_1 = 1 \cdot 10$ kg/m³, $\eta_f = 8 \cdot 10^{-3}$ Pa·s, $R = 5 \cdot 10^{-9}$ m, $c = 1100$ m/s, $\varphi_0 = 0.1$, $K_\perp = 0.65$, ($N_* = 2.3$) and obtain $\eta_{r\perp} = 1.55 \cdot 10^{-3}$ Pa·s. Thus $\eta_{r\perp}$ is about 20 % η_f.

Figure 10.1 shows the dependence of the additional viscosity on the magnetic field strength $\eta_{r\perp}$ (H) in relative units – β_r. The data show that the vibration–rheological the effect is to reduce the

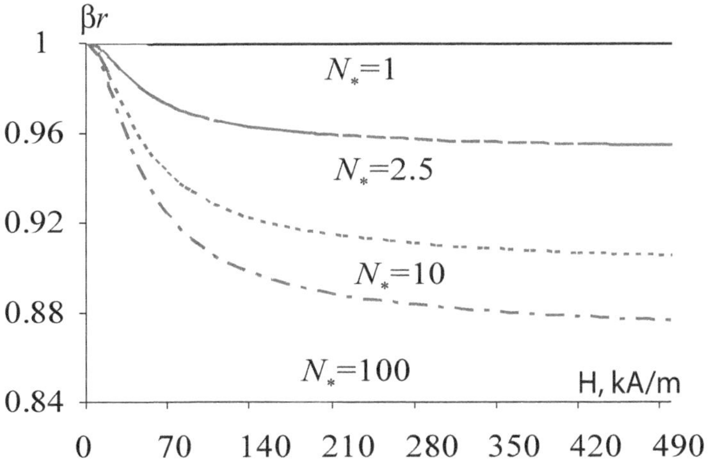

Fig. 10.1. Dependence of additional viscosity on the strength of the magnetic field in relative units.

Table 10.2

d/l	N	$K_{\|}$	$K_{\perp}$	$K \equiv K_{\|}/NK_{\perp}$
0.990	1.010	1.002	0.994	0.998
0.833	1.200	1.040	0.899	0.964
0.769	1.300	1.061	0.860	0.949
0.588	1.700	1.143	0.747	0.900
0.435	2.30	1.265	0.646	0.851
0.313	3.20	1.440	0.561	0.804
0.200	5.0	1.785	0.474	0.753
0.100	10.0	2.647	0.381	0.696
0.067	15.0	3.433	0.342	0.670
0.040	25.0	4.878	0.302	0.646
0.020	50.0	8.117	0.261	0.621
0.010	100	13.859	0.230	0.604
$5 \cdot 10^{-3}$	200	24.280	0.205	0.590
$2 \cdot 10^{-3}$	500	52.022	0.180	0.578
10^{-3}	10^3	93.885	0.165	0.570
10^{-5}	10^5	5695	0.105	0.543
10^{-7}	10^7	408716	0.077	0.531

viscosity of the MF and, therefore, increases the quality factor of the oscillatory system in a magnetic field. Moreover, this effect becomes stronger as the number of particles in the chain N_* increases.

10.3. On a 'non-magnetic' dissipation mechanism of the energy of the oscillatory system

The results of measurements of the damping coefficient of oscillations of the MF-column in the tube in magnitude are several times greater in the absolute value than the values calculated by the formula (10.10) [233]. There is a question about other possible mechanisms of energy dissipation in the discussed oscillatory system. The most probable 'non-magnetic' mechanisms include the mechanism of interphase heat exchange which is not susceptible to external magnetic field and not connected to the viscosity of the MF. The magnetization of the fluid can not affect the contribution of this mechanism to the dissipation of elastic energy of the oscillatory system. However, as will be shown, the interphase heat transfer causes a significant portion of the losses of the energy of the oscillatory system.

I.A. Chaban [228] discussed the damping of the oscillations of a gas bubble in a fluid that is associated with the heat exchange. The Kirchhoff–Rayleigh model [71] is somewhat closer to the situation that we are discussing here and takes into account not only the effect of viscosity but also equally important effects arising from the generation and transfer of heat by heat conduction from the gas to solid walls of the tube and back.

As a result of the adiabatic process of compression and decompression of gas the periodic displacement of the MF column from the equilibrium position is accompanied heat exchange between the gas cavity and the tube walls. Due to the low heat conductivity of the gaseous medium, firstly the heat transfer occurs in a relatively narrow near-wall region and, secondly, is delayed with respect to the oscillations of the bridge. This phase shift causes additional damping of the oscillations in the system.

Some of the assumptions made in the solution of this problem include: air is enclosed inside the cylindrical tube with a circular cross-section; at the base of the tube there is a flat source wave of frequency ω; movement is symmetrical relative to the tube axis; the wall temperature does not change; the coefficient of thermal conductivity of gas is so small that the layer of gas subjected to the direct effect of the pipe walls is only a small fraction of the total amount of gas enclosed in a tube.

Upon excitation of oscillations in the tube filled with the gas a damping sound wave with the damping coefficient α' and the wave number k' will propagate in the tube:

$$\alpha' = \frac{\sqrt{\omega\gamma'}}{\sqrt{2}cr}; k = \frac{\omega}{c} + \frac{\sqrt{\omega\gamma'}}{\sqrt{2}cr},$$

where r is the distance from the tube axis.

$$\gamma' = \sqrt{\frac{\eta_g}{\rho_g}} + \left(\frac{a}{b} - \frac{b}{a}\right)\sqrt{\frac{\chi}{\rho_g C_p}}; a = \sqrt{\frac{\gamma P_0}{\rho_g}}; b = \sqrt{\frac{P_0}{\rho_g}}.$$

Here χ is the coefficient of thermal conductivity, C_p is the specific heat capacity of the gas. If we completely eliminate the thermal conductivity of the gas, i.e. $\chi = 0$, then

$$\alpha' = \frac{1}{cr}\sqrt{\frac{\pi\nu\eta_g}{\rho_g}}.$$

Given the ratio $\beta = \alpha c/2\pi$, we find

$$\beta = \frac{1}{2\pi r}\sqrt{\frac{\pi\nu\eta_g}{\rho_g}}.$$

Selection of r is limited by the conditions that $r < d/2$.

To come back to the original expression for the damping coefficient, determined by viscosity, it is necessary to put $r = d/4\pi$. Then:

$$\alpha' = \frac{4\sqrt{\pi^3\nu}}{cd}\left[\sqrt{\frac{\eta_g}{\rho_g}} + \left(\sqrt{\gamma} - \frac{1}{\sqrt{\gamma}}\right)\sqrt{\frac{\chi}{\rho_g C_p}}\right].$$

According to the Maxwell's kinetic theory:

$$\chi = \frac{1}{3}\rho_g \bar{U}\lambda' C_V;\ \eta_g = \frac{1}{3}\rho_g \bar{U}\lambda'.$$

where λ' is the mean path of the molecule, $\bar{U}$ is the mean speed of thermal motion of the gas molecules.

Taking into account that $\hat{C}_V = \frac{i}{2}R$, and $C_V = \frac{\hat{C}_V}{\mu}$, for the diatomic gases which also include air we obtain $\frac{\chi}{\eta_g} = \frac{5}{2}\frac{R}{\mu}$ or $\chi = \frac{5}{2}\frac{\eta_g R}{\mu}$.

In molecular acoustics there is a slightly different relationship established empirically (Aiken empirical equation) [63]:

$$\chi = C_V \eta_g (9\gamma - 5)/4.$$

Therefore, the sound absorption coefficient that takes into account both mechanisms of energy dissipation can be written in the following form:

$$\alpha = \alpha_\eta + \lambda_T = \frac{4\pi}{cd}\sqrt{\frac{\pi\eta_g\nu}{\rho_g}}\left[1 + \left(\sqrt{\gamma} - \frac{1}{\sqrt{\gamma}}\right)\sqrt{\frac{9\gamma - 5}{4\gamma}}\right].$$

Accordingly for the damping coefficient:

$$\beta''' = \frac{2}{d}\sqrt{\frac{\pi\eta_g \nu}{\rho_g}}\left[1+\left(\sqrt{\gamma}-\frac{1}{\sqrt{\gamma}}\right)\sqrt{\frac{9\gamma-5}{4\gamma}}\right].$$

For the considered oscillatory system in which the gas acts as an elastic member, we can assume in a sufficiently rough approximation that there is no the gas flow in relation to the walls, and the factor of heat transfer due to adiabatic vacuum–compression process is functioning, so it makes sense keep in the expression for β''' only the thermal component:

$$\beta_T = \frac{1}{d}\left(\frac{\gamma-1}{\gamma}\right)\sqrt{\frac{\pi\eta_g(9\gamma-5)\nu}{\rho_g}} \text{ and } \beta_T = \frac{2}{d}\left(\frac{\gamma-1}{\sqrt{\gamma}}\right)\sqrt{\frac{\pi\chi\nu}{\rho_g C_p}}.$$

Using the tabulated values of the physical quantities of air: $\eta_g = 17.2\cdot10^{-6}$ Pa · s, $\rho_g = 1.29\ \frac{\text{kg}}{\text{m}^3}$, $\gamma = 1.4$; and accepting ν = 50 Hz, $d = 10^{-2}$ m, we obtain $\beta_T = 3.7\ \text{s}^{-1}$.

10.4. Comparing the findings of the model theory with experiment

We study the oscillatory system in which a MF column in a tube with the soldered bottom rests on the air cavity [218, 226]. A ring magnet is used to stabilize the bottom surface of the fluid and secure the air cavity. The magnet is disposed coaxially with the tube slightly above its base (the characteristic of the magnetic field of the ring magnet are given in 7.6.3). Glass tubes with an inner diameter of 1.36 cm and 0.95 cm were used. Inside the ring magnet there was a coil in which variable EMF was induced during oscillations at fluctuations of the MF column. The frequency of variable EMF and change in its amplitude are used to define the parameters of the oscillatory process – the frequency ν and damping coefficient β. The oscillation frequency is varied by changing the height of the MF column and (if necessary) reducing the volume of the retained gas cavity.

Tests showed that the measurement error of ν does not exceed 5%. and that at 0.95 confidence the interval of measuring β is 9÷10 %.

Six samples of MF of the 'magnetite in kerosene' type were studied. Table 10.3 shows their physical parameters measured by the methods described in section 9.5.3. The shear

Table 10.3

Sample	ρ, kg/m³	φ, %	φ_m, %	M_s, kA/m	χ	η_s, Pa·s	d_{max}, nm	d_{min}, nm
MF-1	1630	19.1	13	63±1	8.5	$11.6 \cdot 10^{-3}$	17	9.3
MF-2	1499	16.2	12.5	60±1	7.52	$8.1 \cdot 10^{-3}$	16.7	9.3
MF-3	1479	15.8	12.5	60±1	7.5	–	16.6	9.3
MF-4	1444	14.9	–	–	–	$8.3 \cdot 10^{-3}$	–	9.3
MF-5	1330	12.3	–	–	–	$3.8 \cdot 10^{-3}$	–	9.3
MF-6	1294	11.6	10.9	52±1		$3.2 \cdot 10^{-3}$	16.6	9.3

Table 10.4

h, cm	10	15	21	31
v_e, Hz	35.7	27.8	21.5	20.9
v_t, Hz	35.5	29	24.5	20.2

viscosity η_s was measured by the relative capillary method. The reference fluid was distilled water having a viscosity of η_w = $1.004 \cdot 10^{-3}$ Pa·s at 20°C.

The findings discussed in the theoretical model of the formation of the elasticity of the oscillatory the system do not contradict the experimental data obtained.

Table 10.4 presents the theoretical and experimental dependences (ν_t (h) and $\nu_e(h)$) for the oscillatory system in which the inertial-viscous element is the MF-6. ν_t was calculated made by formula (10.3) under the assumption h_g = 2.4 cm, ρ_g = 1.29 kg/m³, ρ = 1294 kg/m³, $M = 35.1 \cdot 10^3$ A/m, $G = 4.6 \cdot 10^6$ A/m².

Importantly, the calculated frequency values differ from the experimental values by less than an order of magnitude. Therefore, formulas (10.2) and (10.3) can be used for approximate calculations.

Figure 10.2 presents experimental values of $\nu \cdot \sqrt{h}$ at different values of h for the samples: MF-1 (triangles), MF-4 (diamonds). MF-5 (squares) and MF-6 (circle in circle). Straight line segments show values of $\langle \nu \cdot \sqrt{h} \rangle$ averaged over all selected values of h.

In accordance with the formulas (10.2) and (10.3), the graphs of the presented dependences $\nu \cdot \sqrt{h}$ on h have the form of horizontal lines.

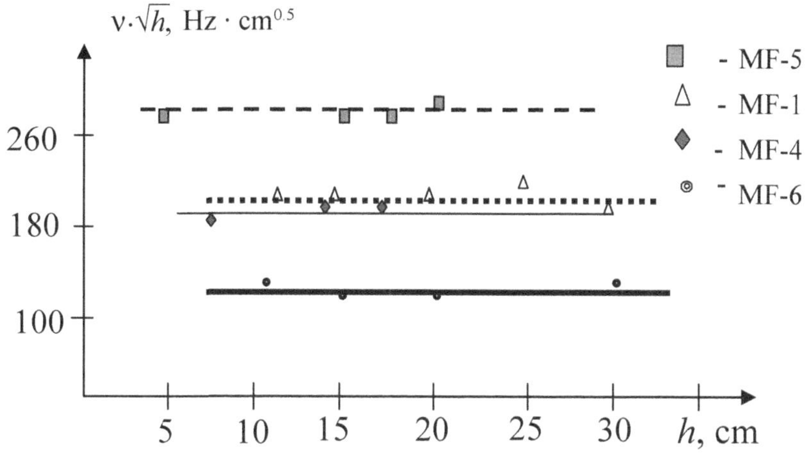

Fig. 10.2. Dependence of $\nu \cdot \sqrt{h}$ on h.

Table 10.5

Specimen No.	h, cm	β, s^{-1}	$\sqrt{\nu}$, $Hz^{0.5}$	$\beta/\sqrt{\nu}$, $s^{-1} \cdot Hz^{-0.5}$	$\beta'' + \beta_T$, s^{-1}	$\frac{\beta'' + \beta_T}{\beta}$
1	5	28.3	9.62	2.94	6.7 + 3.7 = 10.4	0.36
2	10.5	16.5	7.52	2.2	5.2 + 2.9 = 8.1	0.49
3	15	10.8	6.9	1.56	4.8 + 2.62 = 7.42	0.69
4	20	9.4	6.4	1.41	4.4 + 2.43 = 6.83	0.72
5	25	8.9	6.1	1.46	4.3 + 2.54 = 6.84	0.77
6	31.5	7.8	5.7	1.37	4 + 2.2 = 6.2	0.8

Table 10.5 shows the results of measurements of the damping coefficient β of the system in which the inertial–viscous member MF-1. The inner diameter of the tube d = 1.36 cm. In measurements the height of the MF column h was varied which ensured changes in oscillations ν. Since the most likely mechanism of energy dissipation in the studied vibrational system – viscous flow and heat exchange – lead to the dependence of the damping coefficient on frequency in the form $\beta = \gamma \cdot \sqrt{\nu}$, where γ = const, the table includes a column with the value $\beta/\sqrt{\nu}$. Also, Table 10.5 shows the results of calculating the contribution of the main mechanisms of dissipation and the damping coefficient in the absolute $(\beta'' + \beta_T)$ and in relative terms $(\beta'' + \beta_T)/\beta$.

At 'small' magnetic fluid column heights ($h/d < 10$) the deviation of the calculated and experimental values of the damping coefficient was outside the measurement error range. Apparently, this was due to the failure to meet requirements hydrodynamic theory $h/d \gg 1$.

With the increase in the height of the MF column the value $\frac{\beta}{\sqrt{\nu}}$ decreases that indicates an inconsistency in the model theory in comparison with the experiment.

Due to the small difference in the diameter (0.95 and 1.36 cm) and a large measurement error β (≤10 %), the results of experiments performed with pipes of different diameters and the same fluids cannot be used to make unambiguous conclusions regarding the implementation of proportions predicted by the theory $\beta \sim 1/d$.

Table 10.6 compares the experimental values of the damping coefficient β and the corresponding theoretical values $\beta_t = \beta'' + \beta_T$ for oscillatory systems based on four MF samples. β'' was calculated using the formula (10.10) using the data in Table 10.3.

Table 10.6

No.	h, cm	β, s^{-1}	ν, Hz	$\beta'' + \beta_T = \beta_t$, s^{-1}	β_t / β
MF-2	31.5	7.8	32.5	4 + 3.1 = 7.2	0.92
MF-4	17	8.9	43	4.1 + 3.6 = 7.7	0.87
MF-5	25	9.3	47.6	4.34 + 3.8 = 8.14	0.88
MF-6	31	5.5	27	2.13 + 2.9 = 5.0	0.91

Analysis of the data presented in Tables 10.5 and 10.6 shows that β'' and β_T are comparable. At 'large' heights of the MF column ($h/d > 20$) the calculated values of β_T are sufficiently close to the experimental results of β, their quantitative difference does not exceed 20%. However, all calculations are based on an approximate model theory. Therefore. in principle such quantitative correspondence should not be.

Taking into account the vibration–rheological effect by formula (10.20) leads to an increment of β'' by $\leq 0.08\ \beta''$. Therefore, the Helmholtz mechanism and the Kirchhoff–Rayleigh mechanism of interphase heat transfer can be considered the most likely mechanisms of energy dissipation in the study of the oscillatory system.

10.5. Rheology of magnetic fluid with anisotropic properties

V.A. Naletova and Yu.M. Shkel' [259] describe the experimentally observed 'anomalous' increase of the pressure drop in the flow of a magnetic fluid in a pipe in a flow perpendicular to the field [267, 268] using a MF model generalized for the case of a medium with anisotropic properties. The anisotropy of the magnetic fluid is caused by the presence of the dispersed phase along with the individual particles of the aggregates of the ellipsoidal shape fully oriented by a strong magnetic field. It is suggested that the fluid contains at the same time both spherical and elongated particles or aggregates, with the aggregates having the same dimensions, and that all particles of the dispersed system do not interact. It is also assumed that without the field ($H = 0$) the coefficient of viscosity η is calculated according to Einstein's formula:

$$\eta_{H=0} = \eta_0(1+2.5\varphi), \tag{10.22}$$

The relationship of the pressure drop between the points z_1 and z_2 along the fluid column in the longitudinal $\Delta p'_{1.2\|}$ and the transverse magnetic field $\Delta p'_{1.2\perp}$ in the tube with an average flow speed of v_m the tube radius R is described by the equations:

$$\begin{gathered}
\Delta p'_{1.2\perp} = \alpha\eta_0(1+\varphi L_\perp(S)),\ \Delta p'_{1,2\|} = \alpha\eta_0(1+\varphi L_\|(S)),\\
\Delta \rho'_{1,2|H=0} = \alpha\eta_0(1+2.5\varphi),\ \alpha = 8(z_2 - z_1)v_m / R^2,\\
L_\perp = A(S) + 0.5S_2(S) + 0.5(1+2\lambda)S_1(S),\\
L_\| = A(S) + S_2(S) + (1-2\lambda)S_1(S),\ A(S) = 8(S^2-1)/(4S^4 - 10S^2 + 3SL),\\
S_2(S) = 4(S^2-1)/((S^2+1)(4+2S^2-3SL)) - A(S) + \lambda^2 S_1(S),\\
S_1(S) = 2(S^2-1)(S^2+1)/S(L(2S^2-1)-2S),\\
L = ln|(S+(S^2-1)^{1/2})/(S-(S^2-1)^{1/2})/(S^2-1)^{1/2},\\
\lambda = (S^2-1)/(S^2+1),
\end{gathered} \tag{10.23}$$

where S is the geometrical parameter of the nanoaggregate which is the ratio of the larger to the smaller axis of the ellipsoid.

Assuming that the aggregates are stretched enough so that $L_\| \sim 2$, the expression for the pressure drop can be written as:

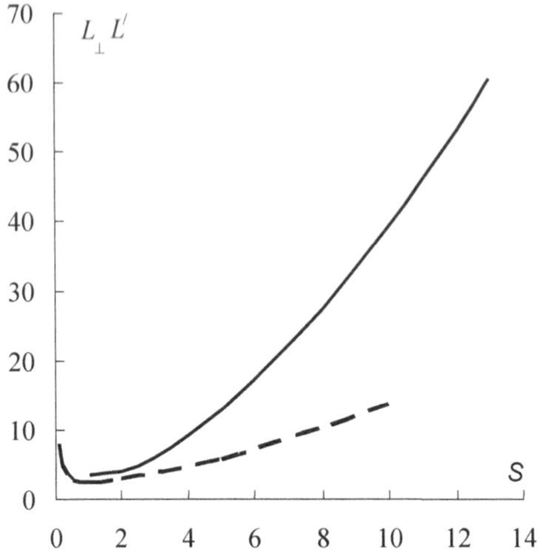

Fig. 10.3. Theoretical dependence $L_\perp(S)$- solid line and $S-1$ the dotted line.

$$\begin{aligned}
&\Delta p'_{1.2\perp} = \alpha\eta_0(1+13(\varphi-\varphi_1)/4+\varphi_1 L_\perp(\mathrm{S})),\\
&\Delta p'_{1.2\|} = \alpha\eta_0(1+4(\varphi-\varphi_1)+2\varphi_1) \qquad (10.24)\\
&\Delta p'_{1.2|H=0} = \alpha\eta_0(1+2.5\varphi),
\end{aligned}$$

where φ_1 – volume fraction of ellipsoidal aggregates.

The system of three equations (10.24) at the parameter values known from the experiment $\Delta p'_{1.2/\perp}$. $\Delta p'_{1.2\|}$, $\Delta p'_{1.2\,|\,H=0}$ may be reduced to two equations with two unknowns φ_1 and S. It is shown that the pressure drop $\Delta p'_{1.2\|}$ in a magnetic field parallel to the fluid flow should be slightly less than without the field. In the perpendicular field the elongated aggregates greatly increase the pressure drop, while spherical particles have no significant effect.

Figure 10.3 shows the theoretical curves of $L_\perp$ (S) and L' (S) with the latter being constructed for the 'regular' (non-magnetic) suspension with ellipsoidal particles using the data in Table 10.1. The graphical dependence $L_\perp$ (S) was obtained by mathematical modelling in MS Excel, starting with values of $S = 2$. This dependence has the character of a monotonically increasing function. A comparison of the curves shows that when the magnetic saturation take place in the magnetic field transverse to the tube with the MF the viscosity of the system increases with the extension of the chain much more than in its absence.

Given the proportionality between the effective viscosity η and the pressure drop $\Delta p'_{1.2\perp}$, the first equation of the system (10.23) can be written in the form

$$\eta = \eta_0(1 + \varphi L_\perp). \tag{10.25}$$

From the equation of the trend line we can find $L_\perp(S)$ at $S \to 1$. In this case $L_\perp (S)_{s\to 1} = 3.44$. which exceeds the corresponding coefficient in the Einstein equation by 0.94 due to the magnetoviscous effect.

We can assume that the dissipation mechanism, described in the [259], is a kind of magnetoviscous effect affecting the ellipsoidal particles rigidly fixed by the magnetic field.

Comparing the findings of the models proposed by V.A. Naletova and V.V. Gogosov in the section of the dependence $\eta_{r\perp}(N_*)$ shows that in the first case $\Delta\eta \sim L_\perp$, and the second $\eta r_\perp \sim 1/K_\perp$.

The above physical mechanisms of elastic energy dissipation in the oscillatory system with a MF inertia member can be divided into two groups, one of which includes mechanisms that are dependent on the strength of the magnetic field (magnetorheological effects) and the other group – independent of it.

The dissipation of elastic energy in the oscillatory system with the inertia member in the form of the MF column (as noted above) is mainly caused by the simultaneous action of the four physical mechanisms :

1. Energy losses in a reciprocating flow of the fluid through the tube as in the shear flow hard magnetic nanoparticles or ellipsoidal nanoaggregates are affected by a moment of force; in the magnetic field the contribution of this mechanism is enhanced by the magnetoviscous effect.
2. Energy losses due to slippage of nanoparticles and nanoaggregates with respect to the liquid matrix.
3. The mechanism of interphase heat exchange of the gas cavity with the container walls.
4. The radiation of elastic waves in structural elements and the environment.

The first mechanism is adequately interpreted on the basis of the concept of shear waves introduced by Stokes. For large values of pipe diameter d and the frequency of v, where $\pi d/2\lambda > 10$, due to the small depth of penetration of the viscous wave σ'' the oscillatory movement of the fluid column is of the 'piston' nature and the fluid flow is concentrated in a thin boundary layer. The active resistance of the tube is given by the Helmholtz formula (10.9) and the damping coefficient is found from the expression (10.10). The contribution to energy dissipation is due to viscous friction in the border area of the wall of the tube and in the framework of the Helmholtz model is proportional to $\sqrt{\nu}$.

The energy loss due to the 'slip' of the particles relative to the liquid matrix leads to the increment of the shear viscosity which tends to decrease with the increase of the magnetic field transverse to the tube with the MF. However, with the increase in the number of particles in the aggregate the contribution to energy dissipation of the oscillatory system, brought by the magnetoviscous effect, much greater than the contribution of the slip of the particles.

As a result of the adiabatic expansion and compression process of the gas the periodic displacement of the fluid column from the equilibrium position heat transfer takes place between the gas cavity on one side and the walls of the tube and the exposed surface of the liquid – on the other side. Due to the low thermal conductivity of the gas medium the heat transfer is delayed in relation to oscillations of the fluid column. This phase shift causes some additional damping of the oscillations.

The fourth of the above energy dissipation mechanisms is associated with the emission elastic vibrations in the tube wall and supporting structural members (holder, stand) into air. Its effectiveness is apparently related to the ratio of the acoustic resistances of structural elements of the oscillatory system and the environment.

The last two mechanisms are not susceptible to the external magnetic field and are not connected with the MF viscosity. Parameters that can have an impact on the physical mechanisms of damping of oscillations of non-viscous origin – acoustic and thermal resistance conductivity of the magnetic colloids are practically independent of the magnetic field strength and the degree of its heterogeneity. The magnetization of the fluid can not affect the contribution of these mechanisms to the dissipation of the elastic energy of the oscillatory system.

Additional energy losses not related to viscosity cause additional damping $\Delta\beta$. The experimental value of the damping coefficient β_{ex} consists of two components $\Delta\beta$ and β:

$$\beta_{ex} = \beta + \Delta\beta. \tag{10.26}$$

After calibration of the measuring device by measuring the viscosity of the sample by another method ($\Delta\beta$ is set at the selected frequency). the value of viscosity of the fluid in a magnetic field is given by:

$$\eta = \frac{\rho d^2(\beta_{ex} - \Delta\beta)^2}{4\pi\nu}. \tag{10.27}$$

Given the proportionality between the effective viscosity η and the pressure drop $\Delta p'_{1,2\perp}$, the first of equations (10.24) can be written as:

$$\eta = \eta_0\left(1 + \frac{13}{4}(\varphi - \varphi_1) + \varphi_1 L_\perp(S)\right). \tag{10.28}$$

In equation (10.28) there are two unknowns φ_1 and S, and the other two parameters η_0 and φ can be directly measured. The effective viscosity, calculated according to the formula (10.28). can be brought in line with the value obtained from the measurements at different combinations of φ_1 and S. The calculation results, obtained in accordance with (10.28), are ambiguous.

For this reason, it would be more correct to use previous results of [259], namely the system of equations (10.23), which allows one to get an 'effective' value of parameter S. Equation (10.28) will be presented in the form of expression

$$\eta = \eta_0(1 + \varphi \mathrm{L}_\perp), \tag{10.29}$$

where φ is the concentration of aggregates with an 'effective' value of parameter S.

The cause of the anomalously strong dependence of the damping coefficient of oscillations in an oscillatory system with the magnetic fluid inertial component on the strength of the magnetic field can be not only the presence of aggregates of magnetic nanoparticles in the dispersed system but also the presence of aggregates of microparticles of the non-magnetic phase in the system.

The presence of non-magnetic particles of micron size in the MF determines the appearance of the magnetorheological effect in the system [265]. Such a particle placed in a magnetizable medium leads to disruption of the homogeneous (on the scale of the particle) distribution of the magnetic field strength. At the poles of the particle the field weakens and at the equator increases. Therefore, adjacent particles in the equatorial region will be repelled and those along the polar axis with be attracted and try to take the position in the region of the minimum values of the magnetic field strength. Increase of the effective viscosity in the field is due to the processes of formation and destruction of non-magnetic structures in the medium as a result of the competing effects of magnetic and hydrodynamic forces.

In [265] an expression is proposed for the increment of the viscous stress in the suspension of non-interacting ellipsoids:

$$\Delta\tau_{max} = \frac{1}{2}\mu_0 M^2 \varphi', \tag{10.30}$$

and for the characteristic orientation relaxation time of the internal structure of the medium:

$$t' = \eta_0 \lambda^2 / \mu_0 M^2 In\lambda, \tag{10.31}$$

In these expressions η_0 is the viscosity of the dispersion medium. φ' is the concentration of the non-magnetic phase, λ is the ratio of major and minor half axes, M is magnetization. The orientation time t' determines the shear rate $\gamma = 1 / t'$ to which the main increase of the viscous stresses in the magnetic field takes place. Using Newton's law $\Delta\eta_{max} = \Delta\tau_{max} / \dot{\gamma}$ and formulas (10.30) and (10.32), we can derive an expression for the increase in the viscosity of the magnetic field associated with the presence of non-magnetic particles

$$\Delta\eta_{max} = \frac{1}{2}\varphi' \eta_0 \frac{\lambda^2}{\ln\lambda}. \tag{10.31}$$

10.6. Measurement procedure

The oscillatory system designed in [256, 257, 262, 266] for the acoustometry of the form of dispersed nano- and microparticles is a glass U-shaped tube partially filled with the MF (inertial member). The elastic member is an isolated air cavity in the tube which is in direct contact with the MF column.

The oscillatory system, used for measuring the rotational viscosity in these studies, is shown schematically in Figure 10.4. The glass U-tube 2, inner diameter d = 10.7 mm, is filled to a certain level in both bends with the magnetic colloid 3. One of the bends is hermetically sealed with a piezoelectric plate – oscillation sensor 4. MF in this case is the inertial–viscous element of the oscillatory system and its elastic element 5 is the air cavity, formed below the piezoelectric plate. The oscillations are excited by the elastic

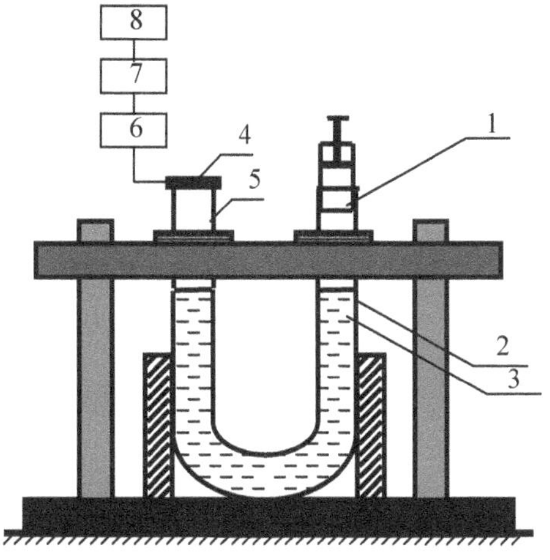

Fig. 10.4. Block diagram of the experimental setup.

tube with the piston 1 installed on the other bend. The tube has an opening designed to release air during movement of the piston to its original position and is covered up at the time of pulling the piston. RF pulsed obtained in the piezoelectric sensor are fed to the storage oscilloscope screen 6. Oscillograms are recorded with the digital camera 7 and transmitted to the computer 8 for further treatment.

It is assumed that both the free fluid surfaces are flat and the overall length of the fluid column is b. The expression for the coefficient of elasticity of the gas cavity for the adiabatic process k_g has the form (7.2).

The effect of fluctuations in the level of the fluid in the tube bends on the elasticity of the system can be neglected, as follows from a comparison of the potential energy of compression (expansion) of the gas cavity

$$E_{pg} = \frac{K_g \Delta h^2}{2}$$

and the increase of potential energy of the fluid column

$$E_{pf} = \frac{\rho g S \Delta h^2}{2}.$$

The ratio $\dfrac{E_{pf}}{E_{pg}} = \dfrac{\rho g h}{\rho_g c^2} = 2 \cdot 10^{-4}$ for h = 3 mm.

Considering the tube wall and the piezoelectric plate absolutely rigid and the fluid incompressible, the formula for the frequency of the free non-damping oscillations has the form:

$$\nu_m = \frac{c}{2\pi}\sqrt{\frac{\rho_g S}{\rho V_0 b}}, \tag{10.33}$$

where ρ is the density of the fluid; b is the length of the fluid.

In this case the tube has a cylindrical shape that allows the isolated air cavity with height h to be described by equation (10.33) in the form:

$$\nu_m = \frac{c}{2\pi}\sqrt{\frac{\rho_g}{\rho h b}}. \tag{10.34}$$

The frequency of free damped oscillation is determined as the inverse of the period T determined from the duration of six or seven full oscillations directly on the oscillogram, with the measurement error of the period and frequency in this method being in the range of 3–5%.

Measurements were carried out on pipes with diameter d = 10.7 mm and d = 15.8 mm at a temperature T = 25 ± 1°C.

The results obtained for β are used to compile a table of experimental data taken from the selected oscillogram using Corel Draw, followed by plotting the graph of the dependence of $\ln\frac{A_r}{A_n}$ on $n \cdot T$, where T is the period of oscillations calculated as $\frac{t_n - t_1}{N}$, N is the number of complete oscillations. The resultant 6÷8 points are connected by a straight line the slope of which is β. The reference value of the amplitude A_r is the amplitude of the 2nd period of the oscillations.

Estimates based on the repeatability of the measurements showed that the relative error of measurements of the damping coefficient of the studied oscillatory system β in experimental conditions is 8–9 % at a confidence level of 0.95.

Discussing the measurement technique, based on the use of the U-shaped tube as a container, it is necessary to consider the impact of bending the tube on the dissipation of oscillatory energy.

We calculate the damping coefficient of the system due to losses in the change in the direction of the fluid flow. i.e., in turning around.

Assume that the cross-section S along the length of the tube is the same. The work of pressure forces applied to sections S on the left and the right is

$$A = p_1 S\Delta l - p_2 S\Delta l = \Delta p \Delta S l,$$

where Δl is the displacement of the fluid column in one of the bends of the U-shaped tube.

The pressure drop due bending of the tube is determined from the Weisbach equation

$$\Delta p = \frac{\xi \rho \upsilon^2}{2}.$$

The loss coefficient ξ can be defined by the empirical formula [291]

$$\xi = 2\left(0.13 + 1.85\left(\frac{r}{R}\right)^{3.5}\right)$$

In this case at the tube radius r = 5.35 mm and the bending radius of the tube R = 15 mm the coefficient ξ = 0.36. In the presence of harmonic oscillations displacement of the fluid column is $\Delta l = A_0 \cos \omega t$, oscillation velocity $v = -A_0 \omega \sin \omega t$.

The energy loss by turning the liquid flow can be found from the expression:

$$\Delta E = -\frac{4}{T}\int_0^{T/4} A dt = -\frac{\xi \rho A_0^3 S \omega^2}{\pi}\int_0^{T/4} \sin^2 \omega t \cos \omega t d\, \omega t = -\frac{\xi \rho A_0^3 S \omega^2}{3\pi}.$$

The damping coefficient of the damping system:

$$\beta = -\frac{1}{2}\frac{\Delta E}{\Delta E_0} v.$$

Since the total mechanical energy of the oscillatory system is

$$\Delta E_0 = \frac{m\omega^2 A_0^2}{2},$$

then

$$\beta = \frac{1}{3}\frac{\xi \rho S A_0 v}{\pi m} = \frac{\xi S A_0 v}{3\pi V}.$$

For the damping coefficient, substituting in the last formula the numerical values of the tube cross-sectional area $S = 9\cdot 10^{-5}\ \text{m}^2$, the fluid volume V = 16 ml, frequency v = 36 Hz, amplitude A_0 = 3 mm, we find $\beta = 2 \cdot 10^{-2}$ s^{-1}. The experimental value is more than an order of magnitude higher than the obtained value of β.

We study the influence of fluid electrical conductivity on the damping coefficient.

When moving the conductor in the magnetic field, the effect of the Lorentz force on the free electrical charge at its ends results in the potential difference e. This process is identical with the Hall effect and thus can be analyzed similarly.

In the magnetic flow transverse to the hydrodynamic flow the effect will be maximized. Assuming that the ferroparticle has the shape of a cube with side $\tilde{d}$, with the direction of movement of the cube coinciding with the normal to one of the faces, we obtain for the condition of the dynamic equilibrium of electric charge fluxes $e/\tilde{d} = \dot{u}B$. The emergence of variable EMF at the ends of the conductor causes the flow of alternating current through the conductor, accompanied by the release of Joule–Lenz heat. The heat released for the selected time t in a single particle is $Q_{T1} \sim I^2Rt$, where R is the ferroparticle resistance, I is the current intensity. Since $R = \rho_e/\tilde{d}$, where ρ_e is the electric resistivity of the ferroparticle material, and according to Ohm's law $I = \dfrac{e}{R} = \dfrac{e_m}{\sqrt{2}R}$, then

$$Q_{T1} = \frac{e_m^2 \tilde{d}}{2\rho_e} t.$$

The heat released during time t in the plane-parallel layer of dispersed medium with area S and thickness dx is: $dQ_T = \dfrac{\mu_m^2 B^2 \varphi \cdot t \cdot dx}{2\rho_e}$.

These losses lower the energy of the system $dQ_d = -dQ_T$.

The total mechanical energy of the oscillatory system:

$$dQ_0 = \frac{\rho S \cdot dx \cdot \dot{u}_m^2}{2}.$$

The logarithmic damping decrement

$$\delta = -\frac{1}{2}\frac{dQ_d}{dQ_0} = \frac{1}{2}\frac{B^2 \varphi \cdot t}{\rho_e \rho}.$$

Since $\beta = \delta v$ and $t = \frac{1}{v}$ then $\beta = \frac{\beta^2 \varphi}{2\rho_e \rho}$. If magnetite is used as the ferromagnetic phase for which $\rho_e = 5 \cdot 10^{-5}$ ohm, then at $B = 0.9$ T, $\varphi = 0.166$, $\rho = 1522$ kg/m^3 we obtain $\beta \approx 0.9$ s^{-1}.

Part of the energy of the oscillatory system is dissipated in the form of Lenz–Joule heat due to the electrical conductivity of the magnetic fluid.

Consider a cube of a magnetic fluid moving in the direction of one of the faces perpendicular to the vector of magnetic induction. Cube edge a, its electrical conductivity σ. In this case, the energy of the system is reduced for the period of oscillations by

$$\Delta Q_d = -\frac{\dot{u}_m^2 B^2 a^3 \sigma T}{2}.$$

Since the total mechanical energy of the oscillatory system is

$$\Delta Q_0 = -\frac{\rho a^3 \dot{u}_m^{-2}}{2},$$

the damping coefficient is

$$\beta = -\frac{1}{2}\frac{\Delta Q_d}{\Delta Q_0} \cdot \frac{1}{T} = \frac{\beta^2 \sigma}{2\rho}.$$

Based on the data of [260], the electrical conductivity of the magnetic liquid with the concentration of the solid phase $\varphi = 16.6\%$, $\sigma = 2 \cdot 10^{-6}$ ohm$^{-1} \cdot$ m^{-1}. Substituting $\rho = 1522$ kg/m^3 and $B = 0.9$ T in the corresponding expression for the damping factor of oscillations. we obtain $\beta = 5 \cdot 10^{-10}$ s^{-1}.

Thus, we can conclude that the contribution of this mechanism to the damping factor is insignificant.

Table 10.7 lists the physical parameters of the samples at 25°C.

The viscosity η of non-magnetized MF samples was measured the Ostwald viscometer with the capillary tube having a diameter 1.31 mm, density was measured in the usual manner using a 10 ml glass pycnometer.

Table 10.7

Fluid	$\eta \cdot 10^{-3}$ Pa·s	ρ, kg/m^3	φ, %	M_s, kA/m	χ	d_{min}, nm	d_{max}, nm
MF-1	1.9	1013	5	20	1.1	10.5	12.9
MF-2	8.9	1522	16.6	55	8.1	14.4	18

Saturation magnetization M_s and initial magnetic susceptibility χ were obtained from the measurements of the magnetization curve.

The maximum and minimum diameters of the dispersed phase particles (excluding the stabilization shell) d_{max}, d_{min} were obtained by the granulometric method [35].

Measuring the vibration damping system with magnetic fluid inertial-viscous element formed on frequency of 36 *Hz* at a temperature of 25°C. The depth of penetration of a viscous waves to the oscillating system with samples MF-1 and MF-2 is respectively σ''_{MF-1} = 0.22 mm. σ''_{MF-2} = 0.20 *mm*.

10.7. Results and analysis

Figure 10.5 shows the dependence β(*H*) for MF-1, MF-2. Figure 10.6

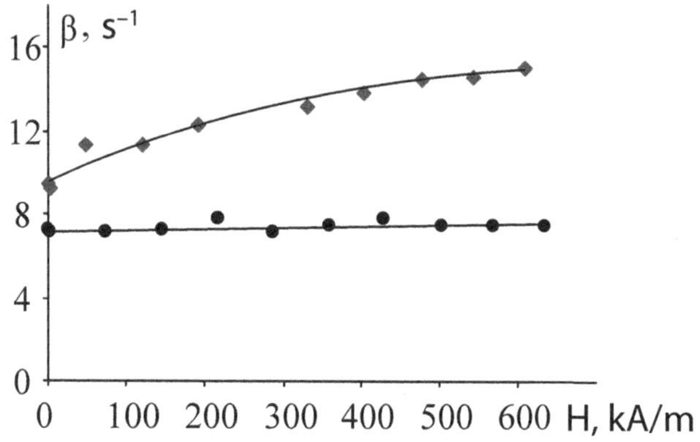

Fig. 10.5. The dependence of the damping coefficient on the strength of the magnetic field: ● – MF-1. ♦ – MF-2.

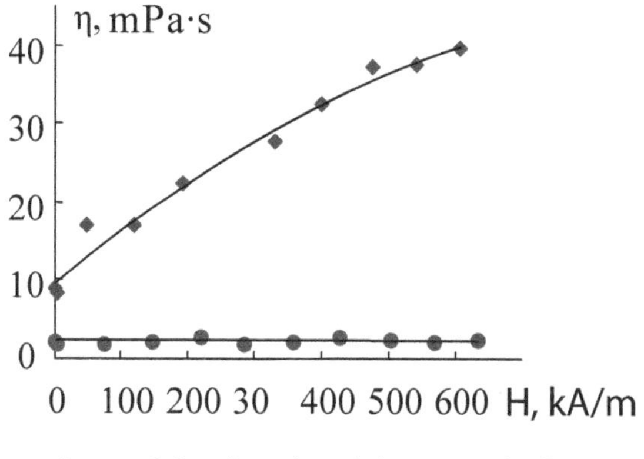

Fig.10.6. The dependence of the viscosity of the magnetic fluid on the strength of the magnetic field. ● – MF-1; ♦ – MF-2.

Table 10.8

Fluid	η, 10^3 Pa·s	ρ, kg/m^3	φ, %	M_s, kA/m	χ	d_{min}, nm	d_{max}, nm
MF-1	8.9	1522	16.6	55	8.1	14.4	18
MF-3	6.1	1378	13.3	55	8.1	14.4	18

shows the dependence $\eta(H)$ for these fluids.

The results of the measurement of the field dependence of the viscosity in the sample with the volume concentration of the solid phase of 5% (MF-1) are consistent with the concept of 'rotational viscosity' of the dilute dispersion of the magnetic nanoparticles. The magnetorheological properties of this sample are characteristic of the 'normal' magnetic colloid and its monoparticle structure is very stable in the investigated range of magnetic field strength.

Noteworthy is the abnormally large growth of the damping coefficient and the viscosity of the MF-2 sample in the magnetic field. As will be shown below, the data obtained for the MF sample with a concentration of 16.6% can be explained by the presence therein of chain aggregates and non-magnetic ellipsoidal particles of the micron size.

Equation (10.27) and the resultant values of β_{ex} are used to calculate the viscosity of the MF samples for different values of H in the investigated magnetic field strength range. Figure 10.5 shows the dependence $\eta(H)$ for MF-1 and MF-2.

According to Fig. 10.6, the maximum viscosity of the MF-1 sample, corresponding to the magnetic saturation of the colloid, is η = 39 mPa·s.

Substituting the value of the parameters η and φ in Table 10.8 in Einstein's formula, we obtain $\eta_0 = 6.3 \cdot 10^{-3}$ Pa·s.

According to the expression (10.29) $L_\perp(S) = 31.3$. The theoretical curve in Fig. 10.3 was used to determine the effective value of $S = 8.7$.

Consequently, if the system is completely aggregated, the chain aggregates contain on average 8–9 magnetic nanoparticles.

However, this estimate was subsequently adjusted as the system was found to contain non-magnetic particles of micron sizes whose presence in the MF also increases viscosity.

The presence in the initial sample of non-magnetic particles of micron sizes was found during further centrifugation of the MF-1 sample TsL 1/3 equipment for one hour. The physical properties were then determined in the decanted upper layer (MF-3 in Table 10.7).

In centrifugal sedimentation, solid particle movement occurs under the action of centrifugal force $\pi d^3 \Delta\rho\omega^2 r/6$ (d – diameter of particles, $\Delta\rho$ – the difference in density between solid and liquid phases, r – distance from the particle to the axis of rotation of the rotor) and the forces of resistance of the fluid $3\pi d\eta_0 V$. The ratio of these forces determines the settling speed V. If the path traversed by the particle in time t is l, then its minimum diameter:

$$d = \sqrt{\frac{18\eta_0 l}{t\Delta\rho\omega^2 r}}. \tag{10.35}$$

Under the experimental conditions the centrifuge rotor speed n = 3000 rpm, i.e. $\omega = 2\pi n$ = 314 rad/s, the length of the MF column in the centrifuge cell l = 4.5 cm, r = 12.3 cm, $\eta_0 = 6.3\cdot10^{-3}$ Pa·s. the typical value $\Delta\rho \sim$ 4420 kg/m^3. Substituting these values into the formula (10.35), we have: $d \sim$ 163 nm. Before centrifuging the MF sample was for a long time (one month) in the stationary state. In that time particles with a diameter of 200 nm settled under the effect of the gravitational force. From these data, the average diameter of the particles removed from the solution was 180 nm. Comparing this value with the data in Table 10.7, we conclude that it is more than an order of magnitude higher than the average particle diameter obtained by magnetic granulometry from the curves in Fig. 10.7.

Using the method described above, the experimental data were used to determine the dependences $\eta(H)$ and $\beta(H)$ for the MF-3. The maximum value of viscosity corresponding to the magnetic saturation colloid is η = 13.9 mPa·s.

Using Einstein' formula, the relation (10.29) and the theoretical curve $L_\perp(S)$ in Fig. 10.3, we get S = 5.5. According to updated data,

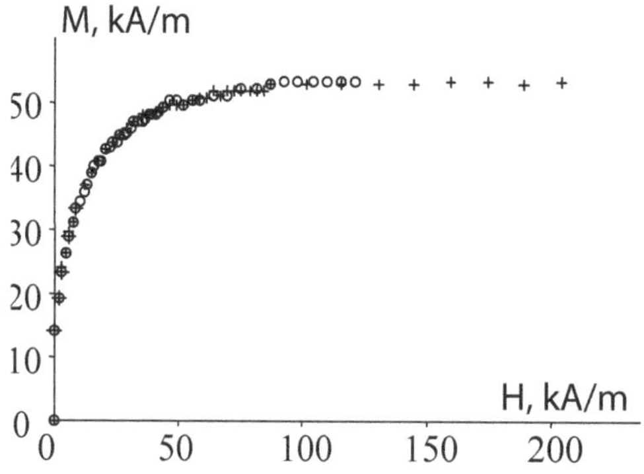

Fig. 10.7. Magnetization curves (+) – MF-1. (∘) – MF-3.

the chain aggregate contains on average 5–6 magnetic particles and the value for MF-1, obtained in the first case, is inflated.

The concentration of the magnetic phase after centrifugation is not changed (as shown by the coincidence of the magnetization curves before and after centrifugation (Fig. 10.7)) and the density and the viscosity decrease. It can be assume that centrifugation resulted in the separation of the non-magnetic phase and that the observed overestimation is due to the presence of non-magnetic particles in the original MF sample. According to experimental data $\Delta\eta_{max}$ = 25.1 MPa·s.

The concentration of the non-magnetic phase is determined by the formula:

$$\varphi' = \varphi_1 - \varphi_2$$

where φ_1 is the concentration of the solid phase of MF-1; φ_2 is the concentration of the solid phase of MF-2 calculated from the density of the samples. In this case $\varphi' = 0.033$.

On the assumption that the viscosity of the dispersed medium is $\eta_0 = 6.3\cdot10^{-3}$ Pa·s, equation (10.32) changed to a transcendental equation of the form:

$$242 ln\lambda = \lambda^2$$

The solution of this equation allows us to set $\lambda = 28.5$. Equating the volume of an ellipsoid of revolution $4\pi ab^2/3$ to the volume of a spherical particle $\pi d^3/6$, we obtain an expression for the semi-minor axis:

$$d = \frac{b}{2}\sqrt[3]{\frac{1}{\lambda}}. \tag{10.36}$$

Substitution of the values of d and λ, calculated earlier, into (10.36) gives $b = 30$ nm. For the semi-major axis: $a = b\lambda = 850$ nm.

For the resultant numerical values of the linear parameters of the magnetic nanoaggregates 'auxiliary' measurement of the size of the magnetic core were taken by the 'non-acoustic' method – by magnetic granulometry. In the previous section we discussed the method of acoustic granulometry based on the use of the acoustomagnetic effect in a magnetic fluid. It is thus possible to obtain the whole set of geometric parameters of the nanoaggregates by acoustometry.

10.8. The expansion of the experimental base vibrorheology of MF on the basis of the magnetic levitation effect

The method of experimental research of the vibrorheology of MFs, described in Section 10.6, can not be preferred in some cases – for example, during a large series of measurements. In the process of obtaining data the cleaning the U-tube and its subsequent filling with the new sample take a long time. In this regard, it is of interest to consider the measurement technique, proposed in [317], to investigate the oscillatory system with an air cavity in the MF held by magnetic levitation.

The air bubble, placed in the MF which is in a magnetic field with the field strength gradient directed vertically upward, is affected by magnetic levitation forces that make the bubble 'heavier', prevent its rising and cause the buble to 'hang' in the liquid medium.

The total force that determines the condition of motion of the air bubble in the magnetized magnetic fluid in the approximation of the 'weakly magnetic' medium can be represented as follows:

$$\vec{F} = -\rho V \vec{g} - \mu_0 M V \nabla H, \tag{10.39}$$

where V is the volume of the air bubble, M and ρ are the magnetization and the density of the magnetic fluid, H is the magnetic field strength. μ_0 is the magnetic constant.

From (10.39) we obtain the levitation (hanging) condition of the air bubble

$$\mu_0 M \nabla H = -\rho \vec{g}.$$

If we ignore the forces of viscous friction in the thin wall layer of the fluid under the condition that

$$\mu_0 M |\nabla H| > |\rho \vec{g}|$$

the magnetic levitation forces cause the air bubble to move down.

Consider the elastic magnetic properties of the MF in the tube with a bottom–air cavity system. Hanging of the cavity and its transport along the column of the magnetic fluid are ensured by magnetic levitation forces. The levitation effect is generated by a ring-shaped magnet magnetized along the axis, inserted on the tube and moving along it. The air cavity is formed by capturing a portion of air from the exposed surface of the magnetic fluid by an inhomogeneous magnetic field as the ring-shaped magnet comes closer from the top to the free surface of the magnetic fluid.

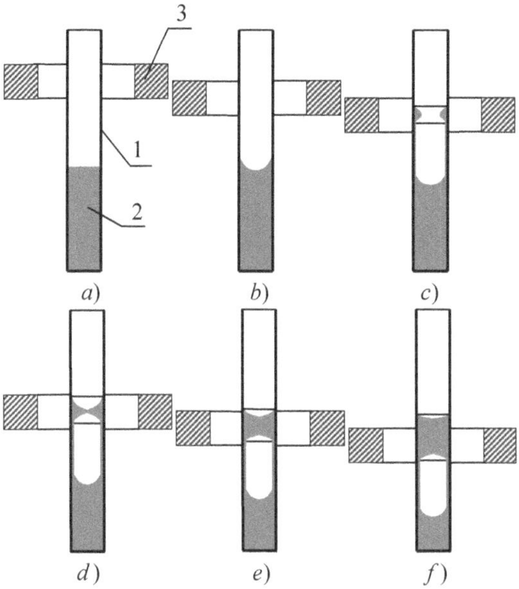

Fig. 10.8. Process of capture of the air bubble by the magnetic fluid.

The 'capture' of the air cavity by the magnetic field is investigated visually recording onto a digital video camera. Figure 10.8 shows the main stages of the process.

The experiments were carried out using a rigidly fixed glass tube 1 partially filled with the MF 2. The annular magnet 3, secured on the kinematic section of the cathetometer, moves down in the tube and the axes of the annular magnet and the tube are aligned with each other.

At the beginning of the experiment (Fig. 10.8 *a*) the free surface of the MF has a flat horizontal shape. With the approach of the magnet to the free surface of the fluid its surface adopts a concave shape (Fig. 10.8 *b*) and then as it approaches the active zone of the magnetic field the ponderomotive forces, holding the liquid to the tube wall, also pull it into the region of the maximum field resulting in the formation of a ring in the symmetry plane of the magnet (Fig. 10.8 *c*).

On further lowering the magnet the ponderomotive forces are significantly higher than the force of gravity, so the MF ring thickens and then covers the section of the tube (Fig. 10.8 *d*). An insulated gas cavity, overlapping the section of the tube, formed under the bridge. The thickness of the bridge then increases due to the fluid flowing from the bottom (Fig. 10.8 *e* and *f*) and the cavity formed under the

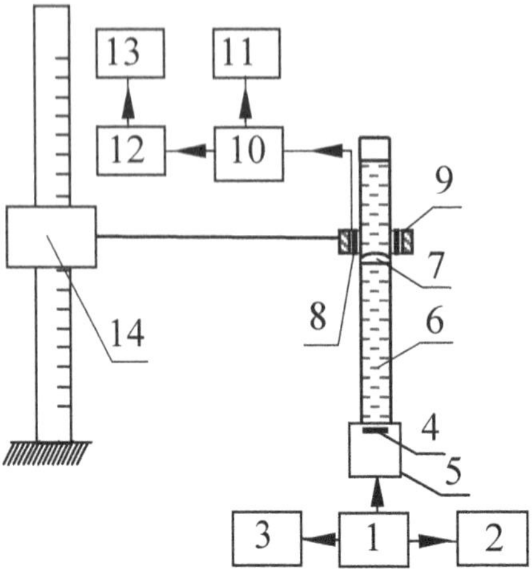

Fig. 10.9. Block diagram of the experimental setup number 1.

action of magnetic levitation is pushed down. It should be noted when the magnet is arrested the flow of the fluid on the tube walls is gradually terminated. Then the inner surface of the tube between the upper and lower columns of the MF becomes completely free from the fluid. Thus, in the steady state the upper and lower MF columns are by the air cavity.

The properties of the system that are of interest to us include the reflective properties of the gas cavity and the vibrational characteristics of the system in which the inertial element is the MF column above the hanging gas cavity, and the elasticity is represented by the elasticity of the air cavity.

The block diagram of the experimental equipment No. 1, intended to identify the opportunities for the use of the transported air cavity as a moving sound wave reflector is shown in Fig. 10.9. The signal from the generator of sound waves 1 is applied in parallel to the frequency meter 2, the voltmeter 3 and the piezoelectric plate 4, pressed to the lid of the acoustic cell 5, the construction of which is described in section 9.5.2. Passing through the column of MF-6 located under the air cavity 7, the sound wave is reflected from its bottom surface. As a result of elastic oscillations of the lower surface of the air cavity the coil 8, mounted in an annular permanent magnet 9, generates variable EMF. After amplification in the selective amplifier 10, the variable EMF is applied in parallel to the oscilloscope 11 and the analog digital converter 12 connected to the computer 13. The magnet with the coil attached to the kinematic section of the cathetometer 14 is moved slowly along the tube axis with MF and at the same time the forces of magnetic levitation move

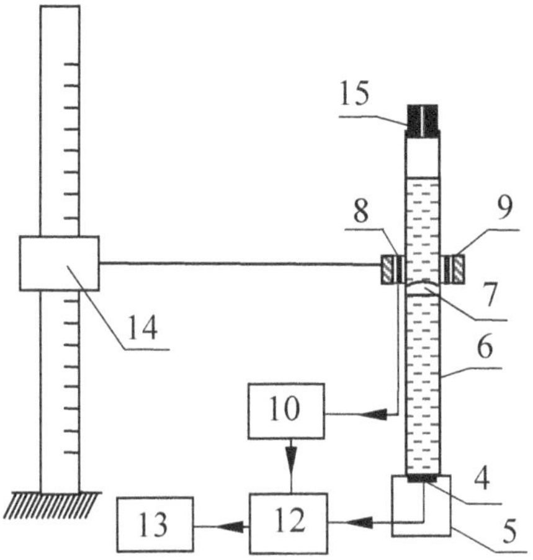

Fig. 10.10 Block diagram of the experimental setup node number 2.

the air cavity 7. Movement of the magnet is fixed with the accuracy of 0.01 mm, and the signal from the coil is taken every 0.5 mm.

The block diagram of the experimental setup No. 2, intended for measuring the parameters of the oscillatory system where the inertial member is the magnetic fluid column situated above the gas cavity, is shown in Fig. 10.10. To avoid repetition with the description of Fig. 10.9, in the description of the block diagram of this installation we list only a few elements used in the task. Piston 15, covering the upper end of the tube, is used to excite oscillations of the MF column in the equilibrium state. In the piston there is a through hole which allows to insert the piston without changing the pressure in the tube. Before pulling our the piston the hole is covered with a finger. The signal, received by the coil 8, is supplied to the broadband amplifier 10, and then to the analog-to-digital converter 12 (ADC) and the computer 13. The ADC also receives a signal from the piezoelectric element 4. The reception and initial processing of the signals from piezoelectric and inductive sensor are carried out using a program developed in the NI LabView environment.

The filling frequency of RF pulses received from the inductor and the piezoelement is measured by determining the duration of 10–15 complete oscillations with subsequent calculation of the period of oscillations and its inverse value.

A ring-shaped magnet, magnetized axially, has the following dimensions: inner diameter 25 mm, outer diameter 50 mm, thickness 5 mm. The axial component of the strength of the magnetic field at the centre of the magnet 91 kA/m.

The volume of the 'trapped' air cavity V_g is defined by the incremental height of the MF column Δh in a tube with an inner diameter d: $V_g = S{\cdot}\Delta h$. When d = 13.5 mm, $S = 1.4{\cdot}10^{-4}$ m^2, $\Delta h = 1{\cdot}10^{-2}$ m. then $V_g = 1.4{\cdot}10^{-6}$ m^3. Before measuring Δh the magnet moves down to a level at which its magnetic field does not affect the curvature of the free liquid surface.

The studies were conducted on a MF sample described in section 9.5.3 (Table 9.2). All measurements were made at temperature of 31 ± 0.2°C.

1. Experience in identifying the possible use of the transported air cavity as a mobile reflector of the sound wave.

Figure 10.11 shows the dependence of the EM induced in the inductor, which is expressed in relative units, on the position of the ring magnet. There are alternating peaks of voltage with the change of the distance by approximately the same value. The pattern is similar to that which can be obtained by the 'normal' interferometric method with insufficiently accurate alignment of the rigid reflector. The unequal peaks indicates the instability of the lower boundary of the air cavity when moving it.

In this experiment, the frequency of sound waves ν = 18.5 kHz, the average distance between the peaks (the length of the standing wave) is ≈ $2.3{\cdot}10^{-2}$ m, which gives a speed of sound of $c \approx 850$ m/s. The significantly reduced value of c compared with the speed in 'unlimited' fluids (according to Fig. 3.2 for the MF with similar physical parameters c = 1160 m/s) is due to yielding of the walls of the tube (section 8.1).

The use of a non-planar reflector with an easily deformable surface increases the measurement error of c is comparison with the magnetic fluid interferometer that uses the MF surface normal to the

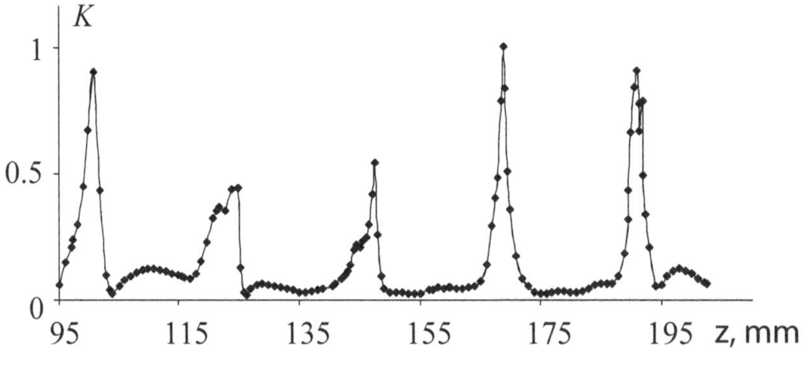

Fig. 10.11. The signal from the coil at a sound wave frequency of 18.5 kHz.

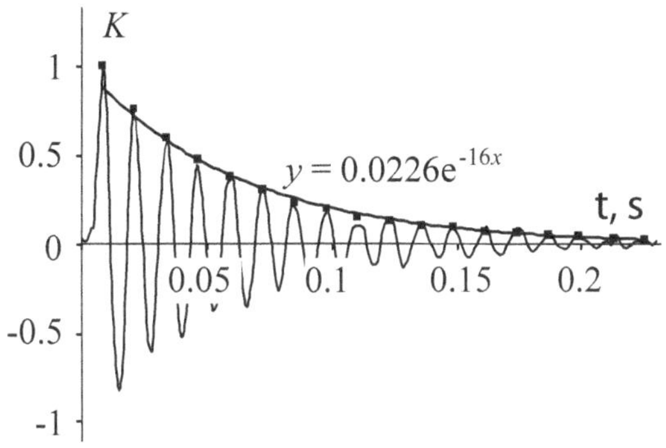

Fig. 10.12. Oscillograms of damped oscillations of the MF column.

wave vector [90]. However, such simple device can be recommended for inclusion in the laboratory training in physics and as a lecture demonstration of the effect of magnetic levitation.

2. Experience with the investigation of the oscillatory parameters of the system with an inertial element in the form of the MF column hanging over the air cavity.

If the MF column, situated under the air cavity, represents a sound waveguide, the MF column in the tube located above the air cavity is an inertial member of the oscillatory system.

The curves of the amplitude–time dependence, obtained from the inductor and the piezoelectric element, are identical to each other. The time dependence of the variable EMF, presented in relative units, $K(t)$ is shown in Fig. 10.12. In this case the signal is taken from the piezoelectric sensor, the MF column height $h = 10.5$ cm. The envelope of this dependence, approximated by the exponential trend line, is also shown.

The role of elasticity in this oscillatory system is performed by the total elasticity of the air cavity with the coefficient of elasticity k_g and the ponderomotive elasticity due to the interaction of the magnetic field with the magnetic fluid in the bottom of the fluid column, with the elasticity coefficient k_p.

The coefficient of elasticity of the system k is determined by the sum (7.1). The expression for k_g for the adiabatic process has the form (7.2). Formula (7.4) was derived for the coefficient of elasticity k_σ of the surface tension of the liquid film overlapping the section of the tube. It can be shown that in the context of the problem being solved the inequality $k_\sigma/k_g \cong 0.05$.

To calculate the parameter k_p we must have very detailed information about the topography of the magnetic field of the ring magnet used. However, it is known that in such systems the ponderomotive elasticity becomes commensurable with the gas elasticity only for sufficiently large values of V_g [29, 30]. Without taking into account the ponderomotive elasticity, the frequency of natural non-damping oscillations of the system v_t is calculated using the formula (10.35).

The results of theoretical calculations of ν_t and experimental data for ν_e differ only slightly in the oscillation frequency (~10%). The observed some excess of v_e over ν_t might be less if the ponderomotive elasticity of the system is taken into account.

The description of the oscillatory system and the results suggest that the system is capable of controlling the frequency both by the mass of the MF column and by the volume of the gas cavity. With decreasing volume of the air cavity the oscillatory system parameters k_g and v_e will have higher and higher numerical values.

Elastic energy dissipation mechanisms in an oscillatory system with the inertial element in the form of the MF column are described in paragraph 10.5.

The coefficient of damping to reciprocating fluid flow through the tube β_r is determined using formula (10.10). Assuming that d = 13.5 mm, $\eta = 4.97 \cdot 10^{-3}$ kg/m·s, $\nu = 59$ Hz, we obtain $\beta_r = 3.9$ s^{-1}.

For the case under consideration, the dissipation of elastic energy due to the interphase heat transfer of the gas cavity with the surrounding fluid is closer the model of the pulsating gas bubble in water as described in [228]. The article gives a formula for the calculation of the dimensionless damping coefficient δ. For approximate evaluation we use the same graphical representation of the dependence of log (δ) on log (ω) presented in [228]. Thus, from the graph (Figure, curve 2) for $\omega = 370$ s^{-1} we have $\delta \approx 0.05$ which at $\nu = 59$ Hz gives for the damping coefficient $\beta_w = \delta \cdot \nu \approx 3$ s^{-1}.

The effectiveness of the mechanism of emission of the energy of elastic waves to structural elements and the environment is associated with the ratio of the acoustic impedances of the tube walls, auxiliary structural elements (holder, stand) and air. It also depends on the area of their contact surface. Theoretical estimation of the contribution of this mechanism to the damping of the oscillations is problematic. One can only suggest that the process of moving the air cavity along the MF column under the effect of levitation forces does not significantly affect the efficiency of the mechanism.

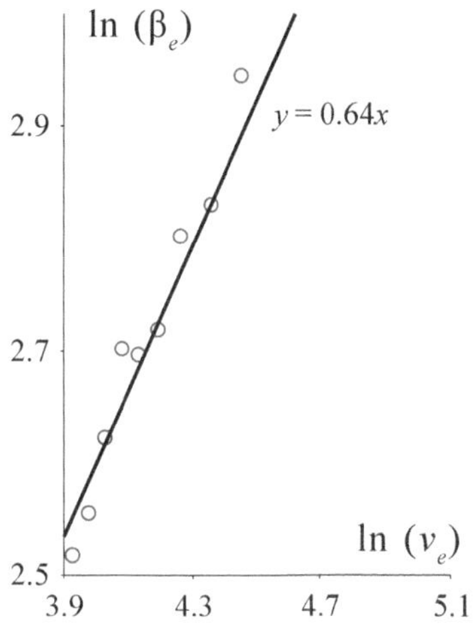

Fig. 10.13. Dependence β_e on ν.

The experimental value of the damping coefficient β_e is obtained on the basis of the exponential approximation of the envelope of the dependence of the amplitude of free damped oscillations on time, shown in Fig. 10.12 by the trend line: $\beta_e = 16\ s^{-1}$.

In view of the above, the contribution of the viscous flow of near-wall layers of MF to β_e is ≈25%. Estimation for the mechanism of interphase heat transfer results in a value of ≈ 20% of β_e. Apparently, most of the energy losses in the vibrating system are connected to the emission mechanism of elastic vibrations.

In Fig. 10.13 the results of measurement of the damping coefficient of oscillations in the frequency range 46–86 Hz are presented as the dependence of ln (β_e) on ln (v_e).

These data suggest that the frequency dependence of the damping coefficient is given by $\beta(\nu) \sim \nu^{0.64}$. However, the mechanism of viscous flow of the near-wall layers predicts the increase of the damping coefficient with frequency in the proportion $\beta(\nu) \sim \nu^{0.5}$. It is appropriate to note that the numerical value of the exponent of the power law is also affected by two other mechanisms of energy dissipation. Possibly, the damping coefficient, associated with the third mechanism, is characterized by the dependence $\beta(\nu) \sim \nu^{\varsigma}$, wherein $\varsigma > 0.5$. Thus, the well-known expression for the transmission coefficient in intensity of the plane sound wave through a planar interface between two media includes only the parameter ε, representing the ratio of impedances of such media and does not include frequency ν [72]: $t_i = 4\varepsilon^2/(1+\varepsilon)^2$. A similar dependence on ε is typical of parameter δ, so the mechanism for the outflow of energy from the system we have $\varsigma = 1$.

In measurements of the frequency dependence of the damping coefficient of oscillations in a system in which the lower surface of the MF column is stabilized with an inhomogeneous magnetic field (section 10.1) the result was almost the same, $\varsigma \approx 0.6$. However, the implementation of this kind of stabilization of the MF–air interface is possible only with the use of concentrated MF samples and has height restrictions for the MF column.

The procedure used to study the oscillatory system with the air cavity in the MF, sustained by the magnetic levitation, is characterized by ease of implementation and the resumption of the measurement process with MF samples of different viscosities. The use of this technique extends the experimental base of MF vibrorheology. The gradual near-wall fluid during capture and movement of the cavity up or down in the tube in various high-speed modes of motion of the magnet opens the possibility of magnetorheological testing of MF samples. The measured parameter may be the volume of the 'trapped' cavity, the critical value of the speed of movement of the magnet for retaining its integrity which is critical for the preservation of its integrity and the time required to obtain the required strength of the received acoustic signal.

Let's notice that the message on stabilization of the bottom surface of a column of MF in a tube by means of an inhomogeneous magnetic field is given in work [59]. However, there is no description in it of mechanisms of formation and movement of an air cavity, properties of the MF column–air cavity oscillatory system, specifics of its inertial and elastic elements are not investigated.

On approach of a magnet to the bottom of a tube, there is a contact of an air cavity with the bottom and appearance of excessive pressure in it. At achievement of the critical pressure difference the bubble comes out of the cavity. The process is accompanied by the electromagnetic radiation, in which the spectrum is present ν – the frequency of fluctuations of a bubble [325, 326]. On the other hand, the frequency of fluctuations of an air bubble of radius of R_0 in a liquid can be received from expression [298]:

$$\nu = \left(2\pi R_0\right)^{-1} \sqrt{3\gamma\left(P_0 + 2\sigma / R_0\right) / \rho}$$

where P_0 – hydrostatic pressure, $\gamma = C_p/C_v$ – ratio of heat capacities of gas, σ and ρ – the coefficient of surface tension and density of a sample of MF.

At σ = 0.028 N/m and $R_0 \approx 10^{-3}$ m, we obtain $2\sigma/R_0$ = 56 Pa. As far as P_0=10^5 Pa, then $2\sigma/R_0 \ll P_0$, that allows to neglect the second member in brackets and to note down the formula for bubble radius: $R_0 = (2\pi\nu)^{-1}\sqrt{3\gamma P_0/\rho}$. In the example under consideration ρ = 1320 kg/m^3, γ = 1.4, experimental value ν = 2.5 kHz, accordingly: R_0 = 1.14 mm. Calculating the mass of gas in a bubble by formula: Δm = 4 $\rho_g \pi R_0^3/3$ and assuming ρ_g = 1.29 kg/m^3, we obtain $\Delta m \approx 8\cdot10^{-9}$ kg.

The results may be of importance for designing a new technique for metering small gas shots to a reactor.

Conclusions

Comments for the first and second chapter. In the propagation of a sound wave in a magnetised magnetic fluid as a result of oscillations of the concentration of the ferroparticles perturbations take place in the magnetisation of the fluid and strength of the magnetic field with the opposite signs. In the absence of compensation between them, the acoustomagnetic effect (AME) starts to operate – induction of the EMF in the conducting circuit. The concentration mechanism of the AME predetermines the nature of its field dependence, the dependence on the amplitude of sound oscillations, and the phase of the wave, the displacement of the maximum of the AME amplitude from the central section of the magnetic fluid cylinder to its base with the increase of the frequency of oscillations.

Comments for the third chapter. The temperature, baric and concentration dependences of the speed of ultrasound in the non-magnetised magnetic fluid (NMMF) are satisfactorily described by the additive model.

Measurements of the speed of ultrasound in magnetisation of the magnetic fluid (MF), exceeding 2 m/s, were found only in individual specimens of the colloids of the third type in which the aggregates at a sufficiently high density have the capacity for the change of the dimensions and shape of the surface in the magnetic field. The absolute values of the increase of the speed do not exceed ~15 m/s, the fluid becomes anisotropic with respect to the given parameter, and the axis of anisotropy is represented by the direction of the magnetic fields. The experimental data for the field dependence of the speed of ultrasound in the magnetic fluid are satisfactorily explained by the dynamic theory.

Comments for the fourth chapter. The results of determination of the additional absorption of ultrasound in the NMMF on the basis of kerosene and magnetite, obtained in experiments and calculated using the equations of the classic theory of microheterogeneous media, are in satisfactory agreement so that the mechanisms of additional absorption – internal heat exchange and the relative motion of the particles of the disperse phase and the fluid – carrier are regarded as the main mechanisms for the magnetic colloids of the given type. At the same time, the application of the conclusions of the classic acoustics of microheterogeneous media for the systems with the dispersed nanoparticles require additional justification because of several reasons. In particular, the theory does not take into account the presence of thermal chaos in the movement of the particles of the dispersed phase, and also the fractal nature of the surface of the nanoparticles on the level of the interatomic distances [237, 253, 270].

Experiments were carried out to determine a large difference in the fabled dependence and the dependence on the angle between H and k of the amplitude of the ultrasound pulse passed through the magnetic fluids with different degrees of aggregate stability, – from almost completely independent to a considerable, non-monotonic and, usually, non-repeated dependence.

The physical nature of this diversity is associated with the special features of the occurrence of structural rearrangement in each specific sample of the magnetic fluid, reflected in the formation of solid or microdroplet aggregates with different packing density and stability, different rates of the rearrangement processes, the possibility of formation of the spatial ordered structure from the filament-like aggregates and the redistribution of the particles of the dispersed phase in the volume as a result of baric and magnetic diffusion. These special features of structure formation in the actual magnetic colloids are capable of causing specific acoustic effects: scattering, diffraction, refraction of the sound beams, dispersion of the speed of propagation and additional absorption of ultrasound.

The absence of the dependence of the amplitude of the pulses, passed through the magnetic fluid, on the strength and gradient of the magnetic field, determined within the measurement error range, is observed in the individual specimens until their structure is similar to modelling considerations of the 'ideal magnetic fluid' in which there is no merger of the ferroparticles into aggregates.

The dynamic theory of the additional absorption of ultrasound in the magnetised magnetic fluid (MMF), taking into account the mechanism of the relative motion of the nanoaggregates in the fluid–carrier, is physically most efficiently substantiated. However, the given theoretical model is, firstly, calculated for low-concentration magnetic fluids and, secondly, does not take into account the processes of internal heat exchange, the scattering of ultrasound waves on the aggregates, with the role of this aggregates changing in the magnetisation process.

Comments for the fifth and sixth chapter. The effect of the ponderomotive mechanism of the magnetoacoustic effect (MAE) – the mechanism of electromagnetic excitation of elastic oscillations in the magnetic fluid has been confirmed by experiments in the kilohertz frequency range.

The investigations of the MAE using a magnetic fluid in the megahertz frequency range determine the following relationships: the curve of the dependence of the relative amplitude of the excited elastic oscillations on the strength of the magnetising field is initially steeper than the magnetisation curve of the given fluid, reaches its maximum value in the magnetic field of ~20 kA/m and subsequently, up to a technical saturation at ~500 kA/m, is characterised by slow fall by ~10%; the dependence of the amplitude of the generated oscillations on the amplitude of the alternating magnetic field is non-linear. Analysis of the resultant relationships indicates the operation of the mechanism of electromagnetic excitation of the elastic oscillations in the magnetised magnetic fluid of the non-ponderomotive nature, which is realised in a homogeneous alternating magnetic field, and its effect influences the entire range of magnetisation of the magnetic colloid, excluding the range of technical saturation. None of the currently available physical models of the mechanisms of electromagnetic excitation of ultrasound in the magnetic fluid (magnetic calorific effect, bulk and linear magnetostriction, dipole–dipole mechanism, ponderomotive mechanism) can be regarded as completed and adequate with respect to the experimental data obtained for the excitation of elastic oscillations in the megahertz frequency range.

The physical mechanisms of the MAE have been discussed up to now using the data for the physical properties of massive ferromagnetics. However, it is well-known that the mechanical, magnetic, dissipative and thermophysical properties of the nanoparticles greatly differ from the properties of macroscopic crystals [238, 253, 270]. Simply, the special features of the MAE in the megahertz frequency range are associated especially with the anomalous magnetoelastic (magnetostriction) characteristics of the nanoparticles.

Comments for the seventh chapter. The investigations of the kinetic and strength properties of the magnetic fluid membrane (MFM) are important because MFM represents firstly the model of magnetic fluid compacting and, secondly, is used as an independent device.

The proposed optical method of studying the kinetic-strength properties of the MFM may be useful for testing the specimens of the magnetic fluids for magnetic fluid compacting, used in the conditions of large pressure drops.

The cavitation model of formation and closing of the orifice in the magnetic fluid–bridge makes it possible to determine the nature of driving forces which cause these processes, and transfer to the level of quantitative estimates.

The results of determination of the critical pressure drop, carried out in different speed conditions, slightly differ from each other also in the conclusions of the model theory wished as not take into account the rheological properties of the magnetic fluid–bridge.

Comments for the eighth chapter. The AME was used in the experimental determination of the spectrum of the oscillation mode is propagating with different phases speeds in the fluid–cylindrical shell system. The data obtained for the dispersion of the speed of sound in the fluid–shell system using shells with a wide range of the geometrical and elasticity parameters have been used to show the efficiency of acoustomagnetic spectroscopy of the modes of oscillations in the fluid–shell system. The proposed experimental procedure may be used for taking measurements in systems of different type, including the systems in which it is difficult to obtain theoretical results.

The comments for the ninth chapter. The new method of investigating the geometrical and magnetic parameters of the nanoparticles, dispersed in the fluid–carrier, acoustic granulometry, has been theoretically and experimentally substantiated.

Investigations were carried out into the process of interaction of the elastic and thermal fields, and also the magnetic and dynamic demagnetising field in the acoustomagnetic effect on the magnetic fluid. The construction of the 'difference' curve is an additional confirmation of the concentration model of the physical mechanism of AME. In other AME mechanism, investigated together with the concentration mechanism, i.e., the 'solid state mechanism', has not as yet been confirmed by the experimental data.

The method has been proposed for the evaluation of the physical parameters of the finest magnetic nanoparticles of the disperse phase of the colloid on the basis of the detailed application of the results of magnetometry and acoustometry.

Comments for the tenth chapter. The anomalously strong dependence of the damping factor of the oscillations in the oscillatory system with the magnetic fluid inertia – viscous element on the strength of the magnetic field may be caused by the presence in the disperse system of aggregates of magnetic nanoparticles and microparticles of the non-magnetic phase. Analysis of the resultant experimental data on the basis of the model of rotational viscosity is used for the evaluation of the geometrical parameters of the magnetic nanoaggregates and non-magnetic microparticles, dispersed in the samples of the actual magnetic colloid.

Appendix

Additional information on the composition and properties of magnetic fluids

Magnetisation of the magnetic fluids

The magnetisation of actual magnetic fluids takes place in accordance with the law which differs from the Langevin law because there is the size distribution of the ferroparticles and the interaction between the particles. For the known function $f(d)$ of the size distribution of the particles and in the absence of interparticle interaction, the magnetisation is described by the equation

$$M(H) = \frac{\pi}{6} M_{SO} \int_0^{\infty} L(\xi) \cdot d^3 f(d) d(d)$$

where d is the particle diameter; M_{SO} is the magnetisation of the particle material.

For approximation of the magnetisation curve of the actual magnetic fluid, the study [220] recommends the equation derived by A.N. Vislovich

$$M = M_S H / (H_T + H)$$

where H_T is the strength of the field at which the magnetisation is $M(H_T) = 0.5\ M_S$.

The limiting concentration of the dispersed phase is close to the concentration in the case of close packing of the spheres which form the magnetic particles together with the adsorption layer. In the currently available high concentration fluids, the magnetite concentration does not exceed $\varphi = 0.35$ [161].

Table A1. The values of maximum magnetisation M_{sm} of magnetite magnetic fluids, produced using various dispersion bases [161] (SAS – surfactant)

Base	Kerosene	Transformer oil	Condenser oil	Medical vaseline oil	Water	Organic silicon fluid
SAS	Oleic acid	Oleic acid	Oleic acid	Oleic acid	Oleic acid	Oleic acid
M_{sm}, kA/m	95–100	85–95	70–75	50–60	40–45	30–40

The physical properties of magnetite

The disperse phase in the magnetic fluids is in most cases magnetite Fe_3O_4 which is associated, firstly, with the simple procedure and technology of producing the magnetite of the required dispersion and, secondly, with the fact that in the case of the colloidal solutions of magnetite in water, hydrocarbons, complex esters, organic silicon and fluorine in organic compounds, surfactants are selected for stabilisation of each system. Magnetite is the iron ferrite with the crystal structure inverse in relation to that of the spinel [240].

Table A2. The main physical parameters of single crystal magnetite at $T = 20°C$

a_0, nm	ρ, kA/m^3	T_k, °C	M_{SO}, kA/m	μ_{00}	ρ_e, ohm·м	κ, W/(m·K)	K_a, J/m^3
0.839	5240	586	477.7	70	$5\cdot10^{-5}$	6	$1.1\cdot10^4$

Comments:

1. Here a_0 is the spacing of the crystal lattice, T_k is the Curie point, M_{SO} is the saturation magnetisation, μ_{00} is the initial magnetic permittivity, ρ_e is the specific electrical resistivity, K_a is the crystallographic anisotropy constant, κ is the heat conductivity coefficient.

2. The magnetic moment of the ferroparticle is expressed through the saturation magnetisation of the ferromagnetic

$$m_* = M_{SO}\cdot V_f,$$

where V_f is the volume of the magnetic part of the particle.

If the 'magnetic core' of the magnetite particle has the form of a sphere with the diameter d, then $m_* = M_{SO}d^3/6 = 2.5\cdot10^5\ d^3$. At d = 10 nm, $m_* = 2.5\cdot10^{-19}$ A·m².

Table A3. Magnetostriction constant of ferrite with the spinel structure (20°C)

Ferrite	$\lambda_{100}\cdot10^6$	$\lambda_{111}\cdot10^6$
Fe_3O_4	–20	+78
$Co_{0.8}\ Fe_{0.2}\ Fe_2O_4$	–590	+120
$Ni_{0.8}\ Fe_{0.2}\ Fe_2O_4$	–36	–4

Table A4. Molar heat capacity of magnetite $\hat{C}_p$ at a pressure of 760 mm Hg at adifferent temperatures [240]

T, K	80	150	250	400	600	1000	1500
$\hat{C}_p, \frac{\text{J}}{\text{mol}\cdot\text{K}}$	37.45	92.76	131.8	172.2	212.5	200.8	200.8

Table A5. The Curie temperature of metals

Substance	T_k, °C		T_k, °C
Gadolinium	20	Magnetite	586
30% Permalloy	70	Electrolytic iron	769
Geisler alloy	200	Iron remelted in hydrogen	774
Nickel	358		
78% Permalloy	550	Cobalt	1140

Surfactants

The most efficient surfactant for the colloidal solutions of magnetite in hydrocarbons is the oleic acid $CH_3(CH_2)_7CH = CH(CH_2)_7COOH$ – the non-saturated monobasic fatty acid.

Table A6. The main physical parameters of oleic acid [51]

Parameter	ρ, kg/m³	η, mPa·s	κ, W/(m·K)	C_p, kJ/(kg·K)	δ_e, ohm⁻¹·cm⁻¹
Numerical value of parameter	898	36.2	0.231	1.848 + 0.447·10⁻² T	$0.5\cdot10^{-10}$
At T_c, °C	18	20	26.5	57–100	15

Comment. Here δ_e is the coefficient of specific bulk electrical conductivity

Physical properties of fluids

Table A7. Dependence of the density of the transformer and condenser oil on pressure and temperature [239]

Fluid	1 atm		200 atm		400 atm		600 atm		800 atm		1000 atm	
	ρ	ρ_T	ρ	ρ_T	ρ	ρ_T	ρ	ρ_T	ρ	ρ_T	ρ	ρ_T
Transformer oil	0.867	6.7	0.880	6.0	0.897	5.5	0.910	5.0	0.920	4.5	0.930	4.0
Condenser oil	0.859	6.5	0.874	6.0	0.887	5.4	0.900	5.2	0.913	5.0	0.923	4.9

Comment. $[\rho]$ = g/cm³, $\rho_T \equiv -\partial\rho / \partial T$, $[\rho_T]$ = g/cm³·K, reference temperature 20°C

Table A8. Density of fluids at 20°C

Substance	ρ, g/m³	Substance	ρ, g/m³
Nitric acid	1.51	Kerosene	0.8–0.82
Aniline	1.02	Sea water	1.01–1.03
Acetone	0.791	Formic acid	1.22
Benzine	0.68–0.8	Oil	0.73–0.94
Benzol	0.879	Nitrobenzene	1.2
Water	0.99823	Nitroglycerine	1.6
Heavy water (D_2O)	1.1086	Mercury	13.55
Hexane	0.660	Sulphuric acid	1.83

Substance	ρ, g/m³	Substance	ρ, g/m³
Heptane	0.684	Hydrochloric acid (38%)	1.19
Glycerine	1.26	Chloroform	1.489
Methyl spirit	0.792	Toluene	0.866
Vaseline oil	0.8	Acetic acid	1.049
Machine oil	0.9	Ethyl alcohol	0.79

Table A9. Viscosity of fluids at 18°C

Substance	$\eta \cdot 10^2$, Pa·s	Substance	$\eta \cdot 10^2$, Pa·s
Aniline	0.46	Purified cylinder oil	0. 109
Acetone	0.0337	Dark cylinder oil	24.0
Benzene	0.0673	Pentane	0.0244
Bromine	0.102	Mercury	0.159
Water	0. 105	Ethyl ester	0.0238
Glycerine	139.3	Ethyl alcohol	0.122
Castor oil	120.0	Toluene	0.0613
Xylol (m)	0.0647	Acetic acid	0.127
Light machine oil	11.3	Chloroform	0.0579
Heavy machine oil	66.0		

Table A10. The coefficient of bulk expansion (for a temperature of approximately 18°C)

Substance	$q \cdot 10^4$, K^{-1}	Substance	$q \cdot 10^4$, K^{-1}
Aniline	8.5	Nitric acid	12.4
Acetone	14.3	Mercury	1.8
Benzene	10.6	Kerosene	10.0
Water at 5–10°	0.53	Oil	9.2
» 10–20°	1.50	Methyl alcohol	11.9
» 20–40°	3.02	Propyl alcohol	9.8
» 40–60°	4.58	Ethyl alcohol	11.0
» 60–80°	5.87	Toluene	10.8
Glycerine	5.0	Ethyl este	16.3

Table A11. Surface tension at 20°C

Substance	σ, mN/m	Substance	σ, mN/m
Nitric acid	59.4	Oil	26
Aniline	42.9	Nitrobenzene	43.9
Acetone	23.7	Sulphuric acid 85%	57.4
Benzene	29.0	Methyl alcohol	22.6
Water	72.8	Propyl alcohol	23.8
Glycerine	59.4	Ethyl alcohol	22.8
Kerosene	28.9 (0°C)	Toluene	28.5
Castor oil	36.4 (18°C)	Acetic acid	27.8
Olive oil	33.06 (18°C)	Ethyl ester	16.9

Table A12. The speed of sound in pure fluids and oils

	T_c, °C	c, m/s	α_T, m/(s·K)
	Pure fluids		
Aniline	20	1656	–4.6
Acetone	20	1192	–5.5
Benzene	20	1326	–5.2
Sea water	17	1510—1550	—
Tap water	25	1497	2.5
Glycerine	20	1923	–1.8
Kerosene	34	1295	—
Mercury	20	1451	–0.46
Methyl alcohol	20	1123	–3.3
Ethyl alcohol	20	1180	–3.6
	Oils		
Peanut oil	31.5	1562	—
Spindle oil	32	1342	—
Transformer oil	32.5	1425	—
Cedar oil	29	1406	—
Linseed oil	31.5	1772	—
Olive oil	32.5	1381	—
Colza oil	30.8	1450	—
Eucalyptus oil	29.5	1276	—

Comment. The speed of sound in fluids decreases with increasing temperature (with the exception of water). The speed can be calculated from the equation $c_t = c + \alpha_T (t - t_0)$, where c is the speed given in the table, α_T is the temperature coefficient indicated for pure fluids in the last graph, t is the temperature at which the speed is measured, t_0 is the temperature given in the table.

Table A13. Absorption of sound in fluids

Fluid	T_c, °C	Frequency range, MHz	$\alpha/\nu^2 \cdot 10^{17}$ s²/cm
Nitrogen	–199	44.5	11
Acetone	25	4–20	50
Benzene	20	1–200	850–900
Water	20	1–200	25
Glycerine	26	4–20	1700
Kerosene	25	6–20	110
Castor oil	18.5	3	11 000
Oil	25	10	~100
Mercury	20	0.5–1000	5.5
Turpentine	25	10	150
Methyl alcohol	20	5–46	43
Ethyl alcohol	20	7–100	52
Ethyl ester	25	10	140

Comment. These values relate to the pressures of 0.1–2 MPa. At these values, absorption is almost completely independent of pressure.

Table A 14. Thermophysical properties of kerosene [161]

T_c, °C	ρ, kg/m³	C_p, kJ/(kg·K)	η, MPa·s	κ, W/(m·K)	$q \cdot 10^4$, K^{-1}	$\sigma \cdot 10^3$, N/m
0	781	1.913	11.640	14.0	–	28.7
15	771	1.982	8.948	13.9	10.65	–
30	760	2.075	6.972	13.8	8.50	25.8
45	752	2.152	5.661	13.7	8.40	–
60	742	2.232	4.899	13.7	–	23.2

Here q is the thermal expansion coefficient, σ is the surface tension coefficient, κ is the heat conductivity coefficient.

Table A15. Dependence of the speed of sound on pressure and temperature [239]

Fluid	1 atm		200 atm		400 atm		600 atm		800 atm		1000 atm	
	c	c_T	c	c_T	c	c_T	c	c_T	c	c_T	c	c_T
Transformer oil	1431	3.7	1520	3.4	1599	3.2	1670	3.0	1748	2.9	1811	2.8
Condenser oil	1432	3.5	1519	3.2	1601	3.0	1673	2.8	1738	–	1800	–

Comment $[c]=\frac{\text{m}}{\text{s}}$, $c_T \equiv -\frac{\partial c}{\partial T}$, $[c_T]=\frac{\text{m}}{\text{s}\cdot\text{K}}$, reference temperature 20°C.

Table A16. The dependence of absorption $\left(\frac{\alpha}{\nu^2}\cdot 10^{17}, \left[\frac{\alpha}{\nu^2}\right]=\text{cm}^{-1}\text{s}^2\right)$ on pressure, temperature and the frequency in the transformer oil [239]

ν, MHz	p, atm / T_c, °C	1	200	400	600	800	1000
4.0	20	565	720	970	1280	1660	2030
	30	440	500	660	780	1030	1380
12.1	12	945	990	1165	1370	1780	2510
	20	560	630	820	1095	1440	1850
	30	415	470	580	720	900	1160
20.1	12	930	940	1100	1310	1685	2160
	20	545	595	755	915	1260	1560
	30	400	430	460	540	670	880
	38	297	310	370	445	555	655

Table A17. Thermophysical properties of organic silicon fluid PES-5 [161]

T_c, °C	ρ, kg/m³	κ, W/(m·K)	C_p, kJ/(kg·K)	ρ_e, ohm·cm	$\nu_*\cdot 10^6$, m²/s
20	998.0	0.157	1.586	$9\cdot 10^{12}$	267.8
40	984.1	0.154	1.622	–	143.6
100	943.0	0.146	1.729	–	32.92
200	874.3	0.133		–	7.284

Comment. Here ν_* is the kinematic viscosity

Table A18. Dependence of density and the speed of sound on pressure and temperature in the organic silicon fluid PMS-5 (reference temperature 20°C) [239]

p, atm	1	200	400	600	800	1000
ρ, g/cm^3	0.917	0.940	0.960	0.990	1.010	1.020
$-\partial\rho/\partial T$, g/cm^3 K	9.0	7.3	6.4	5.7	5.2	5.0
c, m/s	967	1081	1172	1253	1324	1386
$-\partial c/\partial T$, m·s·K	3.1	2.7	2.4	2.3	2.2	2.1

Table A19. The dependence of the absorption of ultrasound $\left(\frac{\alpha}{\nu^2}\cdot 10^{17}, \left[\frac{\alpha}{\nu^2}\right] = \text{cm}^{-1}\text{s}^2\right)$ on pressure, temperature and frequency in PMS-5 organic silicon fluid [239]

ν, MHz	p, atm / T, °C	1	200	400	600	800	1000
12.1	20	215	230	220	230	235	245
	40	190	195	185	165	165	170
20.1	20	205	210	215	220	210	230
	40	170	160	155	140	160	150

Electrical and thermophysical properties of magnetic fluids [51]

The experimental results show that the electrical conductivity of the magnetic fluids is 2–3 orders of magnitude higher than that of the fluid carrier. In the concentration range $0 < \varphi \leq 0.09$ the number of charge carriers increases in proportion to the concentration of the ferromagnetic particles. However, with increasing concentration ($\varphi \geq 0.16$) the mobility of ions decreases as a result of the reduction of the volume fraction of the liquid phase. This results in the formation of a maximum on the curve of the dependence $\delta_e(\varphi)$. The maximum specific electrical conductivity is $\delta_e \approx 2.5\cdot 10^{-6}$ ohm^{-1}·m^{-1}.

The presence of the solid phase in the magnetic fluid increases its heat conductivity because the heat conductivity coefficient of magnetite is on average an order of magnitude higher than the heat conductivity of the liquid base used.

The heat capacity of the magnetic fluid C_p at low and moderate concentrations ($\varphi \leq 0.2$) is calculated on the basis of the mixing rule according to which

$$C_p = \varphi_{m1} C_{p1} + \varphi_{m2} C_{p2} + \varphi_{m3} C_{p3},$$

where φ_{m1}, φ_{m2}, φ_{m3} are the mass fractions of respectively the base, the stabiliser and the magnetic phase; C_{p1}, C_{p2}, C_{p3} are the appropriate heat capacities.

Table A20. Heat conductivity coefficient at T_c = 20°C for the magnetite magnetic fluids based on kerosene (H = 0)

φ, %	0	2	5	10	15	17	20
κ, W/(m·K)	0.138	0.139	0.140	0.158	0.180	0.204	0.260

Table A 21. Physical and mechanical properties of several ferrocolloids [53]

Carrier fluid	Saturation magnetization M_s, kA/m	Density $\rho \cdot 10^{-3}$, kg/m³	Viscosity η, kg/(m·s)	Hardening temperature T, K	Surface tension coefficient σ, N/m	Heat conductivity coefficient λ, W/(m·K)	Bulk heat capacity $c \cdot 10^{-6}$, J/(m³·K)	Thermal expansion coefficient, $q \cdot 10^{4}$, K⁻¹
Diesters	15.9	1.185	0.075	236	—	—	—	—
Hydro-carbons	15.9	1.05	0.003	273	0.028	0.15	1.715	9.0
	31.8	1.25	0.006	281	0.028	0.15	1.840	8.6
Esters	15.9	1.15	0.014	217	0.026	0.31	3.724	8.1
	31.8	1.30	0.030	217	0.026	0.31	3.724	8.1
	47.7	1.40	0.035	217	0.021	0.31	3.724	8.1
Water	15.9	1.18	0.007	273	0.026	1.4	4.184	5.2
	31.8	1.38	0.01	273	0.026	1.4	4.184	5.0

Comment. The solidification point corresponds to the viscosity of 100 kg/(m · s).

Table A 22. Dielectric permittivity of fluids

Substance	T_c, °C						
	0	10	20	25	30	40	50
Acetone	23.3	22.5	21.4	20.9	20.5	19.5	18.7
Benzene	—	2.30	2.29	2.27	2.26	2.25	2.22
Water	87.83	83.86	80.08	78.25	76.47	73.02	69.73
Glycerine	—	—	56.2	—	—	—	—
Kerosene	—	—	2.0	—	—	—	—
Ethyl ester	4.80	4.58	4.33	4.27	4.15	—	—
Ethyl alcohol	27.88	26.41	25.00	24.25	23.52	22.16	20.87

Magnetic fluids with microdroplet aggregates [242]

The magnetic fluids with the microdroplet aggregates relate to highly magnetically sensitive fluids. They are producing by diluting the concentrated magnetic fluid – magnetite in kerosene using solutions of oleic acid in kerosene of different concentration.

The colloidal solution contains the microdroplet aggregates, if the initial fluid contains magnesite with the volume content of solid particles of 7–12% and is dilute by 4–7% solutions of oleic acid in kerosene to the following ratio of the components, wt.%: magnetite 2–3; surfactants 1–2; kerosene 5–6; the balance 4–7% solution of oleic acid in kerosene. This produces a microemulsion containing two liquid phases – concentrated (microdroplet aggregates) and low concentration (initial fluid diluted with the solution of oleic acid in kerosene to the solid phase concentration not higher than 1.3%).

When using a more concentrated magnetic fluid with the content of the solid phase from 15 to 20% as the initial fluid, to produce the microemulsion with the optimum parameters as regards the visualisation of the defects the solvent is represented either by pure kerosene or a solution of oleic acid in kerosene with the concentration not higher than 3%.

References

1. Cary B.B., Fenlon F.H., On the utilization of ferrofluids for transducer applications, J. Acoust. Soc. Amer., 1969, V. 45, No. 5, 1210–1217.
2. Tarapov I.E., Sound waves in the magnetizable medium, PMTF, 1973, No. 1, 15–22.
3. Polunin V.M., Relaxation of the magnetization and the propagation of sound in a magnetic fluid, Akust. Zh., 1983, V. 29, No. 6, 820–823.
4. Polunin V.M., et al., Indication of ultrasound waves in a magnetic fluid, Proceedings of 3 All-Union. school-seminar on magnetic liquids, Moscow, MSU, 1983, 204–205.
5. Polunin V.M., P'yankov E.V., Observation of magnetization perturbation in the propagation of sound in a magnetic fluid, Magn. Gidrodinamika, 1984, No. 1, 126-127.
6. Lukyanov A.E., et al., Wave excitation of magnetization propagation of sound in ferromagnetic fluid, Proc. Conf. on the Physics of magnetic phenomena, Tula: TGPI, 1983.
7. Pirozhkov B.I., et al., The speed of sound in the enzyme fluids, Hydrodynamics. Scientists note, Perm: PSPI, 1976, Vol. 9, 164–166.
8. Polunin V.M., On the speed of ultrasound in the ferromagnetic fluid, Ultrasound and physico-chemical properties of substances, Kursk, 1979, Vol. 13, 151–154.
9. Polunin V.M., Microinhomogeneity of the magnetic fluid and distribution of sound in it, Akust. Zh., 1985, Vol. 31, No. 2, 234–238.
10. Chung D.Y., Isler W.E. Ultrasonic velocity anisotropy in ferrofluids under the influence of a magnetic field, J. Appl. Phys., 1978, V. 49 (3), 1809–1811.
11. Ignatenko N.M., et al. Effect of external magnetic field on the propagation velocity of ultrasonic waves in magnetic fluid, Izv. VUZ, Ser. Physics, 1983, No. 4, 65–69.
12. Pirozhkov B.I., Shliomis M.I., Relaxation absorption of sound in ferrosuspension, Mater. 9 Proc. Acoust. Conf. Section G, Moscow, Nauka, 1977, 123–126.
13. Parsons J.D., Sound velocity in magnetic fluid, J. Phys. D., Appl. Phys., 1975, V. 8, No. 10, 1219–1226.
14. Chung D.Y., Isler W.E., Magnetic field dependence of ultrasonic response times in ferrofluids, IEEE transactions on magnetics, 1978, V. 14, No. 5, 984–986.
15. Polunin V.M., Chernyshova A.A., On the bulk viscosity of the magnetic liquids, \ Magn. gidrodinamika, 1983, No. 1, pp 29-32.
16. Polunin V.M., Chernyshova A.A., Absorption of sound in a magnetic liquid placed in an inhomogeneous magnetic field, Magn. Gidrodinamika, 1984, No. 3, 23Magn. hydrodynamika27.
17. Luk'yanov A.E., et al., Acoustic spectroscopy of magnetic fluids, 11th Riga meeting on magnetohydrodynamics, 3. Magnetic fluids: Salaspils: IF AN Latv. SSR, 1984,

47–50.
18. Gogosov, V.V., Investigation of ultrasound propagation in a magnetic fluid, report number 3236, Institute of Mechanics, Moscow State University, 1985, Number 77066746.
19. Bashtovoy V.G., Krakov, M.S., Resonant excitation of waves in ferromagnetic magnetic fluid by the traveling magnetic field, Ural Conf. on application of magnetohydrodynamics in metallurgy: Proc., Perm, UNTs AN SSSR, 1974, Vol. 1, 136–138.
20. Bashtovoy V.G., Krakov, M.S.,Resonant excitation of sound in ferromagnetic magnetic fluid, Magn. gidrodinamika, 1974, No. 3, 3–7.
21. Baev A.R., Prokhorenko P.P., Resonant excitation of ultrasonic fluctuations in magnetic fluids, DAN BSSR, 1978, Vol. 22, No. 3, 242–243.
22. Polunin V.M., On the question of resonant excitation of oscillations in ferrofluid, Magn. gidrodinamika, 1978, No. 1, 41–143.
23. Polunin V.M., On a method of resonant excitation of ultrasonic oscillations in ferromagnetic fluid, Akust. Zh., 1978, Vol. 24, No. 1, 100–103.
24. Polunin V.M., Some peculiarities of magnetic fluid transducer, Akust. Zh., 1982, Vol. 28, No. 4, 541–546.
25. Polunin V.M.,et al., Audio echo in magnetic fluid, Magn. gidrodinamika, 1981, No. 2, 129–131.
26. Polunin V.M., et al., Magnetic fluid converter of oscillations in the megahertz frequency range, Proc. 15th Conf. on the Physics of magnetic phenomena, Perm: UNTs USSR Academy of Sciences, 1981, Part 2, 105–106.
27. Mace B.R., et al., Wave transmission through structural inserts, J. Acoust. Soc. Amer., 2001, V. 109, No. 4. 1417–1421.
28. Lobova O.V, et al., Elastic properties magnetic fluid sealants, Proc. 11th Session Russ. Acoust. Society, Moscow, GEOS, 2001, Vol. 2, 203–207.
29. Karpova G.V., et al., Resonance properties of magnetic fluid sealants, Magnetohydrodynamics. 2002, Vol. 38, No. 4, 385–390.
30. Karpova G.V., et al. Experimental study of magnetic-resonator, Akust. Zh., 2002, Vol. 48, No. 3, 354–357.
31. Bratukhin Yu.K., Lebedev A.V., Forced oscillations of the magnetic fluid droplets, Zh. Eksp. Tekh. Fiz., 2002, Vol. 121, Vol. 6, 1298–1305.
32. Polunin V.M., et al., Study of the properties magnetic fluid membrane, Akust. Zh., 2005, Vol. 51, No. 6, 778–786.
33. Kameneva Yu. Yu., et al., Kinetic properties of the magnetic fluid membrane, Magnetohydrodynamics, 2005, Vol. 41, 419, No. 1, 87–93.
34. Shliomis M.I., Magnetic fluids, Usp. Fiz. Nauk, 1974, Vol. 112, No. 3, 427–459.
35. Fertman V.E., Magnetic fluid - natural convection and heat exchange, Minsk, Nauka i Tekhnika, 1978.
36. Gogosov V.V., et al., Hydrodynamics of magnetized fluids, Itogi nauki i tekhniki, Ser. Mekh. Zhidkosti i gaza, Moscow, VINITI, 1981.
37. Pat. 3215572 US. Low viscosity magnetic fluid obtained by the colloidal suspension of magnetic particles, Pappell S.S., publ. 1965.
38. Rosenzweig R.E., Ferrohydrodynamics, Usp. Fiz. Nauk, 1967, Vol. 92, No. 2, 339-343.
39. Bibik E.E., Buzunov O.V., Advances in the field of preparation and application of ferromagnetic fluids, Moscow, Central Research Institute of Electronics, 1979, 60 p.
40. Bibik E.E., Preparation of colloidal ferrofluid, Kolloid. Zhurn., 1973, Vol. 35, No. 6, 1141–1142.,
41. Vonsovskii S.V., Magnetism, Moscow, Nauka, 1971.

42. Belov K.P., et al., Rare-earth ferromagnets and antiferromagnets, Moscow, Nauka, 1965..
43. Bibik E.E., Effect of interaction of particles on the properties ferrofluids, The physical properties of magnetic fluids, Sverdlovsk: UNTs AN SSSR, 1983, 3–21.
44. Krakov M.S., Matusevich N.P., On the stability of magnetic colloids and their maximum magnetization, Magnetic fluids: scientific studies and applied research, Minsk: Belorussian Academy of Sciences, ITMO, 1983, 3–11.
45. Hayes Ch.F., Observation of association in a ferromagnetic colloid, J. of Colloid and Interface Science, 1975, V. 52. No. 2, 239–243.
46. Chekanov V.V., On the interaction of particles in magnetic colloids, Hydrodynamics and thermal physics of magnetic fluids, Salaspils, 1980, 69–76.
47. Tsebers S.A., Thermodynamic stability of magnetic fluids, Magn. Gidrodinamika, 1982, No. 2, 42–48.
48. Orlov D.V., et al. Aut. Cert. 516861 USSR. Ferromagnetic magnetic fluid seals. Number 2095965/258; appl. 01/29/75; publ. 1976, Bull. Number 21.
49. Chekanov V.V., et al., The change of the magnetization of the magnetic fluid in the formation of aggregates, Magn. Gidrodinamika, 1984, No. 1, 3–9.
50. Skibin Yu.N., Influence of particle aggregation on extinction and dichroism of magnetic fluids, Physical properties of magnetic fluids, Sverdlovsk, USSR Academy of Sciences, Ufa, 1983, 66–74.
51. Fertman V.E., Magnetic fluids, Minsk, Vysshaya shkola, 1988.
52. Nikitin S.A., et al., Magnetocalorific effect in rare-earth metals and their alloys: technical applications, 16 All-Union. Conf. on the Physics of magnetic phenomena, Tula, TGPI, 1983, 276–277.
53. Bashtovoy V.G., et al., Introduction to thermomechanics of magnetic fluids, Moscow, IVTAN, 1985.
54. Polunin V.M., The electromagnetic effects caused by the elastic deformation of a cylindrical sample of the magnetized fluid, Magn. Gidrodinamika, 1988, No. 3, 43-50.
55. Chechernikov V.I., Magnetic measurements, Moscow, MGU, 1969.
56. Joseph R.J., Ballistic demagnetizing factor in uniformly magnetized cylinders, J. of Applied Physics, 1966, Vol. 37, No. 13, 4639–4643.
57. Landau L.D., Lifshitz E.M., Electrodynamics of continuous media, Moscow, Nauka, 1982.
58. Bloom E.Ya., et al., Magnetic fluids, Riga, Zinatne, 1989.
59. Rosensweig R.E., Ferrohydrodynamics, Cambridge Monographs on Mechanics and Applied Mathematics, New York, 1985.
60. Maiorov M.M., Experimental study of magnetic permeability of the ferrofluid in an alternating magnetic field, Magn. Gidrodinamika, 1979, No. 2, 21–26.
61. Kashevsky B.E., Fertman V.E., The characteristic time of the magnetic interactions in ferromagnetic fluid, Research of convective and wave processes in ferromagnetic fluids, Minsk: Belorussian Academy of Sciences, ITMO, 1975, 46–55.
62. Kondorskii E.I., Nature of high coercive force finely disperse ferromagnets and the theory of a single-domain structure, Izv. AN SSSR, Ser. Fizika, 1952, Vol. 16. No. 4. 398–411.
63. Mikhailov I.G., et al., Fundamentals of molecular acoustics, Moscow, Nauka, 1964.
64. Polunin V.M., Ignatenko N.M. On the elastic properties of ferromagnetic fluids, Magn. Gidrodinamika, 1980, No. 3, 26–30.
65. Polunin V.M., Ignatenko N.M. ,The structure of the magnetic fluid and its elastic properties, Hydrodynamics and thermal physics of magnetic fluids, Proc. Sympo-

sium, Salaspils, 1980, 85–90.

66. Polunin V.M., On the perturbation of the magnetization of a magnetic fluid by sound, Magn. Gidrodinamika, 1984, No. 1, 21–24.
67. Rytov S.M., et al., Propagation of sound in disperse systems, Zh. Eksp. Tekh. Fiz., 1938, Vol. 8. No. 5, 614–626.
68. Vladimirskii V.V., On the theory of sound propagation in dispersed systems, Scientific Collection of MSU students, Ser. Physics, 1939, Vol. 10, 5–30.
69. Frenkel'. Ya.I. The kinetic theory of liquids, Leningrad, Nauka, 1975.
70. Blinova L.P., et al., Acoustic measurements, Moscow, Publishing House of Standards, 1971.
71. Rayleigh J.W., Theory of sound. 2nd ed, Moscow, GITTL, 1955, Vol. 2.
72. Lependin L.F., Acoustics, Moscow, Vysshaya shkola, 1978.
73. Cowley M.D., Rosensweig R.E., The interfacial stability of a ferromagnetic fluid, J. Fluid Mech., 1967, Vol. 80, No. 4, 671–688.
74. Gaititis A., Formation of the hexaganal pattern on the surface of a ferromagnetic fluid in an applied magnetic field, J. Fluid Mech., 1977, Vol. 82, No. 3, 401–413.
75. Polunin V.M., et al., On some peculiarities of perturbation of the magnetization of a magnetic fluid by sound, Magn. Gidrodinamika, 1986, No. 1, 40–44.
76. Allegra J.R., Hawley S.A., Attenuation of sound in suspensions and emulsions: theory and experiments, J. Acoust. Soc. Amer., 1971, Vol. 51, No. 5, 1545–1564; 1972, Vol. 51, No. 5, (part 2), 1545–1563.
77. Polunin V.M., et al., On the magnetoelastic transformation in a magnetic fluid, Magn. Gidrodinamika, 1988, No. 3, 128–130.
78. Besedin A.G., et al., On the AME character in a magnetic liquid poured in a cylindrical container, Magnetohydrodynamics, 2001, Vol. 37, No. 4, 427–431.
79. Dmitriev I.E., Polunin V.M., On the variance of the velocity of sound in the liquid–cylindrical shell system, Akust. Zh.. 1997, Vol. 43, No. 3, 344–349.
80. Polunin V.M., et al., About magnetoelastic transformation in the magnetization of the magnetic fluid, Proc. rep. 5 Proc. Conf. on magnetic fluids, Moscow, IM MGU 1988, Vol. 2, 46–47.
81. Polunin V.M., et al., Study of vibration and acoustic oscillations in the magnetic fluid filling the pipe, Proc. XVI Session of RAO, Moscow, GEOS, 2005, Vol 1, 137-140.
82. Polunin V.M., et al. Experimental study of the magnetic fluid converter,, Proc. 15 Sessions Russian Acoust. Society, Moscow, GELIOS, 2004, Vol. 2, 37–40.
83. Dmitriev I.E., et al., Study of electromagnetic effects in magnetic fluid, Proc. Anniversary Conference KSTU, Kursk, 1995, Part 1, 112–114.
84. Field G.S., Boyle R.W., Dispersion and selective absorption in the propagation of ultrasound in liquid contained in tubes, Canad. Journ. Res., 1932, No. 6, 192.
85. Guelke R.W., Bunn A.E., Transmission line theory applied to sound wave propagation in tubes with compliant walls, Acustica, 1981, Vol. 48, 101–106.
86. Guelke R.W., Bunn A.E., The propagation of sound in liquids confined in tubes with compliant walls, Acustica, 1982, Vol. 52, 131–134.
87. Guelke R.W., Bunn A.E. The influence of wavelength on pressure wave characteristics in fluid-filled compliant tubes, Acustica, 1986, Vol. 59, 247–254.
88. Jacobi W.J. Propagation of sound waves along liquid cylinders, J. Acoust. Soc. Amer., 1949, Vol. 21, No. 2, 120–124.
89. Polunin V.M., et al., AS 1430984 USSR, Training device for physics to demonstrate wave processes. Appl. 03/19/87; publ. 1988, Bull. Number 38.
90. Polunin V.M., On the method of experimental study of normal waves in a thin elastic

cylindrical shell filled with liquid, Akust. Zh., 1989, Vol. 35, No. 3, 557–559.
91. Dmitriev I.E., Polunin V.M., Acoustic dispersion in magnetic fluid interferometer, Magn. Gidrodinamika, 1997, Vol. 33, No. 1, 96–99.
92. Dmitriev I.E., Polunin V.M., Dispersion of sound velosity in a fluid-filled cylindrical shell, Acoustical Physics, Vol. 43, No. 3, 1997, 295–299.
93. Isakovich M.A., Mandel'shtam L.I., Sound propagation in microinhomogeneous media, Usp. Fiz. Nauk, No. 3, 1979, Vol. 129, 531–540.
94. Isakovich M.A., On the propagation of sound in emulsions, Zh. Eksp. Tekh. Fiz., 1948, Vol. 18, No. 10, 907–912.
95. Urick R.J., A sound velocity method for determining the compressibility of finely divided substances, J. Appl. Phys., 1947, Vol. 18, No. 11, 983–987.
96. Urick R.J., Ament W.S., Propagation of sound in composite media, J. Acoust. Soc. Amer., 1949, Vol. 21, No. 2, 115–119.
97. Mikhailov I.G., The propagation of ultrasonic waves in liquids, M.-L, State Publishing House, 1949.
98. Polunin V.M., The use of magnetic fluids in acoustics, Proc. 3rd School-Seminar on magnetic fluids (Pless, 1983), Moscow, IM, Moscow State University, 1983, 200-201.
99. Rykov V.G., Experimental study of the thermal conductivity of magnetic fluid in a magnetic field, Proc. 3rd School-Seminar on magnetic fluids, Moscow, IM, Moscow State University, 1983, 216–218.
100. Polunin V.M., et al., Comparison of various methods for determining the concentration of magnetic fluid, hydrodynamics and thermal physics of magnetic fluids: Proc. Symposium, Salaspils, IF AN Latv. SSR, 1980, 90–96.
101. Dmitriev E.A., Rosliakova L.I., Polunin V.M., et al. Results of measuring the speed of sound in certain magnetic fluids; Kursk Polytekhn. Inst, Kursk, 1984. Dep. VINITI, No. 7212-84.
102. Solodukhin A.D., Fertman V.E., Experimental study of the temperature dependence of the velocity of ultrasound in ferromagnetic fluids, Convection and waves in liquids, Minsk: ITMO Belorussian Academy of Sciences, 1977, 64–68.
103. Polunin V.M., Ignatenko N.M., On the dependence of the velocity of ultrasound in the ferrofluid on the concentration of solids, Ultrasound and physico-chemical properties of matter, Kursk, 1980, Vol. 14, 223–228.
104. Prokhorenko P.P., et al., On the acoustic properties of magnetic ferrofluids applied to ultrasonic flaw detection, Vests. AN BSSR, 1983, No. 1, 88–92.
105. Boelhouwer J.M., Sound velocities in and adiabatic compressibilities of liquid alkanes at various temperatures and pressures, Physica, 1967, Vol. 34, 484–492.
106. Mansurov K.Kh., Sokolov V.V., Acoustic properties of magnetic liquids, Magn. Gidrodinamika, 1987, No. 1, 63–66.
107. Prokhorenko P.P., et al., Study of the acoustic characteristics of magnetic fluids, Vests. AN BSSR, 1981, No. 5, 88–90.
108. Polunin V.M., Rosliakova L.I., On additive compressibility and wave resistance of magnetic fluids, Magn. Gidrodinamika, 1986, No. 3, 136–137.
109. Polunin V.M., Rosliakova L.I., The speed of sound in the water-based magnetic fluid, Proc. 4 Conf. on Magnetic fluids, Ivanovo, IEI, 1985, Vol. 2, 47–48.
110. Berkovskii B.M., et al. Elastic properties of the water-based magnetic liquid, Magn. Gidrodinamika, 1986, No. 1, 67–72.
111. Mikhailov I.G., et al., Speed and absorption of ultrasonic waves in some viscous liquids at pressures up to 1000 atm, Akust. Zh., 1971, Vol. 17, No. 1, 103–109.
112. Dmitriev S.P., Sokolov V.V., The speed of sound in the magnetic fluid at high pres-

sures, 3 Proc. 3RD School-Seminar on magnetic liquids (Pless, 1983), Moscow, IM Moscow State University, 1983, 86–87.
113. Rudnick J. On the attenuation of finite amplitude waves in a liquid, J. Acoust. Soc. Amer., 1958, Vol. 30, No. 6, 564–567.
114. Mikhailov I.G., Shutilov V.A., Nonlinear acoustic properties of aqueous electrolyte solutions, Akust. Zh., 1964, Vol. 10, No. 4, 450–455.
115. Khamzaev B.Kh., Nonlinear acoustic parameters of magnetic liquids, Proc. 5 Conf. on Magnetic fluids, Moscow, MI Moscow State University, 1988, Vol. 2, 122–123.
116. Polunin VM, Rosliakova L.I., On the dependence of the sound velocity in magnetic fluid of the magnetic field and the oscillation frequency, Magn. Gidrodinamika, 1985, No. 4, 59–65.
117. Vinogradov A.N. et al., Acoustic and physico-chemical properties of magnetic fluid, Proc. 5th Conf. on Magnetic fluids, Moscow, MI MSU, 1988, Vol. 1, 57–58.
118. Lin T.C., Morgan G.W., Wave propagation through fluid contained in a cylindrical, elastic shell, J. Acoust. Soc. Amer. 1956, Vol. 28, No. 6, 1165–1176.
119. Chung D.Y., Isler W.E., Sound velocity measurements in magnetic fluids, Physics letters, 1977, Vol. 61 A, No. 6, 373–374.
120. Kaiser R., Magnetic properties of stable dispersions of subdomain magnetic particles, J. Appl. Phys., 1970, Vol. 41, 1064–1072.
121. Narasimham A.V., Direct observation of ultrasound relaxation times in ferrofluids under the action a magnetic field, J. Appl. Phys., 1981, Vol. 19, 1094–1097.
122. Polunin V.M., et al., The influence of the magnetic field on the structural changes and the elastic properties of some magnetic liquids, Magn. Gidrodinamika, 1987, No. 2, 139–141.
123. Tarapov I.E., Hydrodynamics of polarizing and magnetizing media, Magn. Gidrodinamika, 1972, No. 1, 3–11.
124. Tarapov I.E., Simple waves in nonconducting magnetizable medium, Prikl. Matem. Mekh., 1973, Vol. 37, No. 5, 813–821.
125. Patsegon N.F., et al., Investigation of the physical properties of magnetic fluids by the ultrasonic method, Magn. Gidrodinamika, 1983, No. 4, 53–59.
126. Gogosov V.V., et al., Propagation of ultrasound in the magnetic fluid, Magn. Gidrodinamika, 1987, No. 2, 19–27.
127. Lipkin A.I., The acoustic properties of magnetic fluids with aggregates, Magn. Gidrodinamika, 1985, No. 3, 25–30.
128. Gogosov V.V., et al., Propagation of ultrasound in the magnetic fluid, Magn. Gidrodinamika, 1987, No. 3, 15,
129. Lipkin A.I., On the theory of sound propagation in the ferrofluid, Magnetic fluids: Proc. 11th Riga meeting on magnetic hydrodynamics, Salaspils, 1984, IF AN Latv. SSR, 39–42.
130. Gogosov V.V., et al. Investigation of the properties of magnetic fluids by ultrasonic methods, Proc. 4th All-Union. Conf. on Magnetic fluids (Pless, 1985), Ivanovo, IEI, 1985, 90–91.
131. Lipkin A.I., Modulation speed of sound in the colloid by the external electromagnetic field, Akust. Zh., 1986, Vol. 32, No. 3, 340-345.
132. Blum E.Ya, et al., Heat and mass transfer in a magnetic field, Riga, Zinatne, 1980.
133. Brekhovskikh L.M., Waves in layered media, Moscow, 1957.
134. Stashkevich A.P., Acoustics of the sea, Leningrad, Sudostroenie, 1966.
135. Lipkin A.I., Analysis of the experimental results on the measurement of the velocity of sound in the magnetic fluid in a magnetic field, Magn. Gidrodinamika, 1987. No. 4. 123–124.

136. Sokolov V.V., Tolmachev V.V., The anisotropy of the velocity of propagation of sound in a magnetic fluid, Akust. Zh., 1997, Vol. 43, No.1, 106–109.
137. Baev A.R.,et al., AS 794494 USSR. The method of measuring the input angle of ultrasonic vibrations, Claim 10.1.1979, publ. 1981, No. 1, 167.
138. Apsitis L.V., et al., The use of the ferromagnetic fluid to create acoustic contact in studies on the longitudinal and transverse waves, Acoust. Zh., 1982, Vol. 28, No. 6, 721–723.
139. Rosensweig RE, Kaiser R., Miskolezy G. Viscosity of Magnetic Fluid in a Magnetic Field, J. of Colloid and Interface Sience, 1969, Vol. 29, No. 4, 680–686.
140. Shliomis M.I., The effective viscosity of magnetic suspensions, Zh. Eksp. Teor. Fiz., 1971, Vol. 6 (12), 2411–2418.
141. Einstein A, Ann. D Phys., 1906, No. 12, 292.
142. Vand V. Viscosity of solution and suspensions, J. Phys. Coll. Chem., 1948, V. 52, No. 2, 227–299.
143. Buzmakov V.M., Pshenichnikov A.F., About the concentration dependence of the viscosity of magnetic fluids, Magn. Gidrodinamika, 1991, No. 1, 18–22.
144. Varlamov Yu.D, Kaplun A..,B Viscosity measurement of low reactivity magnetic fluids, Magn. Gidrodinamika, 1986, No. 3, 43–49.
145. Zubarev A.Yu., Yushkov A.V., Dynamic properties of moderately concentrated magnetic fluids, Zh. Eksp. Teor. Fiz., 1998, Vol. 114, Vol. 3 (9), 892–909.
146. Bibik E.E., The interaction of particles in ferrofluids, Physically properties and hydrodynamics of disperse ferromagnets, Sverdlovsk, UNTs USSR Academy of Sciences, 1977.
147. De Gennes P.G., Pincus P.A., Pair Correlation in a Ferromagnetic Colloid, Phys. der Konden. Materie, 1970, Vol. 11, No. 3, 189–198.
148. Gilev V.G., Shliomis M.I., Experimental study of flow of magnetic fluid in flat capillaries of varying thickness, Magnetic Liquids, Proc. 11th Riga meeting on magnetohydrodynamics, Salaspils, 1984, 64.
149. Krankalns G.E., et al., Temperature dependence of the physical properties of magnetic fluid, Magn. Gidrodinamika, 1982, No. 2, 38–42.
150. Berkovskii B.M., et al., Viscometric method for magnetic fluids, Magn. Gidrodinamika, 1984, No. 2, 3–10.
151. Shul'man Z.P., Kordonskii V.I. The magnetorheological effect, Minsk, Nauka i tekhnika, 1982.
152. Mayorov M.M., Measuring viscosity of the ferrofluid in a magnetic field, Magn. Gidrodinamika, 1980, No. 4, 11–18.
153. Varlamov Yu.D., Kaplun A.B., Investigation of processes of structurisation in magnetic fluids, Magn. Gidrodinamika, 1983, No. 1, 33–39.
154. Kaplun A.B., Varlamov Yu.D., The study of the relaxation processes in magnetic fluids using vibration viscometer, Physical properties of magnetic fluids, Sverdlovsk, USSR Academy of Sciences, Ufa, 1983, 103–109.
155. Martsenyuk M.A., The viscosity of the suspension of ellipsoidal ferromagnetic particles in a magnetic field, PMTF, 1973, No. 5, 234–236.
156. Ratinskaya I.A., Attenuation of sound in emulsions, Akust. Zh., 1962, Vol. 8, No. 2, 210–215.
157. Vladimirskii V.V., Galanin M.D., Absorption of ultrasound in aqueous emulsions of mercury, Zh. Eksp. Teor. Fiz., 1939, Vol. 9, 233–236.
158. Koltsova I.S., Mikhailov I.G., Attenuation and scattering of ultrasound sound waves in suspensions, Akust. Zh., 1975, Vol. 21, No. 4, 568–575.
159. Luntz G.L., et al., Physical technical handbook, Vol. 1., Mathematics. Physics, Mos-

cow, Fizmatgiz, 1960.
160. Litovits T., Davis K., Structural and shear relaxation in liquids, Physical Acoustics, Vol. 2, Properties of gases, liquids and solutions, Ed. W. Mason, Moscow, Mir, 1968. 288–370.
161. Matusevich N.P., et al., Production and properties of magnetic fluids: Preprint No. 12, Minsk, ITMO AN BSSR, 1985.
162. Kol'tsova I.S., The weakening of the ultrasonic waves in non-magnetic magnetic fluids, Akust. Zh., 1987, Vol. 33, 256–260.
163. Berkovskii B.M., Bashtovoy V.G., Waves in ferromagnetic fluids, Inzh. Fiz. Zh., 1970, Vol. 18, No. 5, 13–21.
164. Polunin V.M., Ignatenko N.M., Dissipation mechanisms of elastic energy in the non-conductive magnetic fluid, Proc. 3rd Meeting on the physics of magnetic fluids (23-25 Sept. 1986)m Stavropol, 1986, 87–88.
165. Postnikov V.S., Internal friction in metals. 2nd ed, Moscow, Metallurgiya, 1974.
166. Taketomi S., The anisotropy of sound attenuation in magnetic fluid under en external magnetic field, J. Phys. Soc. Jap., 1986, Vol. 55, No. 3. 833–844.
167. Bar'yakhtar F.G., et al., Dynamics of the domain structure of magnetic fluids, Physical properties of magnetic fluids, Sverdlovsk, USSR Academy of Sciences, Ufa, 1983, 50–57.
168. Rodionov A.A., Ignatenko N.M., About the absorption associated with the processes of reversible rotation in the magnetic fluid, Izv, VUZ, Ser. Fizika, 1997, No. 7, 14-17.
169. Vinogradov A.N., The spread of ultrasound in polydisperse magnetic fluids, Vestn. Mosk. Univ., Ser. Khimiya, 1999, Vol. 40, No. 2, 90–93.
170. Polunin V.M., et al., Influence of the inhomogeneous magnetic field on the acoustic properties of magnetic fluids, Magnetic fluids, Proc. 11th Riga Meeting on magnetic hydrodynamics, Salaspils, IF AN Latv. SSR, 1984, 43–46.
171. Taketomi S., Tikadzumi S. Magnetic fluids: Trans. from Japanese, Moscow, Mir, 1993.
172. Henjes K. Sound propagation in magnetic fluids, Phys. Rev. E., 1994, Vol. 50, No. 2. 1184–1188.
173. Karelin A.V., Polunin V.M., Modulation of the ultrasonic pulse by the magnetized magnetic colloid, Akust. Zh., 2003, Vol. 49, No. 5, 711–713.
174. Karelin A.V., et al., Self-modulation of ultrasonic pulse in a magnetic fluid, Magnetohidrodynamics. 2004, Vol. 40, No. 2, 161–166.
175. Kuzin B.E., Sokolov V.V., Anisotropy of ultrasound absorption in magnetic fluid, Akust. Zh., 1994, Vol. 40, No. 4, 689.
176. Maximov G.A., Larichev V.A., Propagation of a short pulse in a medium with relaxation resonance. Exact solution, Akust. Zh., 2003, Vol. 49, No. 5, 656–666.
177. Bergman L., Ultrasound and its application in science and technology, Moscow, 1956.
178. Baev A.R., et al. Investigation of generating ultrasonic vibrations in the volume of ferrofluid by the magnetic field, Proc. 9th Riga Meeting on magnetic hydrodynamics, Riga: IF AN Latv. SSR, 1978, Vol. 1, 23–26.
179. Neuringer J.L., Rosensweig R.E. Ferrohydrodynamics, Phys. Fluids, 1964, Vol. 7. No. 12, 1927–1937.
180. Jahnke E., et al., Special functions, Moscow, Nauka, 1968.
181. Dubbelday P.S., Application of ferrofluids as an acoustic transducer material, IEEE Transactions on Magnetics (Second International Conference on Magnetic Fluids). 1980, Vol. 16, No. 2, 372–374.

182. Bashtovoy V.G., Krakov M.S., Excitation of sound in the magnetized liquid, Magn. Gidrodinamika, 1979, No. 4, 3–9.
183. Barkov Yu.I., et al., AS 650663 USSR. Sound radiator. Appl. 27/09/77; publ. 1979, Bull. No.9.
184. Baev A.R., et al., AS 713599 USSR. A method of generating acoustic oscillations. Appl. 02/22/78; publ. 1980, Bull. No. 5, 14.
185. Smirnov V.I., Course of Higher Mathematics, 15th ed., Moscow, Gosizdat, TTL, 1957, Vol. 2, 628.
186. Shutilov V.A., Physics of ultrasound, Leningrad State University, 1980.
187. Bashtovoi V.G., Berkovskii B.M., Thermomechanics of magnetic fluids, Magn. Gidrodinamika, 1973, No. 1, 3–14.
188. Polunin V.M., A function of the magnetocaloric effect in magnetic fluid, Magn. Gidrodinamika, 1985, No. 3, 53–56.
189. Polunin V.M., et al., Some features of excitation of oscillations in the magnetic fluid, Magn. Gidrodinamika, 1982, No. 2, 133–135.
190. Polunin V.M., et al., Some results of experimental studies of magnetic transformation of ultrasonic oscillations, Physical properties of magnetic fluids, Sverdlovsk, UNTs AN SSSR, 1983, 110–114.
191. Ignatenko N.M., Polunin V.M., To the effects of excitation of ultrasonic oscillations in the magnetic fluid, Magn. Gidrodinamika, 1983, No. 3, 142–143.
192. Polunin V.M., Ignatenko N.M., Experimental studies of the properties of magnetic fluid transducer of ultrasonic vibrations, Ultrasound and thermodynamic properties of the substance, Kursk, 1982, 104–108.
193. Polunin V.M., Ignatenko N.M., About magnetic fluid generator of sound vibrations, 8th Intern. Conf. on MHD energy conversion, Moscow, IVTAN, 1984, Vol. 6. Vol. 303–306.
194. Semikhin V.I., Excitation of ultrasonic vibrations of ferromagnetic fluid in a cylindrical resonator, Magn. Gidrodinamika, 1985, No. 1, 127–129.
195. Arenberg D.L. Ultrasonic solid delay lines, J. Acoust. Soc. Amer., 1948,V. 20, No. 1. 1–25.
196. Kolesnikov A.E., Ultrasonic measurements, Moscow, Standartgiz, 1970.
197. Ultrasound, Small Encyclopedia, ed. IP Golyamina, Moscow, Sov. Entsyklopediya, 1979.
198. Bibik E.E., et al., AS 568598 USSR, A method for producing ferrofluids, No. 2303630/26; appl. 12/24/75; publ. 1977, Bull. No. 30.
199. Mc Taque J.P., Magnetoviscosity of magnetic colloids, J. Chem. Phys., 1969, V. 51, No. 1, 133–136.
200. Isler W.E., Chung D.Y., Anomalous attenuation of ultrasound in ferrofluids under the influence of a magnetic field, J. Appl. Phys., 1978, V. 49, No. 3, Pt 2, 1812–1814.
201. Ignatenko N.M., et al. Effect of blocking magnetic particles in a magnetic fluid on the process of transformation of electromagnetic waves to elastic waves, Ultrasound and thermodynamic properties of substances, Proc., Kursk 1985, 162167.
202. Gorelik G.S., Oscillations and waves, Mocow, Leningrad GITTL, 1950..
203. Ignatenko N., Polunin V.M., Measuring the amplitude of elastic waves excited in a magnetic fluid, Proc. 4th Conf. on magnetic fluids, Ivanovo, IEI, 1985, 141–142.
204. Gitis M.B., Electromagnetic excitation of sound in nickel, Fiz. Tverd. Tela, 1972, Vol. 14, No. 12, 3563–3566.
205. Maskaev A.F., Gurevich S.Y., Investigation of the mechanism of excitation of EMA by the method of elastic waves in ferromagnetic materials over a wide temperature range, Physical methods of materials testing, Chelyabinsk, 1974, 51–67.

206. Ignatenko N.M., et al., Excitation of volume magnetostriction ultrasonic oscillations in suspensions of uniaxial ferroparticles, Magn. Gidrodinamika, 1984, No. 3, 19-22.
207. Polunin V.M., Effect of internal heat exchange in the magnetic fluid on its elastic properties, Magn. Gidrodinamika, 1984, No. 2, 138-139.
208. Lipkin A.I., Mechanism for the generation of ultrasonic oscillations in magnetic fluid in a uniform alternating magnetic field, Zh. Tekh. Fiz., 1987, Vol. 57, No. 1, 125–130.
209. Ignatenko N.M., et al., On the physical model of excitation of elastic waves in a magnetic fluid in the electromagnetic field, Proceedings of 3 Proc. School-Seminar on magnetic fluids, Moscow, MI State University, 1983, 112–113.
210. Rodionov A.A.,et al., Magnetostrictive mechanism of generation of elastic waves in a magnetic fluid, Izv. Kursk TU, Kursk, 1998, No. 2, 137–145.
211. Physical Encyclopedic Dictionary, Ed. A.M. Prokhorov, Moscow, Sov. Entsiklopediya, 1983.
212. Polunin V.M., et al., Acoustic phenomena in magnetic colloids, Journal of Magnetism and Magnetic Materials. North-Holland, 1990, No. 85, 141–143.
213. Muller H.W., Liu M., Shear-Excited Sound in Magnetic Fluid, Vol. 89, No. 6, Physcal Review Letters, 5 August 2002, 0672011–0672014.
214. Polunin V.M., et al., The results of experimental study of magnetoelastic properties of magnetic fluid, Coll. tXIII Session of RAO, Moscow, GEOS, 2003, Vol. 1, 193-196.
215. Polunin V.M., et al., The elastic properties of magnetic fluid membrane, Izv. KurskTU, 2003, No. 2 (11), 29–34.
216. Polunin V.M., et al., Dependences of the initial amplitude of the oscillations of the magnetic fluid membrane of the height of the boom, Vibration 2003, Proc. VI Internat. scientific and engineering. conf., Kursk, 2003, 395–398.
217. Polunin V.M., et al., Study of elastic and electrodynamic properties of magnetic fluid membranes, Proc. 11th Intern. Phys Conf. on magnetic fluids, Ivanovo, IGEU, 2004, 101–107.
218. Polunin V.M., et al., On the Dissipation of energy in an oscillating system with magnitc fluid inertial-viscous element, Izv. Kursk. Gos. Tekh. Univ., 2004, No. 1 (12), 54–59.
219. Landau L.D., Lifshitz E.M., Theoretical physics. Hydrodynamics, Moscow, Nauka, 1988, Vol. 6..
220. Orlov D.V.,et al., The magnetic fluids in mechanical engineering, Ed. D.V. Orlov and V.V. Podgorkov, Moscow, Mashinostroenie, 1993, 272 p.
221. Kobelev N.S., et al. patent RF No.2273002, Gas dispenser, No. 2004110653/28; appl. 07.04.2004; publ. 27.03.2006, Bull. No. 9.
222. Rabinovich M.I., Trubetskov D.I., Introduction to the theory of vibrations and waves Moscow, Research Center Regular and chaotic dynamics, 2000.
223. Lobova O.V., et al., The resonance vibrations of magnetic fluid sealant, Izv. Kursk. Gos. Tekh. Univ., 2002. No. 1 (8), 286–294.
224. Landa P.S., Rudenko O.V., About two mechanisms of sound generation, Akust. Zh. 1989, Vol. 35, No. 5, 855–862.
225. Rudenko O.V., Shanin A.V., Non-linear phenomena in systems with finite boundary displacement, Physical and non-linear acoustics, Proc. Seminar of the School of prof. V.A. Krasil'nikov, Moscow, MGU, 2002, 31–46.
226. Karpova G.V., et al., The energy dissipation in magnetic-inert elements, Proc. Scientific. 10th Intern. Conf. on magnetic fluids, Ivanovo: Ivanovo State Power University, 2002, 76–80.

227. Rzhevkin S.N., Lectures on the theory of sound, Moscow, MGU, 1960.
228. Chaban I.A., Damping of oscillations of a gas bubble in a liquid, associated with the heat exchange, Akust. Zh., 1989, Vol. 35, No. 1, 182–183.
229. Lobova O.V., et al., On the speed of sound in the liquid chain system, Ultrasound and thermodynamic. properties of substances, Proc. Kursk State Tekhn. Univ., Kursk, 1998, 80–83.
230. Lobova O.V., et al., Elastic properties of fluid chain, Vibrating machines and technology: 5th Intern. conf. Kursk State Tekhn, Univ, Kursk, 2001, 356–359.
231. Drozdova V.I., et al., Investigation of vibrations of droplets of magnetic fluid, Magn. Gidrodinamika, 1981, No. 2, 17–23.
232. Polunin V.M., Chistyakov M.V., Acoustomagnetic effect on the magnetic fluid in a transverse magnetic field, Proc. Ultrasound and thermodynamic properties of substances, Kursk, KSU, 2005, 28–32.
233. Polunin V.M., et al., Acoustic studies of the structure of real magnetic fluids, Proc. 12th Intern. Conf. on magnetic fluids, Ivanovo: Ivanovo State University, 2006, 68-75.
234. Polunin V.M., Mikhailova Yu.Yu., Investigation of strength properties of magnetic fluid membrane, Proc. XVIII Session of RAO, Moscow, GEOS, 2006, Vol. 1, 55–58.
235. Badescu R., et al., Viscous and thermal effects on acoustic properties of ferrofluids with aggregates. Rom. Journ. Phys., Vol. 55, Nos. 1-2, 173–184, Bucharest, 2010.
236. Gladun A.D., et al., Laboratory workshop on general physics, Textbook, in three volumes, Vol.1, Thermodynamics and molecular physics, Ed. A.D .Gladun, Moscow, MIPT, 2003.
237. Kleman, M., Lavrentovich O.D., Fundamentals of physics of partially ordered environments, Moscow, Fizmatlit, 2007.
238. Zolotukhin I.V., et al., New directions of physical materials science, a textbook, Sci. Ed. B.M. Darinsky, Voronezh, VSU, 2000.
239. Polunin V.M., Study the velocity and absorption of ultrasound in viscous liquids at high pressures, Dissertation, Leningrad, 1971.
240. Tables of physical quantities, handbook, Ed. Acad. I.K. Kikoin, Moscow, Atomizdat 1976.
241. Kovarda V.V., et al., On the strength properties of the magnetic fluid membrane, Magnetohydrodynamics, 2007, Vol. 43, No. 3, 333–344.
242. Shagrova G.V., Methods of control of information on magnetic carriers, Moscow, Fizmatlit, 2005.
243. Baev A.R., Prokhorenko P.P., Physical fundamentals and principles of application of magnetic fluids in the technical acoustics and non-destructive acoustic inspection, Mater. Conf. New smart materials, Minsk: UP Tekhnoprint, 2001, 141–151
244. Tikadzumi S., Physics of ferromagnetism. Magnetic properties of matter, Moscow, Mir, 1983.
245. Polunin V.M., Ferrosuspension as liquid magnet, Magn. Gidrodinamika, 1979, No. 3, 33–37.
246. Polunin V.M., et al., Passage of ultrasound through ferrosuspension, Proc. scientific. 13th Intern. conf. on magnetic fluids, Ivanovo, Ivanovo State Power University, 2008, 94–99.
247. Mikhailova Yu.Yu., et al., On the elasticity of the magnetic fluid membrane, Akust. Zh., 2008, Vol. 54, No. 4, 546–551.
248. Verisokin M.Yu., et al., Non-linear vibrations of the magnetic fluid thin jumper in the ring magnet field, Izv. KGTU, 1, No. (16), 2006, Kursk, 42–45.
249. Erofeev V.I., et al. Effect of different modulus of material on the propagation of

non-linear torsion waves in the rod. Test of materials and structures: Collection of scientific works, ed. S.I. Smirnov and V.I. Erofeev, Nizhny Novgorod, Publisher society IntelService, 2000, Issue 2, 117–137

250. Paukov V.M., et al. Hydroacoustic converters based on dipersed media with magnetic nanoparticles, Proceedings of IX All-Russian Conference Applied Technology of Hydroacoustic and Hydrophysics, St Petersburg, Nauka, 2008, 583–587.
251. Lamb D., Hydrodynamics, Vol. 2, Moscow-Izhevsk, Regular and chaotic dynamics 2003.
252. Odinaev S., Komilov K., The frequency dependence of the velocity and coefficient of absorption of sound waves in magnetic fluids, Akust. Zh., 2008, Vol. 54, No. 6, 920–925.
253. Suzdalev I.P., Nanotechnology: Physics and chemistry of nanoclusters, nanostructures and nanomaterials, Ed. 2nd, rev, Moscow, LIBROKOM, 2009.
254. Polunin V.M., et al., Determination of velocity of sound propagation in a magnetic fluid by the acustomagnetic method, Proc. 53rd conference MIFT, Moscow, 2010, 151–153.
255. Gulamov A.A., et al., Investigation of kinetic-strength properties of magnetic fluid membrane, Nanotekhnika, 2010, No. 1, 10–15.
256. Polunin V.M., Kutuev A.N., Studying the form of dispersed nanoparticles based on the model of the rotational viscosity, Izv. VUZ, Fizika, 2009, No. 8, 10–15.
257. Karpova G.V., et al., On the dissipation processes in the oscillating system with a magneto-liquid element, Magnetohydrodynamics, Vol. 45 (2009), No. 1, 85–94.
258. RF Patent No.2384737. Diaphragm pump [Text], Emelyanov S.G., et al., No. 2009106495/06; appl. 24.02.2009, publ. 20.03.2010, Bull. Number 8.
259. Naletova V.A., Shkel Yu.M., Study of the flow of the magnetic fluid in a tube with anisotropic magnetic fluid in the magnetic field, Mag. Gidrodinamika, 1987, No. 4, 51–57.
260. Polunin V.M., Acoustic effects in magnetic fluids, Moscow, Fizmatlit, 2008.
261. RF Patent No. 2273002. Gas dispenser [Text], S.G. Emelyanov, et al., No. 2008106301/28; appl. 18.02.2008; publ. 10.09.2009. Bull. Number 25.
262. Polunin V.M., et al., An Oscillatory System with a Magnetic_Fluid Viscoinertial Element, Acoustical Physics, 2010. Vol. 56, No. 2, 174–180.
263. Emelyanov S.G., et al., An estimate of physical parameters of magnetic nanoparticles, Akust. Zh., 2010, Vol. 56, No. 3, 316–322.
264. Polunin V.M., et al. An oscillation system with inertial-viscous magnetic fluid element, Akust. Zh., 2010, Vol. 56, No. 2, 197–203.
265. Kashevsky B.E. , et al., Magnetorheological effect in the active suspension carrier liquid., Mag. Gidrodinamika, 1988, No. 1, 35–40.
266. Polunin V.M., et al. Vibratory rheological effect in nanostructured magnetic fluid, Controlled vibration technology and machines, Proc. Ch 1. Kursk. state. tekhn. univ., Kursk, 2010, 300–306.
267. Kamiyama Sh., et al., On the flow of a ferromagnetic fluid in a circular pipe. Report 1. Flow in uniform magnetic field, Bull. ISME, 1979, V. 22, No. 171, 1205–1211.
268. Kamiyama Sh., et al., On the flow of a ferromagnetic fluid in a circular pipe. Report 2. Flow in a non-uniform magnetic field, Sci. Repts. Res. Inst. Tohoku Univ., 1980, V. B41, No. 323, 21–35.
269. Bolotnikov M.F.,et al., Speed of sound, densities and isentropic compressibilities of liquid 1-bromalkanes at temperatures from (243.15 to 423.15) K, J. Chem. Eng. Data, 2009,Vol. 54, No. 6, 1716–1719.
270. Gusev A.I., Nanomaterials, nanostructures, nanotechnology, 2 ed., rev, Moscow,

Fizmatlit, 2009.
271. Korteweg D.J. Uber die Fortpflanzungsgeschwindigkeit des Schalles in elastischen Rohren, Ann. Physik, 1878, No. 5, 525.
272. Skuchik E., Fundamentals of acoustics, Vol. 2, Moscow, Publishing House of Foreign Literature, 1959.
273. El-Raheb M., Wagner P., Harmonic response of cylindrical and toroidal shells to an internal acoustic field, P. 1, Theory, JASA, 1985, Vol. 78, No. 2, 747–757.
274. El-Raheb M., Acoustic propagation in finite length elastic cylinders. P. 1, Axisymmetric excitation, JASA, 1982, Vol. 71, No. 2, 296–306.
275. Fay R.D., Waves in liquid-filled cylinders, JASA, 1952, Vol. 24, No. 5, 459–462.
276. Moodie T.B., Haddow J.B.M Dispersive effects in wave propagation in thin-walled elastic tubes, JASA, 1978, Vol. 64, No. 2, 522–528.
277. Moodie T.B., Haddow J.B., Asymptotic analysis for dispersive fluid filled tubes, JASA, 1980, Vol. 67, No. 2, 446–452.
278. Safaai-Jazi A., et al., Analysis of liquid-core cylindrical acoustic wavegides, JASA, 1987, Vol. 81, No. 5, 1273–1278.
279. Merkulov V.N., et al., Excitation and propagation of normal waves in a thin elastic cylindrical shell filled with a fluid, 1978, Vol. 24, No. 5, 723–730.
280. Kolt'sova I.S., The propagation of ultrasonic waves in heterogeneous media, St. Petersburg University Press, 2007.
281. Ryapolov P.A., Acoustometric analysis of disperse composition of magnetic fluids, Conducting research in the field of industry of nanosystems and materials, Proc. Conf. with Elements of scientific school for young people, Belgorod Univ., BSU, 2009, 144–148.
282. Ryapolov P.A., The study of magnetic nanodispersed magnetic liquids based on the acoustomagnetic effect, Dissertation, Kursk, KSTU Publishing house, 2010.
283. Polunin V.M., et al. Trials of disperse composition of magnetic fluids based on acoustomagnetic effect, Proc. 22th Session of the Russian Acoustical Society and Sessions of Scientific Council on Acoustics, Vol. 1 , Moscow, GEOS, 2010, 74–77.
284. Polunin V.M., et al. Vibratory–rheological effect in nanostructured magnetic fluids, Managed vibration technology and machines, Kursk State Tekhn. University Press, 2010, 300–306.
285. Polunin V.M., et al., Investigation of distribution of magnetic nanoparticles by the acoustomagnetic method, ibid, 2010, 312318.
286. Pshenichnikov A.F., et al., Magnetogranulometric analysis of concentrated ferrocolloids, J. Magn. Magn. Mater., 1996, Vol. 161, 94–162.
287. Ivanov A.O., Kuznetsova O.B., Magnetic properties of dense ferrofluids: an influence of interparticle correlations, Phys. Rev. E., 2001, Vol. 64, 041405-1-041405-12.
288. Ivanov A.O., Kuznetsova O.B., Magnetogranulometric Analysis of Ferrocolloids: Second-Order Modified Mean Field Theory, Kolloidnyi Zhurnal, 2006, Vol. 68, No. 4, 472–483.
289. Kobelev N.S., et al., On the estimation of physical parameters of magnetic nanoparticles in magnetic fluid, Magnetohydrodynamics, Vol. 46 (2010), No. 1, 31–40.
290. Kazakov Yu.B., et al., Sealants based on nanodispersed magnetic fluids and their modeling, ed. Yu.B. Kazakov, VPO, Ivanovo State Power University named after VI Lenin, Ivanovo, 2010.
291. Agroskin I.I., et al., Hydraulics, Energia, Moscow, Leningrad, 1964.
292. Rasa, M.B., Atomic Force Microscopy and Magnetic Force Microscopy Study of Model Colloids. [Text], Journal of Colloid and Interface Science, 2002, V. 250, 303-315.
293. Rykov S.A., Scanning probe microscopy of semiconductor materials and nanostruc-

tures, Nauka, 2001.
294. Golenishtchev-Kutuzov A.V.,et al., Induced domain structures in electrically and magnetically ordered substances, Moscow, Fizmatlit, 2003.
295. Polunin V.M., et al., Study of the kinetic and strength properties of magnetofluid membranes, Magnetohydrodynamics, Vol. 46 (2010), No. 3, 299–308.
296. Rayleigh. On pressure developed in a liquid during the collapse of a spherical cavity, Philos. Mag., 1917, Vol. 34, 94–100.
297. Pernik A.D., Cavitation problems, 2nd ed., Leningrad, Sudostroenie, 1966.
298. Sirotyuk M.G. Acoustic cavitation, Pacific Ocean Institute of Oceanology, Moscow, Nauka, 2008.
299. Mikhailov I.G., Polunin V.M., On the effect of wettability of the surface emitting ultrasound, on separation of bubbles from it, Akust. Zh., 1973, Vol. 19, No. 3, 462-463.
300. Giith W., Zur Entstehung der Stosswellen bei der Kavitation, Acustica, 1956, Vol. 6, No. 6, 526–531.
301. Yemelyanov S.G., et al., Sound speed in a non-uniformly magnetized magnetic fluid, Mag. Gidrodinamika, Vol. 47 (2011), No. 1, 29–39.
302. RF Patent No.2416089, A method for determining the viscosity of the magnetic liquid or magnetic colloid [Text]. S.G. Emelyanov, No. 2010112571/28, appl. 31.03.2010; publ. 10.04.2011, Bull. Number 10.
303. Polunin V.M., et al., Acoustic structural analysis of the nanostructured magnetic fluid, Izv. VUZ, Sek. Fizika, Tomsk, 2011, No. 1, 10–15.
304. Ryapolov P.A., et al., On the estimation of the ferrofluids particle size distribution on the acoustomagnetic effect base, Euromech Colloquium 526 Patterns in Soft Magnetic Matter, Dresden, 2011, 36–37.
305. Skumiel A., Magnetic and acoustic properties of water-based ferrofluids, Molecular and Quantum Acoustics, Vol. 24 (2003), 149–160.
306. Popescua L.B., et al., The application of perturbational statistical theories to the investigation of the static magnetization of magnetic fluids, Journal of Optoelectronics and Advanced Materials Vol. 7, No. 2, April 2005, 753–757.
307. Polunin V.M., et al. Acoustometry of the nanodispersed phase magnetic fluid, Nanotekhnika, 2011, No. 2 (26) 64–69.
308. Shihab M.E., Ferrofluid surface and volume flows in uniform rotating magnetic fields. Massachusetts Institute of Technology, 2006.
309. Kúdelčík J., et al., Acoustic spectroscopy of magnetic fluid based on transformer oil, Acta electrotechnica et informatica, Vol. 10, No. 3, 2010, 90–92.
310. Zieliński B., et al., Determination of magnetic particle size using ultrasonic, magnetic and atomic force microscopy methods, Molecular and Quantum Acoustics, 2005, Vol. 26, 309–316.
311. Sawada T., et al., Influence of a magnetic field on ultrasound propagation in a magnetic fluid, J. Magn. Magn. Mater., 2002, Vol. 252., 186–188.
312. Odenbach S., (Ed.). Colloidal Magnetic Fluids: Basics, Development and Application of Ferrofluids, Lect. Notes Phys, Berlin: Springer, 2009, 430 p.
313. Polunin V.M., et al., On the dynamics of self-restoring of magnetic fluid membranes using a cavitation model, Magnetohydrodynamics, 2011, Vol. 47, No. 3, 303–313.
314. Polunin V.M., Magnetic fluids, Moscow, Great Russian Encyclopedia, Vol.18. Lomonosov - Manizer. 2011, 373–374.
315. Polunin V.M., The acoustic properties of nanodispersed magnetic liquids, Moscow, Fizmatlit, 2012.
316. Storozhenko A.M., et al., Interaction of physical fields under the acoustomagnetic

effect in magnetic fluids, Magnetohydrodynamics, 2011, Vol. 47, No. 4, 345–358.
317. Polunin V.M., et al., Experimental research of oscillatory system with the air cavity kept by forces of the levitation, Magnetohydrodynamics, 2012, Vol. 48, No. 3, 557-566.
318. Dikanskii Yu.I., et al., On the possibility of magnetic ordering in colloidal systems of single-domain particles, Zh. Tekh. Fiz., 2012, V. 82, Vol. 5, 135–139
319. Polunin V.M., Storozhenko A.M., Investigation of the process of interaction of physical fields in the acoustomagnetic effect, Akust. Zh., 2012, Vol. 58, No. 2, 215-221.
320. Polunin V.M., et al., Study of the interaction of physical fields in the acoustomagnetic effect for a magnetic fluid, Russian Physics Journal, 2012. Vol. 55, No. 5, 536-543.
321. Polunin V.M., Tantsyura A.O., Storozhenko A.M., et al., Study of the demagnetizing field induced by a sound wave, Acoustical Physics, 2013, Vol. 59, No. 6, 662–666.
322. Polunin V.M., Storozhenko A.M., Ryapolov P.A., et al., Perturbation of magnetization of a magnetic fluid by ultralow thermal fluctuations accompanying a sound wave, Acoustical Physics, 2014, Vol. 60, No. 5, 476–482.
323. Hanson M., The frequency dependence of the complex susceptibility of magnetic liquids, J. Magn. Magn. Mat., Vol. 96, Issues 1–3, 1 June 1991, 105-113.
324. Storozhenko A.M., Polunin V.M., Ryapolov P.A., Thermal relaxation of magnetization of magnetic fluid in low magnetic field in acoustomagnetic effect, Moscow International Symposium on Magnetism (MISM), Book of Abstracts, 29 June–3 July 2014, Moscow, 690.
325. Boev M.L., Polunin V.M., Ryapolov P.A., et al., Oscillations of a bubble separated from an air cavity under compression caused by magnetic field in a magnetic fluid, Acoustical Physics, 2014, Vol. 60, No. 1, 29–33.
326. Polunin V.M., Storozhenko A.M., Shabanova I.A., et al., Effect of magnetic field perturbation in magnetic fluid with pulsating bubbles, Magnetohydrodynamics, 2014, Vol. 50, No. 4, 431–441.

Index

I

K

L

M